LIGHTWEIGHT MATERIALS

UNDERSTANDING THE BASICS

Edited
by
F.C. Campbell

**The Materials
Information Society**

ASM International®
Materials Park, Ohio 44073-0002
www.asminternational.org

First printing, October 2012

Comments, criticisms, and suggestions are invited, and should be forwarded to ASM International.

Prepared under the direction of the ASM International Technical Book Committee (2011–2012), Bradley J. Diak, Chair.

ASM International staff who worked on this project include Scott Henry, Senior Manager, Content Development and Publishing; Karen Marken, Senior Managing Editor; Victoria Burt, Content Developer; Sue Sellers, Editorial Assistant; Bonnie Sanders, Manager of Production; Madrid Tramble, Senior Production Coordinator; and Patricia Conti, Production Coordinator.

Library of Congress Control Number: 2012940635

ISBN-13: 978-1-61503-849-7
ISBN-10: 0-61503-849-3

SAN: 204-7586

ASM International®
Materials Park, OH 44073-0002
www.asminternational.org

Printed in the United States of America

Contents

CHAPTER 8
Polymer-Matrix Composites 385

CHAPTER 9
Metal-Matrix Composites. 457

CHAPTER 10
Structural Ceramics. 511

Preface

This book deals with the properties, processing, and applications of lightweight materials—aluminum, magnesium, beryllium, titanium, titanium aluminides, engineering plastics, structural ceramics, and composites with polymer, metal, and ceramic matrices. This book is intended primarily for technical personnel who want to learn more about lightweight materials. It is useful to designers, structural engineers, material and process engineers, manufacturing engineers, production personnel, faculty, and students.

The first chapter is an introduction that gives a brief history of lightweight materials development during the 20th century. Somewhat surprisingly, with the exception of ceramics, none of these lightweight materials existed in usable forms until the 20th century. The second part of the introduction gives a survey of each of the lightweight materials and their current applications.

The second through sixth chapters deal with lightweight metals. Each chapter covers the basic metallurgy, properties of the available alloys, processing, and applications. Aluminum is perhaps the most important of the lightweight metals, used in a wide range of commercial, industrial, and defense applications. Magnesium, an even lighter metal than aluminum, is used where extremely light weight is paramount. The other three metals (beryllium, titanium, and titanium aluminides) have important applications in defense and aerospace. As a result of its outstanding corrosion resistance and biocompatibility, titanium has important applications in the chemical processing industry and as implants in the human body.

Plastics are pervasive in our world—we all use plastic items every day. Although the seventh chapter covers plastics in general, the emphasis is on the so-called engineering plastics. While engineering plastics are not near as strong as structural metals, due to their very low densities, they are often competitive on a specific strength (strength/density) and specific stiffness (modulus/density) basis. However, there are certain precautions that must always be followed when using plastics in load bearing applications.

The eighth and ninth chapters cover composite materials having matrices of polymers and metals. Although fiberglass composites have been used since the 1940s, the development of new high performance fibers such as carbon fiber in the early 1960s first led to their widespread use in military aircraft. This technology has been gradually transitioned to the commercial aircraft market where high performance composites will be used for large portions of the airframes of all new commercial aircraft.

Ceramics, covered in the tenth chapter, are not modern materials. They were one of the first materials used by ancient man. However, it was not until the 1950s when structural ceramics were developed that can be used in a much wider range of applications than traditional ceramics. Due to the brittle nature of ceramics, they have been historically confined to applications that place the ceramic in compression, to keep cracks from propagating. This situation is starting to change with the development of the newer toughened ceramic materials and ceramic-matrix composites. Ceramic-matrix composites are the subject of the eleventh chapter.

The twelfth, and final chapter, covers three separate topics. First, some methodologies used in the general materials selection process are discussed. Next, there are some specific guidelines for each of the lightweight materials. The last section covers the importance of the automotive sector to the lightweight materials industries.

I would like to acknowledge the help and guidance of the editorial staff at ASM for their valuable contributions.

F.C. Campbell

CHAPTER 1

Introduction and Uses of Lightweight Materials

DURING THE INDUSTRIAL REVOLUTION of the 18th and 19th centuries, the basic materials of construction were sparse. There was wood, stone, brick and mortar, and iron and steel. However, the irons and steels of that day were vastly inferior to the ones used today. In 1709, Abraham Darby established a coke-fired blast furnace to produce cast iron, and by the end of the 18th century, cast iron began to replace wrought iron, because it was cheaper. The ensuing availability of inexpensive iron was one of the factors leading to the Industrial Revolution. In the late 1850s, Henry Bessemer invented a new steelmaking process, involving blowing air through molten pig iron, to produce mild steel. This made steel much more economical.

With the advent of the 20th century, improved lightweight materials such as aluminum, magnesium, beryllium, titanium, titanium aluminides, engineering plastics, structural ceramics, and composites with polymer, metal, and ceramic matrices began to appear.

Aluminum. Before the Hall-Héroult process was developed in the late 1880s, aluminum was exceedingly difficult to extract from its various ores. This made pure aluminum more valuable than gold. Napoleon III, Emperor of France, is reputed to have given a banquet where the most honored guests were given aluminum utensils, while the others made do with gold. Charles Martin Hall of Ohio in the United States and Paul Héroult of France independently developed the Hall-Héroult electrolytic process that made extracting aluminum from minerals cheaper and is now the principal method used worldwide. In 1888, Hall founded the Pittsburgh Reduction Company, today known as Alcoa.

The next breakthrough in aluminum came in 1901 when Alfred Wilm, while working in a military research center in Germany, accidentally discovered age hardening, in particular age hardening of aluminum alloys.

He made this discovery after hardness measurements on aluminum-copper alloy specimens were found to increase in hardness at room temperature. The increase in hardness was identified after Wilm's measurements were interrupted by a weekend, and when resumed on Monday, the hardness had increased. The economical production of aluminum and the discovery of age hardening allowed the rapid replacement of aluminum for wood and fabric in airplanes.

Magnesium. Magnesium was one of the main aerospace construction materials used for German military aircraft as early as World War I and extensively for German aircraft in World War II. The extraction of magnesium from seawater was pioneered in the United States by Dow. Their magnesium plant became operational in 1941, marking the first time that man had mined the ocean for a metal. After the Japanese attack on Pearl Harbor, the U.S. government asked Dow to increase its magnesium production, so a second plant was constructed. Today, Dow's Texas Operations spreads across more than 5000 acres with more than 75 individual production plants.

Beryllium. The first pure samples of beryllium were produced in 1898. However, it took until World War I (1914 to 1918) before significant amounts of beryllium were produced. Large-scale production started in the early 1930s and rapidly increased during World War II, due to the rising demand for the hard beryllium-copper alloys.

Titanium. The large-scale production and use of titanium was a result of the Cold War. In the 1950s and 1960s, the Soviet Union pioneered titanium use in military and submarine applications. Starting in the early 1950s, titanium began to be used extensively for military aviation purposes, particularly in high-performance jets, the most notable being the Black Bird supersonic spy plane. The Department of Defense realized the strategic importance of the metal and supported early efforts of commercialization. Throughout the Cold War, titanium was considered a strategic material by the U.S. government, and a large stockpile of titanium sponge was maintained by the Defense National Stockpile Center. Today, the world's largest producer, Russian-based VSMPO-Avisma, accounts for an estimated 29% of the world market share.

Engineering Plastics. The first plastic based on a synthetic polymer was phenolic, made from phenol and formaldehyde. In 1907, Leo Baekeland was trying to develop an insulating shellac to coat wires in electric motors and generators. He found that mixtures of phenol and formaldehyde formed a sticky mass when mixed together, and when heated, the mass became an extremely hard solid. After World War I, improvements in chemical technology led to an explosion in new forms of plastics. Among the earliest examples of new plastics were polystyrene and polyvinyl chloride. The real star of the plastics industry in the 1930s was polyamide, far better known by its trade name nylon. Nylon was the first purely synthetic fiber, introduced by DuPont Corporation at the 1939 World's Fair in New York City.

Plastics came of age during World War II. Traditional materials were in short supply due to the war effort and the Pacific supply of rubber was cut off. Plastics became an acceptable substitute for many traditional materials. By 1950, the largest plastic part made was a 19 kg (42 lb) Admiral television. In 1952, engineering plastics came into use, and a door liner for an Admiral refrigerator was produced using injection molding. Since 1976, plastics have become the most used material in the world. The plastics industry in the United States has grown an average of 7.7% per year for the past 40 years. In 1979, the production of plastics in the United States exceeded that of steel. By 2000, the plastics industry was the fourth largest manufacturing segment in the United States, following only transportation, electronics, and the chemical industries.

Structural Ceramics. Ceramics are one of mankind's most ancient materials. Once humans discovered that clay could be dug up and formed into objects by mixing it with water and firing it, the industry was born. As early as 24,000 B.C., animal and human figurines were made from clay and other materials and then fired in kilns partially dug into the ground. Almost 10,000 years later, as settled communities were established, tiles were manufactured in Mesopotamia and India. The first use of functional pottery vessels for storing water and food is thought to be around 9,000 or 10,000 B.C. Fired clay-based products were the only ceramics available until approximately 100 years ago. At that point, new classes of starting materials, such as silicon carbide for abrasives, began to be used. Until the 1950s, the most important ceramic materials were (1) pottery, bricks and tiles, (2) cements, and (3) glass.

The development of structural ceramics has been partially motivated by U.S. government sponsorship of programs during the 1970s and 1980s to develop ceramics for engine applications. Improvements in processing and properties have continued. For engines and more demanding applications, compositions are being developed that are based either on two or more phases of particulates or on matrices reinforced with whiskers or fibers.

Composites. The first known fiberglass-reinforced composite product was a boat hull manufactured in the mid-1930s. Fiberglass-reinforced composites applications have revolutionized the aerospace, sporting goods, marine, electrical, corrosion resistance, and transportation industries. Fiberglass-reinforced composite materials date back to the early 1940s in the defense industry, particularly for use in aerospace and naval applications. The U.S. Air Force and Navy capitalized on their high strength-to-weight ratio and inherent resistance to weather and the corrosive effects of salt air and sea. By 1945, over 7 million pounds of fiberglass were used in military applications. Fiberglass pipe was first introduced in 1948 and was widely used within the corrosion market and the oil industry.

In 1959 and 1962, two processes for manufacturing high-strength and high-modulus carbon fibers from rayon and polyacrylonitrile precursor fibers were invented simultaneously. In 1963, high-modulus carbon fibers made from pitch were invented. High-performance carbon-fiber composites were first used in military aircraft. By the early 1980s, the AV-B Harrier vertical take-off and landing jet contained 28% composites in its airframe. These early successes with composites helped Boeing and Airbus succeed in incorporating advanced composites on commercial aircraft. In 2011, the first Boeing Dreamliner with a 50% composite airframe was introduced into service.

1.1   Today's Lightweight Materials

The distinguishing feature of the materials covered in this book is that they all have low densities. Densities range from as low as 0.80 g/cm^3 (0.030 lb/in.^{3}) for unfilled polymers to as high as 4.5 g/cm^3 (0.160 lb/in.^{3}) for titanium. While the density of titanium is high compared to unfilled polymers, it is significantly lighter than the metals it usually competes with—alloy steel at 7.86 g/cm^3 (0.283 lb/in.^{3}) and superalloys with densities that range from 7.8 to 9.4 g/cm^3 (0.282 to 0.340 lb/in.^{3}). In addition, unfilled polymers have rather low tensile strengths that range from 34 to 103 MPa (5 to 15 ksi), while unidirectional carbon/epoxy can attain tensile strengths as high as 2410 MPa (350 ksi). Some of the lightweight materials can only be used to approximately 66 °C (150 °F), while others maintain useful properties to over 1370 °C (2500 °F). Therefore, as shown in Table 1.1, the lightweight materials covered in this book cover a wide range of properties and, as a result, fulfill a wide range of applications.

1.2   Aluminum Alloys

Aluminum alloys have many outstanding attributes that lead to a wide range of applications, including:

- Good corrosion and oxidation resistance
- High electrical and thermal conductivities
- Low density
- High reflectivity
- High ductility and reasonably high strength
- Relatively low cost

Potential limitations of aluminum alloys are:

- Moderate modulus of elasticity
- Fatigue strength sometimes much lower than static strength
- High-strength 2xxx and 7xxx alloys can be difficult to weld.

Table 1.1 Comparison of lightweight materials

Material	Density		Tensile strength		Modulus of elasticity		Elongation, %	Continuous-use temperature		Relative cost(a)
	g/cm³	lb/in.³	MPa	ksi	GPa	10⁶ psi		°C	°F	
Aluminum alloys	2.77	0.100	207–552	30–80	69–76	10–11	10–30	480	250	3
Magnesium alloys	1.74	0.063	138–345	20–50	45	6.5	3–10	390	200	4
Beryllium alloys	1.85	0.067	345–483	50–70	303	44.0	1–3	2010	1100	8
Titanium alloys	4.43	0.160	827–1241	120–180	103–110	15–16	10–30	1290–2010	700–1100	6
Titanium aluminide alloys	4.21	0.152	448–827	65–120	145–172	21–25	1–4	3000	1650	9
Engineering plastics	0.83–13.8	0.03–0.50	35–103	5–15	2.0–9.0	0.3–1.3	2–700	300–390	150–200	1–4
Polymer-matrix composites										
Discontinuous fibers(b)	1.38	0.050	172	25	9.0	1.3	4	480	250	2
Continuous fibers(c)	1.58	0.057	2413	350	172	25	2	480	250	6
Metal-matrix composites										
Discontinuous fibers(d)	2.93	0.106	276–345	40–50	93–114	13.5–16.5	0.5–1.5	660	350	5
Continuous fibers(e)	3.96	0.144	1379	200	186	27	0.85	2550	1400	8
Structural ceramics	2.49–5.5	0.09–0.2	140–448	20–65	207–414	30–60	<1	3180–4620	1750–2550	5
Ceramic-matrix composites										
Discontinuous fibers(f)	3.60	0.130	903(g)	131(g)	103	15	…	3630	2000	7
Continuous fibers(h)	3.46	0.125	214	31	138	20	<1	4710w	2600	10

(a) 1 is lowest cost; 10 is highest cost. (b) Nylon with 30% glass. (c) Unidirectional carbon/epoxy. (d) Die-cast SiC particles/aluminum. (e) SCS-6 silicon carbide fiber/titanium. (f) Alumina reinforced with 25% SiC whiskers. (g) Bend strength. (h) Woven SiC/SiC

Aluminum is an industrial and consumer metal of great importance. Aluminum and its alloys are used for foil, beverage cans, cooking and food processing utensils, architectural and electrical applications, and structures for boats, aircraft, and other transportation vehicles. As a result of a naturally occurring tenacious surface oxide film (Al_2O_3), a great number of aluminum alloys have exceptional corrosion resistance in many atmospheric and chemical environments. Its corrosion and oxidation resistance is especially important in architectural and transportation applications. On an equal weight and cost basis, aluminum is a better electrical conductor than copper. Its high thermal conductivity leads to applications such as radiators and cooking utensils. Its low density is important for hand tools and all forms of transportation, especially aircraft. Wrought aluminum alloys display a good combination of strength and ductility. Aluminum alloys are among the easiest of all metals to form and machine. The precipitation hardening alloys can be formed in a relatively soft state and then heat treated to much higher strength levels after forming operations are complete. In addition, aluminum and its alloys are not toxic and among the easiest to recycle of any of the structural materials.

Aluminum has a density that is one-third the density of steel. Although aluminum alloys have lower tensile properties than steel, their specific strength (strength/density) is excellent. Aluminum is easily formed, has high thermal and electrical conductivity, and does not show a ductile-to-brittle transition at low temperatures. It is nontoxic and can be recycled with only approximately 5% of the energy needed to make it from alumina, which is why the recycling of aluminum is so successful. The beneficial physical properties of aluminum include its nonferromagnetic behavior and resistance to oxidation and corrosion. However, aluminum does not display a true endurance limit, so failure by fatigue may eventually occur, even at relatively low stresses. Because of its low melting temperature, aluminum does not perform well at elevated temperatures. Finally, aluminum has low hardness, leading to poor wear resistance. Aluminum responds readily to strengthening mechanisms; alloys may be 30 times stronger than pure aluminum. The attractiveness of aluminum is that it is a relatively low-cost, lightweight metal that can be heat treated to fairly high strength levels, and it is one of the most easily fabricated of the high-performance materials, which usually correlates directly with lower costs.

Approximately 25% of the aluminum produced today is used in the transportation industry, another 25% in the manufacture of beverage cans and other packaging, about 15% in construction, 15% in electrical applications, and 20% in other applications.

The dominance of steel in the automotive industry is being challenged by plastics and aluminum. In 2010, approximately 90 kg (200 lb) of aluminum was used in an average car made in the United States. The long-term goal of the aluminum community is an all-aluminum chassis, known as a "body in white." The most notable example of an aluminum chassis is the high-end Audi A8 luxury sedan, which uses a spaceframe construction. This method uses complex hollow extruded sections joined by welding to form a rigid framework. A weight savings of 140 kg (310 lb) relative to a steel body is achieved by covering the spaceframe with prestressed aluminum body panels. The question remains whether car manufacturers and buyers are willing to accept a more expensive, if lighter, product. The present consensus appears to be that except for certain niche products, steel remains the low-cost material of choice. Much will depend on the future price structure for the individual materials, developments in materials processing, and the impetus for further weight-saving in relation to fuel efficiency. In the meantime, the use of higher-strength steels at thinner gage for panels will continue to expand, taking advantage of the very efficient monocoque steel sheet body structure and the high production rates possible with this system, and with increased use of coatings to provide corrosion protection.

Aluminum alloys can be divided into two major groups: wrought and casting alloys (Table 1.2). Wrought alloys, which are shaped by plastic deformation, have compositions and microstructures significantly different

Table 1.2 Designations for aluminum alloys

Series	Aluminum content or main alloying element
Wrought alloys	
1xxx	99.00% min
2xxx	Copper
3xxx	Manganese
4xxx	Silicon
5xxx	Magnesium
6xxx	Magnesium and silicon
7xxx	Zinc
8xxx	Others
9xxx	Unused
Cast alloys	
1xx.0	99.00% min
2xx.0	Copper
3xx.0	Silicon with copper and/or magnesium
4xx.0	Silicon
5xx.0	Magnesium
6xx.0	Unused
7xx.0	Zinc
8xx.0	Tin
9xx.0	Other

from casting alloys, reflecting the different requirements of the manufacturing process. Within each major group, the alloys can be divided into two subgroups: heat-treatable and non-heat-treatable alloys.

The 1xxx, 3xxx, 5xxx, and most of the 4xxx wrought alloys cannot be strengthened by heat treatment. The 1xxx, 3xxx, and 5xxx alloys are strengthened by strain hardening, solid-solution strengthening, and grain-size control. The primary uses of the 1xxx series are in applications that require a combination of extremely high corrosion resistance and formability (e.g., foil and strip for packaging, chemical equipment, tank car or truck bodies, spun hollowware, and elaborate sheet metal work). In the 3xxx alloy series, alloy 3004 and its modification 3104 are the principals for drawn and ironed bodies for beverage cans for beer and soft drinks (Fig. 1.1). As a result, they are among the most-used individual alloys in the aluminum system, in excess of 1.6 billion kg (3.5 billion lb) per year. The 5xxx series of alloys has an outstanding combination of strength, toughness, weldability, and corrosion resistance in saltwater. As a result, they find wide application in building and construction; highway structures including bridges, storage tanks, and pressure vessels; cryogenic tankage and systems for temperatures as low as –270 °C (–455 °F) or near absolute zero; and marine applications. A large welded 5xxx aluminum substructure for a high-speed single-hull ship is shown in Fig. 1.2.

The 2xxx, 6xxx, and 7xxx alloys can be strengthened by precipitation hardening heat treatments. The greatest usage for the high-strength precipitation-hardened 2xxx and 7xxx alloys is in aircraft and spacecraft applications. As shown in Fig. 1.3, 2xxx alloys are used for fuselage

Fig. 1.1 The bodies of beverage cans are alloys 3004 or 3104, making it the largest-volume alloy consumed in the industry. Source: Ref 1.1

construction and 7*xxx* alloys for wing structures. A unique feature of the 6*xxx* alloys is their great extrudability, making it possible to produce in single shapes relatively complex architectural forms, as shown in Fig. 1.4. They can be heat treated to moderate strength levels and have good corrosion resistance.

The highest-usage casting alloys contain enough silicon that the alloys have low melting points, good fluidity, and good castability. Fluidity is the ability of the liquid metal to flow through a mold without prematurely solidifying, and castability refers to the ease with which a good casting can be made from the alloy. Quite intricate structural castings (Fig. 1.5) can be produced using a wide range of casting processes.

Improvements in aluminum manufacturing technology include high-speed machining and friction stir welding. Although higher metal-removal rates are an immediate benefit of high-speed machining, an additional cost-saving is the ability to machine extremely thin walls and webs. This allows the design of weight-competitive high-speed machined assemblies, in which sheet metal parts that were formally assembled with mechanical fasteners can now be machined from a single, or several blocks of, aluminum plate (Fig. 1.6). Another recent development, called friction stir welding, is a new solid-state joining process that has the ability to weld the difficult, or impossible, to fusion weld 2*xxx* and 7*xxx* alloys with less distortion, fewer defects, and better durability than achievable using conventional welding techniques.

(a)

(b)

Fig. 1.2 Aluminum alloy usage in the marine industry. (a) High-speed single-hull ship. (b) 5*xxx* aluminum alloy internal hull stiffener structure. Source: Ref 1.1

(a)

(b)

Fig. 1.3 Aluminum alloys used for commercial aircraft. (a) 2*xxx* alloys for fuselage. (b) 7*xxx* alloys for wings. Source: Ref 1.1

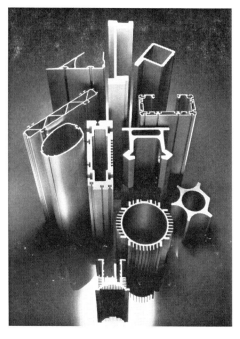

Fig. 1.4 6xxx aluminum alloy structural extrusions. Source: Ref 1.1

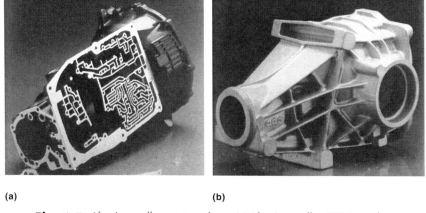

(a)　　　　　　　　　　　　　　　　　(b)

Fig. 1.5 Aluminum alloy cast products. (a) Aluminum alloy 380.0 gearbox casting for passenger car. (b) Aluminum alloy 380.0 rear axle casting. Source: Ref 1.1

1.3　Magnesium Alloys

Magnesium alloys usually compete with aluminum alloys for structural applications. Compared to high-strength aluminum alloys, magnesium alloys are not as strong (tensile strength of 138 to 345 versus 275 to 550 MPa, or 20 to 50 versus 40 to 80 ksi) and have a lower modulus of elasticity (45 versus 69 GPa, or 6.5 versus 10 to 11 $\times$ 10^6 psi). However, magnesium is significantly lighter (1.74 versus 2.77 g/cm^3, or 0.063 versus 0.100 lb/in.^{3})

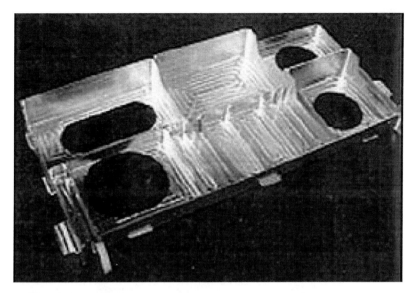

Fig. 1.6 High-speed machined avionics rack. Courtesy of The Boeing Company

than aluminum; therefore, its alloys are competitive on a specific strength and modulus basis. Most magnesium alloys have ratios of tensile strength to density and tensile yield strength to density that are comparable to those of other common structural materials (Fig. 1.7).

Magnesium alloys have very good damping capacity, and castings have found application in high-vibration environments, such as helicopter gear boxes. Magnesium alloys are used in aerospace applications, high-speed machinery, and transportation and materials handling equipment. Although wrought and cast alloys are available, cast alloys are the mostly widely used. An application that takes advantage of the light weight and good damping capacity of magnesium is the main transmission housing for a heavy lift helicopter produced by sand casting shown in Fig. 1.8.

Magnesium has a hexagonal close-packed (hcp) structure that has limited slip planes. Therefore, magnesium alloys have only limited formability at room temperature; most forming operations must be conducted at elevated temperatures. For example, to successfully stretch form the rather simple dome shape shown in Fig. 1.9, the forming had to be conducted at 250 °C (480 °F). This is in contrast to aluminum alloys that can be extensively formed at room temperature. In addition, magnesium alloys are normally more expensive than comparable aluminum alloys.

The biggest obstacle to the use of magnesium alloys is their extremely poor corrosion resistance. In many environments, the corrosion resistance of magnesium approaches that of aluminum; however, exposure to salts, such as those near a marine environment, causes rapid deterioration.

Magnesium occupies the highest anodic position on the galvanic series, and, as such, there is always the strong potential for corrosion. However,

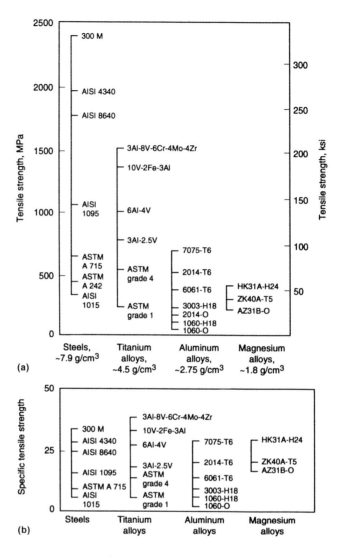

Fig. 1.7 Comparison of structural alloys on the basis of (a) tensile strength and (b) specific tensile strength (tensile strength, in ksi, divided by density, in g/cm³). Source: Ref 1.2

some of the newer alloys have much better corrosion resistance than the older alloys, with some of the newer cast alloys approaching the corrosion resistance of competing aluminum casting alloys.

1.4 Beryllium

Beryllium is a very lightweight metal with an attractive combination of properties. However, beryllium must be processed using powder metallurgy technology that is costly, and beryllium powder and dust are toxic,

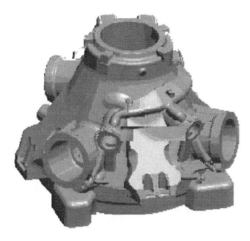

Fig. 1.8 Main transmission housing for a heavy lift helicopter that was sand cast in WE43B magnesium alloy having a T6 temper. Casting weight = 93 kg (206 lb). Courtesy of Fansteel Wellman Dynamics. Source: Ref 1.3

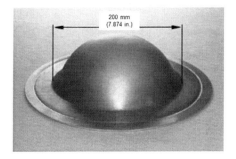

Fig. 1.9 Stretch-formed dome. Material: MgAl3Zn1 (AZ31); initial sheet thickness, 0: 1.3 mm (0.051 in.); forming temperature: 250 °C (480 °F). Source: Ref 1.4

which further increases its cost through the requirement for controlled manufacturing environments and the concern for safety during the repair/service of deployed structures.

Beryllium is lighter than aluminum yet is stiffer than steel, with a modulus of elasticity of 304 GPa (44×10^6 psi). Beryllium alloys, which have yield strengths of 200 to 345 MPa (30 to 50 ksi), have high specific strengths and maintain both strength and stiffness to high temperatures. Instrument-grade beryllium is used in inertial guidance systems where the elastic deformation must be minimal; structural grades are used in aerospace applications; and nuclear applications take advantage of the transparency of beryllium to electromagnetic radiation. The use of beryllium made possible the Hubble Telescope and the building of the James Webb Space Telescope (Fig. 1.10), which is expected to launch in 2012. Beryllium was selected as the mir-

Fig. 1.10 The primary, secondary, and tertiary mirrors of the James Webb Telescope are made of beryllium. The mirrors have been gold coated using physical vapor deposition.

ror technology for its demonstrated track record of operating at cryogenic temperatures on space-based telescopes.

Unfortunately, beryllium is very expensive, brittle, reactive, and toxic. Its production is quite complicated, and hence, the applications of beryllium alloys are very limited. Beryllium oxide (BeO), which is also toxic in a powder form, is used to make high-thermal-conductivity ceramics. Like magnesium, its hcp crystalline structure greatly impairs its formability; elevated temperatures are required for forming operations

1.5 Titanium Alloys

Although titanium has the highest density of the light metals, titanium provides excellent corrosion resistance, high specific strength, and good high-temperature properties. Strengths up to 1400 MPa (200 ksi) provide excellent mechanical properties. An adherent, protective TiO_2 film provides excellent resistance to corrosion. However, even though titanium is the fourth most abundant element in the Earth's crust, due to its high melting point and extreme reactivity, the cost of titanium is high. The high cost includes both the mill operations (sponge production, ingot melting, and primary working) as well as many of the secondary operations conducted by the user.

The primary reasons for using titanium alloys include:

- *Weight savings.* The high strength-to-weight ratio of titanium alloys allows them to replace steel in many applications requiring high strength and fracture toughness. With a density of 4.5 g/cm^3 (0.16 lb/

in.^{3}), titanium alloys are only about half as heavy as steel and nickel-base superalloys, yielding excellent strength-to-weight ratios.

- *Fatigue strength.* Titanium alloys have much better fatigue strength than aluminum alloys and are frequently used for highly loaded bulkheads (Fig. 1.11) and frames in fighter aircraft.
- *Operating temperature capability.* When the operating temperature exceeds approximately 132 °C (270 °F), aluminum alloys lose too much strength, and titanium alloys are often required.
- *Corrosion resistance.* The corrosion resistance of titanium alloys is superior to both aluminum and steel alloys, allowing it to be used in hostile chemical processing environments.
- *Biocompatibility.* The human body accepts titanium much more readily than other metals, including stainless steels. Uses include hip, knee, shoulder, and elbow joint replacements.

Titanium is allotropic, with the hcp crystal structure (α) at low temperatures and a body-centered cubic structure (β) above 882 °C (1620 °F). Alloying elements provide solid-solution strengthening and change the allotropic transformation temperature. The alloying elements can be divided into four groups. Additions such as tin and zirconium provide solid-solution strengthening without affecting the transformation tem-

Fig. 1.11 Alloy Ti-6Al-4V forgings for upper and lower bulkheads used on the F-15 that were produced on a 450 MN (50,000 tonf) hydraulic press using conventional forging methods. Source: Ref 1.5

perature. Aluminum, oxygen, hydrogen, and other α-stabilizing elements increase the temperature at which α transforms to β. Beta stabilizers such as vanadium, tantalum, molybdenum, and niobium lower the transformation temperature, even causing β to be stable at room temperature. Finally, manganese, chromium, and iron produce a eutectoid reaction, reducing the temperature at which the transformation occurs and producing a two-phase structure at room temperature.

The excellent corrosion resistance of titanium allows applications in chemical processing equipment, marine components, and biomedical implants such as hip prostheses. Titanium is an important aerospace material, finding applications as airframe and jet engine components. When it is combined with niobium, a superconductive intermetallic compound is formed; when it is combined with nickel, the resulting alloy displays a shape memory effect; and when it is combined with aluminum, a new class of intermetallic alloys is produced. Titanium alloys are also used for sports equipment such as golf club heads.

Near-net shape processes can lead to savings in materials, machining costs, and cycle times over conventional forged or machined parts. Investment casting, in combination with hot isostatic pressing, can produce aerospace-quality titanium near-net-shaped parts that can offer significant cost-savings over forgings and built-up structure. Examples of investment-cast titanium components are shown in Fig. 1.12. Titanium is also very amenable to superplastic forming (Fig. 1.13), allowing a reduction in the number of detailed parts and fasteners required. It can also be combined with diffusion bonding to produce complex unitized structures. In another process, called directed metal deposition, a focused laser or electron beam melts titanium powder and deposits the melt in a predetermined path on a titanium substrate plate. The deposited preform is then machined to the final part shape.

1.6 Titanium Aluminide Intermetallics

Titanium aluminides represent a new emerging class of alloys that provide a unique set of physical and mechanical properties that can lead to substantial payoffs in the automotive industry, power plant turbines, and aircraft engines. It took more than 20 years of very intensive research to obtain maturity level of TiAl-base alloys that is sufficient to consider this class of materials for critical rotating components in commercial jet engines. The outstanding thermophysical properties of these alloys mainly result from a strongly ordered structure, which provides a high melting point, low density, high elastic modulus, low diffusion coefficient, good structural stability, good resistance against oxidation and corrosion, and high ignition temperature when compared to conventional titanium alloys.

The TiAl-base alloys show superior specific strength-temperature properties when compared to conventional titanium alloys, steels, and

(a)

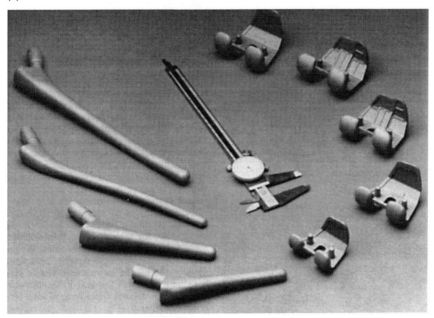

(b)

Fig. 1.12 Titanium alloy investment castings. (a) Investment-cast titanium components for use in corrosive environments. (b) Titanium knee and hip implant prostheses manufactured by the investment casting process. Source: Ref 1.6

nickel-base superalloys in the temperature range from 500 to 900 °C (930 to 1650 °F). Intermetallic TiAl-base alloys exhibit the highest potential for near-term applications in future aircraft engines. While their specific strength is higher than that of competing materials, their room-temperature ductility is poor and is typically about 1% for all TiAl-base alloys. Low ductility is the biggest problem in the application of these alloys as structural components, because 1% ductility is generally accepted as the minimum acceptable level, and cast samples in particular seldom reach even this value. The other major problem is the difficulty in processing to form a component.

General Electric has achieved a major success with TiAl implementation and certification in the new GEnx-1B engine that was developed to power the Boeing 787 Dreamliner that entered service in 2011. The last two stages of the GEnx-1B low-pressure turbine are made from cast gamma TiAl blades. TiAl is also being considered for both intake and exhaust valve applications for automotive engines (Fig. 1.14). The other main area of interest for the application of TiAl in the automotive sector is as a turbocharger wheel material in diesel engines.

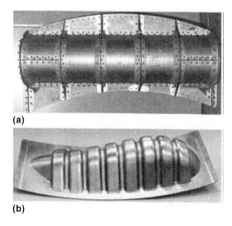

Fig. 1.13 Ti-6Al-4V engine nacelle component for the Boeing 757 aircraft. (a) Part as previously fabricated required 41 detail parts and more than 200 fasteners. (b) Superplastically formed part is formed from a single sheet. Courtesy of Metal Bellows Division of Parker Bertea Aerospace Group. Source: Ref 1.7

Fig. 1.14 Gamma titanium aluminide auto engine component

1.7 Engineering Plastics

Plastics are materials that are composed principally of naturally occurring and modified or artificially-made polymers often containing additives such as fibers, fillers, pigments, and so on that further enhance their properties. Plastics include thermoplastics, thermosets, and elastomers (natural or synthetic).

Engineering plastics offer some unique product benefits. These are usually physical properties, or combinations of physical properties, that allow vastly improved product performance, such as:

* Extremely lightweight
* Potential for low-cost high-volume processing
* Good toughness
* Good electrical and thermal insulation
* Good chemical resistance to some chemicals
* Magnetic inertness
* Some plastics are transparent

Plastics have the following limitations:

* Strength and stiffness are low relative to metals and ceramics.
* Service temperatures are usually limited to only a few hundred degrees because of the softening of thermoplastic plastics or degradation of thermosetting plastics.
* Some plastics degrade when subjected to sunlight, other forms of radiation, and some chemicals.
* Plastics exhibit viscoelastic properties, which can be a distinct limitation in load-bearing applications.

Plastics are used in an amazing number of applications (Fig. 1.15), including clothing, toys, home appliances, structural and decorative items, coatings, paints, adhesives, automobile tires, biomedical materials, car bumpers and interiors, foams, and packaging. Plastics are often used to make electronic components because of their insulating ability and low dielectric constant. More recently, significant developments have occurred in the area of flexible electronic devices based on the useful piezoelectric, semiconducting, optical, and electro-optical properties of some polymers. Polymers such as polyvinyl acetate are water soluble and can be dissolved in water or organic solvents to be used as binders, surfactants, or plasticizers in the processing of ceramics, semiconductors, and as additives to many consumer products. Polyvinyl butyral, a polymer, makes up part of the laminated glass used for car windshields. Polymers are probably used in more technologies than any other class of materials.

Commercial, or standard commodity, polymers are lightweight, corrosion-resistant materials with low strength and stiffness, and they are

Computer mouse

Sunglasses

Toy truck

Tool box

Household wares

Leaf rake

Portable drill

Watch case
and band

Fig. 1.15 Typical plastic parts

not suitable for use at high temperatures. However, these polymers are relatively inexpensive and are readily formed into a variety of shapes, ranging from plastic bags to mechanical gears to bathtubs.

Engineering polymers are designed to give improved strength or better performance at elevated temperatures. These materials are produced in relatively small quantities and often are expensive. Some of the engineering polymers can perform at temperatures as high as 350 °C (660 °F) under no or very low load levels. Others, usually in a fiber form, have strengths that are greater than that of steel.

The properties of both commodity and engineering plastics are often improved with the addition of high-strength fibers—both continuous and discontinuous—to form composites. While discontinuous fibers do not provide the strength and stiffness of continuous fibers, they are amenable to many of the high-production processes used with unfilled plastics. As an example, the strength and stiffness of nylon is significantly increased with the addition of glass fibers (Table 1.3), with only a minimal increase in density.

Polymers also have many useful physical properties. Some polymers, such as acrylics, are transparent and can be substituted for glasses. Although most polymers are electrical insulators, special polymers (such as the acetals) and graphite-fiber polymer-based composites possess useful electrical conductivity. Teflon (E.I. du Pont de Nemours and Company) has a low coefficient of friction and is the coating for nonstick cookware. Polymers also resist corrosion and chemical attack.

1.8 Polymer-Matrix Composites

A composite material can be defined as a combination of two or more materials that results in better properties than when the individual components are used alone. As opposed to metallic alloys, each material retains its separate chemical, physical, and mechanical properties. The two constituents are normally a fiber and a matrix. Typical fibers include glass, aramid, and carbon, which may be continuous or discontinuous. Matrices can be polymers, metals, or ceramics.

The advantages of high-performance composites are lighter weight, the ability to tailor lay-ups for optimum strength and stiffness, improved fatigue life, corrosion resistance, and, with good design practice, reduced assembly costs due to fewer detail parts and fasteners. The specific strength and specific modulus of high-strength fiber composites (Fig. 1.16), especially carbon fibers, are higher than other comparable aerospace metallic alloys. This translates into greater weight savings, resulting in improved performance, greater payloads, longer range, and fuel savings. The physical characteristics of composites and of metals are significantly different. A comparison of some of the properties of composites and metals are given in Table 1.4.

Table 1.3 Effects of glass filler on nylon strength and stiffness

Property	Unfilled nylon	Glass-filled nylon, 35%	Glass-filled nylon, 60%
Density, g/cm^3	1.15	1.62	1.95
Tensile strength, MPa (ksi)	82 (11.9)	200 (29.0)	290 (42.0)
Tensile modulus, GPa (10^6 psi)	2.9 (0.42)	14.5 (2.10)	21.8 (3.16)
Specific tensile strength, σ/ρ	0.082	0.16	0.149
Specific tensile modulus, E/ρ	2.52	8.26	11.18

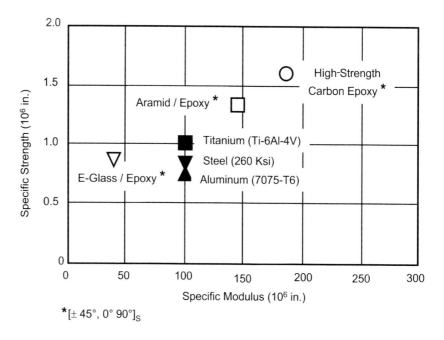

Fig. 1.16 Specific property comparison. * = (±45°, 0°, 90°)$_S$

Table 1.4 Composites versus metals comparison

Condition	Comparative behavior relative to metals
Load-strain relationship	More linear strain to failure
Notch sensitivity	
Static	Greater sensitivity
Fatigue	Less sensitivity
Transverse properties	Weaker
Mechanical property variability	Higher
Fatigue strength	Higher
Sensitivity to hydrothermal environment	Greater
Sensitivity to corrosion	Much less
Damage growth mechanism	In-plane delamination instead of through-thickness cracks

Source: Ref 1.8

Because composites are highly anistropic, the in-plane strength and stiffness are usually high and directionally variable, depending on the orientation of the reinforcing fibers. Properties that do not benefit from this reinforcement, at least for polymer-matrix composites, are comparatively low in strength and stiffness. An example is the through-thickness tensile strength, where the relatively weak matrix is loaded rather than the high-strength fibers. The low through-thickness strength of a typical composite laminate compared to aluminum is shown in Fig. 1.17.

Metals normally have reasonable ductility. When they reach a certain load level, they continue to elongate or compress considerably without picking up more load and without failure. This ductile yielding has two

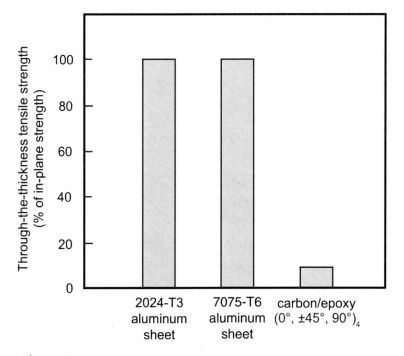

Fig. 1.17 Through-thickness tensile strength comparison. Source: Ref 1.9

important benefits. It provides for local load relief by distributing excess load to adjacent material or structure. Consequently, ductile metals have a great capacity to provide relief from stress concentrations when statically loaded. Second, the ductility of metals provides great energy-absorbing capability, as is indicated by the area under a stress-strain curve. As a result, when impacted, a metal structure will typically deform but not actually fracture. In contrast, composites are relatively brittle. A comparison of typical tensile stress-strain curves for the two materials is shown in Fig. 1.18. The characteristically brittle composite material has poor ability to resist impact damage without extensive internal matrix fracture.

The response of damaged composites to cyclic loading is also significantly different from that of metals. In contrast to the poor composite static strength when it has damage or defects, the ability of composites to withstand cyclic loading is far superior to that of metals. The comparison of the normalized notched specimen fatigue response of a common aircraft metal (7075-T6 aluminum) and a carbon/epoxy laminate is shown in Fig. 1.19. The fatigue strength of the composite is much better relative to its static or residual strength. The static or residual strength requirement for structures is typically much higher than the fatigue requirement. Therefore, because the fatigue threshold of composites is a high percentage of their static or damaged residual strength, they are usually not fatigue-critical. In metal structures, fatigue is typically a critical design consideration.

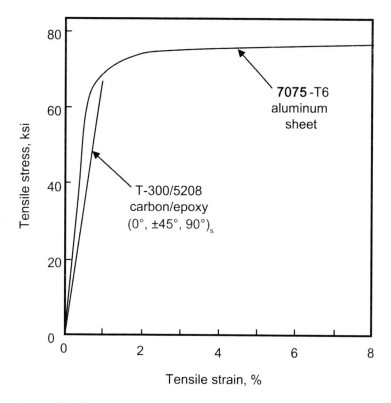

Fig. 1.18 Stress-strain comparison of aluminum and carbon/epoxy. Source: Ref 1.9

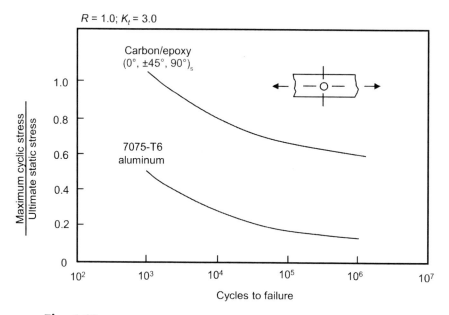

Fig. 1.19 Comparative notched fatigue strength. Source: Ref 1.9

Applications for polymer-matrix composites include aerospace, transportation, construction, marine, sporting goods, and, more recently, infrastructure. In general, high-performance but more costly continuous carbon-fiber composites are used where high strength and stiffness along with light weight are required, and much lower-cost fiberglass composites are used for less demanding applications where weight is not as paramount.

Military aircraft programs (Fig. 1.20) were the earliest users of high-performance continuous carbon-fiber composites, developing much of the technology now used by other industries. Both small and large commercial aircraft rely on composites to decrease weight and increase fuel performance, the most striking example being the 50% composite airframe for the new Boeing 787 (Fig. 1.21). All future Airbus and Boeing aircraft will use large amounts of high-performance composites.

Wind power is the world's fastest growing energy source. The blades for large wind mills (Fig. 1.22) are normally made from composites to improve electrical energy generation efficiency. These blades can be as long as 37 m (120 ft) and weigh up to 5215 kg (11,500 lb). In 2007, nearly

F-15 Eagle
early 1970s
2% composites

AV-8B Harrier
early 1980s
27% composites

F/A-18 Hornet
mid-1970s
10% composites

F/A 18 E/F Hornet
late 1990s
21% composites

Fig. 1.20 Typical fighter aircraft composite applications. Courtesy of The Boeing Company

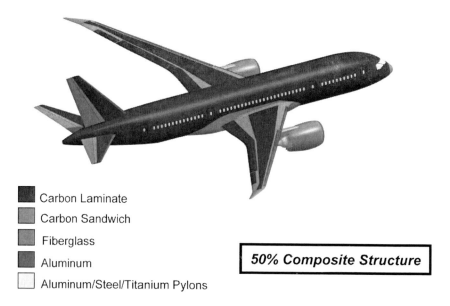

Carbon Laminate
Carbon Sandwich
Fiberglass
Aluminum
Aluminum/Steel/Titanium Pylons

50% Composite Structure

Fig. 1.21 Boeing 787 Dreamliner commercial airplane. Courtesy of The Boeing Company

Fig. 1.22 Composite wind energy application. Composites are being used for wind turbine blades to improve energy generation efficiency and reduce corrosion problems.

50,000 blades for 17,000 turbines were delivered, representing roughly 181 Mg (400 million lb) of composites. The predominant material is continuous glass fibers manufactured by either lay-up or resin infusion.

1.9 Metal-Matrix Composites

Metal-matrix composites offer a number of advantages compared to their base metals, such as higher specific strengths and moduli, higher elevated-temperature resistance, lower coefficients of thermal expansion, and, in some cases, better wear resistance. On the down side, they are more expensive than their base metals and have lower toughness. Metal-matrix composites also have some advantages compared to polymer-matrix composites, including higher matrix-dependent strengths and moduli, higher elevated-temperature resistance, no moisture absorption, higher electrical and thermal conductivities, and nonflammability. However, metal-matrix composites are normally more expensive than even polymer-matrix composites, and the fabrication processes are much more limited, especially for complex structural shapes. Due to their high cost, commercial applications for metal-matrix composites are sparse. However, because they are considered enablers for future hypersonic flight vehicles, both metal- and ceramic-matrix composites remain important materials. Fiber-metal laminates, in particular glass-fiber-reinforced aluminum laminates, are another form of composite material that offers fatigue performance advantages over monolithic aluminum structure.

Aluminum is the most commonly used metal as the matrix in discontinuously reinforced metal-matrix composites. Alumina (Al_2O_3) fibers reinforce the pistons for some diesel engines, and silicon carbide (SiC) fibers and whiskers are used in aerospace applications, including stiffeners and missile fins. Examples of cast discontinuously metal-matrix composite parts are shown in Fig. 1.23. Titanium and titanium aluminides reinforced with continuous SiC fibers are candidates for turbine blades and disks.

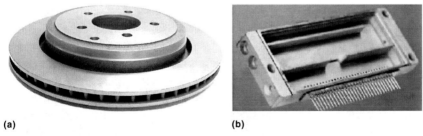

(a)   (b)

Fig. 1.23 Cast discontinuous aluminum-matrix composites. (a) An Al-SiC$_p$ composite brake rotor. (b) Aluminum-SiC microwave radio-frequency packaging for communication satellites. Source: Ref 1.10

However, their extremely high material and fabrication costs currently prohibit their application.

1.10 Structural Ceramics

Ceramics contain many desirable properties, such as high moduli, high compression strengths, high-temperature capability, high hardness and wear resistance, low thermal conductivity, and chemical inertness. However, due to their very low fracture toughness, ceramics are limited in structural applications. They have a very low tolerance to cracklike defects, which can occur either during fabrication or in service. Even a very small crack can quickly grow to critical size, leading to sudden failure. Most ceramics exhibit good strength under compression; however, they exhibit virtually no ductility under tension.

In metals, plastic flow takes place mainly by slip. In metals, due to the nondirectional nature of the metallic bond, dislocations move under relatively low stresses and because all atoms involved in the bonding have an equally distributed negative charge at their surfaces. In other words, there are no positive or negatively charged ions involved in the metallic bonding process. However, ceramics form either ionic or covalent bonds, both of which restrict dislocation motion and slip. One reason for the hardness and brittleness of ceramics is the difficulty of slip or dislocation motion. While ceramics are inherently strong, because they cannot slip or plastically deform to accommodate even small cracks or imperfections, their strength is never realized in practice. They facture in a premature brittle manner long before their inherent strength is approached.

Applications of structural ceramics take advantage of the temperature resistance, corrosion resistance, hardness, chemical inertness, thermal and electrical insulating properties, wear resistance, and mechanical properties. Ceramics offer advantages for structural applications because their density is approximately one-half the density of steel, and they provide very high stiffness-to-weight ratios over a broad temperature range. The high hardness of structural ceramics can be used in applications where mechanical abrasion or erosion is encountered. Structural ceramics are also being developed for turbine engine components.

Ceramics are used in a wide range of technologies, such as refractories, spark plugs, dielectrics in capacitors, sensors, abrasives, and magnetic recording media. Typical ceramic parts are shown in Fig. 1.24. The space shuttle makes use of ~25,000 reusable, lightweight, highly porous ceramic tiles that protect the aluminum frame from the heat generated during re-entry into the Earth's atmosphere. These tiles are made from high-purity silica fibers and colloidal silica coated with a borosilicate glass. The ability to maintain mechanical strength and dimensional tolerances at high temperature makes ceramics suitable for high-temperature use. For electrical applications, ceramics have high resistivity, low dielectric constant, and low

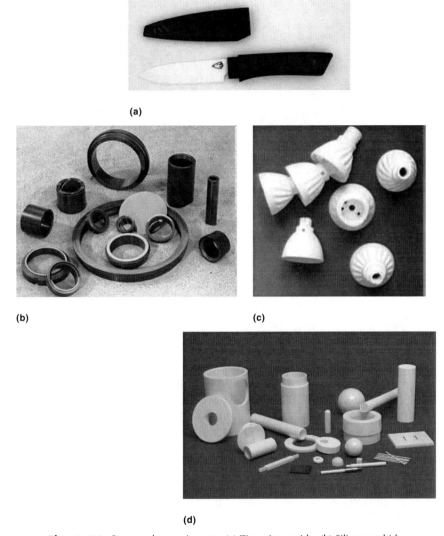

Fig. 1.24 Structural ceramic parts. (a) Zirconium oxide. (b) Silicon carbide. (c) Alumina. (d) Magnesia partially stabilized zirconia

loss factors that, when combined with their mechanical strength and high-temperature stability, make them suitable for extreme electrical insulating applications.

1.11 Ceramic-Matrix Composites

Due to their high costs and the concerns for reliability, there are also very few commercial applications for ceramic-matrix composites. However, carbon-carbon composites have found applications in aerospace for thermal protection systems. Carbon-carbon composites are the oldest and most mature of the ceramic-matrix composites. They were developed in

(a)

(b)

Fig. 1.25 Ceramic-matrix composite structural applications. (a) Ceramic-matrix ceramic exhaust nozzles. (b) Carbon-carbon brakes

the 1950s for use as rocket motor casings, heat shields, leading edges, and thermal protection (Fig. 1.25). The most recognized application is the Space Shuttle leading edges. For high-temperature applications, carbon-carbon composites offer exceptional thermal stability, provided they are protected with oxidation-resistant coatings.

Ceramic-fiber ceramic-matrix composites provide improved strength and fracture toughness compared with monolithic ceramics. While reinforcements such as fibers, whiskers, or particles are used to strengthen polymer- and metal-matrix composites, reinforcements in ceramic-matrix composites are primarily used to increase toughness. Fiber reinforcements improve the toughness of the ceramic matrix in several ways. First, a crack moving through the matrix encounters a fiber; if the bonding between the matrix and the fiber is poor, the crack is forced to propagate around the fiber to continue the fracture process. In addition, poor bonding allows the fiber to begin to pull out of the matrix. Both processes consume energy, thereby increasing fracture toughness. Finally, as a crack in the matrix begins, unbroken fibers may bridge the crack and make it more difficult for the crack to open.

REFERENCES

1.1   J.G. Kaufman, *Introduction to Aluminum Alloys and Tempers*, ASM International, 2000

1.2   Selection and Applications of Magnesium and Magnesium Alloys, *Metals Handbook Desk Edition*, 2nd ed., ASM International, 1998

1.3   K. Savage, Magnesium and Magnesium Alloys, *Casting*, Vol 15, *ASM Handbook*, ASM International, 2008

1.4 K. Droder, "Investigations of Forming of Magnesium Sheet Metal," Ph.D. dissertation, Hanover University, 1999 (in German)

1.5 G.W. Kuhlman, Forging of Titanium Alloys, *Metalworking: Bulk Forming*, Vol 14A, *ASM Handbook*, ASM International, 2005

1.6 M. Guclu, Titanium and Titanium Alloy Castings, *Casting*, Vol 15, *ASM Handbook*, ASM International, 2008

1.7 Forming of Titanium and Titanium Alloys, *Forming and Forging*, Vol 14, *ASM Handbook*, ASM International, 1988

1.8 M.C.Y. Niu, *Composite Airframe Structures*, 2nd ed., Hong Kong Conmilit Press Limited, 2000

1.9 R.E. Horton and J.E. McCarty, Damage Tolerance of Composites, *Composites*, Vol 1, *Engineered Materials Handbook*, ASM International, 1987

1.10 P.K. Rohatgi, N. Gupta, and A. Daoud, Synthesis and Processing of Cast Metal-Matrix Composites and Their Applications, *Casting*, Vol 15, *ASM Handbook*, ASM International, 2008

SELECTED REFERENCES

- M.F. Ashby, *Materials Selection in Mechanical Design*, 3rd ed., Elsevier, 2005
- J.A. Charles, F.A.A. Crane, and J.A.G. Furness, *Selection and Use of Engineering Materials*, 3rd ed., Butterworth-Heinemann, 1997
- G. Davies, *Materials for Automobile Bodies*, Elsevier, 2003
- T. Hall, Special Metals Make Unparalleled Medical Devices Possible, *Adv. Mater. Proc.*, Vol 169 (No. 9), ASM International, 2011, p 34–36
- I.J. Polmear, *Light Alloys—From Traditional Alloys to Nanocrystals*, 4th ed., Elsevier, 2006

CHAPTER **2**

Aluminum Alloys

ALUMINUM is an industrial and consumer metal of great importance. Aluminum and its alloys are used for foil, beverage cans, cooking and food processing utensils, architectural and electrical applications, and structures for boats, aircraft, and other transportation vehicles. As a result of a naturally occurring tenacious surface oxide film (Al_2O_3), a great number of aluminum alloys have exceptional corrosion resistance in many atmospheric and chemical environments. Its corrosion and oxidation resistance is especially important in architectural and transportation applications. On an equal weight and cost basis, aluminum is a better electrical conductor than copper. Its high thermal conductivity leads to applications such as radiators and cooking utensils. Its low density is important for hand tools and all forms of transportation, especially aircraft. Wrought aluminum alloys display a good combination of strength and ductility. Aluminum alloys are among the easiest of all metals to form and machine. The precipitation-hardening alloys can be formed in a relatively soft state and then heat treated to much higher strength levels after forming operations are complete. In addition, aluminum and its alloys are not toxic and are among the easiest to recycle of any of the structural materials. Desirable properties of aluminum and its alloys include:

- Aluminum has a density of only 2.7 g/cm³ (0.10 lb/in.³), approximately one-third as much as steel (7.83 g/cm³, or 0.29 lb/in.³). One cubic foot of steel weighs approximately 220 kg (490 lb); a cubic foot of aluminum, only about 77 kg (170 lb). Such light weight, coupled with the high strength of some aluminum alloys (exceeding that of structural steel), permits the design and construction of strong, lightweight structures that are particularly advantageous for anything that moves—space vehicles and aircraft as well as all types of land and waterborne vehicles.
- Aluminum resists the kind of progressive oxidization that causes steel to rust away. The exposed surface of aluminum combines with oxy-

gen to form an inert aluminum oxide film only a few ten-millionths of an inch thick, which blocks further oxidation. Unlike iron rust, the aluminum oxide film does not flake off to expose a fresh surface to further oxidation. If the protective layer of aluminum is scratched, it will instantly reseal itself. The thin oxide layer itself clings tightly to the metal and is colorless and transparent. The discoloration and flaking of iron and steel rust do not occur on aluminum. Appropriately alloyed and treated, aluminum can resist corrosion by water, salt, and other environmental factors, and by a wide range of other chemical and physical agents.

- Aluminum surfaces can be highly reflective. Radiant energy, visible light, radiant heat, and electromagnetic waves are efficiently reflected, while anodized and dark anodized surfaces can be reflective or absorbent. The reflectance of polished aluminum, over a broad range of wavelengths, leads to its selection for a variety of decorative and functional uses.

- Aluminum typically displays excellent electrical and thermal conductivity, but specific alloys have been developed with high degrees of electrical resistivity. These alloys are useful, for example, in high-torque electric motors. Aluminum is often selected for its electrical conductivity, which is nearly twice that of copper on an equivalent weight basis. The requirements of high conductivity and mechanical strength can be met by use of long-line, high-voltage aluminum steel-cored reinforced transmission cable. The thermal conductivity of aluminum alloys, approximately 50 to 60% that of copper, is advantageous in heat exchangers, evaporators, electrically heated appliances and utensils, and automotive cylinder heads and radiators.

- Aluminum is nonferromagnetic, a property of importance in the electrical and electronics industries. It is nonpyrophoric, which is important in applications involving inflammable or explosive materials handling or exposure. Aluminum is also nontoxic and is routinely used in containers for food and beverages. It has an attractive appearance in its natural finish, which can be soft and lustrous or bright and shiny. It can be virtually any color or texture.

- The ease with which aluminum may be fabricated into any form is one of its most important assets. Often, it can compete successfully with cheaper materials having a lower degree of workability. The metal can be cast by any method known. It can be rolled to any desired thickness down to foil thinner than paper. Aluminum sheet can be stamped, drawn, spun, or roll formed. The metal also may be hammered or forged. Aluminum wire, drawn from rolled rod, may be stranded into cable of any desired size and type. There is almost no limit to the different profiles (shapes) in which the metal can be extruded.

Potential limitations of aluminum alloys include:

- Aluminum has only moderate stiffness. The modulus of elasticity of aluminum alloys is only about 69 GPa (10×10^6 psi), which is approximately two-thirds that of titanium and one-third that of steel.
- The fatigue strength is sometimes much lower than static strength. While precipitation-hardened aluminum alloys can be heat treated to strength levels as high as 517 MPa (75 ksi), the fatigue strength is only about 172 MPa (25 ksi). Increases in static tensile properties have not been accompanied by proportionate improvements in fatigue properties (Fig. 2.1). This lack of fatigue improvement is attributed to two factors: (1) cracks initiating at precipitate-free zones adjacent to grain boundaries, and (2) the re-solution of precipitate particles when they are cut by dislocations. In other words, the cut precipitate particles become smaller than the critical size for thermodynamic stability and redissolve. In addition, aluminum alloys do not have a true endurance limit, meaning that they will fail at even low stress levels if the number of stress cycles becomes great enough.
- High-strength 2xxx and 7xxx alloys can be difficult to fusion weld. The copper content in some of the 2xxx and 7xxx alloys makes them susceptible to weld cracking during fusion welding operations.

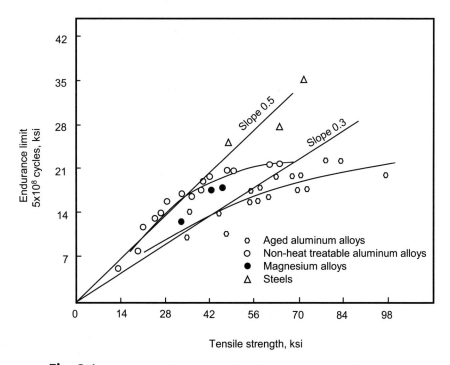

Fig. 2.1 Fatigue strength comparison for aluminum. Source: Ref 2.1

2.1 Applications

Aluminum alloys are economical in many applications. They are used in the automotive industry; aerospace industry; in construction of machines, appliances, and structures; as cooking utensils; as covers for housings for electronic equipment; as pressure vessels for cryogenic applications; and in innumerable other areas. Typical applications for some of the more commonly used wrought and cast alloys are listed in Tables 2.1 and 2.2, respectively.

Table 2.1 Selected applications for wrought aluminum alloys

Alloy	Description and selected applications
1100	Commercially pure aluminum highly resistant to chemical attack and weathering. Low cost, ductile for deep drawing, and easy to weld. Used for high-purity applications such as chemical processing equipment. Also for nameplates, fan blades, flue lining, sheet metal work, spun holloware, and fin stock
1350	Electrical conductors
2011	Screw machine products. Appliance parts and trim, ordnance, automotive, electronic, fasteners, hardware, machine parts
2014	Truck frames, aircraft structures, automotive, cylinders and pistons, machine parts, structurals
2017	Screw machine products, fittings, fasteners, machine parts
2024	For high-strength structural applications. Excellent machinability in the T-tempers. Fair workability and fair corrosion resistance. Alclad 2024 combines the high strength of 2024 with the corrosion resistance of the commercially pure cladding. Used for truck wheels, many structural aircraft applications, gears for machinery, screw machine products, automotive parts, cylinders and pistons, fasteners, machine parts, ordnance, recreation equipment, screws and rivets
2219	Structural uses at high temperature (to 315 °C, or 600 °F). High-strength weldments
3003	Most popular general-purpose alloy. Stronger than 1100 with same good formability and weldability. For general use including sheet metal work, stampings, fuel tanks, chemical equipment, containers, cabinets, freezer liners, cooking utensils, pressure vessels, builder's hardware, storage tanks, agricultural applications, appliance parts and trim, architectural applications, electronics, fin stock, fan equipment, name plates, recreation vehicles, trucks and trailers. Used in drawing and spinning.
3004	Sheet metal work, storage tanks, agricultural applications, building products, containers, electronics, furniture, kitchen equipment, recreation vehicles, trucks and trailers
3105	Residential siding, mobile homes, rain-carrying goods, sheet metal work, appliance parts and trim, automotive parts, building products, electronics, fin stock, furniture, hospital and medical equipment, kitchen equipment, recreation vehicles, trucks and trailers
5005	Specified for applications requiring anodizing; anodized coating is cleaner and lighter in color than 3003. Uses include appliances, utensils, architectural, applications requiring good electrical conductivity, automotive parts, containers, general sheet metal, hardware, hospital and medical equipment, kitchen equipment, name plates, and marine applications
5052	Stronger than 3003 yet readily formable in the intermediate tempers. Good weldability and resistance to corrosion. Uses include pressure vessels, fan blades, tanks, electronic panels, electronic chassis, medium-strength sheet metal parts, hydraulic tube, appliances, agricultural applications, architectural uses, automotive parts, building products, chemical equipment, containers, cooking utensils, fasteners, hardware, highway signs, hospital and medical equipment, kitchen equipment, marine applications, railroad cars, recreation vehicles, trucks and trailers
5056	Cable sheathing, rivets for magnesium, screen wire, zippers, automotive applications, fence wire, fasteners
5083	For all types of welded assemblies, marine components, and tanks requiring high weld efficiency and maximum joint strength. Used in pressure vessels up to 65 °C (150 °F) and in many cryogenic applications, bridges, freight cars, marine components, TV towers, drilling rigs, transportation equipment, missile components, and dump truck bodies. Good corrosion resistance
5086	Used in generally the same types of applications as 5083, particularly where resistance to either stress corrosion or atmospheric corrosion is important
5454	For all types of welded assemblies, tanks, pressure vessels. ASME code approved to 205 °C (400 °F). Also used in trucking for hot asphalt road tankers and dump bodies; also, for hydrogen peroxide and chemical storage vessels
5456	For all types of welded assemblies, storage tanks, pressure vessels, and marine components. Used where best weld efficiency and joint strength are required. Restricted to temperatures below 65 °C (150 °F)

(continued)

Source: Ref 2.2

Table 2.1 continued

Alloy	Description and selected applications
5657	For anodized auto and appliance trim and nameplates
6061	Good formability, weldability, corrosion resistance, and strength in the T-tempers. Good general-purpose alloy used for a broad range of structural applications and welded assemblies including truck components, railroad cars, pipelines, marine applications, furniture, agricultural applications, aircrafts, architectural applications, automotive parts, building products, chemical equipment, dump bodies, electrical and electronic applications, fasteners, fence
6063	Used in pipe railing furniture, architectural extrusions, appliance parts and trim, automotive parts, building products, electrical and electronic parts, highway signs, hospital and medical equipment, kitchen equipment, marine applications, machine parts, pipe, railroad cars, recreation equipment, recreation vehicles, trucks and trailers
7050	High-strength alloy in aircraft and other structures. Also used in ordnance and recreation equipment
7075	For aircraft and other applications requiring highest strengths. Alclad 7075 combines the strength advantages of 7075 with the corrosion-resisting properties of commercially pure aluminum-clad surface. Also used in machine parts and ordnance

Source: Ref 2.2

Table 2.2 Selected applications for aluminum casting alloys

Alloy	Representative applications
100.0	Electrical rotors larger than 152 mm (6 in.) in diameter
201.0	Structural members; cylinder heads and pistons; gear, pump, and aerospace housings
208.0	General-purpose castings; valve bodies; manifolds, and other pressure-tight parts
222.0	Bushings; meter parts; bearings; bearing caps; automotive pistons; cylinder heads
238.0	Sole plates for electric hand irons
242.0	Heavy-duty pistons; air-cooled cylinder heads; aircraft generator housings
A242.0	Diesel and aircraft pistons; air-cooled cylinder heads; aircraft generator housings
B295.0	Gear housings; aircraft fittings; compressor connecting rods; railway car seat frames
308.0	General-purpose permanent mold castings; ornamental grilles and reflectors
319.0	Engine crankcases; gasoline and oil tanks; oil pans; typewriter frames; engine parts
332.0	Automotive and heavy-duty pistons; pulleys, sheaves
333.0	Gas meter and regulator parts; gear blocks; pistons; general automotive castings
354.0	Premium-strength castings for the aerospace industry
355.0	Sand: air compressor pistons; printing press bedplates; water jackets; crankcases. Permanent: impellers; aircraft fittings; timing gears; jet engine compressor cases
356.0	Sand: flywheel castings; automotive transmission cases; oil pans; pump bodies Permanent: machine tool parts; aircraft wheels; airframe castings; bridge railings
A356.0	Structural parts requiring high strength; machine parts; truck chassis parts
357.0	Corrosion-resistant and pressure-tight applications
359.0	High-strength castings for the aerospace industry
360.0	Outboard motor parts; instrument cases; cover plates; marine and aircraft castings
A360.0	Cover plates; instrument cases; irrigation system parts; outboard motor parts; hinges
380.0	Housings for lawn mowers and radio transmitters; air brake castings; gear cases
A380.0	Applications requiring strength at elevated temperature
384.0	Pistons and other severe service applications; automatic transmissions
390.0	Internal combustion engine pistons, blocks, manifolds, and cylinder heads
413.0	Architectural, ornamental, marine, and food and dairy equipment applications
A413.0	Outboard motor pistons; dental equipment; typewriter frames; street lamp housings
443.0	Cookware; pipe fittings; marine fittings; tire molds; carburetor bodies
514.0	Fittings for chemical and sewage use; dairy and food handling equipment; tire molds
A514.0	Permanent mold casting of architectural fittings and ornamental hardware
518.0	Architectural and ornamental castings; conveyor parts; aircraft and marine castings
520.0	Aircraft fittings; railway passenger car frames; truck and bus frame sections
535.0	Instrument parts and other applications where dimensional stability is important
A712.0	General-purpose castings that require subsequent brazing
713.0	Automotive parts; pumps; trailer parts; mining equipment
850.0	Bushings and journal bearings for railroads
A850.0	Rolling mill bearings and similar applications

Source: Ref 2.2

2.2   Aluminum Metallurgy

Pure aluminum and its alloys have a face-centered cubic (fcc) structure, which is stable up to its melting point at 657 °C (1215°F). Because the fcc structure contains multiple slip planes, this crystalline structure greatly contributes to the excellent formability of aluminum alloys. Only a few elements have sufficient solid solubility in aluminum to be major alloying elements, including copper, magnesium, silicon, zinc, and lithium. Important elements with lower solid solubility are the transition metals chromium, manganese, and zirconium, which normally form compounds that help to control the grain structure. Aluminum alloys are normally classified into one of three groups: wrought non-heat-treatable alloys, wrought heat treatable alloys, and casting alloys.

The wrought non-heat-treatable alloys cannot be strengthened by precipitation hardening; they are hardened primarily by cold working. The wrought non-heat-treatable alloys include the commercially pure aluminum series (1xxx), the aluminum-manganese series (3xxx), the aluminum-silicon series (4xxx), and the aluminum-magnesium series (5xxx). While some of the 4xxx alloys can be hardened by heat treatment, others can only be hardened by cold working. Because the wrought non-heat-treatable alloys are hardened primarily by cold working, they are not adequate for load-bearing structural applications at elevated temperatures, because the cold-worked structure could start softening (i.e., recovery) in service.

The wrought heat treatable alloys can be precipitation hardened to develop quite high strength levels. These alloys include the 2xxx series (Al-Cu and Al-Cu-Mg), the 6xxx series (Al-Mg-Si), the 7xxx series (Al-Zn-Mg and Al-Zn-Mg-Cu), and the aluminum-lithium alloys of the 8xxx alloy series. The 2xxx and 7xxx alloys, which develop the highest strength levels, are the main alloys used for metallic aircraft structure.

2.3   Aluminum Alloy Designation

Wrought aluminum alloys are designated by a four-digit numerical system developed by the Aluminum Association (Table 2.3). The first digit defines the major alloying class of the series. The second digit defines variations in the original basic alloy; that digit is always a zero (0) for the original composition, a one (1) for the first variation, a two (2) for the second variation, and so forth. Variations are typically defined by differences in one or more alloying elements of 0.15 to 0.50% or more, depending on the level of the added element. The third and fourth digits designate the specific alloy within the series; there is no special significance to the values of those digits except in the 1xxx series, nor are they necessarily used in sequence.

For cast alloys (Table 2.4), the first digit again refers to the major alloying element, while the second and third digits identify the specific alloy. The zero after a period identifies the alloy as a cast product. If the period

Table 2.3 Designations for aluminum wrought alloys

Series	Aluminum content or main alloying element
1.*xxx*	99.00% min
2.*xxx*	Copper
3.*xxx*	Manganese
4.*xxx*	Silicon
5.*xxx*	Magnesium
6.*xxx*	Magnesium and silicon
7.*xxx*	Zinc
8.*xxx*	Others
9.*xxx*	Unused

Table 2.4 Designations for aluminum casting alloys

Series	Aluminum content or main alloying element
1.*xx*.0	99.00% min
2.*xx*.0	Copper
3.*xx*.0	Silicon with copper and/or magnesium
4.*xx*.0	Silicon
5.*xx*.0	Magnesium
6.*xx*.0	Unused
7.*xx*.0	Zinc
8.*xx*.0	Tin
9.*xx*.0	Other

is followed by the number "1," it indicates an ingot composition that would be supplied to a casting house. A letter prefix is used to denote either an impurity level or the presence of a secondary alloying element. These letters are assigned in alphabetical sequence starting with "A" but omitting "I," "O," "Q," and "X" ("X" is reserved for experimental alloys). For example, the designation A357.0 would indicate a higher purity level than the original alloy 357.0.

The temper designations for aluminum alloys are shown in Table 2.5. Alloys in the as-fabricated condition are designated by an "F"; those in the annealed condition are designated with an "O"; those in the solution-treated condition that have not attained a stable condition are designated with a "W"; and those that have been hardened by cold work are designated with an "H." If the alloy has been solution treated and then aged, either by natural or artificial aging, it is designated by a "T," with the specific aging treatment designated by the numbers "1" through "10."

To redistribute residual stresses after quenching, stress relieving by deformation is often applied to high-strength wrought products. This makes the product less susceptible to warping during machining and improves both the fatigue strength and stress-corrosion resistance. A summary of these stress-relieving tempers is shown in Table 2.6. When stress relieving is used, it is indicated by the number "5" following the last digit for precipitation-hardening tempers (e.g., Tx5x). The number designation "51" indicates stress relief by stretching, while the designation "52" indicates compression stress relief. If the product is an extrusion, a third digit may

Table 2.5 Temper designation for aluminum alloys

Suffix letter F, O, H, T, or W indicates basic treatment condition	First suffix digit indicates secondary treatment used to influence properties	Second suffix digit for condition H only indicates residual hardening
F = As-fabricated	1 = Cold worked only	2 = ¼ hard
O = Annealed; wrought products only	2 = Cold worked and partially	4 = ½ hard
H = Cold-worked, strain hardened	annealed	6 = ¾ hard
	3 = Cold worked and stabilized	8 = hard
		9 = Extra hard
W = Solution heat treated		
T = Heat treated; stable		
T1 = Cooled from an elevated-temperature shaping operation + natural age		
T2 = Cooled from an elevated-temperature shaping operation + cold worked + natural age		
T3 = Solution treated + cold worked + natural age		
T4 = Solution treated + natural age		
T5 = Cooled from an elevated-temperature shaping operation + artificial age		
T6 = Solution treated + artificially aged		
T7 = Solution treated + overaged		
T8 = Solution treated + cold worked + artificial aged		
T9 = Solution treated + artificial aged + cold worked		
T10 = Cooled from an elevated-temperature shaping operation + cold worked + artificial age		

Source: Ref 2.3

Table 2.6 Stress-relieving tempers for aluminum alloys

Stress relieved by stretching:

T-51: Applies to plate and rolled or cold-finished rod or bar, die or ring forgings, and rolled rings when stretched after solution heat treatment or after cooling from an elevated-temperature shaping process. The products receive no further straightening after stretching.

T-510: Applies to extruded rod, bar, profiles, and tubes and to drawn tube when stretched after solution heat treatment or after cooling from an elevated-temperature shaping process

T-51: Applies to extruded rod, bar, profiles, and tubes and to drawn tube when stretched after solution heat treatment or after cooling from an elevated-temperature shaping process. These products may receive minor straightening after stretching to comply with standard tolerances.

These stress-relieved temper products usually have larger tolerances on dimensions than products of other tempers.

Stress relieved by compressing:

T-52: Applies to products that are stress relieved by compressing after solution heat treatment or cooling from an elevated-temperature shaping process to produce a permanent set of 1 to 5%

Stress relieved by combined stretching and compressing:

T-54: Applies to die forgings that are stress relieved by restriking cold in the finish die

For wrought products heat treated from annealed or F temper (or other temper when such heat treatments result in the mechanical properties assigned to these tempers):

T-42: Solution heat treated from annealed or F temper and naturally aged to a substantially stable condition (example: 2024-T42)

T-62: Solution heat treated from annealed or F temper and artificially aged

Source: Ref 2.3

be used. The number "1" for extruded products indicates the product was straightened by stretching, while the number "0" indicates that it was not mechanically straightened.

2.4 Aluminum Alloys

The principal alloying elements in wrought aluminum alloys include copper, manganese, magnesium, silicon, and zinc. Alloys containing copper, magnesium + silicon, and zinc are precipitation hardenable to fairly high strength levels, while those containing manganese or magnesium are hardened primarily by cold working. The strength range for the different series of aluminum alloys is shown in Table 2.7. Note that the heat treatable alloys can obtain higher strengths than the cold-worked non-heat-treatable alloys.

2.5 Wrought Non-Heat-Treatable Alloys

The wrought non-heat-treatable alloys include the commercially pure aluminum alloys (1*xxx*), the aluminum-manganese alloys (3*xxx*), the aluminum-silicon alloys (4*xxx*), and the aluminum-magnesium alloys (5*xxx*). These alloys cannot be hardened by heat treatment and are therefore hardened by a combination of solid-solution strengthening (Fig. 2.2)

Table 2.7 Strength ranges of various wrought aluminum alloys

Aluminum Association series	Type of alloy composition	Strengthening method	Tensile strength range	
			MPa	ksi
1*xxx*	Al	Cold work	70–175	10–25
2*xxx*	Al-Cu-Mg (1–2.5% Cu)	Heat treat	170–310	25–45
2*xxx*	Al-Cu-Mg-Si (3–6% Cu)	Heat treat	380–520	55–75
3*xxx*	Al-Mn-Mg	Cold work	140–280	20–40
4*xxx*	Al-Si	Cold work (some heat treat)	105–350	15–50
5*xxx*	Al-Mg (1–2.5% Mg)	Cold work	140–280	20–40
5*xxx*	Al-Mg-Mn (3–6% Mg)	Cold work	280–380	40–55
6*xxx*	Al-Mg-Si	Heat treat	150–380	22–55
7*xxx*	Al-Zn-Mg	Heat treat	380–520	55–75
7*xxx*	Al-Zn-Mg-Cu	Heat treat	520–620	75–90
8*xxx*	Al-Li-Cu-Mg	Heat treat	280–560	40–80

Source: Ref 2.2

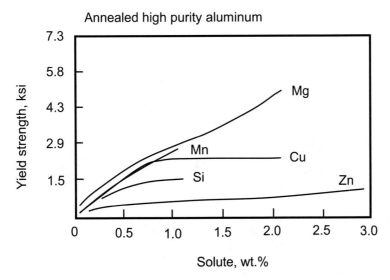

Fig. 2.2 Solid-solution strengthening of aluminum. Source: Ref 2.4

and cold working (Fig. 2.3). The chemical compositions of a number of wrought non-heat-treatable alloys are shown in Table 2.8, and representative mechanical properties are given in Table 2.9.

Commercially Pure Aluminum Alloys (1*xxx*). The major characteristics of the 1*xxx* series are:

- Work hardenable
- Exceptionally high formability, corrosion resistance, and electrical conductivity

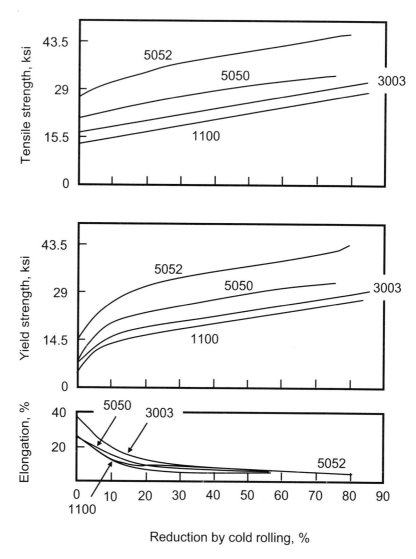

Fig. 2.3 Work-hardening curves for wrought non-heat-treatable aluminum alloys. Source: Ref 2.5

Table 2.8 Compositions of select wrought non-heat-treatable aluminum alloys

Alloy	Alloying element content, wt%			
	Cu	Mn	Mg	Cr
3003	0.12	1.2	...	...
3004	...	1.2	1.0	...
3005	...	1.2	0.4	...
3105	...	0.6	0.5	...
5005	...	...	0.8	...
5050	...	...	1.4	...
5052	...	...	2.5	0.25
5252	...	...	2.5	...
5154	...	...	3.5	0w.25
5454	...	0.8	2.7	0.12
5056	...	0.12	5.0	0.12
5456	...	0.8	5.1	0.12
5182	...	0.35	4.5	...
5083	...	0.7	4.4	0.15
5086	...	0.45	4.0	0.15

Note: All contain iron and silicon as impuritiesSource: Ref 2.6

Table 2.9 Mechanical properties of select wrought non-heat-treatable aluminum alloys

Alloy	Temper	Tensile strength		Yield strength		Elongation in 2 in., %
		MPa	ksi	MPa	ksi	
3003	O	110	16	41	6	30
	H14	124	18	117	17	9
	H18	165	24	152	22	5
3004	O	179	26	69	10	20
	H34	241	35	200	29	9
	H38	283	41	248	36	5
	H19	297	43	283	41	2
3005	O	131	19	55	8	25
	H14	179	26	165	24	7
	H18	241	35	228	33	4
3105	O	117	17	55	8	24
	H25	179	26	159	23	8
	H18	214	31	193	28	3
5005	O	124	18	41	6	25
	H34	159	23	138	20	8
	H38	200	29	186	27	5
5050	O	145	21	55	8	24
	H34	193	28	165	24	8
	H38	221	32	200	29	6
5052	O	193	28	90	13	25
	H34	262	38	214	31	10
	H38	290	42	255	37	7
5252	O	179	26	83	12	23
	H25	234	34	172	25	11
	H28	283	41	241	35	5
5154	O	241	35	117	17	27
	H34	290	42	228	33	13
	H38	331	48	269	39	10
	H112	241	35	117	17	25

(continued)

Source: Ref 2.6

Table 2.9 (continued)

Alloy	Temper	Tensile strength		Yield strength		Elongation in 2 in., %
		MPa	ksi	MPa	ksi	
5454	O	248	36	117	17	22
	H34	303	44	241	35	10
	H111	248	36	124	18	18
	H112	248	36	124	18	18
5056	O	290	42	152	22	35
	H18	434	63	407	59	10
	H38	414	60	345	50	15
5456	O	310	45	159	23	24
	H112	310	45	165	24	22
	H116	352	51	255	37	16
5182	O	276	40	131	19	21

Source: Ref 2.6

- Typical ultimate tensile strength range of 70 to 185 MPa (10 to 27 ksi)
- Readily joined by welding, brazing, and soldering

The 1xxx series of aluminum alloys includes both the superpurity grades (99.99%) and the commercially pure (CP) grades containing up to 1% impurities or minor additions. The last two digits of the alloy number denote the two digits to the right of the decimal point of the percentage of the material that is aluminum. For example, 1060 denotes an alloy that is 99.60% Al.

The 1xxx series of alloys are work hardenable but would not be used where strength is a prime consideration. The primary uses of the 1xxx series are applications in which the combination of extremely high corrosion resistance and formability are required, for example, foil and strip for packaging, chemical equipment, tank car or truck bodies, spun hollowware, and elaborate sheet metal work.

The more prevalent CP grades are available in most product forms and are used for applications such as electrical conductors, chemical processing equipment, aluminum foil, cooking utensils, and architectural products. Because these alloys are essentially free of alloying additions, they exhibit excellent corrosion resistance to atmospheric conditions. The most popular 1xxx alloy is alloy 1100, which has a tensile strength of 90 MPa (13 ksi) that can be increased to 165 MPa (24 ksi) by work hardening.

Electrical applications are one major use of the 1xxx series, primarily 1350, which has relatively tight controls on those impurities that may lower electrical conductivity. As a result, an electrical conductivity of 62% of the International Annealed Copper Standard (IACS) is guaranteed for this material, which, combined with the natural light weight of aluminum, means a significant weight and therefore cost advantage over copper in electrical applications.

Aluminum-Manganese Alloys (3xxx). The major characteristics of the 3xxx series are:

- High formability and corrosion resistance with medium strength
- Typical ultimate tensile strength range of 110 to 285 MPa (16 to 41 ksi)
- Readily joined by all commercial procedures

The 3xxx series of alloys are often used where higher strength levels are required along with good ductility and excellent corrosion resistance. The aluminum-manganese alloys contain up to 1.25% Mn; higher amounts are avoided because the presence of iron impurities can result in the formation of large primary particles of Al_6Mn, which causes embrittlement. Additions of magnesium provides improved solid-solution hardening, as in the alloy 3004, which is used for beverage cans, the highest single usage of any aluminum alloys, accounting for approximately a quarter of the total usage of aluminum. The 3xxx series of alloys are strain hardenable, have excellent corrosion resistance, and are readily welded, brazed, and soldered.

Alloy 3003 is widely used in cooking utensils and chemical equipment because of its superiority in handling many foods and chemicals, and in builders' hardware because of its superior corrosion resistance. Alloy 3105 is a principal for roofing and siding. Because of the ease and flexibility of joining, 3003 and other members of the 3xxx series are widely used in sheet and tubular form for heat exchangers in vehicles and power plants. Typical applications of the 3xxx alloy series include automotive radiator heat exchangers and tubing in commercial power plant heat exchangers.

Aluminum-Silicon Alloys (4xxx). The major characteristics of the 4xxx series are:

- Heat treatable
- Good flow characteristics and medium strength
- Typical ultimate tensile strength range of 175 to 380 MPa (25 to 55 ksi)
- Easily joined, especially by brazing and soldering

The 4xxx series of alloys is not as widely used as the 3xxx and 5xxx alloys. There are two major uses of the 4xxx series, both generated by the excellent flow characteristics provided by relatively high silicon contents. The first is for forgings; the workhorse alloy is 4032, a medium-high-strength, heat treatable alloy used principally in applications such as forged aircraft pistons. The second major application is a weld filler alloy; here, the workhorse is 4043, used for gas metal arc welding (GMAW) and gas tungsten arc welding (GTAW) 6xxx alloys for structural and automotive applications.

Good flow provided by the high silicon content leads to both types of applications. In the case of forgings, this good flow ensures the complete and precise filling of complex dies; in the case of welding, it ensures complete filling of grooves in the members to be joined. For the same reason, other variations of the 4xxx alloys are used for the cladding on brazing sheet, the component that flows to complete the bond. Alloy 4043 is one of the most widely used weld wires in applications such as the automated welding of an autobody structure.

Aluminum-Magnesium Alloys (5xxx). The major characteristics of the 5xxx series are:

- Work hardenable
- Excellent corrosion resistance, toughness, and weldability; moderate strength
- Typical ultimate tensile strength range of 125 to 350 MPa (18 to 51 ksi)

The 5xxx series of alloys have the highest strengths of the non-heat-treatable alloys. They develop moderate strengths when work hardened, have excellent corrosion resistance even in saltwater, and have very high toughness even at cryogenic temperatures to near absolute zero. They are readily weldable by a variety of techniques, at thicknesses up to 20 cm (8 in.). Because aluminum and magnesium form solid solutions over a wide range of compositions, alloys containing magnesium in amounts from 0.8% to approximately 5% are widely used. The 5xxx-series alloys have relatively high ductilities, usually in excess of 25%. The ultimate tensile strength of the 5xxx alloys ranges from a low of 124 MPa (18 ksi) for 5005-O to a high of 310 MPa (45 ksi) for 5456-O.

As a result, 5xxx alloys find wide application in building and construction; highway structures, including bridges, storage tanks, and pressure vessels; cryogenic tankage and systems for temperatures as low as –270 °C (–455 °F) or near absolute zero; and marine applications.

The 5xxx alloys, although still having very good overall corrosion resistance, can be subject to intergranular and stress-corrosion cracking attack. In alloys with more than 3 to 4% Mg, there is a tendency for the β phase (Mn_5Al_8) to precipitate at the grain boundaries, making the alloy susceptible to grain-boundary attack. The precipitation of β occurs slowly at room temperature but can accelerate at elevated temperatures or under highly work-hardened conditions. For this reason, alloys such as 5454 and 5754 are recommended for applications where high-temperature exposure is likely.

A second problem that can be encountered with the 5xxx alloys is one of age softening at room temperature—essentially, over a period of time there is some localized recovery within the work-hardened grains. To avoid this effect, a series of H3 tempers are used in which the alloy is work hardened to a slightly greater level and then given a stabilization aging treatment at 120 to 150 °C (250 to 300 °F). This treatment also helps in reducing the tendency for β precipitation.

Alloys 5052, 5086, and 5083 are the workhorses from the structural standpoint, with increasingly higher strength associated with the increasingly higher magnesium content. Specialty alloys in the group include 5182, the beverage can end alloy and thus among the largest in tonnage; 5754 for automotive body panel and frame applications; and 5252, 5457, and 5657 for bright trim applications, including automotive trim.

High-speed, single-hull ships use 5083-H113/H321 machined plate for hulls, hull stiffeners, decking, and superstructure. Single- or multiple-hull high-speed ferries employ several aluminum-magnesium alloys, 5083, 5383, and 5454, as sheet and plate with all-welded construction.

2.6 Wrought Heat Treatable Alloys

The wrought heat treatable alloys include the aluminum-copper (2xxx) series, the aluminum-magnesium-silicon series (6xxx), the aluminum-zinc (7xxx) series, and the aluminum-lithium alloys of the 8xxx series. These alloys are strengthened by precipitation hardening. The chemical compositions of a number of the wrought heat treatable aluminum alloys are given in Table 2.10, and the mechanical properties of a number of alloys are shown in Table 2.11.

Table 2.10 Compositions of select wrought heat treatable aluminum alloys

	Alloying element content, wt%							
Alloy	Fe	Si	Cu	Mn	Mg	Cr	Zn	Zr
2008	0.40(a)	0.65	0.9	0.3(a)	0.4	...	...	...
2219(b)	0.30(a)	0.20(a)	6.3	0.3	...	...	...	0.18
2519(b)	0.39(a)(c)	0.30(a)(c)	5.8	0.3	0.25	...	...	0.18
2014	0.7(a)	0.8	4.4	0.8	0.5	...	...	...
2024	0.50(a)	0.50(a)	4.4	0.6	1.5	...	...	...
2124	0.30(a)	0.20(a)	4.4	0.6	1.5	...	...	...
2224	0.15(a)	0.12(a)	4.4	0.6	1.5	...	...	...
2324	0.12(a)	0.10(a)	4.4	0.6	1.5	...	...	...
2524	0.12(a)	0.06(a)	4.25	0.6	1.4	...	...	...
2036	0.50(a)	0.50(a)	2.6	0.25	0.45	...	...	...
6009	0.50(a)	0.8	0.4	0.5	0.6	...	...	...
6061	0.7(a)	0.6	0.3	...	1.0	0.2	...	...
6063	0.50(a)	0.4	...	...	0.7	...	...	...
6111	0.4(a)	0.9	0.7	0.3	0.8	...	...	...
7005	0.40(a)	0.35(a)	...	0.45	1.4	0.13	4.5	0.14
7049	0.35(a)	0.25(a)	1.6	...	2.4	0.16	7.7	...
7050	0.15(a)	0.12(a)	2.3	...	2.2	...	6.2	0.12
7150	0.15(a)	0.10(a)	2.2	...	2.4	...	5.4	0.12
7055	0.15(a)	0.10(a)	2.3	...	2.1	...	8.0	0.12
7075	0.50(a)	0.40(a)	1.6	...	2.5	0.25	5.6	...
7475	0.12(a)	0.10(a)	1.6	...	2.2	0.20	5.7	...

(a) Maximum allowable amount. (b) 2219 and 2519 also contain 0.10% V and 0.06% Ti. (c) 0.40% Fe max plus Si. Source: Ref 2.6

Table 2.11 Mechanical properties of select wrought heat treatable aluminum alloys

Alloy	Temper	Product form	Tensile strength		Yield strength		Elongation in 2 in., %
			MPa	ksi	MPa	ksi	
2008	T4	Sheet	248	36	124	18	28
	T6	Sheet	303	44	241	35	13
2014	T6, T651	Plate, forging	483	70	414	60	13
2024	T3, T351	Sheet, plate	448	65	310	45	18
	T361	Sheet, plate	496	72	393	57	13
	T81, T851	Sheet, plate	483	70	448	65	6
	T861	Sheet, plate	517	75	490	71	6
2224	T3511	Extrusion	531	77	400	58	16
2324	T39	Plate	503	73	414	60	12
2524	T3, T351	Sheet, plate	448	65	310	45	21
2036	T4	Sheet	338	49	193	28	24
2219	T81, T851	Sheet, plate	455	66	352	51	10
	T87	Sheet, plate	476	69	393	57	10
2519	T87	Plate	490	71	428	62	10
6009	T4	Sheet	221	32	124	18	25
	T62	Sheet	3303	44	262	38	11
6111	T4	Sheet	283	41	166	24	25
	T6	Sheet	352	51	310	45	10
6061	T6, T6511	Sheet, plate, extrusion, forging	310	45	276	40	12
	T9	Extruded rod	407	59	393	57	12
6063	T5	Extrusion	186	27	145	21	12
	T6	Extrusion	241	35	214	31	12
7005	T5	Extrusion	352	51	290	42	13
7049	T73	Forging	538	78	476	69	10
7050	T74, T745X	Plate, forging, extrusion	510	74	448	65	13
7150	T651, T6151	Plate	600	87	559	81	11
	T77511	Extrusion	648	94	614	89	12
7055	T7751	Plate	641	93	614	89	10
	T77511	Extrusion	669	97	655	95	11
7075	T6, T651	Sheet, plate	573	83	503	73	11
	T73, T735X	Plate, forging	503	73	434	63	13
7475	T7351	Plate		73	434	63	15
	T7651	Plate		66	393	57	15

Source: Ref 2.6

Aluminum-Copper Alloys (2xxx). The major characteristics of the 2xxx series are:

- Heat treatable
- High strength at room and elevated temperatures
- Typical ultimate tensile strength range of 190 to 430 MPa (27 to 62 ksi)
- Usually joined mechanically, but some alloys are weldable

The 2xxx series of alloys are heat treatable and possess in individual alloys good combinations of high strength (especially at elevated temperatures), toughness, and, in specific cases, weldability. They are not as resistant to atmospheric corrosion as several other series and so usually are painted or clad for added protection.

The high-strength 2*xxx* and 7*xxx* alloys are competitive on a strength-to-weight ratio with the higher-strength but heavier titanium and steel alloys and thus have traditionally been the dominant structural material in both commercial and military aircraft. In addition, aluminum alloys are not embrittled at low temperatures and become even stronger as the temperature is decreased without significant ductility losses, making them ideal for cryogenic fuel tanks for rockets and launch vehicles.

The wrought heat treatable 2*xxx* alloys generally contain magnesium in addition to copper as an alloying element. Other significant alloying additions include titanium to refine the grain structure during ingot casting, and transition element additions (manganese, chromium, and/or zirconium) that form dispersoid particles that help control the wrought grain structure. Iron and silicon are considered impurities and are held to an absolute minimum, because they form intermetallic compounds that are detrimental to both fatigue and fracture toughness.

Due to their superior damage tolerance and good resistance to fatigue crack growth, the 2*xxx* alloys are used for aircraft fuselage skins. The 2*xxx* alloys are also used for lower wing skins on commercial aircraft, while the 7*xxx* alloys are used for upper wing skins, where strength is the primary design driver. The superior fatigue performance of 2024-T3 compared to 7075-T6 in the 10^5 cycle range has led to the widespread use of the 2*xxx* alloys in tension-tension applications.

Alloy 2024 has been the most widely used alloy in the 2*xxx* series. It is normally supplied in the T3 temper (i.e., solution heat treated, cold worked, and then naturally aged). Typical cold working operations include roller or stretcher leveling to achieve flatness that introduces modest strains in the range of 1 to 4%. Although 2024-T3 only has a moderate yield strength, it has very good resistance to fatigue crack growth and good fracture toughness. Alloy 2024-T3 sheet is commonly used for fuselage skins where it is Alclad for corrosion protection. The T8 heat treatment is also frequently used with 2*xxx* alloys in which the alloy is solution heat treated, cold worked, and then artificially aged. Cold working before aging helps to nucleate precipitates, decrease the number and size of grain-boundary precipitates, and reduce the aging time required to obtain peak strength. The T8 temper also reduces the susceptibility to stress-corrosion cracking.

Dramatic improvements in aluminum alloys have occurred since they were first introduced in the 1920s. These improvements are a result of a much better understanding of chemical composition, impurity control, and the effects of processing and heat treatment. Improved alloys developed for lower wing skin structure include 2324-T39 plate and 2224-T351 extrusions. Compared to 2024, both compositional and processing changes for these two alloys resulted in improved properties (Fig. 2.4). A lower volume fraction of intermetallic compounds improved fracture toughness. For example, the maximum iron content is 0.12% and silicon is 0.10% in 2224 as compared to 0.50% for both impurities in 2024. The tensile yield strength of the newer plate materials was also increased by increasing the amount of cold work (stretching) after quenching. Note that as the yield strength increases, fracture toughness decreases, a phenomenon observed not only with aluminum alloys but all structural metallic materials.

Although copper is the main alloying element that contributes to the high strength of the 2*xxx* alloys, the corrosion resistance of aluminum alloys is usually an inverse function of the amount of copper in the alloy; therefore, the 2*xxx* alloys, which usually contain approximately 4% Cu, are the least corrosion-resistant alloys. Therefore, 2*xxx* alloys are often Alclad with a thin coating of pure aluminum or an aluminum alloy (e.g., Al-1%Zn) to provide corrosion protection. The Alclad is chosen so that it is anodic to the core alloy and corrodes preferentially instead of the underlying core alloy. The Alclad, which is applied during rolling operations, is usually in the range of 1.5 to 10% of the alloy thickness. Because the Alclad material is usually not as strong as the core material, there is a slight sacrifice in mechanical properties.

The higher-strength 2*xxx* alloys are widely used in truck body (2014) applications, where they generally are used in bolted or riveted construction. Specific members of the series (e.g., 2219 and 2048) are readily joined by GMAW or GTAW and so are used for aerospace applications where that method is the preferred joining method. Alloys 2011, 2017, and 2117 are widely used for fasteners and screw machine stock.

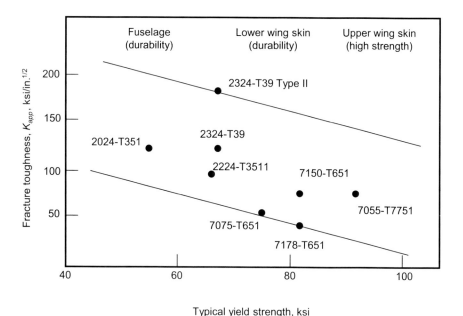

Fig. 2.4 Fracture toughness versus yield strength. Source: Ref 2.7

Aluminum-Magnesium-Silicon Alloys (6xxx). The major characteristics of the 6xxx series are:

- Heat treatable
- High corrosion resistance, excellent extrudability, moderate strength
- Typical ultimate tensile strength range of 125 to 400 MPa (18 to 58 ksi)
- Readily welded by GMAW and GTAW methods

The combination of magnesium (0.6 to 1.2%) and silicon (0.4 to 1.3%) in aluminum forms the basis of the 6xxx precipitation-hardenable alloys. During precipitation hardening, the intermetallic compound Mg_2Si provides the strengthening. Manganese or chromium is added to most 6xxx alloys for increased strength and grain size control. Copper also increases the strength of these alloys, but if present in amounts over 0.5%, it reduces the corrosion resistance.

The 6xxx alloys are heat treatable to moderately high strength levels, have better corrosion resistance than the 2xxx and 7xxx alloys, are weldable, and offer superior extrudability. With a yield strength comparable to that of mild steel, 6061 is one of the most widely used of all aluminum alloys. The highest strengths are obtained when artificial aging is started immediately after quenching. Losses of 21 to 28 MPa (3 to 4 ksi) in strength occur if these alloys are room-temperature aged for one to seven days. Alloy 6063 is widely used for general-purpose structural extrusions because its chemistry allows it to be quenched directly from the extrusion press. Alloy 6061 is used where higher strength is required, and 6071 where the highest strength is required.

Although the 6xxx alloys have not traditionally been able to compete with the 2xxx and 7xxx alloys in applications requiring high strength, a relatively new alloy (6013-T6) has 12% higher strength than Alclad 2024-T3 with comparable fracture toughness and resistance to fatigue crack growth rate, and it does not have to be clad for corrosion protection.

The 6xxx alloys can be welded, while most of the 2xxx and 7xxx alloys have very limited weldability. Some of the 6xxx alloys, such as 6013-T6 and 6056-T6, which contain appreciable copper, tend to form precipitate-free zones at the grain boundaries during precipitation hardening, making them susceptible to intergranular corrosion. However, a new temper (T78) desensitizes 6056 to intergranular corrosion while maintaining a tensile strength close to that of the T6 temper.

A unique feature is their great extrudability, making it possible to produce in single shapes relatively complex architectural forms, as well as to design shapes that put the majority of the metal where it will most efficiently carry the highest tensile and compressive stresses. This feature is a particularly important advantage for architectural and structural members where stiffness-criticality is important.

Alloy 6063 is perhaps the most widely used because of its extrudability; it is not only the first choice for many architectural and structural members, but it has been the choice for the Audi automotive space frame members. A good example of its structural use was the all-aluminum bridge structure in Foresmo, Norway; it was prefabricated in a shop and erected on the site in only a few days. Higher-strength alloy 6061 extrusions and plate find broad use in welded structural members such as truck and marine frames, railroad cars, and pipelines.

Among specialty alloys in the series are 6066-T6, with high strength for forgings; 6070 for the highest strength available in 6xxx extrusions; and 6101 and 6201 for high-strength electrical bus and electrical conductor wire, respectively.

Some of the other most important applications for Al-Mg-Si are structural members of wide-span roof structures for arenas and gymnasiums; geodesic domes, such as the one made originally to house the Spruce Goose, the famous Hughes wooden flying boat, in Long Beach, CA, which was the largest geodesic dome ever constructed, at 300 m (1000 ft) across and 120 m (400 ft) high; an integrally stiffened bridge deck shape, used to produce replacement bridge decks, readily put in the roadway in hours; and a magnetic levitation (Mag-Lev) train in development in Europe and Japan. In addition, aluminum light poles are widely used around the world for their corrosion resistance and crash protection systems, providing safety for auto drivers and passengers.

Aluminum-Zinc Alloys (7xxx). The major characteristics of the 7xxx series are:

- Heat treatable
- Very high strength; special high-toughness versions
- Typical ultimate tensile strength range of 220 to 610 MPa (32 to 88 ksi)
- Mechanically joined or friction stir welded

The 7xxx alloys are heat treatable and, among the Al-Zn-Mg-Cu versions in particular, provide the highest strengths of all aluminum alloys.

The wrought heat treatable 7xxx alloys are even more responsive to precipitation hardening than the 2xxx alloys and can obtain higher strength levels. These alloys are based on the Al-Zn-Mg(-Cu) system. The 7xxx alloys can be naturally aged but are not because they are not stable if aged at room temperature; that is, their strength will gradually increase with increasing time and can continue to do so for years. Therefore, all 7xxx alloys are artificially aged to produce a stable alloy.

The Al-Zn-Mg-Cu versions provide the highest strengths of all aluminum alloys. Some of the 7xxx alloys contain approximately 2% Cu in combination with magnesium and zinc to develop their strength. These alloys, such as 7049, 7050, 7075, 7175, 7178, and 7475, are the strongest but least corrosion resistant of the series. However, the addition of copper does help in stress-corrosion cracking resistance, because it allows precipitation hardening at higher temperatures. The copper-free 7xxx alloys (e.g., 7005 and 7029) have lower strengths but are tougher and exhibit better weldability. However, the majority of the 7xxx alloys are not considered fusion weldable and are therefore joined with mechanical fasteners.

There are several alloys in the 7xxx series that are produced especially for their high fracture toughness, such as 7050, 7150, 7175, 7475, and 7085. Similar to some of the newer 2xxx alloys, controlled impurity levels, particularly iron and silicon, maximize the combination of strength and fracture toughness. For example, while the total iron and silicon content is 0.90% maximum in 7075, the combined total is limited to 0.22% in 7475. Therefore, the size and number of intermetallic compounds that assist crack propagation are much reduced in 7475. The importance of controlling impurity levels is illustrated in Table 2.12. Note the improvement in fracture toughness of 7149 and 7249 compared to the original 7049 composition.

In the peak aged T6 condition, thick plate, forgings, and extrusions of many of the 7xxx alloys are susceptible to stress-corrosion cracking (SCC), particularly when stressed through the thickness (i.e., short transverse). For example, SCC of 7075-T6 has frequently occurred in service. To combat the SCC problem, a number of overaged T7 tempers have been developed for the 7xxx alloys. Although there is usually some sacrifice in strength, these tempers have dramatically reduced the susceptibility to SCC. An additional benefit is that they improve the fracture toughness, especially the through-the-thickness fracture toughness of thick plate. For example, the overaged T73 temper, which was originally developed to reduce the susceptibility to SCC, reduces the yield strength of 7075 by 15% but increases the threshold stress for SCC for 7075 by a factor of 6. Other overaged tempers, such as the T74, T75, and the T77 tempers, were then developed to provide trade-offs in strength and SCC resistance between the T6 and T73 tempers.

Table 2.12 Effect of impurity content on high-strength aluminum extrusions

Alloy and temper	Si max	Fe max	Mn max	Tensile strength		Ultimate tensile strength		Elongation, %	K_{Ic}	
				MPa	ksi	MPa	ksi		MPa$\sqrt{m}$	ksi$\sqrt{in.}$
7049-T7351 1	0.25	0.35	0.35	503	73	552	80	11.6	26	24
7149-T73511	0.15	0.20	0.20	517	75	565	82	13.3	33	30
7249-T73511	0.20	0.12	0.12	531	77	579	84	13.3	37	34

Source: Ref 2.8

Other Aluminum Alloys (8*xxx*). The 8*xxx* series is reserved for those alloys with lesser-used alloying elements, such as iron, nickel, and lithium. Iron and nickel provide strength with little loss in electrical conductivity and so are used in a series of alloys represented by 8017 for conductors. Aluminum-iron alloys have also been developed for potential elevated-temperature applications. The 8*xxx* series also contains some of the high-strength aluminum-lithium alloys that are potential airframe materials.

Aluminum-lithium alloys are attractive for aerospace applications because the addition of lithium increases the modulus of aluminum and reduces the density. Each 1 wt% of lithium increases the modulus by approximately 6% while decreasing the density approximately 3%. However, the early promise of property improvements with aluminum-lithium alloys has not been realized. Even the second-generation alloys that came out in the 1980s and contained approximately 2% Li experienced a number of serious technical problems: excessive anisotropy in the mechanical properties; lower-than-desired fracture toughness and ductility; hole cracking and delamination during drilling; and low SCC thresholds. The anisotropy experienced by these alloys is a result of the strong crystallographic textures that develop during processing, with the fracture toughness problem being one of primarily low strength in the short-transverse direction.

The second-generation alloys included 2090, 2091, 8090, and 8091 in the western world and 1420 in Russia. The aluminum-lithium alloys 2090, 2091, 8090, and 8091 contain 1.9 to 2.7% Li, which results in approximately a 10% lower density and 25% higher specific stiffness than the 2*xxx* and 7*xxx* alloys. To circumvent some of these problems, a third generation of alloys has been developed with lower lithium contents. Alloy 2195 also has a lower copper content and has replaced 2219 for the cryogenic fuel tank on the Space Shuttle, where it provides a higher strength, higher modulus, and lower density than 2219. Other alloys, including 2096, 2097, and 2197, also have lower copper contents but also have slightly higher lithium contents than 2195.

2.7 Melting and Primary Fabrication

Aluminum production starts with the mineral bauxite, which contains approximately 50% alumina (Al_2O_3). In the Bayer process, pure alumina is extracted from bauxite using a sodium hydroxide solution to precipitate aluminum hydroxide, which is then subjected to calcination to form alumina. The Hall-Héroult process is then used to reduce the alumina to pure aluminum. This is an electrolytic process (Fig. 2.5) in which alumina, dissolved in a bath of cryolite, is reduced to pure aluminum by high electrical currents. It should be pointed out that the production of aluminum takes a lot of electrical energy. Because recycling aluminum takes much less energy, a large portion of general-purpose aluminum is currently made from recycled material.

During casting of aluminum ingots, it is important to remove as many oxide inclusions and as much hydrogen gas as possible. Oxides originate primarily from moisture on the furnace charge being melted; therefore, every effort is made to ensure that the materials are dry and free of moisture. Hydrogen gas can cause surface blistering in sheets and is a primary cause of porosity in castings. Fluxing with chlorine, inert gases, and salts removes the oxides and hydrogen from the melt before casting.

Semicontinuous direct chill casting is the primary method for producing ingots for aluminum alloys. In this process (Fig. 2.6), the molten metal is extracted through the bottom of a water-cooled mold, producing fine-grained ingots with a minimum amount of segregation. Low-melting-point alloying elements, such as magnesium, copper, and zinc, are added to the molten charge as pure elements, while high-melting elements (e.g., titanium, chromium, zirconium, and manganese) are added in the form of master alloys. To prevent hot cracking and refine the grain size, inoculants such as titanium and titanium-boron are added to the melt.

The molten aluminum is poured into a shallow, water-cooled, cross-sectional shape of the ingot desired. When the metal begins to freeze in the mold, the false bottom of the mold is slowly lowered, and water is sprayed on the surface of the freshly solidified metal as it comes out of the mold.

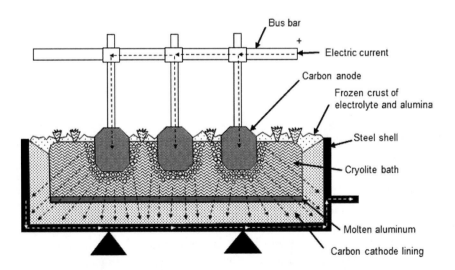

Fig. 2.5 Electrolytic cell used to produce aluminum. Courtesy of Aluminum Company of America

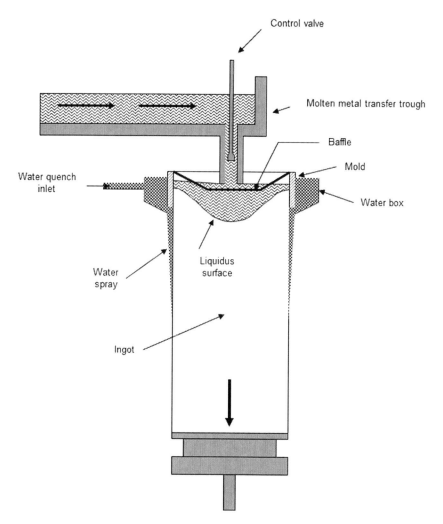

Fig. 2.6 Semicontinuous direct chill casting. Source: Ref 2. 9

The temperature and flow rate of the water are controlled so that it will wet the surfaces and then cascade down the surfaces. Typical casting speeds are in the range of 2.5 to 13 cm/min (1 to 5 in./min).

Because the liquid metal freezing front is almost horizontal and the metal freezes from the bottom to the top of the ingot, the direct chill casting process produces fine-grained ingots with a minimum of segregation. It can also produce fairly large ingots at slow speeds, a necessary requirement for the high-strength alloys to prevent cracking. Also, metal can be transferred to the mold slowly, uniformly, and at relatively low temperatures. Low temperatures, only about 28 °C (50 °F) above the liquidus temperature, are used to minimize hydrogen pickup and oxide formation.

Also, if the temperature of the aluminum is too high, coarse-grained structures will result.

2.8 Rolling Plate and Sheet

Hot rolling is conducted at temperatures above the recrystallization temperature to create a finer grain size and less grain directionality. The upper temperature is determined by the lowest-melting-point eutectic in the alloy, while the lower temperature is determined by the lowest temperature that can safely be passed through the rolling mill without cracking. Hot rolling of as-cast ingots consists of:

1. Scalping of ingots
2. Homogenizing the ingots
3. Reheating the ingots to the hot rolling temperature, if necessary
4. Hot rolling to form a slab
5. Intermediate annealing
6. Cold rolling and annealing for sheet

Scalping of the ingot, in which approximately 6.4 to 9.70 mm (0.25 to 0.38 in.) of material is removed from each surface, is conducted so that surface defects will not be rolled into the finished sheet and plate. Homogenizing of the ingots removes any residual stresses in the ingots and improves the homogeneity of the as-cast structure by reducing the coring experienced during casting. Good temperature control is required during homogenization because the ingots are heated to within 11 to 22 °C (20 to 40 °F) of the lowest-melting eutectic in the alloy. Because homogenization is a diffusion-controlled process, long times are necessary to allow time for the alloying elements to diffuse from the grain boundaries and other solute-rich regions to the grain centers. An important function of homogenization is to remove nonequilibrium low-melting-point eutectics that could cause ingot cracking during subsequent hot working operations. If the alloy is going to be Alclad for corrosion protection, the scalping operation is usually done after homogenization to remove the heavy oxide layer that builds up during the rather long homogenization soaks, making it easier to obtain a good bond between the Alclad and the core during hot rolling.

The ingots can be hot rolled right after removing them from the soaking pits, or, if cooled to room temperature after homogenization, they must be reheated for hot rolling. Ingots as large as 6.1 m (20 ft) long by 1.8 m (6 ft) wide by 0.6 m (2 ft) thick weighing over 20 tons are initially hot rolled back and forth through the rolling mill into plate between 6.4 and 203 mm (0.250 and 8.0 in.) thick. Modern rolling mill facilities (Fig. 2.7) can heat the plate, roll it to the desired thickness, spray quench it to harden it, and then stretch it to relieve stresses. Hot rolling helps to break up the as-cast structure, provide a more uniform grain size, and give a better distribution and size of constituent particles. During hot rolling, the grain structure becomes elongated in the rolling direction (Fig. 2.8). This grain directionally can have a substantial effect on some of the mechanical properties, especially fracture toughness and corrosion resistance, in which the properties are lowest in the through-the-thickness or short-transverse direction.

After the slab has been reduced in thickness, it is removed from the mill, given an intermediate anneal, and then placed in a five-stand, four-high mill to roll to thinner plate or sheet with successive reductions at each station. The cold work put into the aluminum during hot rolling must be sufficient to cause recrystallization during annealing. Intermediate anneals are required after cold reductions in the range of 45 to 85%. Intermediate anneals are required to keep the sheet from cracking during cold rolling; however, the amount of cold work must be sufficient to cause a fine grain size during annealing. If the final product form is sheet, it is annealed and then sent to a four-high cold rolling mill for further reduction. The number of intermediate anneals required during cold rolling depends on the alloy and the final gage required.

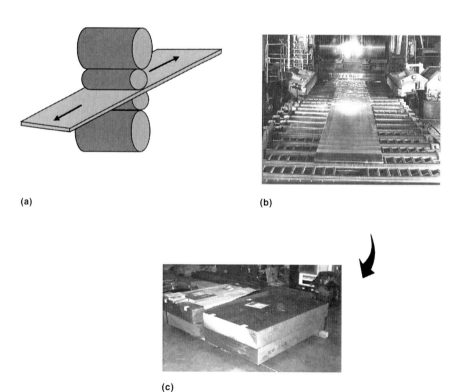

(a)

(b)

(c)

Fig. 2.7 Hot rolling aluminum plate. (a) Four-high mill. (b) Multiple passes. (c) Thick plate product

During spray quenching from the solution heat treating temperature, the surface cools much quicker than the center, resulting in residual stresses. The faster cooling surface develops compressive stresses, while the slower cooling center develops tensile stresses. This residual stress pattern with compressive stresses on the surface helps in preventing fatigue and SCC. However, during machining operations, if the surface material containing the compressive stresses is removed, the interior material with tensile residual stresses is exposed and the part is even more susceptible to warping, fatigue, and SCC. To minimize these problems, aluminum plate and extrusions are often stress relieved by stretching 0.5 to 5%, a temper designated as T*x*5*x* or T*x*5*x*26.

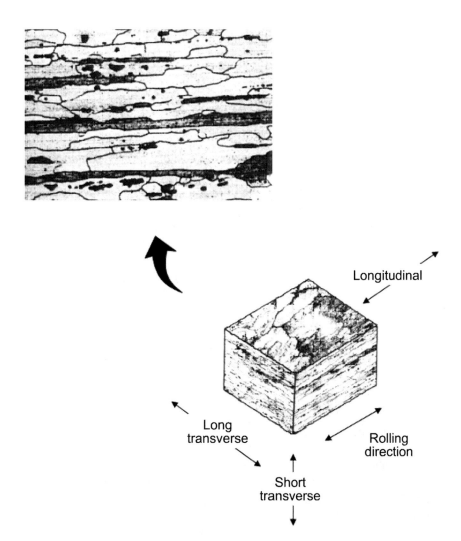

Fig. 2.8 Grain directionality due to rolling. Source: Ref 2.10

Cold working during rolling results in highly directional grain structures that can affect stress-corrosion resistance. The longitudinal direction is the most resistant, followed by the long-transverse direction, with the short transverse being the most susceptible. Thick 7xxx plate is therefore supplied in stress-corrosion-resistant tempers, such as the T73, T74, and T77 tempers, and the 2xxx alloys are given the T6 and T8 tempers. For example, 7075-T6 resists SCC at tensile stresses up to only 48 MPa (7 ksi), while 7075-T73 resists SCC up to 300 MPa (44 ksi) when tested under similar conditions. For thinner sheet, which is not as affected by through-the-thickness effects, the 7xxx alloys can be used in the higher-strength T6 temper, and the 2xxx alloys are given the T3 or T4 tempers.

2.9   Extrusion

Extruded structural sections are produced by hot extrusion, in which a heated cylindrical billet is pushed under high pressure through a steel die to produce the desired structural shape. The extrusion is then fed onto a run-out table where it is straightened by stretching and cut to length. During extrusion, metal flow occurs most rapidly at the center of the ingot, resulting in oxides and surface defects being left in the last 10 to 15% of the extrusion, which is discarded.

The complexity of shapes produced by extrusion is a function of metal-flow characteristics of the process and the means available to control flow. Control of metal flow places a few limitations on the design features of the cross section of an extruded shape that affect production rate, dimensional and surface quality, and costs. Extrusions are classified by shape complexity from an extrusion-production viewpoint into solid, hollow, and semihollow shapes. Each hollow shape—a shape with any part of its cross section completely enclosing a void—is further classified by increasing complexity, as follows:

- *Class 1.* A hollow shape with a round void 25 mm (1 in.) or more in diameter and with its weight equally distributed on opposite sides of two or more equally spaced axes
- *Class 2.* Any hollow shape other than class 1, not exceeding a 125 mm (5 in.) diameter circle and having a single void of not less than 9.5 mm (0.375 in.) diameter or 70 mm^2 (0.110 in.^{2}) area
- *Class 3.* Any hollow shape other than class 1 or 2

Alloy selection is important because it establishes the minimum thickness for a shape and has a basic effect on extrusion cost. In general, the higher the alloy content and the strength of an alloy, the more difficult it is to extrude and the lower its extrusion rate. The relative extrudabilities, as measured by extrusion rate, for several of the more important commercial extrusion alloys are given in Table 2.13. Actual extrusion rate depends on

Table 2.13 Relative extrudability of aluminum alloys

Alloy	Extrudability, % of rate for 6063
1350	160
1060	135
1100	135
3003	120
6063	100
6061	60
2011	35
5086	25
2014	20
5083	20
2024	15
7075	9
7178	8

Source: Ref 2.11

pressure, temperature, and other requirements for the particular shape, as well as ingot quality.

The important shape factor of an extrusion is the ratio of its perimeter to its weight per unit length. For a single classification, increasing shape factor is a measure of increasing complexity. Designing for minimum shape factor promotes ease of extrusion. The size of an extruded shape affects ease of extrusion and dimensional tolerances. As the circumscribing circle size (smallest diameter that completely encloses the shape) increases, extrusion becomes more difficult. In extrusion, the metal flows fastest at the center of the die face. With increasing circle size, the tendency for different metal flow increases, and it is more difficult to design and construct extrusion dies with compensating features that provide uniform metal-flow rates to all parts of the shape.

Ease of extrusion improves with increasing thickness; shapes of uniform thickness are most easily extruded. A shape whose cross section has elements of widely differing thicknesses increases the difficulty of extrusion. The thinner a flange on a shape, the less the length of flange that can be satisfactorily extruded. Thinner elements at the ends of long flanges are difficult to fill properly and make it hard to obtain desired dimensional control and finish. Although it is desirable to produce the thinnest shape feasible for an application, reducing thickness can cause an increase in cost of extrusion that more than offsets the savings in metal cost. Extruded shapes 1 mm (0.040 in.) thick and even less can be produced, depending on the alloy, shape, size, and design.

Size and thickness relationships among the various elements of shape can add to its complexity. Rod, bar, and regular shapes of uniform thickness are easily produced. For example, a bar 3.2 mm (0.125 in.) thick, a rod 25 mm (1 in.) in diameter, and an angle 19 by 25 mm (0.75 by 1 in.) in cross section and 1.6 mm (0.0625 in.) thick are readily extruded, whereas extrusion of a 75 mm (3 in.) bar-type shape with a 3.2 mm (0.125 in.) flange is more difficult.

Semihollow and channel shapes require a tongue in the extrusion die, which must have adequate strength to resist the extrusion force. Channel shapes become increasingly difficult to produce as the depth-to-width ratio increases. Wide, thin shapes are difficult to produce and make it hard to control dimension. Channel-type shapes and wide, thin shapes may be fabricated if they are not excessively thin. Thin flanges or projections from a thicker element of the shape add to the complexity of an extruded design. On thinner elements at the extremities of high flanges, it is difficult to achieve adequate fill to obtain desired dimensions. The greater the difference in thickness of individual elements comprising a shape, the more difficult the shape is to produce. The effect of such thickness differences can be greatly diminished by blending one thickness into the other by tapered or radiused transitions. Sharp corners should be avoided wherever possible because they reduce maximum extrusion speed and are locations of stress concentrations in the die opening that can cause premature die failure. Fillet radii of at least 0.8 mm (0.031 in.) are desirable, but corners with radii of only 0.4 mm (0.015 in.) are feasible.

In general, the more unbalanced and unsymmetrical an extruded-shape cross section, the more difficult that shape is to produce. Despite this, production of grossly unbalanced and unsymmetrical shapes is the basis of the great growth that has occurred in the use of aluminum extrusions, and such designs account for the bulk of extruded shapes produced today.

2.10 Forging

Forgings are often preferred for highly loaded parts because the forging process allows for thinner cross-sectional product forms prior to heat treat and quenching, enabling superior properties. It can also create a favorable grain-flow pattern that increases both fatigue life and fracture toughness when not removed by machining. Also, forgings generally have less porosity than thick plate, and less machining is required.

Aluminum alloys can be forged into a variety of shapes and types of forgings, with a broad range of final part forging design criteria based on the intended application. As a class of alloys, however, aluminum alloys are generally considered to be more difficult to forge than carbon steels and many alloy steels. Compared to the nickel/cobalt-base alloys and titanium alloys, aluminum alloys are considerably more forgeable, particularly in conventional forging-process technology, in which dies are heated to 540 °C (1000 °F) or less.

The relative forgeability of ten aluminum alloys that constitute the bulk of aluminum alloys forging production is illustrated in Fig. 2.9. This arbitrary unit is principally based on the deformation per unit of energy absorbed in the range of forging temperatures typically used for the alloys in question. Also considered in this index is the difficulty of achieving specific degrees of severity in deformation as well as the cracking tendency of the alloy under forging-process conditions. There are wrought aluminum alloys,

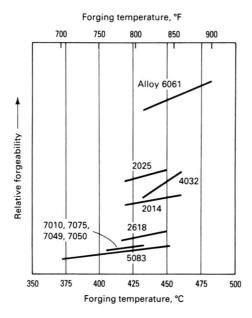

Fig. 2.9 Forgeability and forging temperatures of various aluminum alloys. Source: Ref 11

such as 1100 and 3003, whose forgeability would be rated significantly above those presented; however, these alloys have limited application in forging because they cannot be strengthened by heat treatment.

The 14 aluminum alloys that are most commonly forged, as well as recommended temperature ranges, are listed in Table 2.14. All of these alloys are generally forged to the same severity, although some alloys may require more forging power and/or more forging operations than others. The forging temperature range for most alloys is relatively narrow (generally <55 °C, or <100 °F), and for no alloy is the range greater than 85 °C (155 °F). Obtaining and maintaining proper metal temperatures in the forging of aluminum alloys is critical to the success of the forging process. Die temperature and deformation rates play key roles in the actual forging temperature achieved.

Aluminum alloys are produced by all of the current forging methods available, including open-die (or hand) forging, closed-die forging, upsetting, roll forging, orbital (rotary) forging, spin forging, mandrel forging, ring rolling, and extrusion. Selection of the optimal forging method for a given forging shape is based on the desired forged shape, the sophistication of the forged-shape design, and cost. In many cases, two or more forging methods are combined to achieve the desired forging shape and to obtain a thoroughly wrought structure. For example, open-die forging frequently precedes closed-die forging to prework the alloy (especially when cast ingot forging is used) and to preshape (or preform) the metal to conform to the subsequent closed dies and to conserve input metal.

Table 2.14 Recommended forging temperature ranges for aluminum alloys

Aluminum alloys	Forging temperature range	
	°C	°F
1100	315–405	600–760
2014	420–460	785–860
2025	420–450	785–840
2219	425–470	800–880
2618	410–455	770–850
3003	315–405	600–760
4032	415–460	780–860
5083	405–460	760–860
6061	430–480	810–900
7010	370–440	700–820
7039	380–440	720–820
7049	360–440	680–820
7050	360–440	680–820
7075	380–440	720–820

Source: Ref 2.11

Aluminum alloys can be forged using hammers, mechanical presses, or hydraulic presses. Hammer forging operations can be conducted with either gravity or power drop hammers and are used for both open- and closed-die forgings. Hammers deform the metal with high deformation speed; therefore, it is necessary to control the length of the stroke, the speed of the blows, and the force exerted. Hammer operations are frequently used to conduct preliminary shaping prior to closed-die forging. Both mechanical and screw presses are used for forging moderate-sized parts of modest shapes and are often used for high-volume production runs. Mechanical and screw presses combine impact with a squeezing action that is more compatible with the flow characteristics of aluminum than hammers. Hydraulic presses are the best method for producing large and thick forgings, because the deformation rate is slower and more controlled than with hammers or mechanical/screw presses. The deformation or strain rate can be very fast (>10 s^{-1}) for processes such as hammer forging or very slow (<0.1 s^{-1}) for hydraulic presses. Because higher strain rates increase the flow stress (decrease forgeability) and the 2xxx and 7xxx alloys are even more sensitive than other aluminum alloys, hydraulic presses are usually preferred for forging these alloys. Hydraulic presses are available in the range of 500 to 75,000 tons (biggest press is in Russia) and can produce forgings up to approximately 1360 kg (3000 lb).

The dies are usually heated for forging of aluminum alloys, normally from as low as 93 °C (200 °F) to as high as 316 °C (600 °F). Because the higher temperatures are used for hydraulic press forging, this operation is essentially conducted in the isothermal range in which the dies and part are at or near the same temperature. Aluminum alloys are heated to the forging temperature with a wide variety of heating equipment, including electric furnaces, muffle furnaces, oil furnaces, induction heating units, fluidized beds, and resistance heating units. Regardless of the heating method, it is very important to minimize the absorption of hydrogen, which can result in surface blistering or internal porosity, referred to as bright flakes, in the finished forging.

Open-die forgings, also known as hand forgings, are produced in dies that do not provide lateral restraint during the forging operation. In this process, the metal is forged between either flat or simply shaped dies. This process is used to produce small quantities where the small quantities do not justify the expense of matched dies. Although open-die forgings somewhat improve the grain flow of the material, they offer minimal economic benefit in reduced machining costs. Open-die forging is often used to produce preforms for closed-die forging. Most aluminum forgings are produced in closed dies. Closed-die forgings are produced by forging ingots, plates, or extrusions between a matched set of dies. Closed-die forging uses progressive sets of dies to gradually shape the part to near-net dimensions. Die forgings can be subdivided into four categories from the lowest cost, least intricate to the highest cost, most intricate. A comparison of the relative amount of part definition for these different forging processes is shown in Fig. 2.10.

Blocker forgings may be chosen if the total quantities are small (e.g., 200). Because they have large fillet and corner radii, they require extensive machining to produce a finished part. The fillets are approximately two times the radius, and the corner radii are approximately 1.5 times the radii of conventional forgings. Therefore, a blocker forging costs less than a conventional forging but requires more machining. Examples of large blocker forgings are shown in Fig. 2.11. Finish-only forgings are similar to blocker forgings in that only one set of dies is used. However, because they have one more squeeze applied, they have somewhat better part definition. Fillets are approximately 1.5 times the radius of conventional forgings, with corner radii about the same as conventional forgings. A quantity of approximately 500 may justify the use of finish-only forgings.

Conventional forgings require two to four sets of dies, with the first set producing a blocker-type forging that is subsequently finished in the other sets. This is the most common type of aluminum forging and is usually specified for quantities of 500 or more. Conventional forgings have more definition and require less machining than blocker forgings, but the die cost is higher.

High-definition forgings contain even better definition and tolerance control than conventional forgings, with less machining costs. These forgings are near-net shape forgings produced on multiple die sets. In some applications, some of the forged surfaces may not require machining.

Precision forgings produce the best part definition and highest quality but are, of course, the most expensive. These forgings have tighter tolerances than those produced by even high-definition forgings with better grain flow. Minimal or no machining is required to finish these forgings.

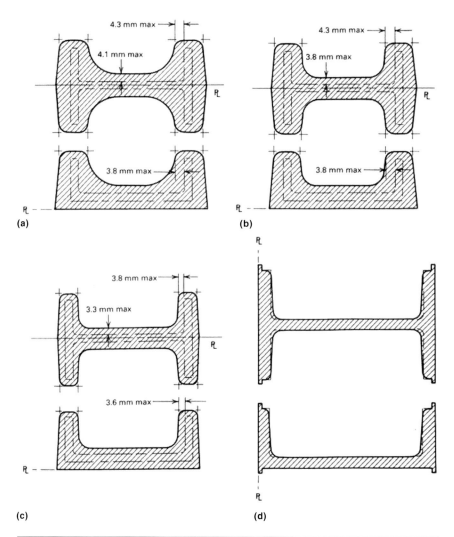

	Tolerance, mm (in.)			
Characteristic	**Blocker-type**	**Conventional**	**High-definition**	**Precision**
Die closure	+2.3, –1.5 (+0.09, –0.06)	+1.5, –0.8 (+0.06, –0.03)	+1.25, –0.5 (+0.05, –0.02)	+0.8, –0.25 (+0.03, –0.01)
Mismatch	0.5 (0.02)	0.5 (0.02)	0.25 (0.01)	0.38 (0.015)
Straightness	0.8 (0.03)	0.8 (0.03)	0.5 (0.02)	0.4 (0.016)
Flash extension	3 (0.12)	1.5 (0.06)	0.8 (0.03)	0.8 (0.03)
Length and width	±0.8 (±0.03)	±0.8 (±0.03)	±0.8 (±0.03)	+0.5, –0.25 (+0.02, –0.01)
Draft angles	5°	5°	3°	1°

Fig. 2.10 Types of aluminum closed-die forgings and tolerances for each. (a) Blocker type. (b) Conventional. (c) High definition. (d) Precision. Source: Ref 2.12

Fig. 2.11 Examples of very large blocker-type aluminum alloy airframe forgings. Source: Ref 2.13

The choice of a particular forging method depends on the shape required and the economics of the number of pieces required, traded off against higher quality and lower machining costs.

2.11 Heat Treating

For aluminum alloys, heat treating usually refers to precipitation hardening of the heat treatable aluminum alloys. Annealing, a process that reduces strength and hardness while increasing ductility, can also be used for both the non-heat-treatable and heat treatable grades of wrought and cast alloys.

Solution Heat Treating and Aging

The importance of precipitation hardening of aluminum alloys can be appreciated by examining the data presented in Fig. 2.12 for naturally aged 2024 and the artificially aged 7075. Note the dramatic increase in strength of both due to precipitation hardening, with only moderate reductions in elongation.

For an aluminum alloy to be precipitation hardened, certain conditions must be satisfied. First, the alloy must contain at least one element or compound in a sufficient amount that has a decreasing solid solubility in aluminum with decreasing temperature. In other words, the elements or compounds must have an appreciable solubility at high temperatures and

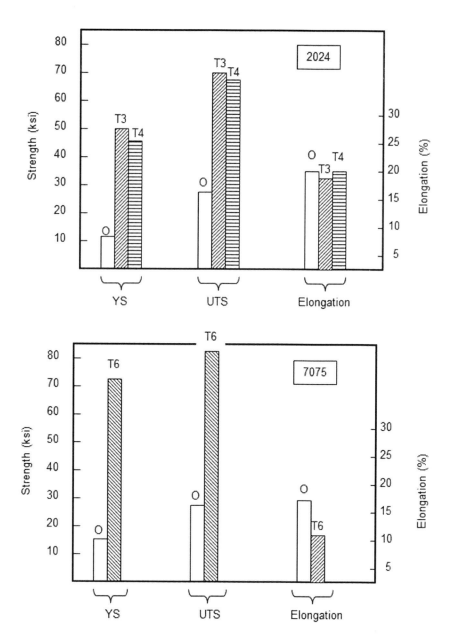

Fig. 2.12 Effect of heat treatment on 2024 and 7075. YS, yield strength; UTS, ultimate tensile strength

only minimal solubility at lower temperatures. Elements that have this characteristic are copper, zinc, silicon, and magnesium, with compounds such as $CuAl_2$, Mg_2Si, and $MgZn_2$. While this is a requirement, it is not sufficient; some aluminum systems that display this behavior cannot be strengthened appreciably by heat treatment. The second requirement is that the element or compound that is put into solution during the solution heat treating operation must be capable of forming a fine precipitate that will produce lattice strains in the aluminum matrix. The precipitation of these elements or compounds progressively hardens the alloy until a maximum hardness is obtained.

Aluminum alloys that satisfy both of these conditions are classified as heat treatable, namely the 2xxx, 6xxx, 7xxx, and some of the 8xxx wrought alloys. The precipitation-hardening process (Fig. 2.13) is conducted in three steps:

1. Heating to the solution heat treating temperature and soaking for long enough to put the elements or compounds into solution

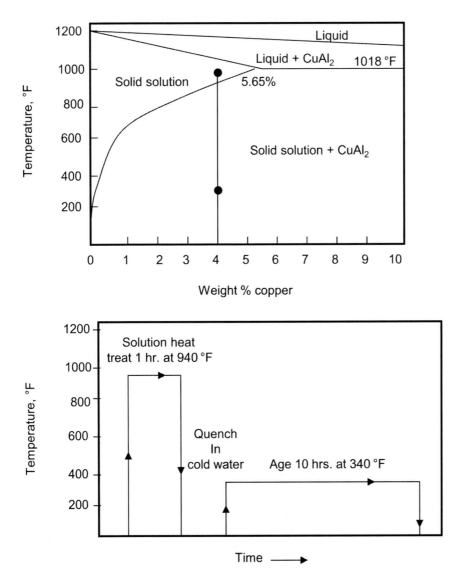

Fig. 2.13 Precipitation hardening of an aluminum-copper alloy. Source: Ref 2.8

2. Quenching to room or some intermediate temperature (e.g., water) to keep the alloying elements or compounds in solution; essentially creating a supersaturated solid solution

3. Aging at either room temperature (natural aging) or a moderately elevated temperature (artificial aging) to cause the supersaturated solution to form a very fine precipitate in the aluminum matrix

The solution heat treating temperature is as high above the solid-solubility curve as possible without melting the lowest-melting-point eutectic constituents. Therefore, close temperature control, normally ±5.6 °C (±10 °F), is required for the furnaces used to heat treat aluminum alloys. If the alloy is heated too high and incipient grain-boundary melting occurs, the part is ruined and must be scrapped. For example, 2024 is solution treated in the range of 488 to 499 °C (910 to 930 °F), while a low-melting eutectic forms at 502 °C (935 °F), only 2.8 °C (5 °F) higher than the upper range of the solution heat treating temperature. On the other hand, if the temperature is too low, solution will be incomplete and the aged alloy will not develop as high a strength as expected. The solution heat treating time should be long enough to allow diffusion to establish an equilibrium solid solution. The product form can determine the time required for solution treating; that is, castings require more time than wrought products to dissolve their relatively large constituents into solution. The time required can vary anywhere from less than a minute for thin sheet to up to 20 h for large sand castings. While longer-than-required soak times are not usually detrimental, caution must be applied if the part is Alclad, because excessive times can result in alloying elements diffusing through the Alclad layer and reducing the corrosion resistance.

After the elements are dissolved into solution, the alloy is quenched to a relatively low temperature to keep the elements in solution. Quenching is perhaps the most critical step in the heat treating operation. The problem is to quench the part fast enough to keep the hardening elements in solution, while at the same time minimizing residual quenching stresses that cause warpage and distortion. In general, the highest strength levels, and the best combinations of strength and toughness, are obtained by using the fastest quench rate possible. Resistance to corrosion and SCC are usually improved by faster quenching rates; however, the resistance to SCC of certain copper-free 7*xxx* alloys is actually improved by slow quenching. While fast quenching rates can be achieved by cold water, slower quenching rates (e.g., hot or boiling water) are often used to sacrifice some strength and corrosion resistance for reduced warpage and distortion.

If premature precipitation during quenching is to be avoided, two requirements must be met. First, the time required to transfer the part from the furnace to the quench tank must be short enough to prevent slow cooling through the critical temperature range where very rapid precipitation takes place. The high-strength 2*xxx* and 7*xxx* alloys should be cooled at rates exceeding 444 °C/s (800 °F/s) through the temperature range of 399 to 288 °C (750 to 550 °F). The second requirement is that the volume of the quenching tank must be large enough so that the quench tank temperature does not rise appreciably during quenching and allow premature precipitation.

Both cold and hot water are common quenching media for aluminum alloys. Cold water, with the water maintained below 29 °C (85 °F), is used with the requirement that the water temperature not rise by more than 11 °C (20 °F) during the quenching operation. The quench rate can be further increased by agitation that breaks up the insulating steam blanket that forms around the part during the early stages of quenching. Parts that distort during quenching require straightening before aging, so hot water quenching, with water maintained between 66 and 82 °C (150 and 180 °F) or at 100 °C (212 °F), is a less drastic quench resulting in much less distortion and is often used for products where it is impracticable to straighten after quenching. Polyalkylene glycol solutions in water are also used to quench aluminum alloys because they produce a stable film on the surface during quenching, resulting in more uniform cooling rates and less distortion.

Aging is conducted at either room temperature (natural aging) or at elevated temperature (artificial aging). Alloys that are not aged sufficiently to obtain maximum hardness are said to be underaged, while those that are aged past peak hardness are said to be overaged. Underaging can be a result of not artificially aging at a high enough temperature or an aging time that is too short, while overaging is usually a result of aging at too high a temperature. A set of aging curves for an Al-4%Cu alloy is shown in Fig. 2.14.

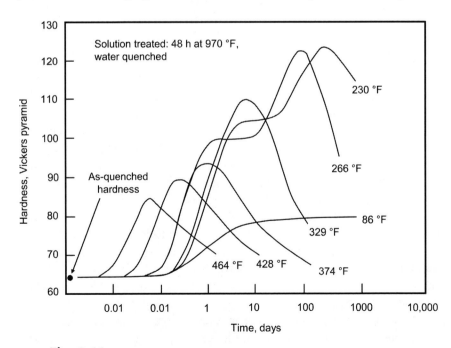

Fig. 2.14 Aging curves for Al-4%Cu alloy. Source: Ref 2.8

The 2*xxx* alloys can be aged by either natural or artificial aging. The 2*xxx* alloys can be naturally aged to obtain full strength after 4 to 5 days at room temperature, with approximately 90% of their strength being obtained within the first 24 h. Natural aging of the 2*xxx* alloys consists of solution heat treating, quenching, and then aging at room temperature to give the T4 temper. Naturally aged alloys are often solution treated and quenched (W temper), refrigerated until they are ready to be formed, and then allowed to age at room temperature to peak strength (T3 temper). To prevent premature aging, cold storage temperature must be in the range of –46 to –73 °C (–50 to –100 °F). It should be noted that artificial aging of the 2*xxx* alloys to the T6 temper produces higher strengths and higher tensile-to-yield strength ratios but lower elongations than natural aging.

The T8 temper (i.e., solution treating, quenching, cold working, and then artificial aging) produces high strengths in many of the 2*xxx* alloys. Alloys such as 2011, 2024, 2124, 2219, and 2419 are very responsive to cold working by stretching and cold rolling; the cold work creates additional precipitation sites for hardening. The T9 temper is similar except that the cold work is introduced after artificial aging (i.e., solution treating, quenching, artificial aging, and cold working). Because the 7*xxx* alloys do not respond favorably to cold working during the precipitation-hardening process, they are not supplied in the T8 or T9 tempers.

Artificial aging treatments are generally low-temperature, long time processes; temperatures range from 116 to 191 °C (240 to 375 °F) for times of 5 to 48 h. The 7*xxx* alloys, although they will harden at room temperature, are all given artificial aging treatments. The 7*xxx* alloys are usually aged at 121 °C (250 °F) for times up to 24 h, or longer, to produce the T6 temper. Many of the thick product forms for the 7*xxx* alloys that contain more than 1.25% Cu are provided in the T7 overaged condition. While overaging reduces the strength properties, it improves the corrosion resistance, dimensional stability, and fracture toughness, especially in the through-the-thickness short-transverse direction. There are a number of T7 tempers that have been developed that trade off various amounts of strength for improved corrosion resistance. Most involve aging at a lower temperature to develop strength properties, followed by aging at a higher temperature to improve corrosion resistance. For example, the T73 aging treatment consists of an aging temperature of 107 °C (225 °F), followed by a second aging treatment at 157 to 177 °C (315 to 350 °F). The T76 temper is similar, with an initial age at 121 °C (250 °F), followed by a 163 °C (325 °F) aging treatment. The T76 temper has a little higher strength than the T73 temper but is also a little less corrosion resistant and produces a lower fracture toughness. The T77 aging treatment developed by Alcoa, which is a variation of a treatment called retrogression and aging, produces the best combination of mechanical properties and corrosion resistance. Although it depends on the specific alloy, in the T77 treatment, the part is solution treated, quenched, and aged. It is then reaged for 1 h at 199 °C (390 °F), water quenched, and finally aged again for 24 h at 121 °C (250 °F).

Verification of heat treatment is usually conducted by a combination of hardness and electrical conductivity.

2.12 Annealing

Cold working results in an increase in internal energy due to an increase in dislocations, point defects, and vacancies. The tensile and yield strengths increase with cold working, while the ductility and elongation decrease. If cold-worked aluminum alloys are heated to a sufficiently high temperature for a sufficiently long time, annealing will occur in three stages: recovery, recrystallization, and grain growth. During recovery, the internal stresses due to cold work are reduced, with some loss of strength and a recovery of some ductility. During recrystallization, new unstrained nuclei form and grow until they impinge on each other to form a new recrystallized grain structure. Although heating for longer times or at higher temperatures will generally result in grain growth, aluminum alloys contain dispersoids of manganese, chromium, and/or zirconium that help to suppress grain growth.

Annealing treatments are used during complex cold forming operations to allow further forming without the danger of sheet cracking. The softest, most ductile, and most formable condition for aluminum alloys is produced by full annealing to the O condition. Strain-hardened products normally recrystallize during annealing, while hot-worked products may or may not recrystallize, depending upon the amount of cold work present. Full annealing treatments for a number of aluminum alloys are given in Table 2.15.

Table 2.15 Typical full annealing treatments for some common wrought aluminum alloys

Alloy	Metal temperature		Approximate time at temperature, h
	°C	°F	
1060	345	650(a)	(a)
1100	345	650	(a)
2024	415(b)	775(b)	2–3
2124	415(b)	775(b)	2–3
3003	415	775	(a)
5050	345	650	(a)
5052	345	650	(a)
5056	345	650	(a)
5154	345	650	(a)
5457	345	650	(a)
5652	345	650	(a)
6061	415(b)	775(b)	2–3
6063	415(b)	775(b)	2–3
6066	415(b)	775(b)	2–3
7050	415(c)	775(c)	2–3
7075	415(c)	775(c)	2–3

(a) Time in the furnace need not be longer than necessary to bring all parts of the load to annealing temperature. Cooling rate is unimportant. (b) These treatments are intended to remove the effects of solution treatment and include cooling at a rate of approximately 30 °C/h (50 °F/h) from the annealing temperature to 260 °C (500 °F). Rate of subsequent cooling is unimportant. Treatment at 345 °C (650 °F), followed by uncontrolled cooling, may be used to remove the effects of cold work or to partly remove the effects of heat treatment. (c) These treatments are intended to remove the effects of solution treatment and include cooling at an uncontrolled rate to 205 °C (400 °F) or less, followed by reheating to 230 °C (450 °F) for 4 h. Treatment at 345 °C (650 °F), followed by uncontrolled cooling, may be used to remove the effects of cold work or to partly remove the effects of heat treatment. Source: Adapted from Ref 2.14

2.13 Forming

Due to their fcc structure and their relatively slow rate of work hardening, aluminum alloys are highly formable at room temperature. High-strength alloys can be readily formed at room temperature, provided the alloy is in either the O or W temper. The choice of the temper for forming depends on the severity of the forming operation and the alloy being formed. Although the annealed or O condition is the most formable condition, it is not always the best choice, because of the potential for warping during subsequent heat treatment. The solution-treated and quenched condition (W temper) is nearly as formable as the O condition and requires only aging after forming to obtain peak strengths, without the potential of warping after forming during the quenching operation. The 2*xxx* and 7*xxx* alloys must be formed immediately after quenching or be refrigerated after heat treating to the W temper prior to forming. Because aluminum has a relatively low rate of work hardening, a fair number of forming operations are possible before intermediate anneals are required. Tools for forming aluminum alloys require good surface finishes to minimize surface marking. The oxide film on aluminum is highly abrasive, and many forming tools are therefore made of hardened tool steels.

Blanking and Piercing. As shown in Fig. 2.15, blanking is a process in which a shape is sheared from a larger piece of sheet, while piercing produces a hole in the sheet by punching out a slug of metal. Both blanking and piercing operations are usually performed in a punch press. The clearance between the punch and die must be controlled to obtain a uniform shearing action. Clearance is the distance between the mating surfaces of the punch and die, usually expressed as a percentage of sheet thickness. The walls of the die opening are tapered to minimize sticking, and the use of lubricants, such as mineral oil mixed with small quantities of fatty oils, also reduces sticking tendencies. A tolerance of 0.13 mm (0.005 in.) is normal in blanking and piercing of aluminum; however, wider tolerances, when permissible, will help in reducing costs.

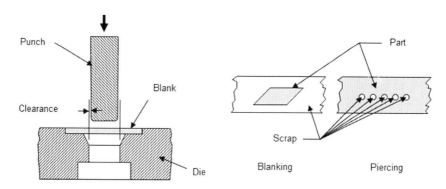

Fig. 2.15 Blanking and piercing

Dull cutting edges on punches and dies have effects similar to excessive clearance, with burrs becoming excessive. With sharp tools and proper clearance, the fractures are clean, without evidence of secondary shearing or excessive burring. When the clearance is too small, secondary shearing can occur, and if the clearance is too large, the sheared edge will have a large radius and a stringy burr. Cast zinc tools, which are much less expensive than steel tools, are often used for runs of up to approximately 2000 parts.

Brake Forming. In brake forming (Fig. 2.16), the sheet is placed over a die and pressed down by a punch that is actuated by the hydraulic ram of a press brake. Springback is the partial return of the part to its original shape after forming. The amount of springback is a function of the yield strength of the material being formed, the bend radius, and the sheet thickness. Springback is compensated by overbending the material beyond the final angle so that it spring backs to the desired angle. The springback allowance (i.e., the amount of overbend) increases with increasing yield strength and bend radius but varies inversely with sheet thickness. Because aluminum sheet tends to develop anisotropy during rolling operations, there is less tendency for cracking during forming if the bend is made perpendicular to the rolling or extrusion direction. The smallest angle that can be safely bent, called the minimum bend radius, depends on the yield strength and on the design, dimensions, and conditions of the tooling. The most severe bends can be made across the rolling direction. If similar bends are to be made in two or more directions, it is best to make all bends at an angle to the direction of rolling. Relatively simple and long parts can usually be press brake formed to a tolerance of 0.75 mm (0.030 in.), while for larger, more complex parts, the tolerance may be as much as 1.6 mm (0.063 in.).

Deep Drawing. Punch presses are used for most deep drawing operations. Typical press setups for deep drawing (Fig. 2.17) include single action, double action, and double action with an inverted die. In a single-action die setup, a punch or male die pushes the sheet into the die cavity

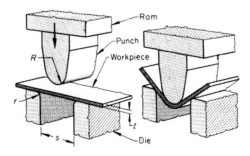

Fig. 2.16 Typical setup for press brake forming in a die with a vertical opening. R, punch radius; r, die radius; s, span width; t, metal thickness. Source: Ref 2.15

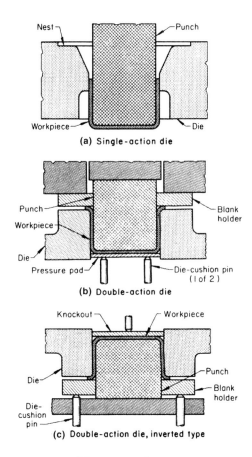

Fig. 2.17 Components of three types of simple dies shown in a setup used for drawing a round cup. Source: Ref 2.16

while it is supported around the periphery by a blankholder. Single-action presses can be operated at 27 to 43 m/min (90 to 140 ft/min), while double-action presses operate at 12 to 30 m/min (40 to 100 ft/min) for mild draws and at less than 15 m/min (50 ft/min) for deep draws. For the higher-strength aluminum alloys, speeds of 6 to 12 m/min (20 to 40 ft/min) are more typical. Clearances between the punch and die are usually equal to the sheet thickness plus an additional 10% per side for the intermediate-strength alloys, while an additional 5 to 10% clearance may be needed for the high-strength alloys. Excessive clearance can result in wrinkling of the sidewalls of the drawn shell, while insufficient clearance increases the force required for drawing and tends to burnish the part surfaces. The draw radius on tools is normally equal to four to eight times the stock thickness. If the punch radius is too large, wrinkling can result, and if the radius is too small, sheet fracture is a possibility. Draw punches and dies should have a surface finish of 16 μin. or less for most applications. Tools are often chrome plated to minimize friction and dirt that can damage the part finish. Lubricants for deep drawing must allow the blank to slip readily and uniformly between the blankholder and die. Stretching and galling during drawing must be avoided. During blank preparation, excessive stock at the corners must be avoided because it obstructs the uniform flow of metal under the blankholder, leading to wrinkles or cracks.

The rate of strain hardening during drawing is greater for the high-strength aluminum alloys than for the low-to-intermediate-strength alloys. For high-strength aluminum alloys, the approximate reductions in diameter during drawing are approximately 40% for the first draw, 15% for the second draw, and 10% for the third draw. Local or complete annealing for 2014 and 2024 is normally required after the third draw. Severe forming operations of relatively thick or large blanks of high-strength aluminum alloys generally must be conducted at temperatures of approximately 316 °C (600 °F), where the lower strength and partial recrystallization aids in forming. This is possible with alloys such as 2024, 2219, 7075, and 7178, but the time at temperature should be minimized to limit grain growth.

Stretch Forming. In stretch forming, the material is stretched over a tool beyond its yield strength to produce the desired shape. Large compound shapes can be formed by stretching the sheet both longitudinally and transversely. In addition, extrusions are frequently stretch formed to moldline curvature. Typical stretch-formed shapes are shown in Fig. 2.18. Variants of stretch forming include stretch draw forming, stretch wrapping, and radial draw forming. Forming lubricants are recommended except when self-lubricating, smooth-faced plastic dies are used. However, the use of too much lubricant can result in workpiece bucking. High-strength alloys can be stretch formed in either the O or W condition. Material properties that help in stretch forming are a high elongation, a large spread between the yield and ultimate strengths (called the forming range), toughness, and a fine grain structure. Alloys with a narrow spread between the yield and ultimate strengths are more susceptible to local necking and failure. For example, 7075-W has a yield strength of 138 MPa (20 ksi), an ultimate strength of 331 MPa (48 ksi), a forming range of 193 MPa (28 ksi); that is, 331 MPa – 138 MPa (48 ksi – 20 ksi) and a stretchability rating of 100, while 7075-T6 has a yield strength of 462 MPa (67 ksi), an ultimate strength of 524 MPa (76 ksi), a forming range of only 62 MPa (9 ksi), and a stretchability rating of only 10.

Rubber Pad Forming. In rubber pad forming, a rubber pad is used to exert nearly equal pressure over the part as it is formed down over a form block, as shown in Fig. 2.19. The rubber pad acts somewhat like a hydraulic fluid, spreading the force over the surface of the part. The pad can either consist of a solid piece or may be several pieces laminated together. The pad is usually in the range of 150 to 300 mm (6 to 12 in.) thick and must be held in a sturdy retainer because the pressures generated can be as high as 138 MPa (20 ksi). Rubber pad forming can often be used to form tighter radii and more severe contours than other forming methods because of the

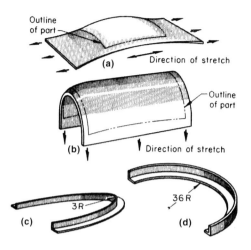

Fig. 2.18 Typical stretch-formed shapes. (a) Longitudinal stretching. (b) Transverse stretching. (c) Compound bend from extrusion. (d) Long, sweeping bend from extrusion. Source: Ref 2.17

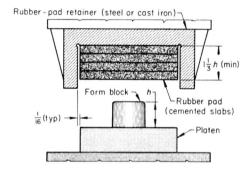

Fig. 2.19 Tooling and setup for rubber pad forming. Dimensions given in inches. Source: Ref 2.18

multidirectional nature of the force exerted on the workpiece. The rubber acts somewhat like a blankholder, helping to eliminate the tendency for wrinkling. This process is very good for making sheet parts with integral stiffening beads. Most rubber pad forming is conducted on sheet 1.6 mm (0.063 in.) or less in thickness; however, material as thick as 16 mm (0.625 in.) thick has been successfully formed. Although steel tools are normally used for long production runs, aluminum or zinc tools will suffice for short or intermediate runs. Fluid cell forming, which uses a fluid cell to apply pressure through an elastomeric membrane, can form even more severe contours than rubber pad forming (Fig. 2.20). Due to the high pressures used in this process, as high as 103 to 138 MPa (15 to 20 ksi), many parts can be formed in one shot with minimal or no springback. However, fluid cell forming presses are usually expensive.

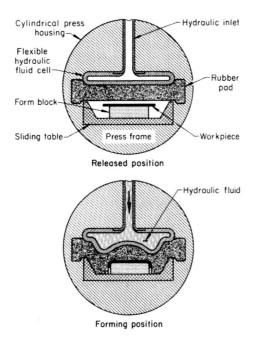

Fig. 2.20 Fluid cell forming. Source: Ref 2.18

Contour Roll Forming. Aluminum alloys are readily shaped by contour roll forming (Fig. 2.21). Operating speeds can be higher for the more ductile aluminum alloys than for most other metals. Speeds as high as 245 m/min (800 ft/min) have been used in mild roll forming of 0.8 mm (0.03 in.) thick alloy 1100-O sections 15 to 30 m (50 to 100 ft) long. Power requirements for roll forming of aluminum alloys are generally lower than for comparable operations on steel, because of the lower yield strength of most aluminum alloys.

Extremely close tolerances are required on tool dimensions. Allowance for springback must be varied with alloy and temper, as well as with material thickness and radius of forming. Final adjustments must be made on the basis of production trials. Tolerances of ±0.127 mm (0.005 in.) are common in contour roll forming, and ±0.05 mm (0.002 in.) can be maintained on small, simple shapes formed from light-gage metals. One or two final sizing stations may be required for intricate contours or when springback effects are great. Lubricants are required in nearly all contour roll forming of aluminum alloys. For high-speed or severe forming operations, the rolls and workpiece may be flooded with a liquid that functions as both a lubricant and a coolant.

Roll-formed aluminum alloy parts made from sheet or coiled strip include furniture parts, architectural moldings, window and door frames, gutters and downspouts, automotive trim, roofing and siding panels, and shelving.

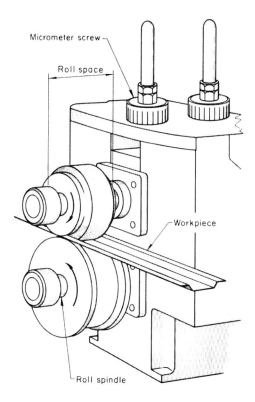

Fig. 2.21 Contour roll forming. Source: Ref 2.19

Tubing in sizes ranging from 19 to 203 mm (0.8 to 8 in.) in outside diameter and from 0.64 to 3.9 mm (0.025 to 0.156 in.) in wall thickness is made in a combined roll forming and welding operation. Linear speeds of 9 to 60 m/min (30 to 200 ft/min) are used in this process. Applications include irrigation pipe, condenser tubing, and furniture parts. Other applications of contour roll forming include the forming of patterned, anodized, or pre-enameled material. Such applications impose stringent requirements on tool design and maintenance, and lubrication sometimes cannot be used because of the nature of the coating or because of end-use requirements.

Superplastic Forming. Superplasticity is a property that allows sheet to elongate to quite large strains without localized necking and rupture. In uniaxial tensile testing, elongations to failure in excess of 200% are usually indicative of superplasticity. Micrograin superplasticity occurs in some materials with a fine grain size, usually less than 10 μm, when they are deformed in the strain range of 0.00005 to 0.01/s at temperatures greater than $0.5T_m$, where T_m is the melting point in degrees Kelvin. Although superplastic behavior can produce strains in excess of 1000%, superplastic forming (SPF) processes are generally limited to approximately 100 to 300%. The advantages of SPF include the ability to make part shapes not possible with conventional forming, reduced forming stresses, improved formability with essentially no springback, and reduced machining costs. The disadvantages are that the process is rather slow, and the equipment and tooling can be relatively expensive.

The main requirement for superplasticity is a high strain-rate sensitivity. In other words, the strain-rate sensitivity, m, should be high, where m is defined as:

$$m = \frac{d(\ln \sigma)}{d(\ln \varepsilon)}$$

where σ is the flow stress, and ε is the strain rate.

The strain-rate sensitivity describes the ability of a material to resist plastic instability or necking. For superplasticity, m is usually greater than 0.5, with the majority of superplastic materials having an m value in the range of 0.4 to 0.8, where a value of 1.0 would indicate a perfectly super-plastic material. The presence of a neck in a material undergoing a tensile strain results in a locally high strain rate and, for a high value of m, to a sharp increase in the flow stress within the necked region; that is, the neck undergoes strain hardening that restricts its further development. Therefore, a high strain-rate sensitivity resists neck formation and leads to the high tensile elongations observed in superplastic materials. The flow stress decreases and the strain-rate sensitivity increases with increasing temperature and decreasing grain size. The elongation to failure tends to increase with increasing m.

Superplasticity depends on microstructure and exists only over certain temperature and strain-rate ranges. A fine grain structure is a prerequisite because superplasticity results from grain rotation and grain-boundary sliding, and increasing grain size results in increases in flow stress. Equiaxed grains are desirable because they contribute to grain-boundary sliding and grain rotation. A duplex structure also contributes to super-plasticity by inhibiting grain growth at elevated temperature. Grain growth inhibits superplasticity by increasing the flow stress and decreasing m.

The original superplastic aluminum alloy, Supral 100 (alloy 2004), was developed especially for its superplastic characteristics. Supral 100 is a medium-strength alloy with mechanical properties similar to 6061 and 2219 and is normally used in lightly loaded or nonstructural applications. Alloy 7475 is a higher-strength aluminum alloy capable of superplastic-ity that derives its superplasticity from thermomechanical processing. In addition, a number of the aluminum-lithium alloys (e.g., 8090) exhibit superplasticity.

In the single-sheet SPF process, illustrated in Fig. 2.22, a single sheet of metal is sealed around its periphery between an upper and lower die. The lower die is either machined to the desired shape of the final part or a die inset is placed in the lower die box. The dies and sheet are maintained at

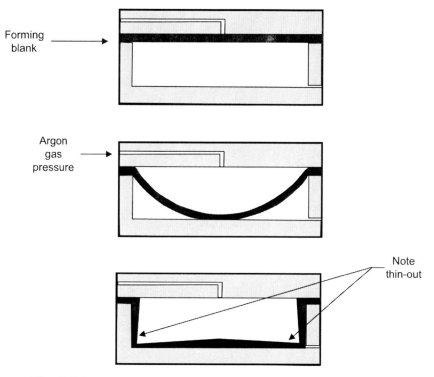

Forming blank

Argon gas pressure

Note thin-out

Fig. 2.22 Single-sheet superplastic forming

the SPF temperature, and gas pressure is used to form the sheet down over the tool. The lower cavity is maintained under vacuum or can be vented to the atmosphere. After the sheet is heated to its superplastic temperature range, gas pressure is injected through inlets in the upper die. This pressurizes the cavity above the metal sheet, forcing it to superplastically form to the shape of the lower die. Gas pressurization is applied slowly so that the strains in the sheet are maintained in the superplastic range, and the pressure is varied during the forming process to maintain the required slow strain rate. As shown in Table 2.16, typical forming cycles for aluminum alloys are 4.8 to 6.2 MPa (700 to 900 psi) at 449 to 524 °C (840 to 975 °F).

During the forming operation, the metal sheet is reduced uniformly in thickness; however, wherever the sheet makes contact with the die, it sticks and no longer thins out. This results in a part with nonuniform thickness. To reduce these thickness variations, overlay forming can be used. In overlay forming, the sheet that will become the final part is cut slightly smaller than the tool periphery. A sacrificial overlay sheet is then placed on top of it and clamped to the tool periphery. As gas is injected into the upper die cavity, the overlay sheet forms down over the lower die, forming the part blank simultaneously with it. While overlay forming does help to mini-

Table 2.16 Superplastic forming parameters for aluminum

Alloy	Forming temperature		Strain rate, s^{-1}	Forming temperature		Back pressure		Strain, %
	°C	°F		MPa	psi	MPa	psi	
2004	449	840	5×10^4	6.0	870	...	...	1000
7475	516	960	2×10^4	4.0	580	3.0	435	500–1000
5083	510	950	1.5×10^4	6.0	870	4.0	580	500
8090	521	970	3×10^4	5.0	725	3.0	435	200–400

Source: Ref 2.7

mize thickness variations, it requires a sacrificial sheet for each run that must be discarded. Two other forming methods, shown in Fig. 2.23, were developed to reduce thickness nonuniformity during forming. However, both of these methods require moving rams within the pressure chamber, which increases capital equipment costs.

The hard particles at the grain boundaries that help control grain growth may contribute to the formation of voids in aluminum alloys, a process called cavitation. Cavitation on the order of 3% can occur after approximately 200% of superplastic deformation. Cavitation can be minimized, or eliminated, by applying a hydrostatic back pressure to the sheet during forming, as shown schematically in Fig. 2.24. Back pressures of 0.69 to 3.4 MPa (100 to 500 psi) are normally sufficient to suppress cavitation.

Gas pressure is an effective pressure medium for SPF for several reasons:

* It permits the application of a controlled uniform pressure.
* It avoids the local stress concentrations that are inevitable in conventional forming where a tool contacts the sheet.
* It requires relatively low pressures (<7 MPa, or 1000 psi).

Forming parameters (time, temperature, and pressure) have traditionally been determined empirically by trial and error; however, there are now a number of finite-element programs that greatly aid in reducing the development time. The disadvantages of SPF are that the process is rather slow and the equipment and tooling can be expensive. For example, a part undergoing 100% strain at 0.0001/s would require almost 3 h at temperature, including the time required for heatup and cooldown.

For titanium alloys, SPF can be combined with diffusion bonding (SPF/DB) to form one unitized structure. Titanium is very amenable to diffusion bonding because the thin protective oxide layer (TiO_2) dissolves into the titanium above 621 °C (1150 °F), leaving a clean surface. However, the aluminum oxide (Al_2O_3) on aluminum does not dissolve and must either be removed, or ruptured, to promote diffusion bonding. Although diffusion bonding of aluminum alloys has successfully been demonstrated in the laboratory, SPF/DB of aluminum alloys is not yet a commercial process.

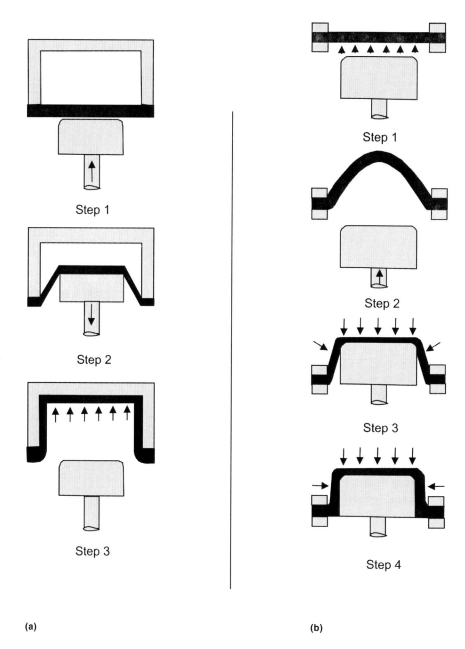

(a)

(b)

Fig. 2.23 Superplastic forming methods for reducing nonuniform thin-out. (a) Plug-assisted forming, female tooling. (b) Snapback forming, male tooling. Source: Ref 2.20

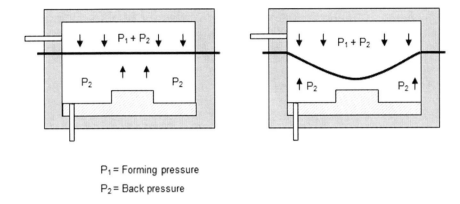

P₁ = Forming pressure

P₂ = Back pressure

Fig. 2.24 Back-pressure forming to suppress cavitation. P_1 = forming pressure; P_2 = back pressure

2.14 Casting

Due to their lower properties and higher variability than wrought product forms, aluminum castings are not used for primary structural applications. However, for lightly loaded secondary structures, castings can offer significant cost-savings by reducing part count and the associated assembly cost. Aluminum casting alloys have different compositions than the wrought alloys; that is, they are tailored to increase the fluidity of the molten metal, be resistant to hot tearing during solidification, and reproduce the details of the mold shape.

2.15 Casting Alloys

In comparison with wrought alloys, casting alloys contain larger proportions of alloying elements such as silicon and copper, which results in a largely heterogeneous cast structure (i.e., one having a substantial volume of second phases). This second-phase material warrants careful study, because any coarse, sharp, and brittle constituent can create harmful internal notches and nucleate cracks when the component is later put under load. The fatigue properties are very sensitive to large heterogeneities. Good metallurgical and foundry practices can largely prevent such defects.

The elongation and strength, especially in fatigue, of most cast products are relatively lower than those of wrought products. This is because current casting practice is as yet unable to reliably prevent casting defects. In recent years, however, innovations in casting processes such as squeeze casting have brought about some significant improvements in the consistency and level of properties of castings, and these should be taken into account in selecting casting processes for critical applications. It should be emphasized that the aluminum-silicon alloys with magnesium and/or copper, the 3xx.x alloys, are by far the most widely used of the aluminum casting alloys. The silicon addition greatly increases liquid metal fluidity and produces superior castings. The order of the alloy series in order of decreasing castability is 3xx.x, 4xx.x, 5xx.x, 2xx.x, and 7xx.x.

Aluminum-Copper Alloys (2xx.x). The major characteristics of the 2xx.x series are:

- Heat treatable sand and permanent mold castings
- High strength at room and elevated temperatures; some high-toughness alloys
- Approximate ultimate tensile strength range of 130 to 450 MPa (20 to 65 ksi)

The strongest of the common casting alloys is heat treated 201.0, which has found important application in the aerospace industry. The castability of the alloy is somewhat limited by a tendency to microporosity and hot tearing, so that it is best suited to investment casting. Its high toughness makes it particularly suitable for highly stressed components in machine tool construction, in electrical engineering (pressurized switchgear castings), and in aircraft construction.

Besides the standard aluminum casting alloys, there are special alloys for particular components, for instance, for engine piston heads, integral engine blocks, or bearings. For these applications, the chosen alloy needs good wear resistance and a low friction coefficient, as well as adequate strength at elevated service temperatures. A good example is the alloy 203.0, which to date is the aluminum casting alloy with the highest strength at approximately 200 °C (400 °F). An example of an application for 2xx.x alloys is an aircraft component that is made of high-strength alloy 201.0-T6.

Aluminum-Silicon plus Copper or Magnesium Alloys (3xx.x). The major characteristics of the 3xx.x series are:

- Heat treatable sand, permanent mold, and die castings
- Excellent fluidity, high strength, and some high-toughness alloys
- Approximate ultimate tensile strength range of 130 to 275 MPa (20 to 40 ksi)
- Readily welded

The 3xx.x series of castings is one of the most widely used because of the flexibility provided by the high silicon content and its contribution to fluidity, plus their response to heat treatment, which provides a variety of high-strength options. In addition, the 3xx.x series may be cast by a variety of techniques ranging from relatively simple sand or die casting to very intricate permanent mold, investment castings, and the newer thixocasting and squeeze casting technologies.

The most widely used aluminum casting alloys are those containing between 9.0 and 13.0 wt% Si. These alloys are of approximately eutectic composition (Fig. 2.25), which makes them suitable as die casting alloys, because their freezing range is small. However, they form a rather coarse eutectic structure (Fig. 2.25a) that is refined by a process known as modification, where small amounts of sodium (~0.01% by weight) are added to the melt just before casting. The sodium delays the precipitation of silicon when the normal eutectic temperature is reached and also causes a shift of the eutectic composition toward the right in the phase diagram. Therefore, as much as 14 wt% Si may be present in a modified alloy without any primary silicon crystals forming in the structure (Fig. 2.25b). It is thought that sodium collects in the liquid at its interface with the newly formed silicon crystals, inhibiting and delaying their growth. Thus, undercooling occurs

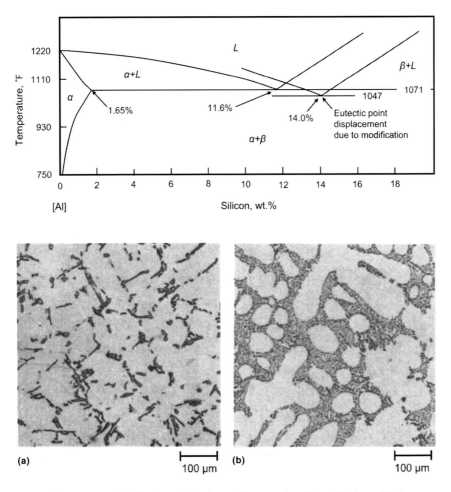

Fig. 2.25 Modification of aluminum-silicon casting alloys. (a) Unmodified. (b) Modified. Source: Ref 2.21

and new silicon nuclei are formed in large numbers, resulting in a relatively fine-grained eutectic structure. Modification raises the tensile strength and the elongation in the manner shown in Fig. 2.26. The relatively high ductility of this cast eutectic alloy is due to the solid-solution phase in the eutectic constituting nearly 90% of the total structure. Therefore, the solid-solution phase is continuous in the microstructure and acts as a cushion against much of the brittleness arising from the hard silicon phase. More recently, modification strontium is replacing sodium, because it has the advantage that there is less loss during melting due to oxidation or evaporation. Overmodification (too much addition) is not a problem, because it produces the innocuous compound $SrAl_3Si_3$, and strontium suppresses the formation of large primary silicon particles in hypereutectic alloys.

Among the workhorse alloys are 319.0 and 356.0/A356.0 for sand and permanent mold casting; 360.0, 380.0/A380.0, and 390.0 for die casting; and 357.0/A357.0 for many types of casting, including, especially, the relatively newly commercialized squeeze/forge cast technologies.

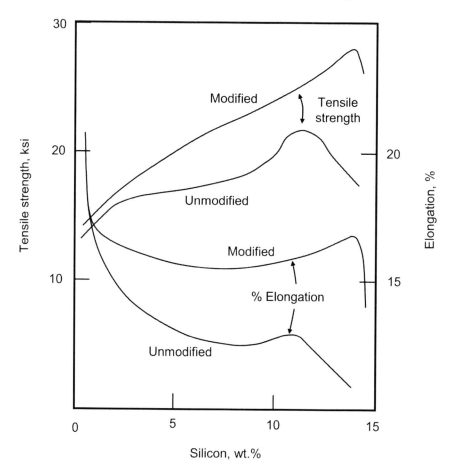

Fig. 2.26 Effects of silicon content and modification on aluminum casting alloys

Alloy 332.0 also is one of the most frequently used aluminum casting alloys, because it can be made almost exclusively from recycled scrap.

Aluminum-Silicon Alloys (4xx.x). The major characteristics of the 4xx.x series are:

* Non-heat-treatable sand, permanent mold, and die castings
* Excellent fluidity; good for intricate castings
* Approximate ultimate tensile strength range of 120 to 175 MPa (17 to 25 ksi)

Alloy B413.0 is notable for its very good castability and excellent weldability, which are due to its eutectic composition and low melting point of 700 °C (1292 °F). It combines moderate strength with high elongation before rupture and good corrosion resistance. The alloy is particularly suitable for intricate, thin-walled, leakproof, fatigue-resistant castings.

These alloys have found applications in relatively complex cast parts for typewriter and computer housings and dental equipment and also for fairly critical components in marine and architectural applications.

Aluminum-Magnesium Alloys (5xx.x). The major characteristics of the 5xx.x series are:

* Non-heat-treatable sand, permanent mold, and die castings
* Tougher to cast; provides good finishing characteristics
* Excellent corrosion resistance, machinability, and surface appearance
* Approximate ultimate tensile strength range of 120 to 175 MPa (17 to 25 ksi)

The common feature of this group of alloys is good resistance to corrosion.

Alloys 512.0 and 514.0 have medium strength and good elongation and are suitable for components exposed to seawater or to other similar corrosive environments. These alloys often are used for door and window fittings, which can be decoratively anodized to give a metallic finish or provide a wide range of colors. Their castability is inferior to that of the aluminum-silicon alloys because of its magnesium content and, consequently, long freezing range. For this reason, it tends to be replaced by 355.0, which has long been used for similar applications.

For die castings where decorative anodizing is particularly important, alloy 520.0 is quite suitable.

Aluminum-Zinc Alloys (7xx.x). The major characteristics of the 7xx.x series are:

* Heat treatable sand and permanent mold castings (harder to cast)
* Excellent machinability and appearance
* Approximate ultimate tensile strength range of 210 to 380 MPa (30 to 55 ksi)

Because of the increased difficulty in casting 7xx.x alloys, they tend to be used only where the excellent finishing characteristics and machinability are important. Representative applications include furniture, garden tools, office machines, and farming and mining equipment.

Aluminum-Tin Alloys (8xx.x). The major characteristics of the 8xx.x series are:

- Heat treatable sand and permanent mold castings (harder to cast)
- Excellent machinability
- Bearings and bushings of all types
- Approximate ultimate tensile strength range of 105 to 210 MPa (15 to 30 ksi)

As with the 7xx.x alloys, 8xx.x alloys are relatively hard to cast and tend to be used only where their combination of superior surface finish and relative hardness are important. The prime example is for parts requiring extensive machining and for bushings and bearings.

Premium-quality castings provide higher quality and reliability than conventional cast products. Important attributes include high mechanical properties determined by test coupons machined from representative parts, low porosity levels as determined by radiography, dimensional accuracy, and good surface finishes. Premium casting alloys include A201.0, A206.0, 224.0, 249.0, 354.0, A356.0, A357.0, and A358.0. The requirements for premium-quality castings are usually negotiated through special specifications.

Aluminum Casting Control

Molten aluminum alloys are extremely reactive and readily combine with other metals, gas, and sometimes with refractories. Molten aluminum dissolves iron from crucibles; therefore, aluminum is usually melted and handled in refractory-lined containers. For convenience in making up the charge, and to minimize the chance of error, most foundries use standard prealloyed ingot for melting, rather than doing their own alloying. Most alloying elements used in aluminum castings, such as copper, silicon, manganese, zinc, nickel, chromium, and titanium, are not readily lost by oxidation, evaporation, or precipitation. Alloying elements that melt at temperatures higher than the melting temperature of aluminum, such as chromium, silicon, manganese, and nickel, are added to the molten metal as alloy-rich ingots or master alloys. Some elements, such as magnesium, sodium, and calcium, which are removed from the molten bath by oxidation and evaporation, are added in elemental form to the molten bath, as required, to compensate for loss.

Because of its reactivity, molten aluminum is easily contaminated. The principal contaminants are iron, oxides, and inclusions. When the iron content exceeds 0.9 wt%, an undesirable acicular grain structure develops in the thicker sections of the casting. When the iron content exceeds 1.2 wt% in the higher-silicon alloys, sludging is likely to occur, particularly if the temperature drops below 649 °C (1200 °F). To prevent sludging, the quantity wt% Fe + 2(wt% Mn) + 3(wt% Cr) should not exceed 1.9 wt%. When this quantity exceeds 1.9 wt%, the castings are likely to contain hard spots that impair machining and may initiate stress cracks in service. Iron also causes excessive shrinkage in aluminum castings and becomes more severe as the iron content increases beyond 1 wt%.

Oxides must be removed from the melt; if they remain in the molten metal, the castings will contain harmful inclusions. Magnesium is a strong oxide former, making magnesium-containing alloys difficult to control when melting and casting. Oxides of aluminum and magnesium form quickly on the surface of the molten bath, developing a thin, tenacious skin that prevents further oxidation as long as the surface is not disturbed. Molten aluminum also reacts with moisture to form aluminum oxide, releasing hydrogen. Oxidation is also caused by excessive stirring, overheating of the molten metal, pouring from too great a height, splashing of the metal, or disturbing the metal surface with the ladle before dipping. Oxides that form on the surface of the molten bath can be removed by surface-cleaning fluxes. These fluxes usually contain low-melting-point ingredients that react exothermically on the surface of the bath. The oxides separate from the metal to form a dry, powdery, floating dross that can be skimmed. Some denser oxides sink to the bottom and are removed by gaseous fluxing or through a drain hole located in the bottom of the furnace.

Hydrogen is the only gas that dissolves to any extent in molten aluminum alloys and, if not removed, will result in porosity in the castings. As shown in Fig. 2.27, the solubility of hydrogen is significantly higher in the liquid than in the solid state. During cooling and solidification, the solubility decreases, and hydrogen is precipitated as porosity. Hydrogen is introduced in the molten aluminum by moisture and dirt in the charge and by the products of combustion. Degassing fluxes to remove hydrogen are used after the surface of the bath has been fluxed to remove oxides. The degassing fluxes also help to lift fine oxides and particles to the top of the bath. Removal of hydrogen by degassing is a mechanical action; hydrogen gas does not combine with the fluxing gases. Degassing fluxes include chlorine gas, nitrogen-chlorine mixtures, and hexachloroethane.

The grain size of aluminum alloy castings can be as small as 0.13 mm (0.005 in.) in diameter to as large as 13 mm (0.5 in.) in diameter. Fine-grained castings are desired for several reasons. First, while any porosity is undesirable, coarse porosity is the most undesirable. Because the coarseness of porosity is proportional to grain size, porosity in fine-grained castings is finer and less harmful. Second, shrinkage and hot cracking are usually associated with coarse-grained structures. A finer grain size minimizes shrinkage, resulting in sounder castings. Third, the mechanical

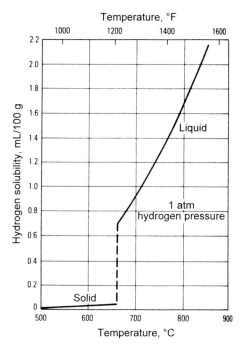

Fig. 2.27 Solubility of hydrogen in aluminum. Source: Ref 2.22

properties, such as tensile strength and ductility, are better for fine-grained castings than those of coarse-grained castings.

The grain size of aluminum alloy castings is influenced by pouring temperature, solidification rate, and the presence or absence of grain refiners. For all aluminum alloys, the grain size increases as the pouring temperature is increased. This is the main reason that aluminum alloy castings should be poured at the lowest temperature that will produce a sound casting. Rapid solidification rates produce finer grain sizes; therefore, castings done in steel dies will have finer grain sizes due to the faster solidification rates than those produced by sand casting. Grain-refining elements, such as titanium, boron, and zirconium, are helpful in producing finer-grained castings.

Casting Processes

Aluminum alloys can be cast using almost any casting process. Six of the major processes include sand casting, plaster mold casting, evaporative pattern casting, investment casting, permanent mold casting, and die casting. In addition, two emerging processes, thixocasting and rheocasting, are briefly discussed. Sand, permanent mold, and die casting are the most prevalent aluminum alloy casting processes. Factors in their selection are listed in Table 2.17.

Table 2.17 Factors affecting selection of casting process for aluminum alloys

	Casting process		
Factor	Sand casting	Permanent mold casting	Die casting
Cost of equipment	Lowest cost if only a few items required	Less than die casting	Highest
Casting rate	Lowest rate	11 kg/h (25 lb/h) common; higher rates possible	4.5 kg/h (10 lb/h) common; 45 kg/h (100 lb/h) possible
Size of casting	Largest of any casting method	Limited by size of machine	Limited by size of machine
External and internal shape	Best suited for complex shapes where coring is required	Simple sand cores can be used, but more difficult to insert than in sand castings	Cores must be able to be pulled because they are metal; undercuts can be formed only by collapsing cores or loose pieces
Minimum wall thickness	3.0–5.0 mm (0.125–0.200 in.) required; 4.0 mm (0.150 in.) normal	3.0–5.0 mm (0.125–0.200 in.) required; 3.5 mm (0.150 in.) normal	1.0–2.5 mm (0.100–0.040 in.); depends on casting size
Types of cores	Complex baked sand cores can be used	Reusable cores can be made of steel, or nonreuseable baked cores can be used	Steel cores; must be simple and straight so they can be pulled
Tolerance obtainable	Poorest; best linear tolerance is 300 mm/m (300 mils/in.)	Best linear tolerance is 10 mm/m (10 mils/in.)	Best linear tolerance is 4 mm/m (4 mils/in.)
Surface finish	6.5–12.5 µm (250–500 µin.)	4.0–10 µm (150–400 µin.)	1.5 µm (50 µin.); best finish of the three casting processes
Gas porosity	Lowest porosity possible with good technique	Best pressure tightness; low porosity possible with good technique	Porosity may be present
Cooling rate	0.1–0.5 °C/s (0.2–0.9 °F/s)	0.3–1.0 °C/s (0.5–1.8 °F/s)	50–500 °C/s (90–900 °F/s)
Grain size	Coarse	Fine	Very fine on surface
Strength	Lowest	Excellent	Highest, usually used in the as-cast condition
Fatigue properties	Good	Good	Excellent
Wear resistance	Good	Good	Excellent
Overall quality	Depends on foundry technique	Highest quality	Tolerance and repeatability very good
Remarks	Very versatile in size shape, and internal configurations	…	Excellent for fast production rates

Source: Ref 2.23

Sand casting is perhaps the oldest casting process known. The molten metal is poured into a cavity shaped inside a body of sand and allowed to solidify. Advantages of sand casting are low equipment costs, design flexibility, and the ability to use a large number of aluminum casting alloys. It is often used for the economical production of small lot sizes and is capable of producing fairly intricate designs. The biggest disadvantages are that the process does not permit close tolerances, and the mechanical properties are somewhat lower due to larger grain sizes as a result of slow cooling rates. However, the mechanical properties are improving as a result of improvements in casting materials and procedures. The steps involved in sand casting are shown in Fig. 2.28 and consist of:

1. Fabricate a pattern, usually wood, of the desired part and split it down the centerline.
2. Place the bottom half of the pattern, called the drag, in a box called a flask.
3. Apply a release coating to the pattern, fill the flask with sand, and then compact the sand by ramming.

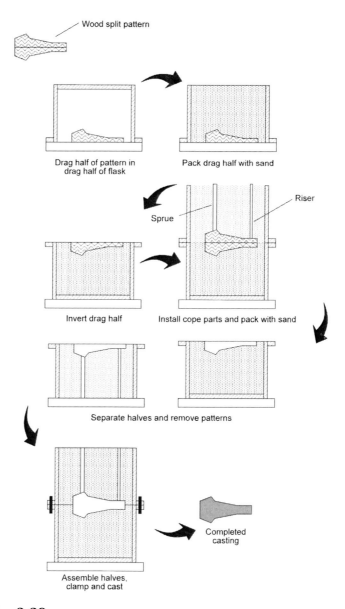

Fig. 2.28 Sand casting process

4. Turn the drag half of the mold over and place the top half of the flask on top of it. The top half of the pattern, called the cope, is then placed over the drag half of the pattern and release coated.

5. Risers and a sprue are then installed in the cope half of the flask. The sprue is where liquid metal enters the mold. In a complex casting, the sprue is usually gated to different positions around the casting. The risers are essentially reservoirs for liquid metal that keep the cast-

ing supplied with liquid metal as the metal shrinks and contracts on freezing.
6. The cope half is then packed with sand and rammed.
7. The two halves are separated and the patterns are removed. If hollow sections are required, a sand core is placed in the drag half of the mold. A gating system is then cut into the sand on the cope half of the mold.
8. The two halves are reassembled and clamped or bolted shut for casting.

Molding sands usually consist of sand grains, a binder, and water. The properties that are important are good flowability or the ability to be easily worked around the pattern, sufficient green strength, and sufficient permeability to allow gas and steam to escape during casting. Sand cores for molded-in inserts can be made using either heat-cured binder systems or no-bake binder systems. No-bake binder systems are usually preferred because they provide greater dimensional accuracy, have higher strengths, are more adaptable to automation, and can be used immediately after fabrication. The no-bake systems typically consist of room-temperature-curing sodium silicate sands, phenolic-urethanes, or furan acids combined with sand.

Gates are used to evenly distribute the metal to the different locations in the casting. The objective is to have progressive solidification from the point most distant from the gate toward the gate; that is, the metal in the casting should solidify before the metal in the gates solidifies. Normally, the area of the gates and runner system connecting the gates should be approximately four times larger than the sprue. When feeding needs to be improved, it is better to increase the number of gates rather than increase the pouring temperature.

Plaster and Shell Molding. Plaster mold casting is basically the same as sand casting, except gypsum plasters replace the sand in this process. The advantages are very smooth surfaces, good dimensional tolerances, and uniformity due to slow uniform cooling. However, as a result of the slow solidification rates, the mechanical properties are not as good as with sand castings. In addition, because plaster can absorb significant moisture from the atmosphere, it may require slow drying prior to casting.

Shell molding can also be used in place of sand casting when a better surface finish or tighter dimensional control is required. Surfaces finishes in the range of 6.4 to 11.4 μm (250 to 450 μin.) are typical with shell molding. Because it requires precision metal patterns and more specialized equipment, shell molding should be considered a higher-volume process than sand casting. Shell molding, shown in Fig. 2.29, consists of:

1. A fine silica sand coated with a phenolic resin is placed in a dump box that can be rotated.
2. A metal pattern is heated to 204 to 260 °C (400 to 500 °F), mold released, and placed in the dump box.

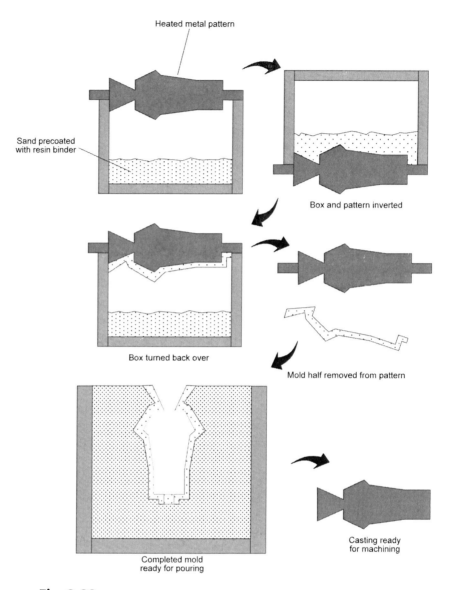

Fig. 2.29 Shell molding process

3. The pattern and sand are inverted, allowing the sand to coat the heated pattern. A crust of sand fuses around the part as a result of the heat.
4. The dump box is turned right side up; the pattern with the shell crust is removed and cured in an oven at 343 to 399 °C (650 to 750 °F).
5. The same process is repeated for the other half of the mold.
6. The two mold halves are clamped together and placed in a flask supported with either sand or metal shot.

In permanent mold casting, liquid metal is poured into a metal mold and allowed to solidify. This method is second only to die casting in the number of aluminum castings produced annually. However, due to the tooling costs, it is usually reserved for high-volume applications. The castings produced are normally small compared to sand casting and rather simple in shape. The process produces fairly uniform wall thicknesses but, unless segmented dies are used, is not capable of undercuts. Compared to sand castings, permanent mold castings are more uniform and have better dimensional tolerances, superior surface finishes (7 to 12.7 µm, or 275 to 500 µin., are typical), and better mechanical properties due to the faster solidification rates. Mold materials include gray cast iron and hot work die steels such as H11 and H13. When a disposable sand or plaster core is used with this process, it is referred to as semipermanent mold casting. Another variant of the permanent mold process is low-pressure permanent mold casting. Here, the casting is done inside a pressure vessel, and an inert gas is used to apply 34 to 69 kPa (5 to 10 psi) pressure on the liquid metal. This results in shorter cycle times and excellent mechanical properties.

Die casting is a permanent mold casting process in which the liquid metal is injected into a metal die under high pressure. It is a very high rate-production process using expensive equipment and precision-matched metal dies. Because the solidification rate is high, this process is amenable to high-volume production. It is used to produce very intricate shapes in the small-to-intermediate part size range. Characteristics of the process include extremely good surface finishes and the ability to hold tight dimensions; however, die castings should not be specified where high mechanical properties are important, because of the inherently high porosity level. The high-pressure injection creates a lot of turbulence that traps air, resulting in high porosity levels. In fact, die cast parts are not usually heat treated because the high porosity levels can cause surface blistering. To reduce the porosity level, the process can be done in a vacuum (vacuum die casting), or the die can be purged with oxygen just prior to metal injection. In the latter process, the oxygen reacts with the aluminum to form an oxide dispersion in the casting.

Investment casting is used where tighter tolerances, better surface finishes, and thinner walls are required than can be obtained with sand casting. Although the investment casting process is covered in greater detail in Chapter 5, "Titanium Alloys," in this book, a brief description of the process is that investment castings are made by surrounding, or investing, an expendable pattern, usually wax, with a refractory slurry that sets at room temperature. The wax pattern is then melted out, and the refractory mold is fired at high temperatures. The molten metal is cast into the mold, and the mold is broken away after solidification and cooling.

Evaporative pattern casting is used quite extensively in the automotive industry. Part patterns of expandable polystyrene are produced in metal dies. The patterns may consist of the entire part, or several patterns may be assembled together. Gating patterns are attached, and the completed pattern is coated with a thin layer of refractory slurry that is allowed to dry. The slurry must still be permeable enough to allow mold gases to escape during casting. The slurry-coated pattern is then placed in a flask supported by sand. When the molten metal is poured, it evaporates the polystyrene pattern. This process is capable of producing very intricate castings with close tolerances (Fig. 2.30), but the mechanical properties are low due to the large amounts of entrapped porosity.

Thixocasting and Rheocasting. Emerging casting processes include thixocasting and rheocasting. Both are variants of semisolid metal casting. Both variants are capable of creating a nondendritic microstructure from a wide range of alloys. Hypoeutectic aluminum-silicon alloys will typically display a microstructure in which the α-aluminum primary phase is present in the form of noninterconnected matrix surrounded by the aluminum-silicon eutectic. Semisolid metal (SSM) microstructures typical of these processes are shown in Fig. 2.31. Evolution in this area has proceeded from thixocasting (electromagnetically or mechanically stirred direct chill cast billet that is reheated until the eutectic is liquefied) to slurry-on-demand processes that create SSM slugs directly from liquid in the case of rheocasting. The resultant SSM slugs serve as feedstock to either vertical or horizontal high-pressure die casting machines. The

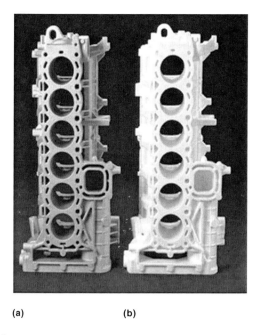

(a)   (b)

Fig. 2.30 Evaporative pattern casting. (a) Casting. (b) Pattern

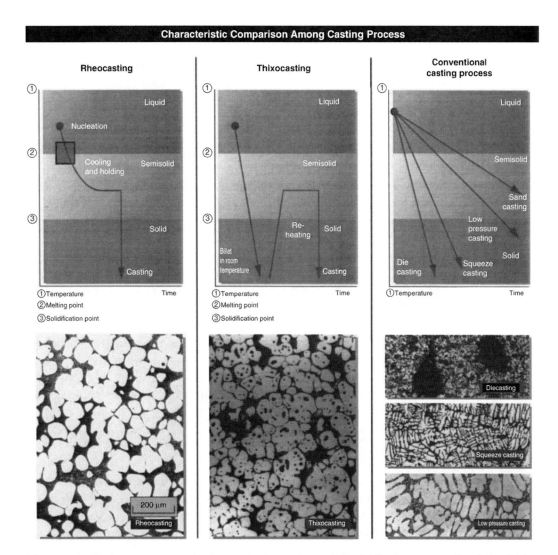

Fig. 2.31 Casting processes can be characterized according to their solidification rate as well as their solidification profile. Conventional liquid-based processes display a dendritic microstructure whose scale is dependent on the time to solidify. The faster the freezing rate, the smaller the dendrite arm spacing. Thixocasting and rheocasting display nondendritic microstructures in which the alpha-aluminum phase is present as nonconnected spheroids. Source: Ref 2.24

semisolid slurry flows thixotropically; that is, the viscosity decreases as the applied shear rate increases. This is different from liquid metal, and it helps to prevent many of the turbulence- and entrained air-related defects inherent in the conventional die casting process. Because an SSM melt is already typically half solid, only half as much energy must be removed to complete the solidification process. This gives process benefits of faster cycle times as well as reduced thermal shock to the dies, with a resultant increased tooling life. In addition, the high chill rates associated with this process may allow T5 properties close to those of T6 properties in sand or permanent mold to be achieved if the parts are quenched straight out of the dies. Alloys used for SSM parts today are primarily A356/357 alloys, but applications using 2*xx* and wrought 6*xxx* compositions are also being investigated.

Casting Heat Treatment

The major differences between heat treating cast aluminum alloys, as compared with wrought alloys, are the longer soak times during solution heat treating and the use of hot water quenches. Longer soak times are needed because of the relatively coarse microconstituents present in castings that do not have the benefit of the homogenization treatments given to wrought products before hot working. Boiling water is a common quenchant to reduce distortion for castings that normally contain more complex configurations than wrought products. Cast aluminum alloys are usually supplied in the T6, T7, or T5 tempers. The T6 temper is used where maximum strength is required. If low internal stresses, dimensional stability, and resistance to SCC are important, then the casting can be overaged to a T7 temper. The T5 temper is produced by aging the as-cast part without solution heat treating and quenching. This treatment is possible because most of the hardening elements are retained in solid solution during casting; however, the strengths obtained with the T5 temper will be lower than those with the T6 heat treatment.

Casting Properties

Because the mechanical properties of castings are not as consistent as wrought products, it is normal practice to use a casting factor for aluminum castings. The casting factor usually ranges from 1.0 to 2.0, depending on the end usage of the casting. For example, if the casting factor is 1.25 and the material has a yield strength of 207 MPa (30 ksi), the maximum design strength would be 207 MPa/1.25 = 165 MPa (30 ksi/1.25 = 24 ksi). In addition, sampling is used during production in which a casting is periodically selected from the production lot and cut up for tensile testing. The sampling plan depends on the criticality of the casting. All premium castings are subjected to both radiographic and penetrant inspection. Premium castings can also be hot isostatic pressed (HIPed) to help reduce internal porosity. Hot isostatic pressing is usually conducted using argon pressure at 103 MPa (15 ksi) and temperatures in the range of 482 to 527 °C (900 to 980 °F). Hot isostatic pressing usually results in improved mechanical properties, especially fatigue strength, but of course it adds to the cost and cycle time. The improvement in fatigue life for A201.0-T7 as a result of HIP is shown in Fig. 2.32. Surface defects in casting are normally repaired by GTAW using filler wire cast from the same alloy as the casting.

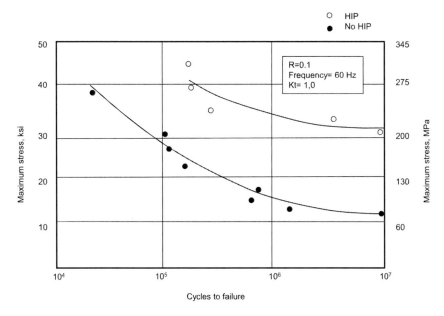

Fig. 2.32 Effect of hot isostatic pressing (HIP) on fatigue life of A201.0-T7 casting. Source: Ref 2.25

2.16 Machining

Aluminum exhibits extremely good machinability. Its high thermal conductivity readily conducts heat away from the cutting zone, allowing high cutting speeds. Cutting speeds for aluminum alloys are high, with speeds approaching or exceeding 305 surface m/min (1000 surface ft/min), common. In fact, as the cutting speed is increased from 30 to 60 m/min (100 to 200 ft/min), the probability of forming a built-up edge on the cutter is reduced; the chips break more readily, and the surface finish on the part is improved. The depth of cut should be as large as possible to minimize the number of cuts required. During roughing, depths of cuts range from 6.4 mm (0.25 in.) for small parts to as high as 38 mm (1.5 in.) for medium and large parts, while finishing cuts are much lighter, with depths of cuts less than 0.64 mm (0.025 in.) commonplace. Feed rates for roughing cuts are in the range of 0.15 to 2 mm/rev (0.006 to 0.080 in./rev), while lighter cuts are used for finishing, usually in the range of 0.05 to 0.15 mm/rev (0.002 to 0.006 in./rev).

Because aluminum alloys have a relatively low modulus of elasticity, they have a tendency to distort during machining. Also, due to the high coefficient of thermal expansion of aluminum, dimensional accuracy requires that the part be kept cool during machining; however, the high thermal conductivity of aluminum allows most of the heat to be removed with the chips. The flushing action of a cutting fluid is generally effective in removing the remainder of the heat. The use of stress-relieved tempers, such as the Tx51 tempers, stress relieved by stretching, also helps to minimize distortion during machining.

Excessive heat during machining can cause a number of problems when machining aluminum alloys. Friction between the cutter and the workpiece can result from dwelling, dull cutting tools, lack of cutting fluid, and heavy end mill plunge cuts rather than ramping the cutter into the workpiece. Inadequate backup fixtures, poor clamping, and part vibration can also create excessive heat. Localized overheating of the high-strength grades can even cause soft spots, which are essentially small areas that have been overaged due to excessive heat experienced during machining. These often occur at locations where the cutter is allowed to dwell in the work; for example, during milling in the corners of pockets.

Standard high-speed tool steels, such as M2 and M7 grades, work well when machining aluminum. For higher-speed machining operations, conventional C-2 carbides will increase tool life, resulting in less tool changes and allowing higher cutting speeds. For example, typical peripheral end milling parameters for the 2xxx and 7xxx alloys are 122 to 244 m/min (400 to 800 ft/min) with high-speed tool steel and 244 to 396 m/min (800 to 1300 ft/min) for carbide tools. For higher-speed machining operations, large cuttings, such as inserted end mills, should be dynamically balanced. Diamond tools are often used for the extremely abrasive hypereutectic aluminum-silicon casting alloys.

High-speed machining is somewhat an arbitrary term. It can be defined for aluminum as:

* Machining conducted at spindle speeds greater than 10,000 rpm
* Machining at 760 m/min (2500 ft/min) or higher, where the cutting force falls to a minimum
* Machining at speeds in which the impact frequency of the cutter approaches the natural frequency of the system

In end milling aluminum alloys, a respectable and easily achievable metal-removal rate using conventional machining parameters is 655 cm^3/min (40 in.^{3}/min). However, with the advent of high-speed machining of aluminum, even higher metal-removal rates are obtainable. It should be emphasized that while higher metal-removal rates are good, another driver for developing high-speed machining of aluminum is the ability to machine extremely thin walls and webs. For example, the minimum machining gage for conventional machining may be 1.5 to 2 mm (0.060 to 0.080 in.) or higher without excessive warpage, while with high-speed machining, wall thicknesses as thin as 0.5 to 0.76 mm (0.020 to 0.030 in.) without distortion are readily achievable. This allows the design of weight-competitive, high-speed-machined "assemblies" in which sheet metal parts that were formally assembled with mechanical fasteners can now be machined from a single or several blocks of aluminum plate. For example, as shown in Fig. 2.33, an aircraft assembly that originally consisted of 44 formed sheet metal parts was replaced by a single high-speed-machined part and five other pieces, resulting in a 73% cost reduction. The final part weighs 3.9 kg (8.5 lb), needs only five tools, and takes only 38.6 h to manufacture, as opposed to the original built-up design that weighed 4.3 kg (9.5 lb), required 53 individual tools, and took 1028 h to manufacture.

Successful high-speed machining requires an integrated approach between the cutter, workpiece, machine tool, and cutting strategy. For example, step cutting, as shown in Fig. 2.34, uses the workpiece to help

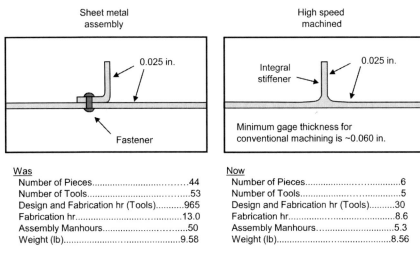

Sheet metal assembly — 0.025 in. — Fastener

High speed machined — Integral stiffener — 0.025 in. — Minimum gage thickness for conventional machining is ~0.060 in.

Was
Number of Pieces....................................44
Number of Tools.....................................53
Design and Fabrication hr (Tools)...........965
Fabrication hr...13.0
Assembly Manhours...............................50
Weight (lb)..9.58

Now
Number of Pieces..6
Number of Tools..5
Design and Fabrication hr (Tools).........30
Fabrication hr..8.6
Assembly Manhours...............................5.3
Weight (lb)...8.56

Fig. 2.33 Comparison between assembled and high-speed-machined assembly

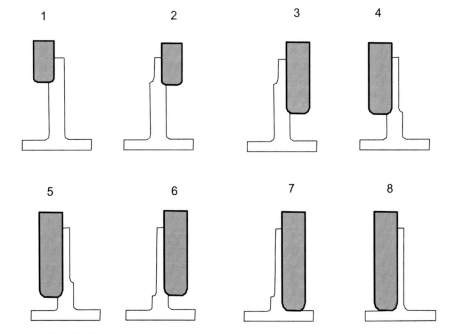

1   2   3   4

5   6   7   8

Fig. 2.34 Step-cutting methodology

provide rigidity during the machining process. For thin ribs (i.e., side cutting), this means using a large radial depth of cut and a small axial depth of cut, while for thin floor webs (i.e., end cutting), a large axial depth of cut should be combined with a small radial depth of cut. Although the depths of cut are small compared to conventional machining, the feed rates used in high-speed machining are so much higher that they allow higher metal-removal rates. Another fundamental associated with step cutting is that all cutters have a maximum depth of cut before they chatter; therefore, taking small depths of cut with the step-cutting technique helps to prevent chatter; that is, the cutter remains in the stability zone where chatter does not occur.

The comparison between conventional and high-speed machining in Table 2.18 illustrates the small depths of cuts used in high-speed machining. Again, because the feed rates are so much higher in high-speed machining, the metal-removal rate is still much higher. Attempts to cut extremely thin-wall structures using conventional machining are generally unsuccessful. In conventional machining, cutter and part deflection result in cutting through the thin walls. Typical high-speed machining operations are shown in Fig. 2.35.

2.17 Chemical Milling

Shallow pockets are sometimes chemically milled into aluminum skins for weight reduction. The process is used mainly for parts having large surface areas requiring small amounts of metal removal. Rubber maskant is applied to the areas where no metal removal is desired. In practice, the maskant is placed over the entire skin and allowed to dry. The maskant is then scribed according to a pattern and the maskant removed from the areas to be milled. The part is then placed in a tank containing sodium hydroxide heated to 91 ± 2.3 °C (195 ± 5 °F), with small amounts of triethanolamine to improve the surface finish. The etchant rate is in the range of 0.02 to 0.03 mm/min (0.0008 to 0.0012 in./min). Depths greater than 3.2 mm (0.125 in.) are generally not cost-competitive with conventional machining, and the surface finish starts to degrade. After etching, the part is washed in fresh water and the maskant is stripped.

Table 2.18 Comparison of conventional and high-speed machining

	Conventional machining	High-speed machining
Cutter (diam × length), mm (in.)	38×51 (1.5×2)	25×51 (1.0×2)
Speed, rpm	3600	20,000
Feed, cm/min (in./min)	102 (40)	710 (280)
Axial depth of cut, mm (in.)	25 (1)	10 (0.4)
Radial depth of cut, mm (in.)	25 (1)	25 (1)
Metal-removal rate, cm^3/min ($in.^3$/min)	655 (40)	1835 (112)

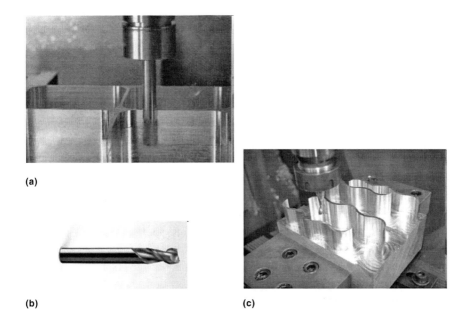

(a)

(b) (c)

Fig. 2.35 High-speed-machined thin-wall structure. (a) Step cutting. (b) Two-flute carbide end mill. (c) Deep-web machining

2.18 Joining

Aluminum alloys can be joined by a variety of commercial methods, including welding, brazing, soldering, adhesive bonding, and mechanical fastening.

Welding

Weldability is defined as the ability to produce a weld free of discontinuities and defects that results in a joint with acceptable mechanical properties, either in the as-welded condition or after a postweld heat treatment. Although aluminum has a low melting point, it can be rather difficult to weld for several reasons:

- The stable surface oxide must be removed by either chemical methods or, more typically, by thoroughly wire brushing the joint area.
- The high coefficient of thermal expansion of aluminum can result in residual stresses, leading to weld cracking or distortion.
- The high thermal conductivity of aluminum requires high heat input during welding, further leading to the possibility of distortion or cracking.
- The high solidification shrinkage of aluminum with a wide solidification range also contributes to cracking.
- The high solubility of aluminum for hydrogen when in the molten state leads to weld porosity.

- The highly alloyed, high-strength 2*xxx* and 7*xxx* alloys are especially susceptible to weld cracking.

The 2*xxx* alloys that can be fusion welded include 2014, 2195, 2219, and 2519. These alloys have lower magnesium contents, reducing their propensity for weld cracking. Alloy 2219 is readily weldable, and 2014 can also be welded to a somewhat lesser extent. Alloy 2319 is also commonly used as a filler metal when welding 2219. Although the weldability of the aluminum-lithium alloy 2195 approaches that of 2219, it tends to crack more than 2195 during repair welding. The 6*xxx* alloys can also be prone to hot cracking but are successfully welded in many applications. Postweld heat treatments can be used to restore the strength of 6*xxx* weldments. For the 7*xxx* alloys, those with a low copper content, such as 7004, 7005, and 7039, are somewhat weldable, while the remainder of the 7*xxx* series are not fusion weldable due to weld cracking and excessive strength loss.

The ability to fusion weld aluminum is often defined as the ability to make sound welds without weld cracking. Two types of weld cracking can be experienced: solidification cracking and liquation cracking. Solidification cracking, also called hot tearing, occurs due to the combined influence of high levels of thermal stresses and solidification shrinkage during weld solidification. Solidification cracking occurs in the fusion zone, normally along the centerline of the weld or at termination craters. Solidification cracking can be reduced by minimal heat input and by proper filler metal selection. The 4*xxx* alloys, with their narrow solidification range, are often used as filler metals when welding the 2*xxx* and 7*xxx* alloys. Liquation cracking occurs in the grain boundaries next to the heat-affected zone (HAZ). Highly alloyed aluminum alloys typically contain low-melting eutectics that can melt in the adjacent metal during the welding operation. Similar to solidification cracking, liquation cracking can be minimized by lower heat input and proper filler wire selection.

Due to its high solubility in molten aluminum and low solubility in solid aluminum, hydrogen can enter the molten pool and, with its decreasing solubility during freezing, form porosity in the solidified weld. Hydrogen is approximately 20 times more soluble in the liquid than the solid. Hydrogen normally originates from three sources:

- Hydrogen from the base metal
- Hydrogen from the filler metal
- Hydrogen from the shielding gas

Hydrogen from the base metal and filler wire can be minimized by ensuring that there is no moisture present and that all hydrocarbon residues and the surface oxide are thoroughly removed prior to welding.

The reduction of strength and hardness in a fusion-welded HAZ is illustrated in Fig. 2.36. The degradation of properties within the HAZ usually

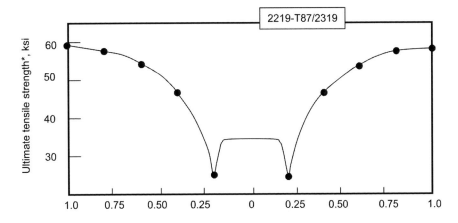

* Estimated from hardness readings.

Distance from centerline, in.

Fig. 2.36 Strength across fusion weld joint. Ultimate tensile strength values are estimated from hardness readings. Source: Ref 2.26

dictates joint strength. High heat inputs and preheating prior to welding increase the width of the HAZ and the property loss. The property loss within the HAZ can be minimized by the elimination of preheating, minimizing heat input, using high welding speeds where possible, and by using multipass welding. Postweld heat treatments can also be employed to improve properties, either by complete solution heat treating and aging or by postweld aging only. Solution heat treating and aging will restore the highest properties, but water quenching may result in excessive distortion; therefore, polymer quenchants that produce slower cooling rates are often used for postweld heat treatments. While postweld aging at moderate temperatures does not achieve as high properties as full heat treating, it is often used because it does not result in excessive distortion or warpage. It should be noted that both full heat treatment and postweld aging result in decreased joint ductility.

Welding processes that produce a more concentrated heat source result in smaller HAZs and lower postweld distortions; however, the capital cost of the equipment is roughly proportional to the intensity of the heat source. The nature of welding in the aerospace industry is characterized by low unit production, high unit cost, extreme reliability, and severe service conditions; therefore, the more expensive and more concentrated heat sources, such as plasma arc, laser beam, and electron beam welding, are often selected for welding of critical components.

Gas metal arc welding, as shown in Fig. 2.37, is an arc welding process that creates the heat for welding by an electric arc that is established between a consumable electrode wire and the workpiece. The consumable electrode wire is fed through a welding gun that forms an arc between

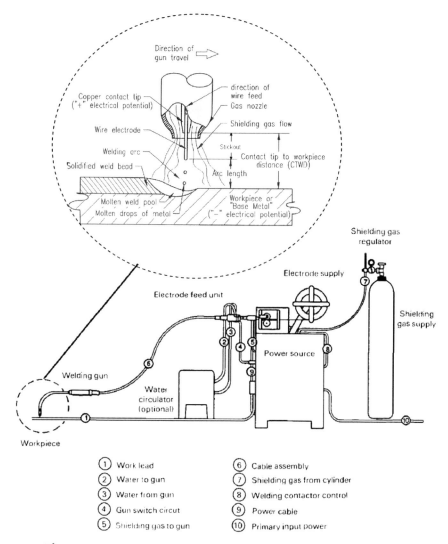

Fig. 2.37 Schematic of gas metal arc welding process. Source: Ref 2.27

the electrode and the workpiece. The gun controls the wire feed, the current, and the shielding gas. In GMAW, the power supply is direct current with a positive electrode. The positive electrode is hotter than the negative weld joint, ensuring complete fusion of the wire in the weld joint. In addition, the direct current electrode positive (DCEP) arrangement provides cathodic cleaning of the oxide layer during welding. When the electrode is positive and a direct current is used, there is a flow of electrons from the workpiece to the electrode, with ions traveling in the opposite direction and bombarding the workpiece surface. The ion bombardment breaks up and disperses the oxide film to create a clean surface for welding. The DCEP arrangement also provides good arc stability, low spatter, a good weld bead profile, and the greatest depth of penetration. A shielding gas, such as argon, protects the liquid metal fusion zone; however, the addition of helium to argon provides deeper penetration. GMAW has the advantage of good weld metal deposit per unit time.

Gas tungsten arc welding uses a nonconsumable tungsten electrode to develop an arc between the electrode and the workpiece. A schematic of the GTAW process is shown in Fig. 2.38. Although it has lower metal-deposition rates than GMAW, it is capable of higher-quality welds. However, when the joint thickness exceeds 9.5 mm (0.375 in.), GMAW is probably a more cost-effective method. For welding aluminum with GTAW, an alternating current arrangement is used, which, like the DCEP arrangement for GMAW, cleans the oxide layer during the welding process. The alternating

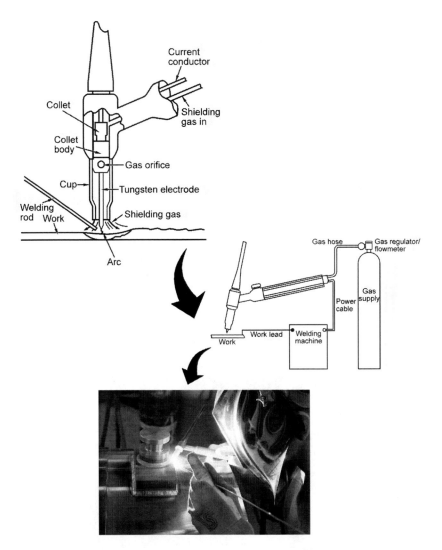

Fig. 2.38 Key components of the gas tungsten arc welding process. Source: Ref 2.28

current causes rapid reversing of the polarity between the workpiece and the electrode at 60 Hz. For this welding arrangement, tungsten electrodes and argon shielding gas are used. In general, material less than 3.2 mm (0.125 in.) thick can be welded without filler wire addition if solidification cracking is not a concern.

Plasma Arc Welding. Automated variable polarity plasma arc (VPPA) welding is often used to weld large aluminum alloy fuel tank structures. Plasma arc welding, shown in Fig. 2.39, is a shielded arc welding process in which heat is created between a tungsten electrode and the workpiece. The arc is constricted by an orifice in the nozzle to form a highly collimated arc column, with the plasma formed through the ionization of a portion of the argon shielding gas. The electrode positive component of the VPPA process promotes cathodic etching of the surface oxide, allowing good flow characteristics and consistent bead shape. Pulsing times are in the range of 20 ms for the electrode negative component and 3 ms for the electrode positive polarity. A keyhole welding mode is used in which the arc fully penetrates the workpiece, forming a concentric hole through the thickness. The molten metal then flows around the arc and resolidifies behind the keyhole as the torch traverses through the workpiece. The keyhole process allows deep penetration and high welding speeds while minimizing the number of weld passes required. VPPA welding can be used for thicknesses up to 13 mm (0.50 in.) with square-grooved butt joints and even thicker material with edge beveling. While VPPA welding produces high-integrity joints, the automated equipment used for this process is expensive and maintenance-intense.

Laser Welding. There is considerable interest in laser beam welding of high-strength aluminum alloys, particularly in Europe, where limited aerospace production has been announced. The process is attractive because

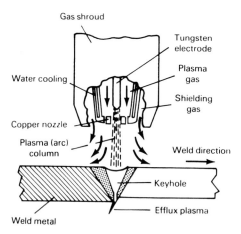

Fig. 2.39 Plasma arc welding process, showing constriction of the arc by a copper nozzle and a keyhole through the plate. Source: Ref 2.29

it can be conducted at high speeds with excellent weld properties. No electrode or filler metal is required, and narrow welds with small HAZs are produced. Although the intensity of the energy source is not quite as high as that in electron beam welding (EBW), EBW must be conducted in a vacuum chamber. The coherent nature of the laser beam allows it to be focused on a small area, leading to high energy densities. Because the typical focal point of the laser beam ranges from 0.1 to 1.0 mm (0.004 to 0.040 in.), part fit-up and alignment are more critical than conventional welding methods. Both high-power continuous-wave carbon dioxide (CO_2) and neodymium-doped yttrium aluminum garnet (Nd:YAG) lasers are being evaluated. The wavelength of light from a CO_2 laser is 10.6 μm, while that of Nd:YAG laser is 1.06 μm. Because the absorption of the beam energy by the material being welded increases with decreasing wavelength, Nd:YAG lasers are better suited for welding aluminum. In addition, the solid-state Nd:YAG lasers use fiber optics for beam delivery, making it more amenable to automated robotic welding. Laser welding produces a concentrated high energy density heat source that results in very narrow HAZs, minimizing both distortion and loss of strength in the HAZ. In 2 mm (0.080 in.) sheet, speeds up to 2 m/min (6.5 ft/min) are achievable with a 2 kW Nd:YAG laser and 4.9 to 6.1 m/min (16 to 20 ft/min) with a 5 kW CO_2 laser.

Resistance welding can produce excellent joint strengths in the high-strength heat treatable aluminum alloys. Resistance welding is normally used for fairly thin sheets where joints are produced with no loss of strength in the base metal and without the need for filler metals. In resistance welding, the faying surfaces are joined by heat generated by the resistance to the flow of current through workpieces held together by the force of water-cooled copper electrodes. A fused nugget of weld metal is produced by a short pulse of low-voltage, high-amperage current. The electrode force is maintained while the liquid metal rapidly cools and solidifies. In spot welding, as shown in Fig. 2.40, the two parts to be joined are pressed together between two electrodes during welding. In seam welding, the two electrodes are replaced with wheels. While the 2*xxx* and 7*xxx* alloys are easy to resistance weld, they are more susceptible to shrinkage cracks and porosity than lower-strength aluminum alloys. Alclad materials are also more difficult to weld due to the lower electrical resistance and higher melting point of the clad layers. Removal of the surface oxide is important to produce good weld quality. Both mechanical and chemical methods are used, with surface preparation being checked by measuring the surface resistivity.

Friction Stir Welding. A relatively new welding process that has the potential to revolutionize aluminum joining has been developed by The Welding Institute (TWI) in Cambridge, United Kingdom. Friction stir welding (FSW) is a solid-state process that operates by generating frictional heat between a rotating tool and the workpiece, as shown sche-

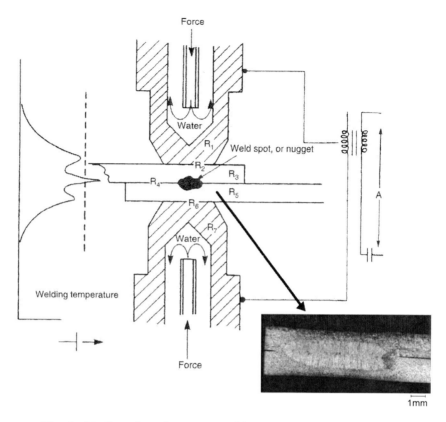

Fig. 2.40 Setup for resistance spot welding. Cross section shows shape of nugget and position of nugget relative to inner and outer surfaces of workpieces. Source: Ref 2.30, 2.31

matically in Fig. 2.41. The welds are created by the combined action of frictional heating and plastic deformation due to the rotating tool.

A tool with a knurled probe of hardened steel or carbide is plunged into the workpiece, creating frictional heating that heats a cylindrical column of metal around the probe, as well as a small region of metal underneath the probe. As shown in Fig. 2.42, a number of different tool geometries have been developed, which can significantly affect the quality of the weld joint. The threads on the probe cause a downward component to the material flow, inducing a counterflow extrusion toward the top of the weld, or an essentially circumferential flow around the probe. The rotation of the probe tool stirs the material into a plastic state, creating a very fine-grained microstructural bond. The tool contains a larger-diameter shoulder above the knurled probe, which controls the depth of the probe and creates additional frictional heating on the top surface of the workpiece. It also prevents the highly plasticized metal from being expelled from the joint. Prior to welding, the workpieces must be rigidly fixed with the edges butted to each other and must have a backing plate to withstand

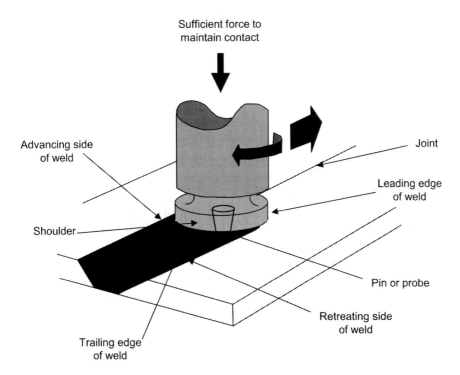

Fig. 2.41 Friction stir welding process. Source: Ref 2.32

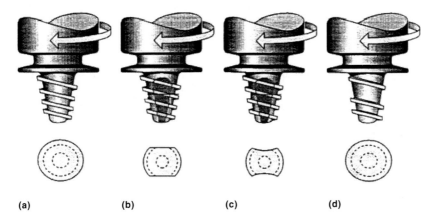

Fig. 2.42 Sample friction stir welding tool geometries. (a) Oval shape. (b) Paddle shape. (c) Re-entrant. (d) Changing spiral form. Courtesy of The Welding Institute

the downward forces exerted by the tool. A typical welding operation is shown in Fig. 2.43.

The larger the diameter of the shoulder, the greater is the frictional heat it can contribute to the process. After the shoulder makes contact, the

Direction

Weld

Fig. 2.43 Friction stir welding. Courtesy of ESAB Welding Equipment AB

thermally softened metal forms a frustum shape corresponding to the tool geometry, with the top portion next to the shoulder being wider and then tapering down to the probe diameter. The maximum temperature reached is of the order of 0.8 of the melting temperature. Material flows around the tool and fuses behind it. As the tool rotates, there is some inherent eccentricity in the rotation that allows the hydromechanically incompressible plasticized material to flow more easily around the probe. Heat-transfer studies have shown that only about 5% of the heat generated in FSW flows into the tool, with the rest flowing into the workpieces; therefore, the heat efficiency in FSW is very high relative to traditional fusion welding processes where the heat efficiency is only about 60 to 80%.

After the tool has penetrated the workpieces, the frictional heat caused by the rotating tool and rubbing shoulder results in frictional heating and plasticization of the surrounding material. Initially, the material is extruded at the surface, but as the tool shoulder contacts the workpieces, the plasticized metal is compressed between the shoulder, workpieces, and backing plate. As the tool moves down the joint, the material is heated and plasticized at the leading edge of the tool and transported to the trailing edge of the probe, where it solidifies to form the joint.

The advantages of FSW include:

- The ability to weld butt, lap, and T-joints
- Minimal or no joint preparation
- The ability to weld the difficult-to-fusion-weld 2xxx and 7xxx alloys

- The ability to join dissimilar alloys
- The elimination of cracking in the fusion and HAZs
- Lack of weld porosity
- Lack of required filler metals
- In the case of aluminum, no requirement for shielding gases

In general, the mechanical properties are better than for many other welding processes. For example, the static properties of 2024-T351 are between 80 and 90% of the parent metal, and the fatigue properties approach those of the parent metal. In a study of lap shear joints, FSW joints were 60% stronger than comparable riveted or resistance spot welded joints.

The weld joint does not demonstrate many of the defects encountered in normal fusion welding, and the distortion is significantly less. A typical weld joint, as shown in Fig. 2.44, contains a well-defined nugget with flow contours that are almost spherical in shape but are somewhat dependent on tool geometry. TWI has recommended the microstructural classification shown in Fig. 2.44 be used for friction stir welds. The fine-grained recrystallized weld nugget and the adjacent unrecrystallized but plasticized material is referred to as the thermomechanically affected zone (TMAZ); therefore, the TMAZ results from both thermal exposure and

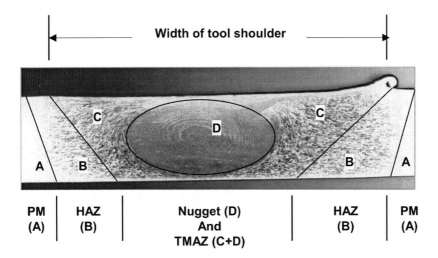

A- Parent metal
B- Heat affected zone (HAZ)
C- Unrecrystallized area
D- Recrystallized nugget
C + D- Thermomechanically affected zone (TMAZ)

Fig. 2.44 Friction stir fusion weld. A = parent metal (PM); B = heat-affected zone (HAZ); C = unrecrystallized area; D = recrystallized nugget; C + D = thermomechanically affected zone (TMAZ). Courtesy of The Welding Institute

plastic deformation and extends from the width of the shoulder at the top surface to a narrower region on the backside. A series of concentric rings, called onion rings, are frequently observed within the weld nugget, possibly as a result of the swirling motion of the plasticized material behind the advancing tool probe. The unrecrystallized portion of the TMAZ, which has also undergone thermal exposure and plastic deformation, has a grain size similar to that of the parent metal. The HAZ is typically trapezoidal in shape for a single-pass weld, with a greater width at the tool shoulder due to the heat generated between the shoulder and the top of the workpieces. The HAZ results primarily from thermal exposure with little or no plastic deformation present.

The FSW process has already been adapted to a number of industrial applications. In 1999, the fuel tanks on the Boeing Delta II rocket were launched with friction stir longitudinal welds. Based on this early success, Boeing purchased a large friction stir welder for its Delta IV fuel tanks. In addition, Lockheed Martin and National Aeronautics and Space Administration (NASA) Marshall Space Flight Center have developed and implemented FSW on the longitudinal welds of the 2195 aluminum-lithium liquid hydrogen and liquid oxygen barrel segments of the external tank for the Space Shuttle (Fig. 2.45). Another early adapter of FSW has been the marine industry, as shown in the welded ship skin in Fig. 2.46. Construction of high-speed trains is another important potential application for FSW.

Adhesive Bonding

Adhesive bonding is a materials joining process in which an adhesive (usually a thermosetting or thermoplastic resin) is placed between the faying surfaces of the parts or bodies called adherends. The adhesive then solidifies or hardens by physical or chemical property changes to produce a bonded joint with useful strength between the adherends.

Fig. 2.45 Friction stir weld process development tool at the Marshall Space Flight Center (MSFC) shown with an 8.2 m (27 ft) diameter barrel segment of the 2195 aluminum-lithium Space Shuttle. National Aeronautics and Space Administration (NASA) Michoud Assembly Facility in New Orleans. Courtesy of NASA MSFC. Source: Ref 2.33

Fig. 2.46 Large friction stir welder being used in the marine industry. Courtesy of ESAB Welding Equipment AB

Adhesive-bonded joints are used extensively in aircraft components and assemblies where structural integrity is critical. The structural components are not limited to aircraft applications; they can be translated to commercial and consumer product applications as well. Adhesive bonding is very competitive when compared to other joining methods in terms of production cost, ability to accommodate manufacturing tolerances and component complexity, facility and tooling requirements, reliability, and repairability.

Some typical aircraft applications for adhesive-bonded structures include:

* Metal-to-metal bonded structures that are locally reinforced by bonded doubler plates or some other type of reinforcement (Fig. 2.47)
* Metal-to-metal bonded multiple laminations where each layer progressively increases the cross-sectional area of the component (e.g., for stringers and spar caps)
* Bonded joints between rather thin metal sheet and a low-density core material, called honeycomb or sandwich structures (Fig. 2.47). Core materials include paper dipped in a phenolic resin, fiberglass, aluminum alloy foil, Nomex (aramid fiber paper dipped in phenolic resin), graphite-reinforced plastic, and Kevlar (E.I. du Pont de Nemours and Company) fabric.

Aluminum alloys that are commonly adhesive bonded include 2024 (T3, T6, and T86), 3xxx-series alloys, 5052-H39, 5056-H39, and 7075-T6. These alloys may be bonded to themselves, each other, other metals, and many

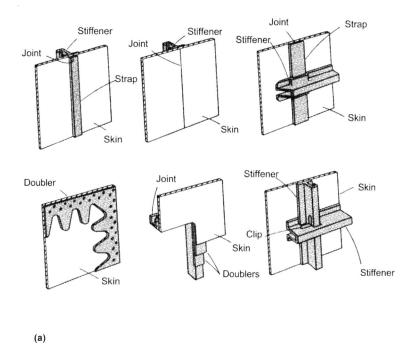

(a)

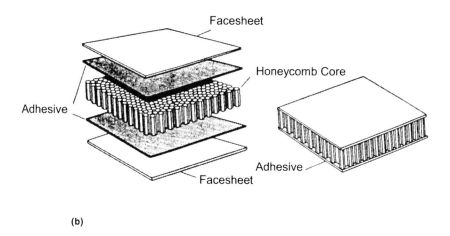

(b)

Fig. 2.47 (a) Examples of adhesively bonded joint configurations. (b) Honeycomb bonded assembly. Source: Ref 2.34

nonmetals, including all forms of paper products, insulation board, wood-particle board, plaster board, plywood, fiberglass, and various polymers and organic-matrix composites. Successful adhesive bonding requires stringent cleaning procedures, and the cleaned surfaces must be protected from contamination until they are bonded.

Because an adhesive can transmit loads from one member of a joint to another, it allows a more uniform stress distribution than is obtained using mechanical fasteners. Thus, adhesives often permit the fabrication of structures that are mechanically equivalent or superior to conventional assemblies and have cost and weight benefits. For example, adhesives can join thin metal sections to thick sections so that the full strength of the thin section is used. In addition, adhesives can produce joints with high strength, rigidity, and dimensional precision in light metals, such as aluminum, that may be weakened or distorted by welding.

Because the adhesive in a properly prepared joint provides full contact with mating surfaces, it forms a barrier so fluids do not attack or soften it. An adhesive may also function as an electrical and/or thermal insulator in a joint. Its thermal insulating efficiency can be increased, if necessary, by forming an adhesive with the appropriate cell structure in place. On the other hand, electrical and thermal conductivity can be raised appreciably by adding metallic fillers. Oxide fillers, such as alumina, increase only thermal conductivity. Electrically conductive adhesives, filled with silver flake, are available with specific resistivities less than 50 times that of bulk silver.

Adhesives can also prevent electrochemical corrosion in joints between dissimilar metals. They may also act as vibration dampers. The mechanical damping characteristics of an adhesive can be changed by formulation. However, changing such a property in an adhesive generally changes other properties of the joint, such as tensile or shear strength, elongation, or resistance to peel or cleavage. A property somewhat related to the ability to damp vibration is resistance to fatigue. A properly selected adhesive can generally withstand repeated strains induced by cyclic loading without the propagation of failure-producing cracks.

Adhesives usually do not change the contours of the parts they join. Unlike screws, rivets, or bolts, adhesives give little or no visible external evidence of their presence. They are used to join skins to airframes, and they permit the manufacture of airfoils, fuselages, stabilizers, and control surfaces that are smoother than similar conventionally joined structures and that consequently have better aerodynamic efficiency. These structures also have greater load-bearing capability and higher resistance to fatigue than conventionally joined structures.

2.19 Corrosion of Aluminum and Aluminum Alloys

Aluminum and most aluminum alloys have good corrosion resistance in natural atmospheres, fresh waters, seawater, many soils, many chemicals and their solutions, and most foods. This resistance to corrosion is the result of the presence of a very thin, compact, and adherent film of aluminum oxide on the metal surface. Whenever a fresh surface is created by cutting or abrasion and is exposed to either air or water, a new film forms rapidly, growing to a stable thickness. The film formed in air at ambient temperatures is ~5 nm thick. The thickness increases with increasing temperature and in the presence of water. The oxide film is soluble in alkaline solutions and in strong acids, with some exceptions, but is stable over a pH range of ~4.0 to 9.0.

There are different types of corrosion and various interactions with induced or imposed stresses, so that the effects can range from unimportant to highly damaging. For some types of applications, a distinction should be made between appearance and durability. The surface can become unattractive because of roughening by shallow pitting and can darken with dirt retention, but these conditions may have no effect on durability or function. On the other hand, SCC or highly localized, severe corrosion due to heavy metal ions in solutions, stray electrical currents, or galvanic couples with more-anodic metals can be quite damaging. Good design and application practices must be observed to avoid these conditions. This includes selection of alloys appropriate for the conditions of the application.

Effects of Alloy Composition. Among the principal alloying elements (magnesium, silicon, copper, manganese, and zinc), the element that has the greatest effect on general corrosion resistance is copper. Copper reduces resistance because it replates from solution as minute metallic particles forming highly active corrosion couples. The effects are apparent at copper contents exceeding a few tenths of one percent. In the heat treatable 2xxx or 2xx.x alloys with several percent copper, the state of solution or precipitation affects the type of corrosion attack as well as susceptibility to SCC. These effects are recognized in the alloy/temper ratings for resistance to general corrosion and to SCC of wrought alloys (Table 2.19) and for casting alloys (Table 2.20).

Among the common impurity elements, iron is probably the most important, degrading general corrosion resistance by increasing the volume fraction of cathodic and surface-film-weakening intermetallic-phase microconstituents.

The 7xxx Al-Zn-Mg alloys without copper have high general corrosion resistance. The alloys of this group that contain more than 1% Cu are less resistant to general corrosion. Appropriate tempers should be used to avoid SCC.

The 3xxx alloys are generally among those having the highest general corrosion resistance, as are those of the 5xxx group, which outperform any of the others in marine exposures. The 6xxx alloys also have high resistance.

Corrosion in Atmospheres. Alloys other than those with the higher copper contents have excellent resistance to atmospheric corrosion (often called weathering) and, in many outdoor applications, require no protection or maintenance. Products widely used under such conditions include electrical conductors, outdoor lighting poles, bridge railings, and ladders. These often retain a bright metallic appearance for many years but may

Table 2.19 Relative ratings of resistance to general corrosion and to stress-corrosion cracking (SCC) of wrought aluminum alloys

Alloy No.	Temper	General(a)	SCC(b)
1100	All	A	A
2011	T3, T4, T451	D(c)	D
	T8	D	B
2014	O	…	…
	T3, T4, T451	D(c)	C
	T6, T651, T6510, T6511	D	C
2024	O	…	…
	T4, T3, T351, T3510, T3511, T361	D(c)	C
	T6, T861, T81, T851, T8510, T8511	D	B
	T72	…	…
3003	All	A	A
4032	T6	C	B
5052	All	A	A
5056	O, H11, H12, H32, H14, H34	A(d)	B(d)
	H18, H38	A(d)	C(d)
	H192, H392	B(d)	D(d)
6053	O	…	…
	T6, T61	A	A
6061	O	B	A
	T4, T451, T4510, T4511	B	B
	T6, T651, T652, T6510, T6511	B	A
6063	All	A	A
7001	O	C(c)	C
7075	T6, T651, T652, T6510, T6511	C(c)	C
	T73, T7351	C	B
7178	T6, T651, T6510, T6511	C(c)	B

(a) Ratings are relative and in decreasing order of merit, based on exposure to sodium chloride solution by intermittent spraying or immersion. Alloys with A and B ratings can be used in industrial and seacoast atmospheres without protection. Alloys with C, D, and E ratings generally should be protected, at least on faying surfaces. (b) Stress-corrosion cracking ratings are based on service experience and on laboratory tests of specimens exposed to alternate immersion in 3.5% NaCl solution. A, no known instance of failure in service or in laboratory tests; B, no known instance of failure in service; limited failure in laboratory tests of short-transverse specimens; C, service failures when sustained tension stress acts in short-transverse direction relative to grain structure; limited failures in laboratory tests of long-transverse specimens; D, limited service failures when sustained stress acts in longitudinal or long-transverse direction relative to grain structure. (c) In relatively thick sections, the rating would be E. (d) This rating may be different for material held at elevated temperatures for long periods. Source: Ref 2.35

Table 2.20 Relative ratings of resistance to general corrosion and to stress-corrosion cracking (SCC) of cast aluminum alloys

Alloy No.	Temper	General(a)	SCC(b)
Sand castings			
208.0	F	B	B
A242.0	T75	D	C
C355.0	T6	C	A
356.0	T6, T7, T71, T51	B	A
A356.0	T6	B	A
443.0	F	B	A
535.0	F	A	A
771.0	T6	C	C

(continued)

(a) Relative ratings of general corrosion resistance are in decreasing order of merit, based on exposures to NaCl solution by intermittent spray or immersion. (b) Relative ratings of resistance to SCC are based on service experience and on laboratory tests of specimens exposed to alternate immersion in 3.5% NaCl solution. A, no known instance of failure in service when properly manufactured; B, failure not anticipated in service from residual stresses or from design and assembly stresses below approximately 45% of the minimum guaranteed yield strength given in applicable specifications; C, failures have occurred in service with either this specific alloy/temper combination or with alloy/temper combinations of this type; designers should be aware of the potential SCC problem that exists when these alloys and tempers are used under adverse conditions. (c) For electric motor rotors. Source: Ref 2.35

Table 2.20 (continued)

No.	Temper	General(a)	SCC(b)
Alloy		**Resistance to corrosion**	
Permanent mold castings			
242.0	T571, T61	D	C
355.0	All	C	A
356.0	All	B	A
A357.0	T61	B	A
358.0	T6	B	A
359.0	All	B	A
513.0	F	A	A
705.0	T5	B	B
852.0	T5	C	B
Die castings			
A360.0	F	C	A
364.0	F	C	A
380.0	F	E	A
A380.0	F	E	A
392.0	F	E	A
413.0	F	C	A
518.0	F	A	A
Rotor metal(c)			
100.1	...	A	A

(a) Relative ratings of general corrosion resistance are in decreasing order of merit, based on exposures to NaCl solution by intermittent spray or immersion. (b) Relative ratings of resistance to SCC are based on service experience and on laboratory tests of specimens exposed to alternate immersion in 3.5% NaCl solution. A, no known instance of failure in service when properly manufactured; B, failure not anticipated in service from residual stresses or from design and assembly stresses below approximately 45% of the minimum guaranteed yield strength given in applicable specifications; C, failures have occurred in service with either this specific alloy/temper combination or with alloy/temper combinations of this type; designers should be aware of the potential SCC problem that exists when these alloys and tempers are used under adverse conditions. (c) For electric motor rotors. Source: Ref 2.35

darken with mild surface roughening caused by shallow pitting and with accumulation of dirt.

An important characteristic of weathering, as well as corrosion of aluminum under many other environmental conditions, is that the rate of corrosion decreases with time. Figure 2.48 shows typical curves of the average changes that occurred in exposures of two generic types, seacoast and industrial, over a 30-year period with sheet specimens of 1100, 3003, and 3004 alloys, all in H14 temper. The curve shapes are similar, whether the amount of corrosion is measured by weight loss, depth of pitting, or loss in strength. Leveling-off usually occurs in 6 months to 2 years, after which the rate becomes approximately linear at a low rate. In rural atmospheres the weight-loss rate may be less than 0.025 µm/year (0.001 mil/year). In industrial locations, rates vary from ~0.75 to 2.75 µm/year (0.03 to 0.11 mil/year), with a few particularly aggressive sites at which rates up to 12.5 µm/year (0.5 mil/year), based on maximum pit depth, were observed. The atmospheric corrosion rates for wrought aluminum alloy 1100-H14 with that of commercially pure copper, lead, and zinc are compared in Table 2.21.

Castings of 4xx.x and low-copper 3xx.x alloys have been used for many years in such applications as bridge guardrail supports and lighting pole

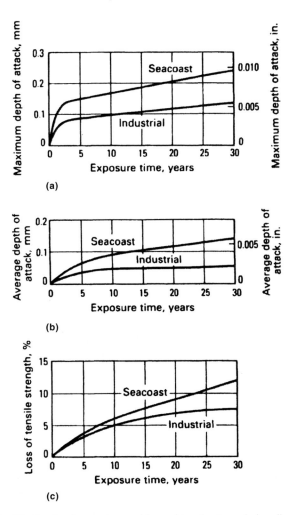

Fig. 2.48 Depth of corrosion and loss of tensile strength for alloys 1100, 3003, and 3004 (shown in graphs a, b, and c, respectively). Data are given for the average performance of the three alloys, all in the H14 temper. Seacoast exposure was at a severe location (Point Judith, RI); industrial exposure was at New Kensington, PA. Tensile strengths were computed using original cross-sectional areas, and loss in strength is expressed as a percentage of original tensile strength. Source: Ref 2.35

bases with little adverse effect from corrosion. The 5xx.x casting alloys with 4 to 8% Mg are particularly suitable for cast parts that are used in marine and seawater exposure sites. Copper-containing alloys such as 295.0, 333.0, 380.0, and even 355.0 require surface protection for satisfactory use in corrosive marine or industrial atmospheres. Stress-corrosion cracking has been encountered occasionally with some of the 7xx.x alloys and alloy 520.0-T4 in atmospheric exposures of stressed parts. Applications for these alloys must be carefully engineered to avoid such problems.

Table 2.21 Atmospheric corrosion rates for aluminum and other nonferrous metals at several exposure sites

Location	Type of atmosphere	Aluminum(b) 10 yr	Aluminum(b) 20 yr	Copper(c) 10 yr	Copper(c) 20 yr	Lead(d) 10 yr	Lead(d) 20 yr	Zinc(e) 10 yr	Zinc(e) 20 yr
Phoenix, AZ	Desert	0.000	0.076	0.13	0.13	0.23	0.10	0.25	0.18
State College, PA	Rural	0.025	0.076	0.58	0.43	0.48	0.30	1.07	1.09
Key West, FL	Seacoast	0.10	...	0.51	0.56	0.56	...	0.53	0.66
Sandy Hook, NJ	Seacoast	0.20	0.28	0.66	...	...	...	1.40	...
La Jolla, CA	Seacoast	0.71	0.63	1.32	1.27	0.41	0.53	1.73	1.73
New York, NY	Industrial	0.78	0.74	1.19	1.37	0.43	0.38	4.8	5.6
Altoona, PA	Industrial	0.63	...	1.17	1.40	0.69	...	4.8	6.9

Header: Depth of metal removed per side(a), in μm/yr, during exposure of indicated length for specimens of

(a) Calculated from weight loss, assuming uniform attack, for 0.89 mm (0.035 in.) thick panels. (b) Aluminum 1100-H14. (c) Tough pitch copper (99.9% Cu). (d) Commercial lead (99.92% Pb). (e) Prime western zinc (98.9% Zn). Source: Ref2. 35

Corrosion in Waters. Corrosion resistance of aluminum in high-purity water, distilled or deionized, and in steam condensate is so high that these fluids are regularly contained and handled in aluminum equipment. Resistance is also high in most natural fresh waters. Soft waters have the least pitting tendency. Components of natural waters that increase pitting are copper ions, bicarbonate, chloride, sulfate, and oxygen. Thus, harder waters with more bicarbonate have a higher pitting tendency.

Service experience with aluminum alloys in marine and coastal applications, including structures, buoys, pipelines, lifeboats, motor launches, cabin cruisers, patrol boats, barges, and larger vessels, has demonstrated their good resistance and long life under conditions of partial, intermittent, and total immersion. Wrought alloys of the 3xxx, 5xxx, and 6xxx groups are used. Those alloys of the 5xxx aluminum-magnesium group are most resistant and most widely used because of favorable strength and good weldability. The rate of corrosion based on weight loss does not exceed ~5 μm/year (0.2 mil/year), which is <5% of the rate for unprotected low-carbon steel in seawater. Corrosion is mainly pitting, decelerating with time from rates of 2.5 to 5 μm/year (0.1 to 0.2 mil/year) in the first year to average rates over a 10 year period of 0.75 to 1.5 μm/year (0.03 to 0.06 mil/year). The curve of maximum depth of pitting versus time follows an approximate cube-root law, from which it follows that doubling material thickness increases time to perforation by a factor of 8.

Casting alloys of the 5xx.x group with up to 8% Mg have high resistance to seawater corrosion and are used for fittings. Aluminum-silicon and Al-Si-Mg alloys are less resistant, although 443.0 and 356.0-T6 are sometimes used. Resistance of 2xxx, 2xx.x, 7xxx, and 7xx.x alloys is distinctly inferior, and their use without cladding or metallizing is not recommended.

Corrosion in Soils. Soils differ widely in mineral content, texture, permeability, moisture, pH, and electrical conductivity as well as aeration, organic matter, and microorganisms. With this variability, corrosion performance of unprotected buried aluminum, like that of other metals, varies considerably. However, in many cases where carbon steel requires protective coatings, unprotected aluminum alloys have performed well. In most cases, protection is recommended for buried applications. Use of 2xxx or 7xxx alloys or their cast counterparts is generally not recommended. Stray current effects and contact with more cathodic metals should be avoided. Successful applications include pipelines employing alloys 3003, 6061, and 6063 and culverts made of Alclad 3004.

Exposure to Foods. Aluminum alloys of the 3xxx and 5xxx groups are resistant to most foods and beverages. Aluminum products constitute a substantial share of the domestic cooking utensil market and are used extensively for commercial handling and processing of foods. Large quantities of foil, foil laminated to plastics or paper, and cans are used for packaging and marketing of foods and beverages. Beverage can bodies are generally produced from alloy 3004, food can bodies from 5352 or 5050, and can ends from 5182. These cans generally have internal and external organic coatings, not for corrosion protection but for decoration and prevention of effects on product taste.

Exposure to Chemicals. Aluminum alloys are used to store, process, handle, and package a variety of chemical products. They are compatible with most dry inorganic salts. Within the passive pH range, approximately 4 to 9, they resist corrosion in solutions of most inorganic chemicals but are subject to pitting in aerated solutions, particularly of halides. Aluminum alloys are not suitable for containing or handling mineral acids, with the exceptions of nitric acid in concentrations over 82 wt% and sulfuric acid in concentrations from 98 to 100 wt%.

Aluminum alloys resist most alcohols; however, some may cause corrosion when extremely dry and at elevated temperatures. Similar characteristics are associated with phenol. Aldehydes have little or no corrosive effects. Care must be taken in using aluminum with halogenated organic compounds because under some conditions when moisture is present, they may hydrolyze and react violently.

Resistance of aluminum and aluminum alloys to many foods and chemicals, representing practically all classifications, has been established in laboratory tests and, in many cases, by service experience. Data are readily available from handbooks, proprietary literature, and trade association publications.

Exposure to Nonmetallic Building Materials. Aluminum alloys are not corroded seriously by long-time embedment in portland cement concrete, standard or lime brick mortar, hard-wall plaster, or stucco. Superficial etching of the aluminum occurs while these products are setting, but after curing, further attack is minimal. Special consideration should be given to protection where crevices between the concrete and the metal can entrap environmental contaminants. For example, highway railings and streetlight standards or stanchions usually are coated with a sealing compound where they are fastened to concrete, to prevent entry of salt-laden road splash into crevices.

Absorbent materials, such as paper, wood refuse, and wallboard in contact with aluminum under conditions where it may become wet, will cause corrosion. Composite-bonded insulated aluminum panels have a moisture barrier on the inside to prevent condensation and wetting of the insulation. Some insulating materials, such as magnesia, are alkaline and quite aggressive to aluminum. Magnesium oxychloride, a flooring compound sometimes used in subway and railway passenger cars and ship decks, is very corrosive and should not be used with aluminum. Although wood usually is not corrosive, it can become corrosive if moisture content exceeds 18 to 20%. Wood preservatives containing copper are detrimental, and those containing mercury should not be used where aluminum is involved.

Forms of Corrosion

Form or type of corrosion can be categorized by the morphology, which may or may not be related to the microstructure of the metal, or by the conditions causing the corrosion. Uniform attack or dissolution, which may occur if the surface oxide film is soluble in the corroding medium, is infrequent in service. Most corrosion in service is localized in one way or another. When the oxide film is insoluble in the corroding medium, corrosion is localized at weak spots in the film, which can result from microstructural features such as the presence of microconstituents. Local cells are formed by such nonuniformities in the metal as well as environmental nonuniformities, such as those created by differential aeration cells or by heavy metals plated out on the surface. Localized corrosion in a microscopic sense results from galvanic coupling and stray-current effects.

Pitting is the most common form of localized corrosion and frequently is difficult to associate with specific metallographic features. Pit shape can vary from shallow depressions to cylindrical or roughly hemispherical cavities. These shapes distinguish pitting from intergranular or exfoliation corrosion. Superpurity aluminum has the highest resistance to pitting, and, among the 1xxx and 1xx.x aluminums, resistance improves with purity. Among commercial alloys, those of the 5xxx group have the lowest pitting probability and penetration rates, followed by alloys of the 3xxx group.

Intergranular corrosion is a selective attack of grain boundaries. The mechanism is electrochemical, resulting from local cell action in the boundaries. Microconstituents precipitated in grain boundaries have a corrosion potential differing from that of adjacent solid-solution and transition precipitate structure and form cells with it. In alloys of the 5xxx and 7xxx groups, the precipitates (Al_8Mg_5, $MgZn_2$, and $Al_2Mg_3Zn_3$) are anodic to the matrix. In 2xxx alloys, the precipitates (Al_2Cu and Al_2CuMg) are cathodic. Intergranular corrosion can occur with either type. Susceptibility depends on the extent of intergranular precipitation, which is controlled by fabricating or heat treating parameters.

In 2xxx alloys, grain-boundary precipitation is caused by an inadequate cooling rate during the quenching operation of heat treatment. Thick-section products cannot be cooled sufficiently rapidly to completely avoid susceptibility to intergranular corrosion in T3- and T4-type tempers. Resistance to this type of attack is much higher in T6- and T8-type tempers. Alloys of the 6xxx group with a balanced magnesium/silicon ratio (Mg_2Si proportions) show little tendency toward intergranular corrosion; susceptibility is higher in those with excess silicon over the Mg_2Si ratio. Alloys of the 7xxx group may corrode intergranularly; overaging to T7-type tempers provides high resistance.

Because intergranular corrosion is involved in SCC of aluminum alloys, it is often presumed to be more deleterious than pitting or general (uniform) corrosion. However, in alloys that are not susceptible to SCC, for example, the 6xxx-series alloys, intergranular corrosion is usually no more severe than pitting corrosion, tends to decrease with time, and, for equal depth of corrosion, the effect on strength is no greater than that of pitting corrosion, although fatigue cracks can be more likely to initiate at areas of intergranular corrosion than at random pits.

Exfoliation corrosion is selective attack that proceeds along multiple subsurface paths roughly parallel to the surface. It can be intergranular but also is associated with striated insoluble microconstituents and dispersoid bands aligned parallel to the product surface. It is most common in thin-section products with highly worked, flattened, and elongated metallurgical structures. Leafing or delamination accompanied by swelling caused by expansion of corrosion products is characteristic and is apparent in metallographic section (Fig. 2.49). Exfoliation frequently proceeds from sheared edges and may be initiated at pit surfaces. It is not accelerated by applied stress but is intensified by slightly acidic solutions and by galvanic coupling.

In 2xxx and copper-containing 7xxx alloys, exfoliation is considerably affected by section thickness and corresponding microstructure and by temper. The 2xxx alloys are susceptible only in T3- and T4-type tempers and are resistant in T6- or T8-type tempers. In the copper-containing 7xxx-group alloys, resistance is greatly improved by overaging beyond peak strength (T6-type tempers), so that these materials are resistant in T7-type tempers. In extrusions of these heat treated alloys, the recrystallized peripheral zone near surfaces is often highly resistant to exfoliation, while the underlying unrecrystallized portion may be vulnerable to this type of attack.

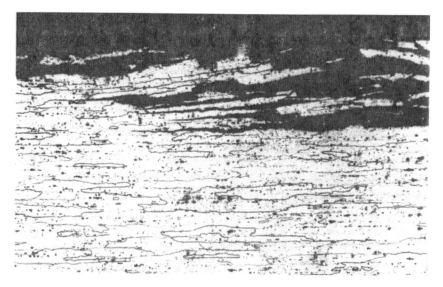

Fig. 2.49 Exfoliation corrosion in an alloy 7178-T651 plate exposed to a seacoast environment. Cross section of the plate shows how exfoliation develops by corrosion along boundaries of thin, elongated grains. Source: Ref 2.35

The 1*xxx* aluminums and 3*xxx* alloys are highly resistant to exfoliation in all tempers. Exfoliation has been encountered in highly cold-worked, high-magnesium 5*xxx* alloys such as 5456-H321 boat hull plate. Improved tempers (H116 and H117) with high resistance have been established.

Galvanic Corrosion. Most forms of corrosion discussed previously are electrochemical in nature and involve cells formed by microstructural or environmental features. A number of other conditions establish potential differences that intensify and localize corrosion. The accelerated corrosion resulting from electrical contact with a more noble metal or with a nonmetallic conductor such as graphite is termed galvanic corrosion. The most common examples occur when aluminum alloys are joined to steel or copper and are exposed to wet saline environments. In such situations, the aluminum is more rapidly corroded than it would be in the absence of the dissimilar metal.

For each environment, metals can be arranged in a galvanic series from most to least active. Table 2.22 lists potentials based on measurements in sodium chloride solution. The rate of corrosive attack when two metals are coupled depends on several factors: the potential difference, the electrical resistance between the metals, the conductivity of the electrolyte, the cathode/anode area ratio, and the polarization characteristics of the metals. Although the corrosion potential can be used to predict which metal will be attacked galvanically, the extent of attack cannot be predicted, because of polarization. For example, the potential difference between aluminum

Table 2.22 Electrode potentials of representative aluminum alloys and other metals

Aluminum alloy(a) or other metal	Potential(b), V
Chromium	+0.18 to −0.40
Nickel	−0.07
Silver	−0.08
Stainless steel (300 series)	−0.09
Copper	−0.20
Tin	−0.49
Lead	−0.55
Mild carbon steel	−0.58
2219-T3, -T4	−0.64(c)
2024-T3, -T4	−0.69(c)
295.0-T4 (S or PM)	−0.70
2014-T6, 355.0-T4 (S or PM)	−0.78
355.0-T6 (S or PM)	−0.79
2219-T6, 6061-T4	−0.80
2024-T6	−0.81
2219-T8, 2024-T8, 356.0-T6 (S or PM), 443.0-F (PM), cadmium	−0.82
1100, 3003, 6061-T6, 6063-T6, 7075-T6(c), 443.0-F (S) 1060, 1350, 3004, 7050-T73(c),	−0.83
7075-T73(c)	−0.84
5052, 5086	−0.85
5454	−0.86
5456, 5083	−0.87
7072	−0.96
Zinc	−1.10
Magnesium	−1.73

(a)The potential of an aluminum alloy is the same in all tempers wherever the temper is not designated. S is sand cast; PM is permanent mold cast. (b) Measured in an aqueous solution of 53 g of sodium chloride and 3 g of H_2O_2 per liter at 25 °C; 0.1N calomel reference electrode. (c) Potential varies ±0.01 to 0.02 V with quenching rate. Source: Ref 2.35

and stainless steel exceeds that between aluminum and copper; however, the galvanic effect of stainless steel on aluminum, because of polarization, is much less than that of copper, which shows little polarization.

In natural environments, including saline solutions, zinc is anodic to aluminum and corrodes preferentially, giving protection to the aluminum. Magnesium is also protective, although in severe marine environments, it can cause corrosion of aluminum because of an alkaline reaction. Cadmium is neutral to aluminum and can be used safely in contact with it. Copper and copper alloys, brass, bronze, and Monel are the most harmful, followed closely by carbon steel in saline environments. Nickel is less aggressive than copper, approaching stainless steel in effect, as does chromium electroplate. Lead can be used with aluminum, except in marine environments.

Stray Current Corrosion. Whenever an electric current is conducted from aluminum to an environment such as water, soil, or concrete, the aluminum is corroded in the area of anodic reaction in proportion to the current. At low current densities, the corrosion may be in the form of pitting, while at higher current densities, considerable metal destruction can occur at rates that do not diminish with time.

In soils, stray current corrosion can be caused by close proximity to other buried metal systems, which are protected by an impressed current cathodic-protection system. The ground current can leak onto a buried aluminum structure at one point, then off at another where the corrosion occurs, taking a lower resistance path between the driven buried anode and the nearby structure being protected. Common bonding of all buried metal systems in close proximity is the usual way to avoid such attack.

Deposition corrosion is a special form of galvanic corrosion that causes pitting. It occurs when particles of a more cathodic metal are plated out of solution on the aluminum surface, setting up local galvanic cells. The ions aggressive to aluminum are copper, lead, mercury, nickel, and tin, often referred to as heavy metals. Their effects are greater in acidic solutions and are much less severe in alkaline solutions, in which their solubility is low.

Copper ions most commonly cause this type of corrosion in applications of aluminum. For example, rain runoff from copper roof flashing can cause corrosion of aluminum gutters with no electrical contact between the two metals. Very small amounts of copper in solution (as low as 0.05 ppm) can be detrimental. The inferior general corrosion resistance of alloys containing copper is attributed to deposition corrosion by copper replated from the dissolved corrosion products.

Mercury is the ion most aggressive to aluminum, and even traces can cause serious problems. Liquid mercury does not wet aluminum, but if the natural oxide film on the aluminum surface is broken, aluminum dissolves in the mercury, forming amalgam, and the corrosion reaction becomes catastrophic. In corrosive solutions, any concentration of mercury greater than a few parts per billion should be cause for concern.

Crevice Corrosion. If an electrolyte is present in a crevice formed between two facing aluminum surfaces or between an aluminum surface and a nonmetallic material such as a gasket, localized corrosion in the form of pits or etch patches can occur. This is the result of formation of a concentration cell or differential aeration cell. Staining that occurs on interwrap surfaces of coiled sheet or foil or in packages of flat sheet or circles is a result of the same mechanism and can be preliminary to more severe corrosion that will make separation difficult. Such damage can be prevented by ensuring that the product is initially dry and by avoiding ingress of moisture by protecting it against condensation, rain, and other sources of contamination.

Filiform corrosion, sometimes termed "worm-track" corrosion, occurs on aluminum when it is coated with an organic coating and exposed to warm, humid atmospheres. The corrosion appears as threadlike filaments that initiate at defects in the organic coating, are activated by chlorides, and grow along the metal/coating interface at rates to 1 mm/day (0.04 in./day). The moving end of the filament is called the head, and the remainder of the track is called the tail (Fig. 2.50a). The occurrence of filiform corrosion on painted surfaces in aircraft (Fig. 2.50b) exposed to marine and other high-humidity environments has been controlled by use of chemical conversion coatings, anodizing, or application of chromate-inhibited primers prior to final coating.

Stress-Corrosion Cracking. Time-dependent cracking under the combined influence of sustained tensile stress and a corrosive environment is labeled SCC. In aluminum products, SCC, which is characteristically intergranular in nature, has been experienced only in higher-strength alloys and tempers of the 2*xxx*, 7*xxx*, and 5*xxx* types (with more than 3% Mg) and of the 6*xxx* type (with excess silicon). The relative resistances of aluminum products made of such alloys in various tempers, and with respect to the direction of tensile stress, are indicated by the ratings listed in Table 2.23. No SCC problems have been encountered in service with 1*xxx* aluminum or with 3*xxx*, 6*xxx* (Mg_2Si ratio), or 5*xxx* (containing 3% Mg or less) alloys. The alloy 6061-T6, which is included, is a balanced-ratio alloy.

In general, high tensile stress is a prerequisite to cracking when the stress direction is parallel to either the longitudinal or the long-transverse direction. When the tensile stress is in the short-transverse direction (perpendicular to the surfaces of plate or across the flash plane of die forgings), SCC can occur in susceptible alloy/temper combinations at relatively low stresses. Cracking is accelerated by aggressive, chloride-containing environments but can occur in humid air.

Because of the orientation dependence of SCC, it is important to minimize stresses in the most susceptible direction. In addition to the

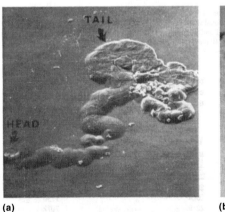
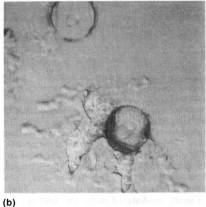

(a) (b)

Fig. 2.50 Examples of filiform corrosion of aluminum. (a) Aluminum foil coated with polyvinyl chloride showing the advancing head and cracked tail section of a filiform cell. Scanning electron microscopy. Original magnification: 80×. (b) Filiform corrosion of a painted aluminum aircraft skin. Source: Ref 2.35

Table 2.23 Relative stress-corrosion cracking ratings for wrought products of high-strength aluminum alloys

Alloy and temper(a)	Test direction(b)	Rolled plate	Rod and bar(c)	Extruded shapes	Forgings
2011-T3, -T4	L (d)	(d)	B	(d)	(d)
	LT	(d)	D	(d)	(d)
	ST	(d)	D	(d)	(d)
2024-T3, -T4	L	A	A	A	(d)
	LT	B(e)	D	B(e)	(d)
	ST	D	D	D	(d)
2024-T6	L	(d)	A(d)	A	¼
	LT	(d)	B	(d)	A(e)
	ST	(d)	B	(d)	D
2124-T851	L	A	(d)	(d)	(d)
	LT	A	(d)	(d)	(d)
	ST	B	(d)	(d)	(d)
6061-T6	L	A	A	A	A
	LT	A	A	A	A
	ST	A	A	A	A
7050-T76	L	A	A	A	(d)
	LT	A	B	A	(d)
	ST	C	B	C	(d)
7075-T6	L	A	A	A	A
	LT	B(e)	D	B(e)	B(e)
	ST	D	D	D	D
7075-T76	L	A	(d)	A	(d)
	LT	A	(d)	A	(d)
	ST	C	(d)		
	C	(d)			
7175-T736	L	(d)	(d)	(d)	A
	LT	(d)	(d)	(d)	A
	ST	(d)	(d)	(d)	B

Note: Resistance ratings are as follows: A, very high, no record of service problems, stress-corrosion cracking not anticipated in general applications. B, high, no record of service problems, stress-corrosion cracking not anticipated at stresses of the magnitude caused by solution heat treatment. Precautions must be taken to avoid high sustained tensile stresses (exceeding 50% of the minimum specified yield strength) produced by any combination of sources, including heat treatment, straightening, forming, fit-up, and sustained service loading. C, intermediate, stress-corrosion cracking not anticipated if total sustained tensile stress is maintained below 25% of minimum specified yield strength. This rating is designated for the short-transverse direction in products used primarily for high resistance to exfoliation corrosion in relatively thin structures, where appreciable stresses in the short-transverse direction are unlikely. D, low. Failure from stress-corrosion cracking is anticipated in any application involving sustained tensile stress in the designated test direction. This rating is currently designated only for the short-transverse direction in certain products. (a) Ratings apply to standard mill products in the types of tempers indicated and also in Tx5x and Tx5xxx (stress-relieved) tempers and may be invalidated in some instances by use of nonstandard thermal treatments or mechanical deformation at room temperature by the user. (b) Test direction refers to orientation of direction in which stress is applied relative to the directional grain structure typical of wrought alloys, which, for extrusions and forgings, may not be predictable on the basis of the cross-sectional shape of the product: L, longitudinal; LT, long transverse; ST, short transverse. (c) Sections with width-to-thickness ratios equal to or less than two, for which there is no distinction between LT and ST properties. (d) Rating not established because product not offered commercially. (e) Rating is one class lower for thicker sections: extrusions, 25 mm (1 in.) and thicker; plate and forgings, 40 mm (1.5 in.) and thicker. Source: Ref 2.35

stresses imposed by service loading, the residual stresses from quenching or forming and any resulting from interference fits or assembly misfits must be taken into account. Minimizing these stresses in the short-transverse direction greatly reduces the probability of SCC failure of susceptible alloy/temper combinations. Use of resistant tempers is also recommended.

For thin-section products used under conditions that induce little or no stress in the short-transverse direction, resistance of 2xxx alloys in T3- or T4-type tempers or of 7xxx alloys in T6-type tempers is often satisfactory.

For rolled, extruded, or forged thick-section products, resistance in the short-transverse direction or across the flash plane of die forgings usually controls their use. Overaging stress-relief treatments that sacrifice some strength are very effective in providing high resistance to SCC in the copper-containing 7xxx alloys (7075, 7175, 7475, 7049, and 7050) in the T73, T736, and T74 tempers. Overaging of the premium-strength 2xx.x casting alloys to T63 to T7 tempers provides good resistance to SCC, and 2219 in the T6 temper is similarly overaged with respect to strength to achieve high SCC resistance.

Medium-strength copper-free and low-copper alloys of the 7xxx group tend to be susceptible to SCC. Successful use of 7039, which has poor resistance, in armor plate applications requires control of short-transverse stresses and weld overlays. Alloys 7016, 7021, and 7029 with copper contents up to 1% have good formability and finishing properties for automotive applications such as bumpers. Maximum SCC resistance is obtained by forming in the freshly quenched (W) temper followed by two-step aging. Casting alloys of the 7xx.x group have compositions similar to those of the aforementioned wrought alloys, and some SCC problems have occasionally been encountered in their applications.

Corrosion Fatigue. Fatigue strengths of aluminum alloys are lower in such corrosive environments as seawater and other salt solutions than in air, especially when evaluated by low-stress, long-duration tests. Such corrosive environments produce smaller reductions in fatigue strength in the more corrosion-resistant alloys, such as the 5xxx and 6xxx series, than in the less-resistant alloys, such as 2xxx and 7xxx series.

Like SCC, corrosion fatigue requires the presence of water. In contrast to SCC, however, corrosion fatigue is not appreciably affected by test direction, because the fracture that results from this type of attack is predominantly transgranular.

Hydrogen Embrittlement. Only recently has it been determined that hydrogen embrittles aluminum. For many years, all environmental cracking of aluminum and aluminum alloys was represented as SCC; however, testing in specific hydrogen environments has revealed that aluminum is susceptible to hydrogen damage. Hydrogen damage in aluminum alloys may take the form of intergranular or transgranular cracking or blistering. Blistering is most often associated with the melting or heat treatment of aluminum, in which reaction with water vapor produces hydrogen. Blistering due to hydrogen is frequently associated with grain-boundary precipitates or the formation of small voids. Blister formation is different from that in ferrous alloys because in aluminum it is more common to have a multitude of near-surface voids that coalesce to produce a large blister.

In a manner similar to the mechanism in iron-base alloys, hydrogen diffuses into the aluminum lattice and collects at internal defects. This occurs most frequently during annealing or solution treating in air furnaces prior to age hardening.

Most of the work on hydrogen embrittlement of aluminum has been on the 7xxx alloys; therefore, the full extent of hydrogen damage in aluminum alloys has not been determined and the mechanisms have not been established.

Erosion-Corrosion. In noncorrosive environments, such as high-purity water, the stronger aluminum alloys have the greatest resistance to erosion-corrosion because resistance is controlled almost entirely by the mechanical components of the system. In a corrosive environment, such as seawater, the corrosion component becomes the controlling factor; thus, resistance may be greater for the more corrosion-resistant alloys even though they are lower in strength. Corrosion inhibitors and cathodic protection have been used to minimize erosion-corrosion, impingement, and cavitation on aluminum alloys.

Corrosion at Joints

Joints fastened by mechanical fasteners (bolts, screws, cold-headed rivets, and threaded connections) or by pressure or adhesive bonding require special attention to avoid galvanic effects, crevice corrosion, and assembly stresses in products susceptible to SCC. Methods of joining that involve heat (welding, brazing, and soldering) alter metallurgical structures of the parent material adjacent to the fusion zones and introduce composition differences between the joint proper and the parent metal. These can be a source of galvanic effects, particularly in heat treatable alloys, and are a factor of importance in selection of appropriate joining materials.

Alclad Products

In Alclad aluminum products, the difference in solution potential between the core alloy and the cladding alloy provides cathodic protection to the core. These products, primarily sheet and tube, consist of a core coated on one or both surfaces with a metallurgically bonded layer of an alloy that is anodic to the core alloy. The thickness of the cladding layer usually is less than 10% of the overall thickness of the product. Cladding alloys generally are of the non-heat-treatable type, although, for higher strength, heat treatable alloys sometimes are used. Composition relationships of core and cladding alloys generally are designed so that the cladding is 80 to 100 mV more anodic than the core. Several core alloy/cladding alloy combinations for common Alclad products are listed in Table 2.24. Because of the cathodic protection provided by the cladding, corrosion progresses only to the core/cladding interface and then spreads laterally. This is highly effective in eliminating perforation of thin products.

Table 2.24 Combination of aluminum alloys used in some Alclad products

Core alloy	Cladding alloy
2014	6003 or 6053
2024	1230
2219	7072
3003	7072
3004	7072 or 7013
6061	7072
7075	7072, 7008, or 7011
7178	7072

Source: Ref 2.35

Protective Coatings

Paints and other coatings are applied to aluminum alloy products for decorative as well as protective purposes. Almost any type of paint for metals (acrylic, alkyl, polyester, vinyl, etc.) is suitable; the performance of a particular paint on aluminum, when applied properly, is better than the performance on steel. As with any metal, surface preparation is important. Conversion coatings, of either the chromate or the phosphate type, are recommended for preparation of aluminum alloys. For milder environments, the paint can be applied to the conversion coating; for more aggressive environments, such as those containing chlorides, a chromated primer should be applied first. Aluminum alloys are especially amenable to waterborne paints, which are increasingly used because of environmental considerations. Many products precoated in a variety of colors for agricultural, industrial, and residential applications are available commercially.

Although more expensive, and restricted to in-plant application, anodized coatings provide excellent protection for aluminum alloys. They are also sometimes used as bases for paints. The many monumental buildings with outer walls of anodized aluminum alloys attest to the durability of these materials under conditions of weathering. Anodized coatings also provide a variety of colorations, most commonly shades of gray and bronze, produced by selection of both alloy and anodizing process.

Anodized coatings are produced by an electrolytic process in which the surface of an alloy that is made the anode is converted to aluminum oxide; this oxide is bound to the alloy as tenaciously as the natural oxide film but is much thicker. Coatings used to provide corrosion resistance range in thickness from 5 to 30 µm (0.2 to 1.2 mils); little or no additional protection is provided by thicker coatings. As with the alloys themselves, anodized coatings are not resistant to most environments with pH values outside the range from 4 to 9. Within this range, resistance to corrosion can be improved by an order of magnitude or more; in atmospheric weathering tests, the number of pits that developed in the base metal was found to decrease exponentially with coating thickness.

ACKNOWLEDGMENTS

Portions of the text for this chapter came from "Introduction to Aluminum and Aluminum Alloys" and "Corrosion Resistance of Aluminum and Aluminum Alloys," both in *Metals Handbook Desk Edition,* 2nd ed., ASM International, 1998, and *Introduction to Aluminum Alloys and Tempers* by J.G. Kaufman, ASM International, 2000.

REFERENCES

2.1 P.C. Varley, *The Technology of Aluminum and Its Alloys*, Newnes-Butterworths, London, 1970

2.2 Introduction to Aluminum and Aluminum Alloys, *Metals Handbook Desk Edition*, 2nd ed., ASM International, 1998

2.3 J.R. Kissell, Aluminum Alloys, *Handbook of Materials for Product Design*, McGraw-Hill, 2001, p 2.1–2.178

2.4 R.E. Sanders et al., *Proceedings of the International Conference on Aluminum Alloys—Physical and Mechanical Properties* (Charlottesville, VA), Engineering Materials Advisory Services, Warley, U.K., 1941, 1986

2.5 W.A. Anderson, in *Aluminum*, Vol 1, American Society for Metals, 1967

2.6 R.N. Caron and J.T. Staley, Aluminum and Aluminum Alloys: Effects of Composition, Processing, and Structure on Properties of Nonferrous Alloys, *Materials Selection and Design*, Vol 20, *ASM Handbook*, ASM International, 1997

2.7 B. Smith, The Boeing 777, *Adv. Mater. Process.*, Sept 2003, p 41–44

2.8 W.F. Smith, Precipitation Hardening of Aluminum Alloys, Lesson 9, *Aluminum and Its Alloys*, ASM Course, American Society for Metals, 1979

2.9 Aluminum Casting Principles, Lesson 5, *Aluminum and Its Alloys*, ASM Course, American Society for Metals, 1979

2.10 K.R. Van Horn, Ed., *Aluminum*, Vol 1, *Properties, Physical Metallurgy and Phase Diagrams*, American Society for Metals, 1967

2.11 J.W. Bray, Aluminum Mill and Engineered Wrought Products, *Properties and Selection: Nonferrous Alloys and Special-Purpose Materials*, Vol 2, *ASM Handbook*, ASM International, 1990

2.12 Forging of Specific Metals and Alloys, *Metals Handbook Desk Edition*, 2nd ed., ASM International, 1998

2.13 G.W. Kuhlman, Forging of Aluminum Alloys, *Metalworking: Bulk Forming*, Vol 14A, *ASM Handbook*, ASM International, 2005

2.14 Heat Treating of Nonferrous Alloys, *Metals Handbook Desk Edition*, 2nd ed., ASM International, 1998

2.15 Press-Brake Forming, *Forming and Forging*, Vol 14, *ASM Handbook*, ASM International, 1993

2.16   Deep Drawing, *Forming and Forging*, Vol 14, *ASM Handbook*, ASM International, 1993

2.17   Forming of Aluminum Alloys, *Forming and Forging*, Vol 14, *ASM Handbook*, ASM International, 1993

2.18   Rubber-Pad Forming, *Forming and Forging*, Vol 14, *ASM Handbook*, ASM International, 1993

2.19   Contour Roll Forming, *Forming and Forging*, Vol 14, *ASM Handbook*, ASM International, 1993

2.20   C.H. Hamilton and A.K. Ghosh, Superplastic Sheet Forming, *Forming and Forging*, Vol 14, *ASM Handbook*, ASM International, 1988

2.21   M. Warmuzek, Metallographic Techniques for Aluminum and Its Alloys, *Metallography and Microstructures*, Vol 9, *ASM Handbook*, ASM International, 2004

2.22   E.L. Rooy, Aluminum and Aluminum Alloys, *Casting*, Vol 15, *ASM Handbook*, ASM International, 1988

2.23   Aluminum Foundry Products, *Metals Handbook Desk Edition*, 2nd ed., ASM International, 1998

2.24   J.F. Major, Aluminum and Aluminum Alloy Castings, *Casting*, Vol 15, *ASM Handbook*, ASM International, 2008

2.25   J.R. Davis, *ASM Specialty Handbook: Aluminum and Aluminum Alloys*, ASM International, 1993, p 108

2.26   R.P. Martukanitz, Selection and Weldability of Heat-Treatable Aluminum Alloys, *Welding, Brazing, and Soldering,* Vol 6, *ASM Handbook*, ASM International, 1993

2.27   C. Conrardy, Gas Metal Arc Welding, *Welding Fundamentals and Processes*, Vol 6A, *ASM Handbook*, ASM International, 2011

2.28   L.E. Allgood, Gas-Tungsten Arc Welding, *Welding Fundamentals and Processes*, Vol 6A, *ASM Handbook*, ASM International, 2011

2.29   I.D. Harries, Plasma Arc Welding, *Welding Fundamentals and Processes*, Vol 6A, *ASM Handbook*, ASM International, 2011

2.30   R.W. Messler, Overview of Welding Processes, *Welding Fundamentals and Processes*, Vol 6A, *ASM Handbook*, ASM International, 2011

2.31   Metallography and Microstructures of Weldments, *Metallography and Microstructures*, Vol 9, *ASM Handbook*, ASM International, 2004

2.32   W.M. Thomas and E.D. Nicholas, Friction Stir Welding for the Transportation Industries, *Mater. Des.*, Vol 18 (No. 4/6), 1997, p 269–273

2.33   R.S. Mishra and M.W. Mahoney, *Friction Stir Welding and Processing*, ASM International, 2007

2.34   Joining, *Metals Handbook Desk Edition*, 2nd ed., ASM International, 1998

2.35   Corrosion Resistance of Aluminum and Aluminum Alloys, *Metals Handbook Desk Edition*, 2nd ed., ASM International, 1998

SELECTED REFERENCES

- Alloy and Temper Designation Systems for Aluminum, *Metals Handbook Desk Edition*, 2nd ed., ASM International, 1998
- Aluminum Casting Principles, Lesson 5, *Aluminum and Its Alloys*, ASM Course, American Society for Metals, 1979
- Aluminum Wrought Products, *Metals Handbook Desk Edition*, 2nd ed., ASM International, 1998
- H.K.D.H. Bhadeshia, Joining of Commercial Aluminum Alloys, *Proceedings of the International Conference on Aluminum*, 2003, p 195–204
- Y.J. Chao, X. Qi, and W. Tang, Heat Transfer in Friction Stir Welding—Experimental and Numerical Studies, *Trans. ASME*, Vol 125, 2003, p 138–145
- R.W. Fonda, J.W. Bingert, and K.J. Colligan, Development of Grain Structure during Friction Stir Welding, *Scr. Mater.*, Vol 51, 2004, p 243–248
- R.F. Gaul, Hot and Cold Working Aluminum Alloys, Lesson 7, *Aluminum and Its Alloys*, ASM Course, American Society for Metals, 1979
- C.H. Hamilton and A.K. Ghosh, Superplastic Sheet Forming, *Forming and Forging*, Vol 14, *ASM Handbook*, ASM International, 1988
- Heat Treating of Aluminum Alloys, *Heat Treating*, Vol 4, *ASM Handbook*, ASM International, 1991
- J.G. Kaufman, Aluminum Alloys, *Handbook of Materials Selection*, John Wiley & Sons, Inc., 2002, p 89–134
- J.G. Kaufman, *Introduction to Aluminum Alloys and Tempers*, ASM International, 2000
- J.R. Koelsch, High-Speed Machining: A Strategic Weapon, *Mach. Shop Guide*, Nov 2001
- G.W. Kuhlman, Forging of Aluminum Alloys, *Forming and Forging*, Vol 14, *ASM Handbook*, ASM International, 1988
- S. Li and S.L. Gobbi, Laser Welding for Lightweight Structures, *J. Mater. Process. Technol.*, Vol 70, 1997, p 137–144
- G. Mathers, Chap. 8, Other Welding Processes, *The Welding of Aluminum and Its Alloys*, Woodhead Publishing Limited and CRC Press, 2002
- P.F. Mendez and T.W. Eager, Welding Processes for Aeronautics, *Adv. Mater. Process.*, May 2001, p 39–43
- I.J. Polmear, *Light Alloys—Metallurgy of the Light Metals*, 3rd ed., Butterworth Heinemann, 1995
- E. Schubert et al., Light-Weight Structures Produced by Laser Beam Joining for Applications in Automobile and Aerospace Industry, *J. Mater. Process. Technol.*, Vol 115, 2001, p 2–8
- B. Smith, The Boeing 777, *Adv. Mater. Process.*, Sept 2003, p 41–44

- E.A. Starke and J.T. Staley, Application of Modern Aluminum Alloys to Aircraft, *Prog. Aerosp. Sci.*, Vol 32, 1996, p 131–172
- W.M. Thomas and E.D. Nicholas, Friction Stir Welding for the Transportation Industries, *Mater. Des.*, Vol 18 (No. 4/6), 1997, p 269–273
- J.C. Williams and E.A. Starke, Progress in Structural Materials for Aerospace Systems, *Acta Mater.*, Vol 51, 2003, p 5775–5799

CHAPTER **3**

Magnesium Alloys

MAGNESIUM has the lowest density (1.74 g/cm^3, or 0.063 lb/in.^{3}) of the common structural metals, with a density of approximately two-thirds that of aluminum and one-quarter that of steel. Although magnesium alloys have only moderate tensile strengths, in the range of 138 to 345 MPa (20 to 50 ksi), and a modulus of elasticity of only 45 GPa (6.5 × 10^6 psi), due to their low densities, they exhibit favorable specific strengths (tensile strength/density) and specific moduli (modulus/density) comparable to other structural metals. Magnesium has relatively good electrical conductivity (38.6% IACS) and thermal conductivity (154.5 W/m · K, or 89.2 Btu/ft · h · °F) values.

The majority of the annual production of magnesium is used for alloying elements in aluminum alloys, with only about 15% of the annual production being used for structural applications, with the majority of these being castings. Magnesium and its alloys are used in a wide variety of structural applications, such as automotive, industrial, materials handling equipment, kitchen appliances, handheld tools, luggage frames, computer housings, and ladders. Magnesium is relatively inexpensive; easy to cast, machine, and weld; and its conductivity and heat capacity are relatively high. Magnesium alloys have very good damping capacity, and castings have found applications in high-vibration environments.

3.1 Magnesium Metallurgy

Pure magnesium has a melting point of 650 °C (1202 °F) and a hexagonal close-packed (hcp) crystalline structure. Because the hcp structure restricts slip to the basal planes, magnesium is difficult to plastically deform at room temperature; that is, the work-hardening rate at room temperature is high and the ductility is low. At elevated temperatures, other slip planes become operative, so magnesium alloys are normally formed at temperatures greater than 227 °C (440 °F), usually in the range of 343 to 510 °C (650 to 950 °F). Another consequence of the hcp structure is the mechanical property anisotropy (or directionality) in cold-rolled sheet due to its crystallographic texture. For example, a mildly rolled sheet has a tensile strength and tensile elongation in the rolling direction (241 MPa, or 35 ksi, and 2% elongation) that differ from those transverse to the rolling direction (262 MPa, or 38 ksi, and 8% elongation). The yield strength in compression of wrought products is only about 40 to 70% of that in tension. Because of the difficulty of cold forming magnesium alloys, castings are the more prevalent product form than wrought products. One consequence of its rather low melting point is its susceptibility to creep at moderately elevated temperatures. However, alloys with improved creep performance have been developed.

Two major magnesium alloy systems are available. The first includes alloys that contain 2 to 10% Al combined with minor additions of zinc and manganese. These alloys are widely available at moderate cost, and their mechanical properties are good to 95 to 120 °C (200 to 250 °F). Beyond this, the properties deteriorate rapidly with increasing temperature. The second group consists of magnesium alloyed with various elements (rare earths, zinc, thorium, silver, and silicon) except aluminum, all containing a small but effective zirconium content that imparts a fine-grained structure and improves mechanical properties. These alloys generally possess better elevated-temperature properties, but their more costly elemental additions, combined with the specialized manufacturing technology required, result in higher cost.

Aluminum and zinc provide solid-solution strengthening. Aluminum, in addition to providing strength and hardness, widens the freezing range and makes the alloy easier to cast. Aluminum is alloyed with magnesium because it increases strength, castability, and corrosion resistance. Because aluminum has a maximum solid solubility in magnesium of 12.7% at 427 °C (800 °F) that decreases to approximately 2% at room temperature (Fig. 3.1), it would at first appear that this system could be strengthened by precipitation hardening. However, the resulting precipitate is rather coarse and results in only moderate hardening.

However, when zinc is added to the composition, it refines the precipitate and increases the strength by a combination of solid-solution strengthening and precipitation hardening (Fig. 3.2). Zinc is the second most important alloying element. Zinc can also be used in combination with zirconium, rare earths (REs), thorium, or silver to produce precipitation-hardening alloys. With tensile strengths in the range of 214 to 241 MPa (31 to 35 ksi) and elongations of 1 to 8%, the Mg-Al-Zn alloys are not particularly strong or ductile but have low densities and are easy to cast.

Magnesium alloys are limited to a total aluminum and zinc level of less than 10%; at higher levels, the ductility is drastically reduced due to the formation of intermetallic compounds. Thus, if the zinc content of an Mg-Al-Zn alloy is raised to 3%, the aluminum content must be reduced to

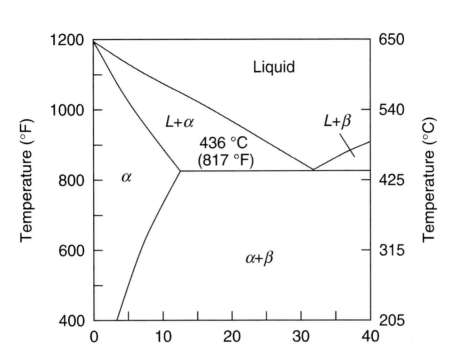

Fig. 3.1 Magnesium-rich portion of the magnesium-aluminum phase diagram

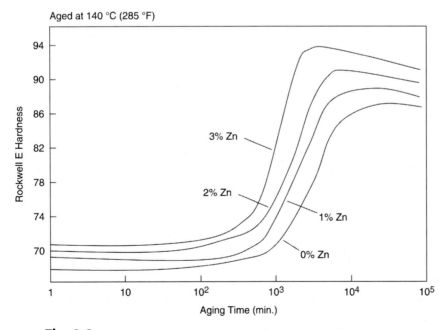

Fig. 3.2 Aging curves for Mg-9%Al alloy with various zinc additions

approximately 6%, as in AZ63. However, as the zinc contents increase in the Mg-Al-Zn alloys, there is an increase in microporosity and shrinkage.

Zirconium is used in casting alloys for grain refinement. The powerful grain-refining effect of zirconium is shown in Fig. 3.3. However, zirconium is not used in alloys containing aluminum because brittle compounds are formed. Grain refining by the addition of zirconium is extremely efficient; in fact, it is the most efficient grain-refining additive ever reported for metallic alloys. The grain structure observed in sand or permanent mold castings consists of spherical grains, with a size depending on the cooling rate. It is typically possible to achieve grain sizes in the range of 10 to 100 μm, depending on the cooling conditions. The effect of zirconium refining on pure magnesium is shown in Fig. 3.4. Both strength and ductility increase dramatically with increasing zirconium contents up to approximately 0.6 wt%. The spherical grain structure promotes extremely good flow characteristics.

Manganese additions improve the corrosion resistance in seawater by removing iron from solution. Silicon increases fluidity for casting alloys but decreases the corrosion resistance if iron is present. Thorium and yttrium improve creep resistance; however, the use of thorium, which is

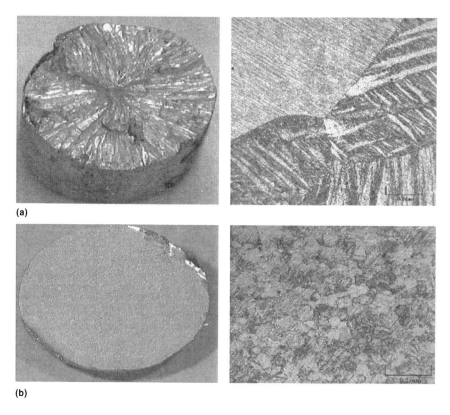

(a)

(b)

Fig. 3.3 Grain refinement with zirconium. (a) Pure magnesium. (b) Pure magnesium plus zirconium. Source: Ref 3.1

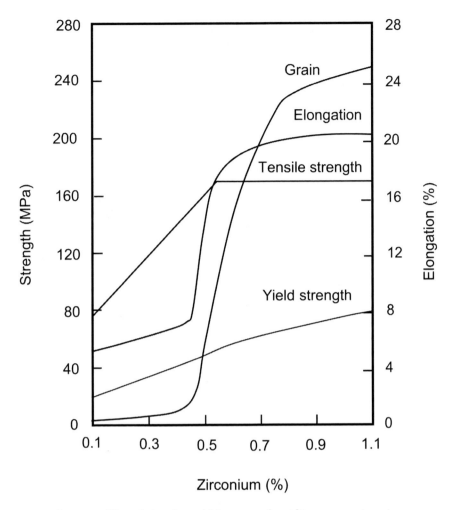

Fig. 3.4 Effect of zirconium additions to sand-cast binary magnesium-zirconium alloys on mechanical properties and grain size. Source: Ref 3.2

mildly radioactive, has decreased due to increasing regulations on its use. Although much less soluble than aluminum and zinc, the RE elements are potent solid-solution strengtheners. The REs are usually added as natural mixtures of either mischmetal or as didymium. Mischmetal contains approximately 50% Ce, with the remainder being mainly lanthanum and neodymium, while didymium contains approximately 85% Nd and 15% Pr. Rare earths and cerium also provide precipitation-hardening capability. As little as 1% RE additions increase strength and reduce the tendency for weld cracking.

Magnesium occupies the highest anodic position on the galvanic series, and as such, there is always the strong potential for corrosion. A severely corroded magnesium part is shown in Fig. 3.5. The impurity elements nickel, iron, and copper must be held to low levels to minimize corrosion. The effect of iron on the corrosion susceptibility of pure magnesium is shown in Fig. 3.6. More corrosion-resistant Mg-Al-Zn alloys were developed in the mid-1980s by using higher-purity starting materials and by limiting the amounts of iron (<0.005%), nickel (<0.001%), and copper (<0.015%). The low levels of nickel and copper are controlled by the

Fig. 3.5 Severely corroded magnesium part

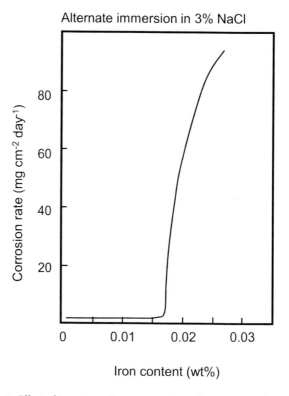

Fig. 3.6 Effect of iron impurities on corrosion of pure magnesium. Source: Ref 3.3

purity of the starting materials, while the low iron levels are controlled with $MnCl_2$ additions. For example, the high-purity alloy AZ91E, due to its lower iron content, has improved corrosion resistance compared to the earlier alloy AZ91C (Fig. 3.7). As shown in Fig. 3.8, some of the newer casting alloys approach the corrosion resistance of competing aluminum casting alloys.

Magnesium alloys are produced in both the wrought and cast conditions. Some alloys are strengthened by cold working, while others can be pre-

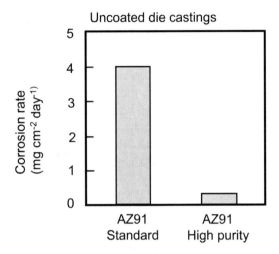

Fig. 3.7 Effect of purity on corrosion resistance of AZ91 alloy. Source: Ref 3.4

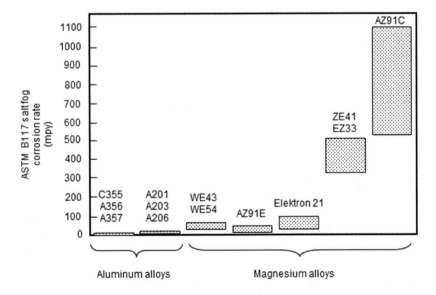

Fig. 3.8 Corrosion comparison of aluminum and magnesium casting alloys. Source: Ref 3.1

cipitation hardened by heat treatment. The tensile properties of magnesium alloys generally range from 69 to 344 MPa (10 to 50 ksi) yield strength and 138 to 379 MPa (20 to 55 ksi) tensile strength with elongations of 1 to 15%.

3.2 Magnesium Applications

Magnesium and magnesium alloys are used in a wide variety of structural and nonstructural applications. Magnesium compounds, primarily magnesium oxide, are used mainly as a refractory material in furnace linings for producing iron and steel, nonferrous metals, glass, and cement. More than 75% of the magnesium compounds produced in the United States are used as refractories.

Structural uses, mostly die castings, account for ~25% of total reported consumption. Applications include automotive, industrial, materials handling, commercial, and aerospace equipment. The automotive applications include instrument panel support beams, brake and clutch pedal brackets, air intake grilles, steering column support brackets, steering wheels, seat back frames and seat bottoms, and battery cases (for electric vehicles). AZ91 is the most commonly used magnesium-base alloy for die casting, with excellent castability and good strength. It is typically used for automobile and computer parts, mobile telephones, sporting goods, housings and covers, brackets, chain saw parts, handheld tools, and household equipment. AM60 and AM50 are alloys with outstanding ductility and energy-absorbing properties combined with good strength and castability. Typical uses are automotive seat frames, steering wheels, instrument panels, brackets, and fans. AM20 is an alloy known for its ductility and impact strength. Typical uses are automotive safety parts where the highest possible ductility is required. In industrial machinery, such as textile and printing machines, magnesium alloys are used for parts that operate at high speeds and thus must be lightweight to minimize inertial forces. Materials handling equipment includes dock boards, grain shovels, and gravity conveyors. Commercial applications include handheld tools, computer housings, and mobile phone cases. Magnesium alloys are valuable for aerospace applications because they are lightweight and exhibit good strength and stiffness at both room and elevated temperatures. Aerospace components produced from magnesium castings include main transmission housings for helicopters and gearboxes and gearbox housings for commercial and military aircraft.

Aluminum beverage cans have become a significant factor in the supply-demand relationship for magnesium. In the mid-1970s, aluminum alloys began to be used in substantial quantities for beverage cans, and they have gained nearly 100% of the beverage can market. This has had a marked impact on magnesium consumption because aluminum cans contain ~2.5% Mg. The advent of the aluminum can also had a significant effect on secondary magnesium. As aluminum cans were recycled by the consumer, some of the magnesium content was also recycled; consequently, recycled magnesium has become an essential portion (~35%) of U.S. production, with ~80% of the recycled magnesium originating from magnesium-containing aluminum scrap.

3.3 Magnesium Alloy Designation

Magnesium alloys are designated by a combination of letters and numbers. The first two letters indicate the two major alloying elements in the alloy, while the following two numbers give the approximate amounts for the first and second alloying elements, respectively. For example, the alloy AZ91 contains approximately 9% Al and 1% Zn. There is also a letter that follows the basic alloy designation; A is the original composition, B is the second modification, C is the third modification, D indicates a high-purity version, and E is a corrosion-resistant composition. In the aforementioned example, AZ91C would indicate the third modification to AZ91. The magnesium alloy designation system is shown in Table 3.1. The heat treatment temper designations for magnesium alloys are the same as for aluminum

Table 3.1 ASTM International designation for magnesium alloys

First part	Second part	Third part	Fourth part
Indicates the two principal alloying elements	Indicates the amounts of the two principal alloying elements	Distinguishes between different alloys with the same percentages of the two principal alloying elements	Indicates condition (temper)
Consists of two code letters representing the two main alloying elements arranged in order of decreasing percentage (or alphabetically if percentages are equal)	Consists of two numbers corresponding to rounded-off percentages of the two main alloying elements and arranged in the same order as alloy designations in the first part	Consists of a letter of the alphabet assigned in order as compositions become standard	Consists of a letter followed by a number (separated from the third part of the designation by a hyphen)
• A, aluminum • B, bismuth • C, copper • D, cadmium • E, rare earth • F, iron • G, magnesium • H, thorium • K, zirconium • L, lithium • M, manganese • N, nickel • P, lead • Q, silver • R, chromium • S, silicon • T, tin • W, yttrium • Y, antimony • Z, zinc	• Whole numbers	• A, first compositions, registered ASTM International • B, second compositions, registered ASTM International • C, third compositions • D, high purity, registered ASTM International • E, high corrosion resistant, registered ASTM International • X1, not registered with ASTM • International	• F, as-fabricated • O, annealed • H10 and H11, slightly strain hardened • H23, H24, and H26, strain hardened and partially annealed • T4, solution heat treated • T5, artificially aged only • T6, solution heat treated and artificially aged • T8, solution heat treated, cold worked, and artificially aged

Source: Ref 3.5

alloys; however, because most applications use cast magnesium alloys, the most prevalent tempers are the T4, T5, and T6 tempers.

3.4 Magnesium Casting Alloys

Magnesium castings are used in structural applications because of their low weight and good damping characteristics. Magnesium alloys have a very low viscosity, allowing the metal to flow long distances and fill narrow mold cavities. Their relatively low melting points allow the use of hot chamber die casting, and their minimal reactivity with steel below 704 °C (1300 °F) allows the use of inexpensive steel crucibles and molds. The chemical compositions of magnesium castings alloys are given in Table 3.2, and their mechanical properties are shown in Table 3.3.

The choice of casting method is determined primarily by size, shape, quantity, cost, and desired mechanical properties. Die castings, sand, and

Table 3.2 Nominal compositions of magnesium casting alloys

Alloy		Composition, wt%						
ASTM No.	UNS No.	Al	Mn(a)	Zn	Th	Zr	Rare earths	Other
Sand and permanent mold castings								
AM100A	M10100	10.0	0.1	0.3	...	...	...	...
AZ63A	M11630	6.0	0.15	3.0	...	...	...	...
AZ81A	M11810	7.6	0.13	0.7	...	...	...	...
AZ91C	M11914	8.7	0.13	0.7	...	...	...	...
AZ91E	M11918	8.7	0.13	0.7	...	...	...	0.005 Fe(b)
AZ92A	M11920	9.0	0.10	2.0	...	...	...	...
EQ21A	M12210	...	...	...	...	0.7	2.25(c)	1.5 Ag
EZ33A	M12330	...	...	2.55	...	0.75	3.25	...
HK31A	M13310	...	...	0.3	3.25	0.7	...	...
HZ32A	M13320	...	...	2.1	3.25	0.75	0.1	...
K1A	M18010	...	...	...	...	0.7	...	...
QE22A	M18220	...	...	...	...	0.7	2.15(c)	2.5 Ag
QH21A	M18210	...	...	0.2	1.1(d)	0.7	1.05(c)(d)	2.5 Ag
WE43A	M18430	...	0.15	0.2	...	0.7	3.4(e)	4.0 Y
WE54A	M18410	...	0.15	0.2	...	0.7	2.75(e)	5.0 Y
ZC63A	M16331	...	0.25	6.0	...	...	...	2.7 Cu
ZE41A	M16410	...	0.15	4.25	...	0.7	1.25	...
ZE63A	M16630	...	...	5.75	...	0.7	2.55	...
ZH62A	M16620	...	...	5.7	1.8	0.75	...	...
ZK51A	M16510	...	...	4.55	...	0.75	...	...
ZK61A	M16610	...	...	6.0	...	0.8	...	...
Die castings								
AM60A	M10600	6.0	0.13	0.22	...	...	...	0.5 Si; 0.35 Cu
AS41A	M10410	4.25	0.20	0.12	...	...	...	1.0 Si
AS41B	M10412	4.25	0.35	0.12	...	...	...	1.0 Si
AZ91A	M11910	9.0	0.13	0.7	...	...	...	0.5 Si
AZ91B	M11912	9.0	0.13	0.7	...	...	...	0.5 Si; 0.35 Cu
AZ91D	M11916	9.0	0.15	0.7	...	...	...	...
AM60B	M10602	6.0	0.24	0.22	...	...	...	...
AM50A	M10500	4.9	0.26	0.22	...	...	...	...

(a) Minimum. (b) If iron exceeds 0.005%, the iron-to-manganese ratio should not exceed 0.032. (c) Rare earth elements are in the form of didymium (a mixture of rare earth elements made chiefly of neodymium and praseodymium). (d) Thorium and didymium total is 1.5 to 2.4%. (e) Rare earths are 2.0 to 2.5% and 1.5 to 2.0% Nd for WE43A and WE54A, respectively, with the remainder being heavy rare earths. Source: Ref 3.5

Table 3.3 Minimum mechanical properties for magnesium casting alloys

Alloy-temper	Tensile strength		Yield strength		Elongation in 50 mm (2 in.), %	Hardness, HB
	MPa	ksi	MPa	ksi		
Sand and permanent mold castings						
AM100A-T6	241	35	117	17	...	69
AZ63A-T6	234	34	110	16	3	73
AZ81A-T4	234	34	76	11	7	55
AZ91C-T6	234	34	110	16	3	70
AZ91E-T6	234	34	110	16	3	70
AZ92A-T6	234	34	124	18	1	81
EQ21A-T6	234	34	172	25	2	78
EZ33A-T5	138	20	96	14	2	50
HK31A-T6	186	27	89	13	4	66
HZ32A-T5	186	27	89	13	4	55
K1A-F	165	24	41	6	14	...
QE22A-T6	241	35	172	25	2	78
QH21A-T6	241	35	186	27	2	...
WE43A-T6	221	32	172	25	2	85
WE54A-T6	255	37	179	26	2	85
ZC63A-T6	193	28	125	18	2	60
ZE41A-T5	200	29	133	19.5	2.5	62
ZE63A-T6	276	40	186	27	5	...
ZH62A-T5	241	35	152	22	5	70
ZK51A-T5	234	34	138	20	5	65
ZK61A-T6	276	40	179	26	5	70
Die castings						
AM50A	200	29	110	16	10	...
AM60A and B	220	32	130	19	8	...
AS41A and B	210	31	140	20	6	...
AZ91A, B, and D	230	34	160	32	3	...

Source: Ref 3.5

permanent mold castings are more widely used than investment castings. The disadvantages of the investment process are the cost of the casting in terms of both capital equipment and cost per casting. There is also a much greater restriction on the size of the casting than can be produced.

Refinement of the grain and dendrite structure caused by rapid cooling is extremely beneficial for mechanical properties. This is fully exploited in thin-walled die-cast parts, where cooling rates are typically in the range of 100 to 1000 °C/s (180 to 1800 °F/s). The influence of cooling rate on tensile properties is shown in Fig. 3.9. In real castings, various defects caused by improper feeding or impurities will occur, reducing the ductility. With reference to Fig. 3.9, it should be noted that even if a thin-walled die-cast part may contain more casting defects than an optimally cast permanent mold casting, the die-cast part will show superior properties due to the finer grain size. As the wall thickness increases, the advantage of die casting diminishes, and for heavy wall thickness, permanent mold castings may compare favorably.

Die castings made of magnesium alloys may be selected in preference to aluminum die castings of the same design because of the weight savings.

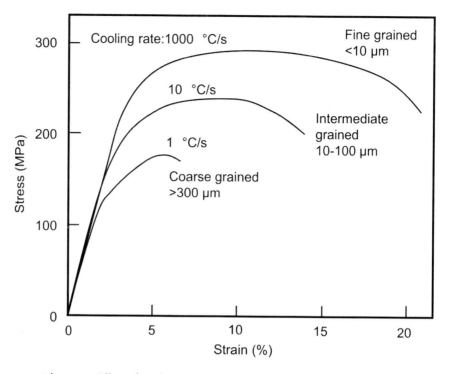

Fig. 3.9 Effect of cooling rate on magnesium casting tensile stress-strain curves. Source: Ref 3.2

Magnesium die castings can also replace assembled steel parts, zinc die castings, and plastics. Service requirements and size may govern whether a magnesium alloy is selected for use in a die casting, but quantity is the most important factor, because die castings are high-production items.

Magnesium alloys are cast by the permanent mold process when the number of parts required justifies the high mold cost. The mechanical properties of sand and permanent mold castings are comparable, but the permanent mold process normally provides closer control of dimensions and produces better surface finishes.

The cost of magnesium alloy castings is governed largely by ingot price, alloy castability, and required heat treatment. Ingot price increases with additions of RE metals, zirconium, silver, and thorium. Small changes in composition can affect cost of heat treatment. Casting cost is also influenced by required tolerances, mold and die costs, and machining costs. The quantity of a part to be produced is an important factor affecting cost and must be considered in selecting the most economical method of production. Magnesium alloy die castings, like castings in general, are always priced and purchased on a per piece basis. Cost per pound varies, depending primarily on complexity of design, wall thickness, number of cavities in the mold, and quality level.

3.5 Die Casting Alloys

Die casting is a permanent mold casting process in which the liquid metal is injected into a metal die under high pressure. It is a very high-rate production process using expensive equipment and precision-matched metal dies. Two methods, the hot and cold chamber die casting, are shown in Fig. 3.10. A large number of machines of both the hot and cold chamber type are currently in use. The cold chamber method is often preferred because of the higher injection pressure that can be used, allowing larger parts to

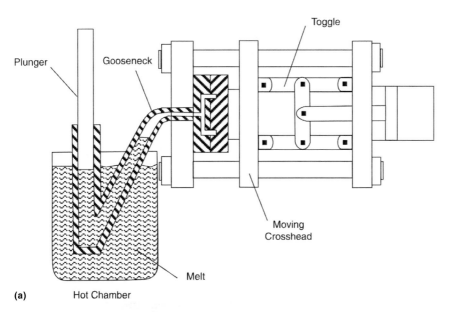

(a)   Hot Chamber

– Higher production rate, 15 cycles per minute for small parts
– Metal injected directly from melt zone

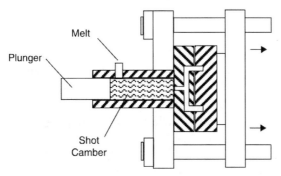

(b)   Cold Chamber

– Melt is poured into cylinder, which is then shot into chamber

Fig. 3.10 (a) Hot and (b) cold chamber die casting

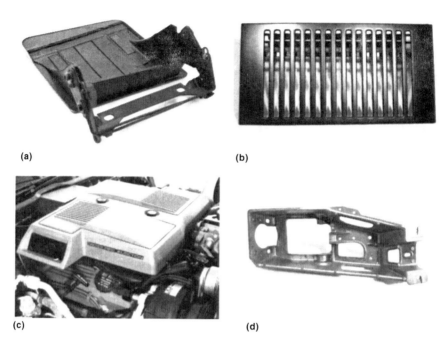

Fig. 3.11 AZ91 die-cast magnesium alloy used in automotive applications. (a) Door frame for hidden headlight assembly weighing 0.370 kg (0.816 lb). (b) Air intake grille weighing 3.240 kg (7.143 lb). (c) Air cleaner cover (shown mounted on a vehicle engine) weighing 2.307 kg (5.086 lb). (d) Brake and clutch pedal bracket weighing 0.637 kg (1.40 lb). Courtesy of International Magnesium Association. Source: Ref 3.6

be fabricated. The machines must be equipped with a means for ensuring that no water comes into contact with the magnesium.

Die casting is by far the most common production method, accounting for ~70% of the magnesium used. Typical automotive parts are shown in Fig. 3.11. The process is well understood with a high rate of production. Die casting is ideally suited for intricately shaped and thin-walled components, which necessitates the use of low-viscosity alloys. The low latent heat of fusion in magnesium allows rapid solidification in the die, with resultant high production rates. Molten magnesium does not react with or solder to die steels, resulting in longer die life and increased productivity (dies last two to three times longer than with aluminum).

Characteristics of die casting include extremely good surface finishes and the ability to hold tight dimensions; however, die castings should not be specified where high mechanical properties are important, because of an inherently high porosity level. The high-pressure injection creates a lot of turbulence that traps air, resulting in high porosity levels. In fact, die-cast parts are not usually heat treated because the high porosity levels can cause surface blistering. The tensile properties decrease significantly with increasing wall thickness, due to a coarser microstructure and higher porosity. To reduce the porosity level, the process can be done in a vacuum (vacuum die casting).

There are three systems of magnesium alloys used commercially for high-pressure die casting: Mg-Al-Zn-Mn (AZ), Mg-Al-Mn (AM), and Mg-Al-Si-Mn (AS). Die castings are used in the as-cast condition.

The most commonly used magnesium die-casting alloy is AZ91D. Alloy AZ91D exhibits good mechanical and physical properties in combination with excellent castability and saltwater corrosion resistance. The excellent corrosion resistance of high-purity AZ91D is essentially the result of controlling the impurity level of three critical contaminants: iron, nickel, and copper. High-purity AZ91D alloy has replaced less-pure AZ91B as the workhorse for die casting. However, alloy AZ91B is still used because it can be easily produced from scrap or secondary metal. This cost-saving alternative can be used for applications in which corrosion resistance is not an important consideration, for example, painted parts in a noncorrosive environment.

For applications requiring greater ductility than available with AZ91D, high-purity die-cast alloys AM50A and AM60B are used. AM50A and AM60B have better elongation and toughness than AZ91D. Despite the decrease in aluminum content, the tensile and yield strengths of AM50A and AM60B are only slightly lower than those of AZ91D. AM60B is used in the production of die-cast automobile wheels and in some archery and other sports equipment. AM50A is used where even higher ductility than AM60B is required. As with AM50A and AZ91D, AM60B exhibits excellent saltwater corrosion resistance.

Die-cast alloy AS41A has much better creep strength than AZ91D, AM50A, or AM60B alloys at temperatures up to 175 °C (350 °F). It also has good elongation, yield strength, and tensile strength. AS41A has been used in crankcases of air-cooled automotive engines. A high-purity version of AS41A, which exhibits excellent saltwater corrosion resistance, is AS41B.

Besides being readily cast directly from a liquid melt, magnesium alloys are also formed from semisolid slugs. The thixomolding process, which combines the features of die casting and injection molding, is being used to produce alloy AZ91D battery cases for electric vehicles.

3.6 Sand and Permanent Mold Casting Alloys

Several systems of magnesium alloys are available for sand and permanent mold castings:

- Mg-Al-Mn with and without zinc (AM and AZ)
- Mg-Zr (K)
- Mg-Zn-Zr with and without REs (ZK, ZE, and EZ)
- Mg-Th-Zr with and without zinc (HK, HZ, and ZH)

- Mg-Ag-Zr with REs or thorium (QE and QH)
- Mg-Y-RE-Zr (WE)
- Mg-Zn-Cu-Mn (ZC)

Sand casting is often used when the parts are too large for sound die casting and the number of parts required is less, as shown in the military applications of Fig. 3.12.

Magnesium-Aluminum Casting Alloys. The magnesium sand and permanent mold casting alloys that contain aluminum as the primary alloying ingredient (AM100A, AZ63A, AZ81A, AZ91C, AZ91E, and AZ92A) exhibit good castability, good ductility, and moderately high yield strength at temperatures up to ~120 °C (250 °F). Of these alloys, AZ91E has become prominent; it has almost completely replaced AZ91C because it has superior corrosion performance. In AZ91E, the iron, nickel, and copper contaminants are controlled to very low levels. As a result, it exhibits excellent saltwater corrosion resistance. In any of the Mg-Al-Zn alloys, an increase in aluminum content raises yield strength but reduces ductility for comparable heat treatment. Final selection of the specific composition may be based on tests of the finished castings. Alloy KIA is used primarily where high damping capacity is required. It has low tensile and yield strength.

High-Zinc-Content Alloys. Magnesium alloys that contain high levels of zinc (ZK51A, ZK61A, ZE63A, ZH62A, and ZC63A) develop the highest yield strengths of the casting alloys and can be cast into compli-

Fig. 3.12 Magnesium alloy sand castings. (a) Main transmission housing for a heavy lift helicopter that was sand cast in WE43B magnesium alloy having a T6 temper. Casting weight = 206 lb (93 kg). Courtesy of Fansteel Wellman Dynamics. (b) Gearbox housing for a military fighter aircraft composed of ZE41A magnesium alloy of T5 temper. Courtesy of Haley Industries Ltd. Source: Ref 3.6

cated shapes. However, these grades are more costly than the alloys of the AZ series. Therefore, these alloys are used where exceptionally good yield strengths are required. They are intended primarily for use at room temperature.

Because ZK61A has a higher zinc content, it has significantly greater strength than ZK51A. Both alloys maintain high ductility after an artificial aging treatment (T5). The strength of ZK61A can be further increased (3 to 4%) by solution treatment plus artificial aging (T6), without impairing ductility. Both of these alloys have fatigue strengths equal to those of the Mg-Al-Zn alloys, but they are more susceptible to microporosity and hot cracking and are less weldable. Addition of either thorium or RE metals overcomes these deficiencies. The strength properties of ZE63A are equivalent to those of ZK61A, and those of ZH62A are equivalent to or better than those of ZK51A.

ZE63A is a high-strength grade with excellent tensile strength and yield strength; these superior properties are obtained by heat treating in a hydrogen atmosphere. Because hydriding proceeds from the surface, heat treating time, wall thickness, and penetrability are limiting factors. This alloy also has excellent casting characteristics.

ZE41A was developed to meet the growing need for an alloy with medium strength, good weldability, and improved castability in comparison with AZ91C and AZ92A. It has good fatigue and creep properties and maximum freedom from microshrinkage. Unlike the AZ alloys, there is a very close relationship between separately cast test bar properties and those obtained from the casting itself, even where relatively thick cast sections are involved. ZE41A is used at temperatures up to 160 °C (320 °F) in such applications as aircraft engines, helicopter and airframe components, and wheels and gear boxes.

ZC63A alloy exhibits good castability, and it is pressuretight and weldable. No grain refining or hardeners are required to obtain the properties, but a heat treatment must be used to achieve the full properties. It has attractive properties at room temperature and at moderately elevated temperatures. The corrosion resistance of the alloy is similar to that of AZ91C, but it is less than that of AZ91E.

The Mg-RE-Zr alloys are used at temperatures between 175 and 260 °C (350 and 500 °F). Because their high-temperature strengths exceed those of the Mg-Al-Zn alloys, thinner walls can be used, and a weight-savings is possible.

The Mg-RE-Zr alloy EZ33A has good strength stability when exposed to elevated temperatures. (Strength stability is the ability to resist deterioration of strength from extended exposure to elevated temperatures.) This alloy is more difficult to cast in some designs than Mg-Al-Zn alloys. Castings of EZ33A have excellent pressure tightness. ZE41A is similar to EZ33A but has higher tensile and yield strengths because of higher zinc content. There is some sacrifice in castability and weldability in ZE41A to obtain the higher mechanical properties.

The Mg-Th-Zr alloys HK31A and HZ32A are intended primarily for use at temperatures of 200 °C (400 °F) and higher; at these temperatures, properties superior to those of EZ33A are required. However, these alloys have lost favor because of environmental considerations and are generally considered obsolete. In the United Kingdom, for example, alloys containing as little as 2% of thorium are classified as radioactive materials that require special handling. Many thorium-containing magnesium parts, however, are still in use, and some replacement castings are still being produced.

For full development of properties, HK31A requires a T6 treatment, whereas HZ32A, which contains zinc, requires only the T5 treatment. Castings of HK31A and HZ32A have been used at temperatures as high as 345 to 370 °C (650 to 700 °F) in a few applications. The Mg-Zn-Th-Zr alloy ZH62A differs from other Mg-Th-Zr alloys in that it is intended primarily for use at room temperature.

The Mg-Th-Zr alloys are more difficult to cast than EZ33A because they are more susceptible to the formation of inclusions and defects as a result of gating turbulence. The tendency for inclusions to form in the Mg-Th-Zr alloys is particularly marked in thin-walled parts that require rapid pouring rates. These alloys have adequate castability for production of complex parts of moderate-to-heavy wall thickness.

At 260 °C (500 °F) and slightly higher, HZ32A is equal to or better than HK31A in short-time and long-time creep strength. HK31A has higher tensile, yield, and short-time creep strengths up to 370 °C (700 °F). However, HZ32A has greater strength stability at elevated temperatures and has much better foundry characteristics than does HK31A.

Magnesium-Silver Casting Alloys. The addition of 2.5% Ag and 2.5% REs produces better precipitation-hardening response with good tensile properties up to 204 °C (400 °F) in QE22, which has tensile strength of 241 MPa (35 ksi) in the T6 condition. The microstructure of sand-cast QE22A-T6 is shown in Fig. 3.13. The presence of silver improves the room-temperature strength of magnesium alloys. When RE elements or thorium are present, along with the silver, elevated-temperature strength is also increased. The QE22A and EQ21A grades are high-tensile-strength and yield-strength alloys with fairly good properties at temperatures up to 205 °C (400 °F). Alloy QH21A has similar properties to QE22A and EQ21 at room temperature, but it exhibits superior properties at temperatures from 205 to 260 °C (400 to 500 °F). The alloys QE22A, EQ21A, and QH21A have good castability and weldability. They require solution and aging heat treatments to achieve the higher mechanical properties. The QE22A and QH21A alloys are relatively expensive because of their silver contents. EQ21A, which has a lower silver content, is less expensive.

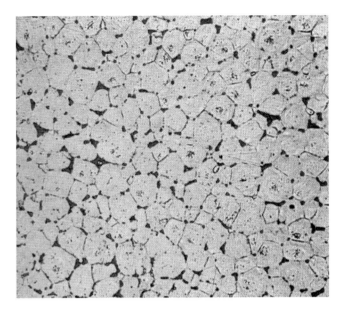

Fig. 3.13 Microstructure of sand-cast QE22A-T6 magnesium alloy. Source: Ref 3.7

The Mg-Ag-RE casting alloys with approximately 4 to 5% yttrium have also been developed that have better elevated-temperature properties. For example, alloy WE43 has a room-temperature tensile strength of 248 MPa (36 ksi) when heat treated to the T6 condition. This alloy maintains a tensile strength of 248 MPa (36 ksi) after long-term aging (5000 h) at 204 °C (400 °F). The effect of 204 °C (400 °F) exposure on the room-temperature strength of WE43 is shown in Fig. 3.14. A relatively new alloy, Elektron 21, offers many of the advantages of WE43, because the cost is lower and the castability is better. Instead of using yttrium, neodymium and gadolinium are used along with zinc and zirconium.

WE54A and WE43A have high tensile strengths and yield strengths, and they exhibit good properties at temperatures up to 300 and 250 °C (570 and 480 °F), respectively. WE54A retains its properties at high temperature for up to 1000 h, whereas WE43A retains properties at high temperature in excess of 5000 h. Both WE54A and WE43A have good castability and weldability, but they require solution and aging heat treatments to optimize their mechanical properties. They are relatively expensive because of their yttrium content. Both alloys are corrosion resistant, with corrosion rates similar to those of the common aluminum-base casting alloys.

The investment (lost wax) process is used for casting complex components, parts with very thin-walled sections, when fine surface finishes are required, and when improvement in dimensional accuracy is required. The same alloys can be cast by this process as with sand casting.

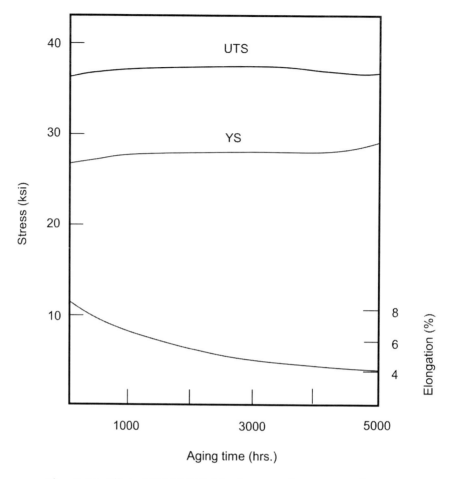

Fig. 3.14 Effect of 204 °C (400 °F) aging on tensile properties of WE43A-T6. UTS, ultimate tensile strength; YS, yield strength. Courtesy of Magnesium Electron, Ltd.

3.7   Wrought Magnesium Alloys

Wrought magnesium alloys are used less often than castings, and a relatively limited number of alloys are available in wrought form. Because magnesium is somewhat more expensive than aluminum, and aluminum is much more easily cold formed, aluminum has a decided cost advantage over wrought alloys. Consequently, wrought magnesium alloys have rather limited usage. Wrought alloys are available as bars, billets, sheet, plate, and forgings. The principal sheet and plate alloy is AZ31. Because AZ31 is strengthened by a combination of solid-solution strengthening, grain size control, and cold working, it is not really a candidate for elevated-temperature structural applications due to the possibility of softening (i.e., recovery) at elevated temperature. Alloy AZ31 is available in several tempers, but all are limited to approximately 93 °C (200 °F). For higher-temperature applications, the thorium-containing alloys HK31 and HM21 are available. For maximum creep resistance, HK31 requires a T6 heat treatment, while HM21 is cold worked prior to aging (T8 temper). In addition, due to the hcp structure, all but the most mild forming operations must be done at elevated temperature.

Wrought magnesium alloys are produced as bars, billets, shapes, wires, sheet, plate, and forgings. Nominal compositions of these alloys are given in Table 3.4, and tensile properties for wrought alloys are listed in Table 3.5.

Due to anisotropy, or texture, produced by mechanical working, the compression yield strength of wrought magnesium alloys can be appreciably less than the tensile yield strength. The compression yield strength varies from approximately 0.4 to 0.7 of the tension yield strength. Because castings do not develop texture, the compression yield strength of castings is approximately equal to the tensile yield strength.

Table 3.4 Nominal compositions of wrought magnesium alloys

Alloy		Product	Composition, wt%					
ASTM No.	UNS No.	form(a)	Al	Mn, Min	Zn	Th	Zr	Other
AZ31B	M11310	F, S, E	3.0	0.20	1.0	...	...	...
AZ31C	M11312	S, E	3.0	0.15	1.0	...	...	...
AZ61A	M11610	F, E	6.5	0.15	0.95	...	...	...
AZ80A	M11800	F, E	8.5	0.12	0.5	...	...	...
HK31A	M13310	S	3.0	...	0.3	3.25	0.7	...
HM21A	M13210	F, S	...	0.45	...	2.0	...	...
LA141A	M14141	S	1.25	0.15	...	...	...	14 Li
M1A	M15100	E	...	1.6	...	...	...	0.3 Ca
ZE10A	M16100	S	...	0.15	1.25	...	...	0.17 RE(b)
ZK40A	M16400	E	...	...	4.0	...	0.45	...
ZK60A	M16600	F, E	...	...	5.5	...	0.45	...

(a) F, forging; S, sheet and plate; E, extruded bar, shape, tube, and wire. (b) RE, rare earths. Source: Ref 3.5

Table 3.5 Minimum mechanical properties for wrought magnesium alloys

Alloy-temper	Tensile strength		Yield strength		Elongation in 50 mm (2 in.), %	Hardness, HB
	MPa	ksi	MPa	ksi		
Extruded bars, rods, and shapes						
AZ31B-F	220–240	32–35	140–150	20–22	7	...
AZ61A-F	260–275	38–40	145–165	21–24	7–9	...
AZ80A-F	290–295	42–43	185–195	27–28	4–9	...
AZ80A-T5	310–325	45–47	205–230	30–33	2–4	...
M1A-F	200–205	29–32	...	...	2–3	...
ZK40A-T5	275	40	255	37	4	...
ZK60A-F	295	43	215	31	4–5	...
ZK60A-T5	295–310	43–45	215–250	31–36	4–6	...
			(continued)			

Source: Ref 3.5

Table 3.5 (continued)

Alloy-temper	Tensile strength		Yield strength		Elongation in 50 mm (2 in.), %	Hardness, HB
	MPa	ksi	MPa	ksi		
Forgings						
AZ31B-F	234	34	131	19	6	...
AZ61A-F	262	38	152	22	6	...
AZ80A-F	290	42	179	26	5	...
AZ80A-T5	290	42	193	28	2	...
HM21A-T5	228	33	172	25	3	...
ZK60A-T5	290	42	179	26	7	...
ZK60A-T6	296	43	221	32	4	...
Sheet and plate						
AZ31B-O	221	32	...	...	9–12	...
AZ31B-H26	241–269	35–39	145–186	21–27	6	...
HK31A-O	200–207	29–30	97–124	14–18	12	...
HK31A-H24	228–234	33–34	172–179	25–26	4	...
HM21A-T81	234	34	172	25	4	...
LA141A-T7	124–131	18–19	103	15	10	...
ZE10A-O	200–207	29–30	83–124	12–18	12–15	...
ZE10A-H24	214–248	31–36	138–172	20–25	6	...

Source: Ref 3.5

3.8 Sheet and Plate

Sheet and plate are rolled from Mg-Al-Zn (AZ and photoengraving, or PE, grade) alloys, Mg-Th (HK and HM) alloys, and Mg-Li-Al (LA) alloys.

AZ31B is the most widely used alloy for sheet and plate and is available in several grades and tempers. It can be used at temperatures up to 100 °C (200 °F). HK31A and HM21A are suitable for use at temperatures up to 315 and 345 °C (600 and 650 °F), respectively. However, HM21A has superior strength and creep resistance.

Alloys with very low density and good formability have been made by adding lithium to magnesium. For example, LA141A, which contains 14% Li, has a density of 1.35 g/cm^3 (0.049 lb/in.^{3}), or only 78% that of pure magnesium. Magnesium-lithium alloys have found limited use in aerospace applications because of their low density.

Alloy PE (a modified version of AZ31B containing 3.3% Al and 0.7% Zn) is a special-quality sheet with excellent flatness, corrosion resistance, and etchability. It is used in photoengraving.

3.9 Extrusions

For normal strength requirements, one of the Mg-Al-Zn (AZ) alloys is usually selected, such as AZ31, AZ61, or AZ80. The strength of these alloys increases as aluminum content increases. AZ31B is a widely used moderate-strength grade with good formability; it is used extensively for cathodic protection. AZ31C is a lower-purity commercial variation of AZ31B for lightweight structural applications that do not require maximum corrosion resistance. AZ10A has a low aluminum content and thus is of lower strength than AZ31B, but it can be welded without subsequent stress relief. AZ61A and AZ80A alloys can be artificially aged for additional strength (with a sacrifice in ductility); however, AZ80A is not available in hollow shapes. M1A and ZM21A can be extruded at higher speeds than AZ31B, but they have limited use because of their lower strength.

The banded structure produced by hot extrusion of ZK60 is shown in Fig. 3.15. Because the extrusion process is carried out at approximately the solution heat treating temperature, and the extrusion cools fairly rapidly in air, it is only necessary to age these alloys to produce the T5 temper. For example, ZK60 in the T5 condition is often specified where higher strength and toughness is required. ZK40A is of lower strength and is more readily extrudable than ZK60A; it has had limited use in hollow tubular strength requirements.

Other high-strength extrusion alloys include ZK61 and ZCM711. For high-temperature applications, the alloys HK31 and HM21 can be specified. An important factor in extrusion is symmetry, preferably around both axes. The optimum width-to-thickness ratio for magnesium extrusions is normally less than 20. HM31A is of moderate strength. It is suitable for use in applications requiring good strength and creep resistance at temperatures in the range of 150 to 425 °C (300 to 800 °F).

Many shapes can be extruded economically. Extrusion dies are relatively inexpensive, and dimensions can be held closely enough so that machining often is unnecessary. Magnesium alloys are extruded as round rods and a variety of bars, tubes, and shapes. A wide variety of special shapes

Fig. 3.15 Section of hot-extruded ZK60 magnesium alloy. Source: Ref 3.7

also can be extruded. Extrusion is selected as a means of producing certain shapes when several small extrusions or a combination of extrusions and sheet can be joined to form an assembly, shapes are desired that are uneconomical to machine from castings, and pieces cut from extrusions can replace individually cast or forged parts.

The extrusion process offers many design possibilities not economically attainable by other production methods. These include re-entrant angles and undercuts, thin-walled tubing of large diameter, and variations in section thickness, almost without restriction. Probably the most important factor in determining whether a magnesium alloy shape will extrude well is good symmetry, preferably around both axes.

Very thin and wide sections with large circumscribing circles should be avoided. The optimum width-to-thickness (w/t) ratio for magnesium extrusions normally is less than 20. Parts with higher ratios can be extruded but require more generous tolerances. A thick section tapering to a thin wedge must always be modified by rounding the edge, or the die may not fill properly. A thin leg attached to a thick body of an extrusion should be limited to a length not exceeding ten times the leg thickness. Semiclosed shapes requiring long, thin die tongues should be avoided. For best extrudability, the length of the tongue should not exceed three times the width, although it is possible to extrude lengths five times the width. Similarly, shapes requiring unbalanced die tongues do not constitute good extrusion design. Hollow shapes that contain unsymmetrical voids, or voids separated by sections of inadequate thickness, are undesirable.

Sharp outside corners result in excessive stress concentration and die breakage and should be avoided. Inside corners should be filleted to reduce stress concentrations in the part and to ensure complete filling of the die during extrusion. Regardless of the shape being extruded, it is difficult to hold distances between thin sections to close tolerances.

Impact extrusions are tubular parts of symmetrical shape. The impact type of hot extrusion is particularly applicable when:

* It is not practical to make the part by any other method, such as with parts requiring very thin walls, where thin walls having high strength are essential, and where irregular profiles must be incorporated in the part.
* High production rates are required, where scrap loss from machining would be excessive if the part were made by other means, where strength requirements cannot be met by die castings, where the number of manufacturing operations or the number of parts in an assembly can be reduced by the use of impact extrusions, where portions of the part require zero draft, and where closer tolerances are required.

In designing impact extrusions, the following factors should be taken into consideration:

- A wide variety of symmetrical shapes is possible.
- Variations in wall thickness are possible (thin sidewall and thick bottom or thick sidewall and thin bottom).
- Ribs, flanges, bosses, and indentations can be incorporated.
- Length-to-diameter ratios may range from 1.5:1 to 15:1. Ratios from 6:1 to 8:1 are considered good working ratios.
- Reduction in area varies with the alloy being extruded and is limited by the size of available equipment. Parts with reduction in area up to 95% have been made. In general, extrusions of M1A and AZ31B can have thinner walls than AZ61A, AZ80A, and ZK60A extrusions.
- Sharp corner radii are possible in some areas of impact extrusions. This is not true of other product forms.
- Average properties of impact extrusions are slightly higher than typical properties of the hot-extruded stock from which the parts are made.

3.10 Forgings

Forgings are made of AZ31B, AZ61A, AZ80A, HM21A, and ZK60A. ZK60 has slightly better forgeability than the other alloys. HM21A and AZ31B can be used for hammer forgings (whereas the other alloys are almost always press forged). AZ80A alloy has greater strength than AZ61A and requires the slowest rate of deformation of the Mg-Al-Zn alloys. ZK60A has essentially the same strength as AZ80A but with greater ductility. To develop maximum properties, both AZ80A and ZK60A are heat treated to the artificially aged (T5) condition. AZ80A may be given the T6 solution heat treatment, followed by artificial aging to provide maximum creep stability. HM21A is given the T5 temper. It is useful at elevated temperatures up to 370 to 425 °C (700 to 800 °F) for applications in which good creep resistance is needed.

Magnesium forgings can be produced in the same variety of shapes and sizes as can forgings of other metals. Maximum size is limited primarily by size of available equipment. Tolerances can be held to the same values as in normal forging of other metals and vary somewhat with forging size and design. Magnesium alloys are heated to 340 to 510 °C (650 to 950 °F) for forging.

Forgings have the best combination of strength characteristics of all forms of magnesium. They are used where light weight coupled with rigidity and high strength are required. Magnesium forgings are sometimes used because of their pressure tightness, machinability, and lack of warpage rather than because of their high strength-to-weight ratio.

Forging is used for parts to be produced in quantities sufficient to amortize die costs and for parts requiring high strength and ductility and greater uniformity and soundness than can be obtained with castings. For small quantities, hand forgings can be used, but die forgings have better mechanical properties and are less expensive in larger quantities.

The ease with which magnesium can be worked greatly reduces the number of forging operations needed to produce finished parts. Many of the steps commonly required in forging brass, bronze, and steel (such as punching, planishing, drawing and ironing, sizing and coining, and edging and rolling) are unnecessary in forging magnesium. Bending, blocking, and finishing are the principal steps used in forging magnesium.

3.11 Forming

Wrought magnesium alloys, like other alloys with the hcp structure, are much more formable at elevated temperatures than at room temperature. Wrought alloys are usually formed at elevated temperatures; room-temperature forming is used only for mild deformations around generous radii. The approximate formability of magnesium alloy sheet is indicated by the ability to withstand 90° bending over a mandrel without cracking. The minimum size of the mandrel (minimum radius) over which the sheet can be bent without cracking depends on alloy composition and temper, material thickness, and temperature (Table 3.6). The methods and equipment used in forming magnesium alloys are the same as those commonly employed in forming alloys of other metals, except for differences in tooling and technique that are required when forming is done at elevated temperatures.

Forming magnesium alloys at elevated temperatures has several advantages:

- Forming operations can usually be conducted in one step without the need for intermediate anneals.
- Parts can be made to closer tolerances with less springback.
- Hardened steel dies are not necessary for most forming operations.

The formability depends on composition and temper, material thickness, and forming temperature. With correct temperatures and forming parameters, all magnesium sheet alloys can be deep drawn to approximately equal reductions. Maximum forming temperatures and times for various wrought magnesium alloys are given in Table 3.7.

Table 3.6 Recommended minimum radii for 90° bends in magnesium sheet

Alloy and temper	Forming Temperature(a)							
	20 °C (70 °F)	95 °C (200 °F)	150 °C (300 °F)	205 °C (400 °F)	260 °C (500 °F)	315 °C (600 °F)	370 °C (700 °F)	425 °C (800 °F)
AZ31B-O	5.5t	5.5t	4t	3t	2t	...	...	...
AZ31B-H24	8t	8t	6t	3t	2t	...	...	...
HK31A-O	6t	6t	6t	5t	4t	3t	2t	1t
HK31A-H24	13t	13t	13t	9t	8t	5t	3t	...
HM21A-T8	9t	9t	9t	9t	9t	8t	6t	4t

(a) The numerical values for bend radii are given as multiples of sheet thickness. Source: Ref 3.5

Table 3.7 Maximum forming temperatures and times for wrought magnesium alloys

Alloy	Temperature		Time(a)
	°C	°F	
Sheet			
AZ31B-O	288	550	1 h
AZ31B-H24	163	325	1 h
HK31A-H24	343	650	15 min
	371	700	5 min
	399	750	3 min
Extrusions			
AZ61A-F	288	550	1 h
AZ31B-F	288	550	1 h
M1A-F	371	700	1 h
AZ80A-F	288	550	½ h
AZ80A-T5	193	380	1 h
ZK60A-F	288	550	½ h
ZK60A-T5	204	400	½ h

(a) Maximum time the alloy can be held at temperature without adverse effects on properties. Source: Ref 3.5

3.12 Magnesium Heat Treating

Wrought magnesium alloys can be annealed by heating to 288 to 454 °C (550 to 850 °F) for 1 to 4 h to produce the maximum anneal practical. Because most forming operations are done at elevated temperature, the need for full annealing is less than with many other metals. Stress relieving is used to remove or reduce residual stresses in wrought magnesium alloys produced by cold and hot working, shaping and forming, straightening, and welding. Stress relieving is generally conducted at 149 to 427 °C (300 to 800 °F) for times ranging from 15 to 180 min. Castings are also stress relieved for a variety of reasons:

- To prevent stress-corrosion cracking for castings containing more than 1.5% Al, especially if the casting has been weld repaired
- To allow precision machining of castings to close dimensional tolerances
- To avoid warpage and distortion in service

Although magnesium castings do not normally contain high residual stresses, even moderate residual stresses can cause large elastic strains due to the low modulus of elasticity of magnesium. Residual stresses can result from nonuniform contraction during solidification, nonuniform cooling during heat treatment, machining operations, and weld repair. Welded Mg-Al-Zn castings that do not require solution heat treatment after welding should be stress relieved 1 h at 260 °C (500 °F) to eliminate the possibility of stress-corrosion cracking. Likewise, Mg-Al-Zn wrought alloys require stress relieving after cold forming to prevent stress-corrosion cracking.

Although magnesium alloys do not attain the high strengths that aluminum alloys experience during precipitation hardening, there is some strength benefit to heat treatment for a number of the casting alloys. The solution heat treatment helps to reduce or eliminate the brittle interdendritic networks in the as-cast structure. Thus, solution-treated castings have better ductility than as-cast alloys, with some increase in strength. The most prevalent precipitation-hardening treatments for cast magnesium alloys are solution treating and naturally aging (T4), naturally aging only after casting (T5), and solution treating and artificially aging (T6).

For solution heat treatment, the parts are usually placed in a preheated furnace (260 °C, or 500 °F) and slowly heated to 390 to 527 °C (735 to 980 °F). To prevent excessive surface oxidation during solution heat treating, protective atmospheres of sulfur hexafluoride, sulfur dioxide, or carbon dioxide are used. The furnaces are also equipped to handle a fire in case the furnace malfunctions and overheating occurs. In the event of a fire, boron trifluoride gas can be pumped into the furnace. Although there are exceptions, slow heating to the solution-treating temperature is recommended to avoid melting of eutectic compounds, with the subsequent formation of grain-boundary voids. Castings are held at the solution heat treating temperature for times in the range of 16 to 24 h. These hold times are long because the solution treatment also serves the purpose of homogenizing the cast structure. Castings often require support fixtures during solution heat treating to prevent them from sagging under their own weight. Some magnesium alloys are subject to excessive grain growth during solution heat treating; however, there are special heat treatments available to minimize grain growth.

Magnesium is normally quenched in air following solution treatment. Still air is usually sufficient, but forced-air cooling is recommended for dense loads or parts that have thick sections. Hot water quenching is used for the alloys QE22 and QH21 to develop the best mechanical properties. Glycol quenchants can also be used to help prevent distortion. Artificial aging consists of reheating to 168 to 232 °C (335 to 450 °F) and holding for 5 to 25 h. Hardness cannot be used for verification of heat treatment. For cast products, tensile test specimens must either be cut from a portion of the casting or cast as separate tensile test bars.

3.13 Joining Methods

Magnesium alloys can be joined using most industrial joining processes, including welding, adhesive bonding, and mechanical fastening.

Welding

Magnesium alloys are welded readily by gas metal arc welding and by resistance spot welding. Rods of approximately the same composition as the base metal generally are satisfactory. Butt and fillet joints are preferred in magnesium because they are the easiest to make by arc welding, and they provide more consistent results than other types of joints. Lap joints sometimes are used but generally are less satisfactory than butt joints for load-carrying applications.

Arc-welded joints in annealed magnesium alloy sheet and plate have room-temperature tensile strengths less than 10% lower than those of the base metal (joint efficiencies greater than 90%). However, as a result of the annealing effect from the welding heat, tensile strengths of arc welds in hard-rolled material are significantly lower than those of the base metal (joint efficiencies of only 60 to 85%). Consequently, room-temperature strengths of arc-welded joints in magnesium alloy sheet and plate are approximately the same regardless of the temper of the base metal. There are no appreciable differences in properties between welds made with alternating current and those made with direct current.

Arc welds in some magnesium alloys, specifically the Mg-Al-Zn series and alloys containing >1% Al, are subject to stress-corrosion cracking, and thermal treatment must be used to remove the residual stresses that cause this condition. The parts are placed in a jig or clamping plate and heated at the temperatures for the specified times shown in Table 3.8.

Table 3.8 Times and temperatures for stress relieving arc welds in magnesium alloys

Alloy	Temperature		Time, min
	°C	°F	
Sheet			
AZ31B-H24(a)	150	300	60
AZ31B-O(a)	260	500	15
Extrusions(b)			
AZ31B-F(a)	260	500	15
AZ61A-F(a)	260	500	15
AZ80A-F(a)	260	500	15
AZ80A-T5(a)	204	400	60
HM31A-T5	425	800	60
ZK60A-F(c)	260	500	15
ZK60A-T5(c)	150	300	60
Castings			
AM100A	260	500	60
AZ63A	260	500	60
AZ81A	260	500	60
AZ91C	260	500	60
AZ92A	260	500	60
EZ33A	260	480	600
HZ32A	350	660	120
ZE41A	350	625	120
ZH62A	350	625	120

(a) Postweld stress relief is required to prevent possible stress-corrosion cracking in this alloy. Postweld heat treatment of other alloys is used primarily for straightening or for stress relieving prior to machining. (b) When extrusions are welded to sheet, distortion can be minimized by using a lower stress-relieving temperature and a longer time, for example, 60 min at 150 °C (300 °F) instead of 15 min at 260 °C (500 °F). (c) ZK60 has limited weldability. Source: Ref 3.5

After heating, the parts are cooled in still air. The use of jigs is sometimes necessary so that relief of stresses does not result in warpage of the assembly. The other types of magnesium alloys, including those containing manganese, REs, thorium, zinc, or zirconium, are not sensitive to stress corrosion and normally do not require stress relief after welding.

Spot welds in magnesium have good static strength, but fatigue strength is lower than for either riveted or adhesive-bonded joints. Spot-welded assemblies are used mainly for low-stress applications and are not recommended where joints are subject to vibration.

Seam welds of the continuous and intermittent types have strength properties comparable to those of spot welds. Shear strengths of ~19.2 to 40.2 kg/linear mm (1075 to 2250 lb/linear in.) of welded seam can be obtained in AZ31B sheet from 1.0 to 3.2 mm (0.040 to 0.12 in.) thick.

The cost of weldments is less likely to vary significantly with quantity than the cost of other methods of fabrication. Therefore, weldments are used most often where quantities are small or where fabrication of specific designs is impractical or impossible by other methods.

Adhesive Bonding

The fatigue characteristics of adhesive-bonded lap joints are better than those of other types of joints. The probability of stress-concentration failure in adhesive-bonded joints is minimal. Adhesive bonding permits the use of thinner materials than can be effectively riveted. The adhesive fills the spaces between the contacting surfaces and thus acts as an insulator between any dissimilar metals in the joint. It also permits manufacture of assemblies having surfaces smoother than those associated with riveting.

Adhesive bonding is limited almost exclusively to lap joints. The following are a few of the general factors that should be considered when designing adhesively bonded joints:

- Joint strengths vary with lap width, metal thickness, direction in which loads are applied, and type of adhesive used.
- The joint should be designed so that it provides a sufficiently large bonded area.
- The adhesive layer should be uniform in thickness.
- The adhesive layer should be as thin as possible yet applied in sufficient quantity so that no joints are "starved."
- Joints should be designed so that pressure and heat can be readily applied.
- The curing temperatures of the common structural adhesives are below the temperatures at which the properties of hard-rolled magnesium sheet are affected, and thus, they do not significantly reduce the properties of magnesium alloys.

Riveting

Essentially the same procedures employed in riveting other materials are used to rivet magnesium alloys. Standard procedures are used for drilling and countersinking holes. Both dimpling and machine countersinking are used in flush riveting. With machine countersinking, it is desirable to have a cylindrical land with a minimum depth of 0.38 mm (0.015 in.) at the bottom of the hole. Thus, machine countersinking is limited to sheet thick enough to permit lands of this depth with a given size of rivet. Dimpling of magnesium alloy sheet is a hot forming operation; to prevent reduction of properties during dimpling, the sheet must not be heated to excessively high temperatures or for long periods.

Only aluminum rivets should be used if galvanic incompatibility is to be minimized, and those up to 8 mm ($^5/_{16}$ in.) in diameter can be driven cold. The ease of driving rivets of alloy 5056 will vary with the temper. Quarter-hard temper (5056-H32) is satisfactory for all normal riveting, although rivets of 6053-T61 or 6061-T6 can sometimes be substituted.

3.14 Machinability

Magnesium and magnesium alloys can be machined at extremely high speeds using greater depths of cut and higher rates of feed than can be used in machining other structural metals. There are no significant differences in machinability among magnesium alloys. Therefore, a specific magnesium alloy rarely, if ever, is selected in place of another magnesium alloy solely on the basis of machinability. The power required to remove a given amount of metal is lower for magnesium than for any other commonly machined metal. Because of the free-cutting characteristic of magnesium, chips produced in machining it are well broken. Dimensional tolerances of approximately ±0.1 mm (a few thousandths of an inch) can be obtained using standard operations.

An outstanding machining characteristic of magnesium alloys is their ability to acquire an extremely fine finish. Often, it is unnecessary to grind and polish magnesium to obtain a smooth finished surface. Surface smoothness readings of ~0.1 µm (3 to 5 µin.) have been reported for machined magnesium and are attainable at both high and low speeds, with or without cutting fluids.

In machining of magnesium alloys, cutting fluids provide far smaller reductions in friction than they provide in machining of other metals and thus are of little use in improving surface finish and tool life. Cutting fluids are sometimes used for cooling the work, but until recently, most machining of magnesium was done dry. Machining coolants that repress hydrogen formation, however, are now available and are paving the way to more wet machining applications, especially in high-volume automotive applications.

Although less heat is generated during machining of magnesium alloys than during machining of other metals, higher cutting speeds and the low heat capacity of magnesium and relatively high thermal expansion characteristics of magnesium may make it necessary to dissipate the small amount of heat that is generated. Generation of heat can be minimized by use of correct tooling and machining techniques, but sometimes cutting fluids are needed to reduce the possibilities of distortion of the work and ignition of fine chips. Because they are used primarily to dissipate heat, cutting fluids are referred to as coolants when used to machine magnesium alloys. Numerous mineral oil cutting fluids of relatively low viscosity are satisfactory for use as coolants in machining magnesium. Suitable coolants represent a compromise between cooling power and flash point. Additives designed to increase wetting power are usually beneficial. Only mineral oils should be used as coolants; animal and vegetable oils are not recommended. Water-soluble oils, oil/water emulsions, or water solutions of any kind should not be used on magnesium. Water reduces the scrap value of magnesium turnings and introduces potential fire hazards during shipment and storage of machine-shop scrap. Distortion of magnesium parts during machining occurs rarely and usually can be attributed to excessive heating or improper chucking or clamping.

Safe Practice. The possibility of chips or turnings catching fire must be considered when magnesium is machined. Chips must be heated close to their melting point before ignition can occur. Roughing cuts and medium finishing cuts produce chips too large to readily ignite during machining. However, fine finishing cuts produce fine chips that can ignite by a spark. Stopping the feed and letting the tool dwell before disengagement and letting the tool or tool holder rub on the work produce extremely fine chips and should be avoided.

Factors that increase the probability of chip ignitions are extremely fine feeds, dull or chipped tools, improperly designed tools, improper machining techniques, and sparks caused by tools hitting iron or steel inserts. Feeds <0.02 mm (<0.001 in.) per revolution and cutting speeds higher than 5 m/s (1000 ft/min) increase the risk of fire. Even under the most adverse conditions, with dull tools and fine feeds, chip fires are very unlikely at cutting speeds below 3.5 m/s (700 ft/min).

3.15 Magnesium Corrosion

The corrosion resistance of magnesium or a magnesium alloy part depends on many of the same factors that are critical to other metals. However, because of the electrochemical activity of magnesium (Table 3.9), the relative importance of some factors is greatly amplified. This section discusses the effects of heavy-metal impurities, the type of environment (rural atmosphere, marine atmosphere, and elevated temperatures), the

Table 3.9 Standard reduction potentials

Electrode	Reaction	Potential, V
Li,Li$^+$	Li$^+$ + e^- → Li	–3.02
K,K$^+$	K$^+$ + e^- → K	–2.92
Na,Na$^+$	Na$^+$ + e^- → Na	–2.71
Mg,Mg^{2+}	Mg^{2+} + e^- → Mg	–2.37
Al,Al^{3+}	Al^{2+} + e^- → Al	–1.71
Zn,Zn^{2+}	Zn^{2+} + e^- → Zn	–0.76
Fe,Fe^{2+}	Fe^{2+} + e^- → Fe	–0.44
Cd,Cd^{2+}	Cd^{2+} + e^- → Cd	–0.40
Ni,Ni^{2+}	Ni^{2+} + e^- → Ni	–0.24
Sn,Sn^{2+}	Sn^{2+} + e^- → Sn	–0.14
Cu,Cu^{2+}	Cu^{2+} + e^- → Cu	0.34
Ag,Ag$^+$	Ag$^+$ + e^- → Ag	0.80

Source: Ref 3.8

surface condition of the part (such as as-cast, treated, and painted), and the assembly practice. In some environments, a magnesium part can be severely damaged unless galvanic couples are avoided by proper design or surface protection.

Magnesium alloys, when properly made and applied, are corrosion resistant and are used successfully in a wide variety of commercial, industrial, and aerospace applications. The corrosion encountered usually is a result of improper design or application or inadequate protective finish.

Metallurgical Factors

Chemical Composition. Figure 3.16 shows the effects of 14 elements on the saltwater corrosion performance of magnesium in binary alloys with increasing levels of the individual elements. Most of the elements included in Fig. 3.16 (aluminum, manganese, sodium, silicon, tin, and lead, plus thorium, zirconium, beryllium, cerium, praseodymium, and yttrium) are known to have little if any deleterious effect on the basic saltwater corrosion performance of pure magnesium when present at levels exceeding their solid solubility or up to a maximum of 5%. Four elements in Fig. 3.16 (cadmium, zinc, calcium, and silver) have mild-to-moderate accelerating effects on corrosion rates, whereas four others (iron, nickel, copper, and cobalt) have extremely deleterious effects because of their low solid-solubility limits and their ability to serve as active cathodic sites for the reduction of water at the sacrifice of elemental magnesium. Although cobalt is seldom encountered at detrimental levels and cannot be introduced even through the long immersion of cobalt steels in magnesium melts, iron, nickel, and copper are common contaminants that can be readily introduced through poor molten metal handling practices. These elements must be held to levels under their individual solubility limits (or their activity moderated through the use of alloying elements such as manganese or zinc) to obtain good corrosion resistance.

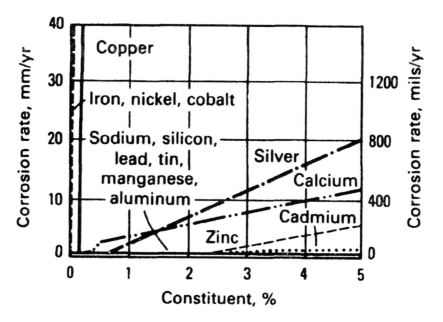

Fig. 3.16 Effect of alloying and contaminant metals on the corrosion rate of magnesium as determined by alternate immersion in 3% NaCl solution. Source: Ref 3.8

Figure 3.17 illustrates the effect of increasing iron, nickel, and copper contamination on the standard ASTM International salt spray performance of die-cast AZ91 test specimens as compared to the range of performance observed for cold-rolled steel and die-cast aluminum alloy samples. Such results have led to the definition of the critical contaminant limits for two magnesium-aluminum alloys in both low- and high-pressure cast form and the introduction of improved high-purity versions of the alloys. Some of the critical contaminant limits are listed in Table 3.10. The iron tolerance for the magnesium-aluminum alloys depends on the manganese present. For AZ91 with a manganese content of 0.15%, this means that the iron tolerance would be 0.0048% (0.032 × 0.15%). Therefore, higher-purity versions of AZ91, such as AZ91D (Fe 0.001% max, Ni 0.001% max, Cu 0.015% max, and Mn 0.17% min), were developed for those die castings requiring high corrosion resistance.

It should also be noted that the nickel tolerance depends strongly on the cast form, which influences grain size, with low-pressure cast alloys showing just a 10 ppm tolerance for nickel in the as-cast (F) temper. Therefore, alloys intended for low-pressure cast applications should be of the lowest possible nickel level. The low tolerance limits for the contaminants in AM60 alloy when compared to AZ91 alloy can be related to the absence of zinc. Zinc is thought to improve the tolerance of magnesium-aluminum alloys for all three contaminants, but it is limited to 1 to 3% because of

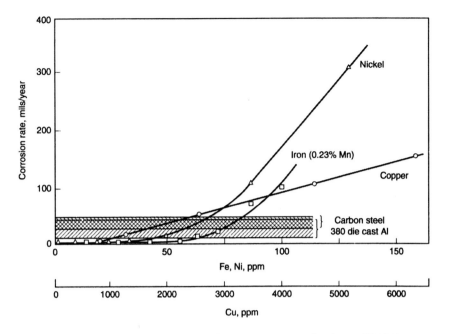

Fig. 3.17 Effect of iron, nickel, and copper contaminant levels on ASTM B117 salt spray corrosion rates in AZ91 alloy versus cold-rolled steel and 380 die-cast aluminum. Source: Ref 3.8

Table 3.10 Known contaminant tolerance limits in high- and low-pressure cast forms

		Critical contaminant limit, %		
Alloy/form	Grain size, μm	Fe	Ni	Cu
Unalloyed magnesium	...	0.015	0.0005	0.1
AZ91/high pressure	5–10	0.032 Mn(a)	0.0050	0.040
AZ91/low pressure	100–200	0.032 Mn(a)	0.0010	0.040
AM60/high pressure	5–10	0.021 Mn(b)	0.0030	0.010
AM60/low pressure	100–200	0.021 Mn(b)	0.0010	0.010
AZ63/low pressure	...	0.003(c)	0.0040	>0.45
K1A/low pressure	...	>0.003	0.003	...

(a) Iron tolerance equals manganese content of alloy times 0.032. (b) Iron tolerance equals manganese content of alloy times 0.021. (c) Magnesium content of AZ63 reported as 0.2%. Source: Ref 3.8

the detrimental effects on microshrinkage porosity and the accelerating effect on corrosion.

For the RE, thorium, and zinc alloys containing zirconium, the normal saltwater corrosion resistance is only moderately reduced when compared to high-purity magnesium-aluminum alloys—0.5 to 0.76 mm/year (20 to 30 mils/year) as opposed to less than 0.25 mm/year (10 mils/year) in 5% salt spray, but contaminants must again be controlled. Zirconium is effective because it serves as a strong grain refiner for magnesium alloys, and it precipitates the iron contaminant from the alloys before casting. However, if alloys containing more than 0.5 to 0.7% Ag or more than 2.7 to 3% Zn are used, a sacrifice in corrosion resistance should be expected (Fig. 3.17). Nevertheless, when properly finished, these alloys provide excellent service in harsh environments.

Heat Treatment and Cold Work. With controlled-purity Mg-Al-Mn alloys containing 0 to 1% Zn, there is little if any difference in corrosion rate between as-fabricated material and solution heat treated and aged material, provided that the heat treated material is air cooled from the solution heat treating temperature. With similar cooling, controlled-purity alloys containing 2 to 3% Zn corrode at slightly higher rates in the solution heat treated or solution heat treated and aged condition than in the as-cast condition. In alloys containing iron above the tolerance limit, solution heat treatment will increase the corrosion rate by a factor of 2 to 5. This factor is lower if aging follows solution treatment. The effect of heat treatment on corrosion rate in 3% NaCl is particularly noticeable adjacent to welds in alloys containing aluminum and zinc (severe attack may occur in the rapidly cooled heat-affected zone).

Cold working of magnesium alloys, such as stretching or bending, has no appreciable effect on corrosion rate. Corrosion rate can be increased by shot or grit blasting because of surface contamination. Acid pickling to a depth of 0.01 to 0.05 mm (0.0004 to 0.002 in.) often is used to remove the corrosion-active contamination, but reprecipitation of the contaminant is possible with acid pickling. Therefore, fluoride anodizing is used instead when complete removal of the contaminant is essential.

Environmental Factors

Atmospheres. A clean, unprotected magnesium alloy surface exposed to indoor or outdoor atmospheres free from salt spray will develop a gray film that protects the metal from corrosion while causing only negligible losses in mechanical properties. Materials handling equipment is an example of successful application of unfinished magnesium. Chlorides, sulfates, and foreign materials that hold moisture on the surface will promote corrosion and pitting unless the metal is protected by properly applied coatings.

The surface film that ordinarily forms on magnesium alloys exposed to the atmosphere does not protect the metal from further attack. However, if the film is tight and adherent, it decreases the rate of further attack. Unprotected magnesium and magnesium alloy parts are resistant to rural atmospheres and moderately resistant to industrial and mild marine atmospheres, provided that they do not contain joints or recesses that entrap water and are not in contact with dissimilar metals. Magnesium alloys generally do lose strength in industrial and mild marine atmospheres, whether exposed in a stressed or an unstressed condition. However, except for exposure close to seawater, the conductivity of any entrapped water present during atmospheric exposure is low. Consequently, the rate of corrosion is low and is less influenced by alloy composition and impurities.

Corrosion of magnesium alloys increases with relative humidity. At 9.5% relative humidity (RH), neither pure magnesium nor any magnesium alloys show evidence of surface corrosion after 18 months. At 30% RH, a small amount of an amorphous phase and slight corrosion are evident. At 80% RH, an amorphous phase is present over ~30% of the surface, and the surface exhibits considerable corrosion. Crystalline magnesium hydroxide is formed only when the RH is at or above 93%.

In marine atmospheres heavily loaded with salt spray, magnesium alloys require protection for prolonged survival. Magnesium alloys are less resistant to atmospheric corrosion than aluminum alloys but considerably more resistant than low-carbon steel. Table 3.11 compares a magnesium alloy, an aluminum alloy, and a low-carbon steel on the basis of loss in weight and loss in tensile strength in three different atmospheres. None of the metals was given surface protection.

Fresh Water. In stagnant distilled water at room temperature, magnesium alloys rapidly form a protective film that prevents further corrosion. Small amounts of dissolved salts in water, particularly chlorides or heavy-metal salts, will break down the protective film locally, which usually results in pitting.

Dissolved oxygen plays no major role in the corrosion of magnesium in either freshwater or saline solutions. However, agitation or any other means of destroying or preventing the formation of a protective film leads to corrosion. When magnesium is immersed in a small volume of stagnant water, the corrosion rate is negligible. When the water is constantly replenished so that the solubility limit of magnesium hydroxide ($Mg(OH)_2$) is never reached, the corrosion rate may increase.

Table 3.11 Results of a 2.5-year atmospheric exposure of an aluminum alloy, a magnesium alloy, and a low-carbon steel

Alloy	Corrosion rate		Change in tensile strength, %
	mm/year	mils/year	
Marine atmosphere(a)			
Aluminum, 2024-T3	2	0.06	2.5
Magnesium, AZ31B-H24	18	0.70	7.4
Low-carbon steel (0.27% Cu)	150	5.91	75.4
Industrial atmosphere(b)			
Aluminum, 2024-T3	2	0.08	1.5
Magnesium, AZ31B-H24	27.7	1.09	11.2
Low-carbon steel (0.27% Cu)	25.4	1.00	11.9
Rural atmosphere(c)			
Aluminum, 2024-T3	0.1	0.005	0.4
Magnesium, AZ31B-H24	13	0.53	5.9
Low-carbon steel (0.27% Cu)	15	0.59	7.5

(a) At Kure Beach, NC. (b) At Madison, IL. (c) Near Midland, MI. Source: Ref 3.8

The corrosion of magnesium alloys by pure water increases substantially with temperature. At 100 °C (212 °F), the AZ alloys corrode typically at 0.25 to 0.50 mm/year (10 to 20 mils/year). Pure magnesium and ZK60A corrode excessively at 100 °C (212 °F), with rates up to 25 mm/year (1000 mils/year). At 150 °C (302 °F), all alloys corrode excessively.

Salt Solutions. Severe corrosion can occur in neutral solutions of salts of heavy metals, such as copper, iron, and nickel. Such corrosion occurs when the heavy metal, the heavy-metal basic salts, or both plate out to form active cathodes on the anodic magnesium surface.

Chloride solutions are corrosive because chlorides, even in small amounts, usually break down the protective film on magnesium. Fluorides form insoluble magnesium fluoride and consequently are not appreciably corrosive. Oxidizing salts, especially those containing chlorine or sulfur atoms, are more corrosive than nonoxidizing salts, but chromates, vanadates, phosphates, and many others are film forming and thus retard corrosion, except at elevated temperatures.

Acids and Alkalis. Magnesium is rapidly attacked by all mineral acids except hydrofluoric acid (HF) and chromic acid (H_2CrO_4). Hydrofluoric acid does not attack magnesium to an appreciable extent, because it forms an insoluble, protective magnesium fluoride film on the magnesium; however, pitting develops at low acid concentrations. With increasing temperature, the rate of attack increases at the liquid line but increases to a negligible extent elsewhere.

Pure H_2CrO_4 attacks magnesium and magnesium alloys at a very low rate. However, traces of chloride ion in the acid will markedly increase this rate. A boiling solution of 20% H_2CrO_4 in water is widely used to remove corrosion products from magnesium alloys without attacking the base metal. Magnesium resists dilute alkalis, and 10% caustic solution is commonly used for cleaning at temperatures up to the boiling point.

Organic Compounds. Aliphatic and aromatic hydrocarbons, ketones, ethers, glycols, and higher alcohols are not corrosive to magnesium and magnesium alloys. Ethanol causes slight attack, but anhydrous methanol causes severe attack. The rate of attack in the latter is reduced by the presence of water. Gasoline-methanol fuel blends in which the water content equals or exceeds ~0.25 wt% of the methanol content do not attack magnesium.

Pure halogenated organic compounds do not attack magnesium at ambient temperatures. At elevated temperatures or if water is present, such compounds may cause severe corrosion, particularly those compounds having acidic final products. Dry fluorinated hydrocarbons, such as the freon refrigerants, usually do not attack magnesium alloys at room temperature, but when water is present, they may cause significant attack. At elevated temperatures, fluorinated hydrocarbons may react violently with magnesium alloys. In acidic foodstuffs, such as fruit juices and carbonated beverages, attack of magnesium is slow but measurable. Milk causes attack, particularly when souring. At room temperature, ethylene glycol solutions produce negligible corrosion of magnesium that is used alone or galvanically connected to steel; at elevated temperatures, such as 115 °C (240 °F), the rate increases, and corrosion occurs unless proper inhibitors are added.

Gases. Dry chlorine, iodine, bromine, and fluorine cause little or no corrosion of magnesium at room or slightly elevated temperature. Even when it contains 0.02% H_2O, dry bromine causes no more attack at its boiling temperature (58 °C, or 136 °F) than at room temperature. The presence of a small amount of water causes pronounced attack by chlorine, some attack by iodine and bromine, and negligible attack by fluorine. Wet chlorine, iodine, or bromine below the dewpoint of any aqueous phase causes severe attack. Dry, gaseous sulfur dioxide causes no attack at ordinary temperatures. If water vapor is present, some corrosion may occur. Wet (below dewpoint) sulfur dioxide gas is severely corrosive due to the formation of sulfurous and sulfuric acids. Ammonia, wet or dry, causes no attack at ordinary temperatures. Dry, gaseous sulfur dioxide (SO_2) or ammonia causes no attack at ordinary temperatures; however, some corrosion may occur if water vapor is present.

Water vapor in air or in oxygen sharply increases the rates of oxidation of magnesium and magnesium alloys above 100 °C (212 °F), but boron trifluoride (BF_3), SO_2, and sulfur hexafluoride (SF_6) are effective in reducing oxidation rates. The presence of BF_3 or SF_6 in the ambient atmosphere is particularly effective in suppressing high-temperature oxidation up to and including the temperature at which the alloy normally ignites.

The oxidation rate of magnesium in oxygen increases with temperature. At elevated temperature (approaching melting), the oxidation rate is a linear function of time. Cerium, lanthanum, calcium, and beryllium in the metal reduce the oxidation rate below that of pure magnesium. Beryllium additions have the most striking effects, protecting some alloys at temperatures up to the melting point over extended periods of time. However, structural applications of magnesium alloys at elevated temperature are usually limited by creep strength rather than by oxidation.

Soils. Except when used as galvanic anodes, magnesium alloys have good corrosion resistance in clay or nonsaline sandy soils but have poor resistance in saline sandy soils.

Stress-Corrosion Cracking

Magnesium alloys containing more than ~1.5% Al are susceptible to stress-corrosion cracking (SCC), and the tendency increases with aluminum content. Alloys in wrought form appear to be more susceptible than castings. The stress sources likely to promote cracking are weldments and inserts. Welded structures of these alloys require stress-relief annealing.

While there is little documented record of SCC failures of castings in service, magnesium castings have been shown to fail in laboratory tests under tensile loads as low as 50% of yield strength in environments causing negligible general corrosion. The apparent low incidence of SCC service failures of castings is attributable to low stresses actually applied or to stress relaxation by yielding or creep when a fixed deflection is imposed.

Galvanic Corrosion

Insufficient attention to galvanic corrosion has been one of the major obstacles to the growth of structural applications of magnesium alloys. Serious galvanic problems occur mainly in wet saline environments. Outstanding improvements in general saltwater corrosion resistance of magnesium alloys have been achieved by reducing the internal corrosion currents through strict limitations on the critical impurities nickel, iron, and copper, as well as on the iron-to-manganese ratio. These improvements have no significant effect on galvanic corrosion, because the electromotive force (emf) for corrosion now comes from an external source, the dissimilar metal coupled to magnesium. Prevention of galvanic damage thus requires consideration of a combination of measures involving:

- Design to prevent access and entrapment of saltwater at the dissimilar-metal junction
- Selection of the most compatible dissimilar metals
- Introduction of high resistance into the metallic portion of the circuit through insulators or into the electrolytic portion of the circuit by increasing the length of the path the electrolytic current must follow
- Protective coating of the full assembly

Relative Compatibility of Metals. All structural metals are cathodic to (more noble than) magnesium (Table 3.9). The degree to which the corrosion of magnesium is accelerated under a given set of exposure conditions depends partly on the relative positions of the two metals in the emf series but more importantly on how rapidly the effective potential of the couple is reduced by polarization as galvanic current flows. The principal polarization mechanism in a magnesium couple in saltwater is the resistance to the formation and liberation of hydrogen gas at the cathode. Therefore, metals of low hydrogen overvoltage, such as nickel, iron, and copper, constitute efficient cathodes for magnesium and cause severe galvanic corrosion. Metals that combine active potentials with higher hydrogen overvoltages, such as aluminum, zinc, cadmium, and tin, are much less damaging, although not fully compatible with magnesium.

Data were compiled in tests at Kure Beach, NC, in which sheets of dissimilar metals were fastened to panels of AZ31B and AZ61A. The dissimilar metals were divided into five groups based on observed gradations of galvanic damage to magnesium at the 24.4 and 244 m (80 and 800 ft) stations. These ratings are summarized as follows:

- Group 1 (least effect)
  a. Aluminum alloy 5052
  b. Aluminum alloy 5056
  c. Aluminum alloy 6061
- Group 2
  a. Aluminum alloy 6063
  b. Alclad alloy 7075
  c. Aluminum alloy 3003
  d. Aluminum alloy 7075
- Group 3
  a. Alclad alloy 2024
  b. Aluminum alloy 2017
  c. Aluminum alloy 2024
  d. Zinc
- Group 4
  a. Zinc-plated steel
  b. Cadmium-plated steel
- Group 5 (greatest effect)
  a. Low-carbon steel
  b. Stainless steel
  c. Monel
  d. Titanium
  e. Lead
  f. Copper
  g. Brass

Effects of Anode and Cathode Areas. The relative areas of the magnesium anode and the dissimilar-metal cathode have an important effect on the corrosion damage that occurs. A large cathode coupled with a small area of magnesium results in rapid penetration of the magnesium, because the galvanic current density at the small magnesium anode is very high, and anodic polarization in chloride solutions is very limited. This explains why painted magnesium should not be coupled with an active cathodic metal if the couple will be exposed to saline environments. A small break in the coating at the junction results in a high concentration of galvanic current at that point unmitigated by any polarization. Unfavorable area effects can also be experienced in the behavior of some proprietary coatings using aluminum or zinc powder. When used as a coating on a steel bolt attached to magnesium, the metallic pigment can present a very large effective surface area, which may be more detrimental than bare steel. Galvanic action is further accelerated if the metallic pigment contains impurities such as iron.

Effects of Minor Constituents on Compatibility of Aluminum with Magnesium. Aluminum alloys containing small percentages of copper (7000 and 2000 series and 380 die-casting alloy) may cause serious galvanic corrosion of magnesium in saline environments. Very pure aluminum is quite compatible, acting as a polarizing cathode, but when the iron content exceeds 200 ppm, cathodic activity becomes significant (apparently because of the depolarizing effect of the intermetallic compound $FeAl_3$), and galvanic attack of magnesium increases rapidly with increasing iron content. The effect of iron is diminished by the presence of magnesium in the alloy (Fig. 3.18). This agrees with the relatively compatible behavior of aluminum alloys 5052, 5056, and 6061.

Cathodic Corrosion of Aluminum. The compatibility of aluminum alloys with magnesium alloys is complicated by the fact that aluminum can be attacked by the strong alkali generated at the cathode when magnesium corrodes sacrificially in static NaCl solutions. Such attack destroys compatibility in alloys containing significant iron contamination, apparently by exposing fresh, active sites with low overvoltage. The aluminum alloys having substantial magnesium content (5052 and 5056) are more resistant to this effect but not completely so. The essential requirement for a fully compatible aluminum alloy, as indicated in Fig. 3.18, would be met by a 5052 alloy with a maximum of 200 ppm Fe or a 5056 alloy with a maximum of 1000 ppm Fe. Commercially produced 5052 alloy is permitted by

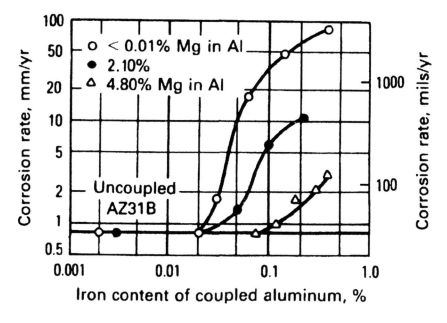

Fig. 3.18 Corrosion rates in 3% NaCl solution of magnesium alloy AZ31B coupled with aluminum containing varying amounts of iron and magnesium. The corrosion rate of uncoupled AZ31B is shown for comparison. Source: Ref 3.8

specification to have a maximum (iron + silicon) content of 0.45% and may typically contain 0.3% Fe. In a severe exposure, such as 5% NaCl immersion, this iron content, combined with the cathodic corrosion caused by the current from the magnesium, can render the 5052 alloy incompatible with magnesium. However, in most real situations, this extreme condition would not exist, and a 5052 washer under the head of a plated steel bolt in a magnesium assembly would reduce galvanic attack of the magnesium. For maximum effect of the washer, the linear distance along the aluminum from the bolt should be approximately 4.8 mm ($^3/_{16}$ in.).

Cathodic corrosion of aluminum is much less severe in seawater than in NaCl solution, because the buffering effect of magnesium ions reduces the equilibrium pH from 10.5 to approximately 8.8. The compatibility of aluminum with magnesium is accordingly better in seawater and is less sensitive to iron content.

Cathodic Damage to Coatings. Hydrogen evolution and strong alkalinity generated at the cathode can damage or destroy organic coatings applied to fasteners or other accessories coupled to magnesium. Alkaliresistant resins are necessary, but under severe conditions such as salt spray or salt immersion, the coatings may be simply blown off by hydrogen, starting at small voids or pores. Because of its severity, the salt spray test can lead to rejection of some fastener coatings that may provide useful benefits in real service environments. Salt spray should not be relied upon exclusively to evaluate these coatings.

Compatibility of Plated Steel. Zinc, cadmium, or tin plating on steel all reduce galvanic attack of magnesium substantially compared to bare steel. This agrees with the more compatible potentials and/or the higher hydrogen overvoltages of the plated deposits. The relative merit of the three electroplates is generally considered to be (in decreasing order) tin, cadmium, and zinc.

3.16   Corrosion Protection of Magnesium

A dissimilar metal in contact with magnesium will not by itself result in galvanic corrosion. For corrosion to occur, both surfaces must also be wetted by a common electrolyte. The degree to which precautions against galvanic corrosion are taken will depend on many factors, of which the operating environment is of primary importance.

For indoor use, where condensation is not likely, no protection is necessary. Even in some sheltered outdoor environments, magnesium components can give good service lives without special precautions against galvanic attack, provided other mitigating factors are present. These may include design elimination of water traps, good ventilation, component warmth, or the presence of an oil film.

For continuous outdoor use, during which magnesium assemblies may be wetted or subjected to salt splash or spray, precautions against galvanic attack must be taken. Although corrosive attack from any source can jeopardize the satisfactory performance of magnesium components, attack resulting from galvanic corrosion is probably the most detrimental. Under corrosive conditions, use of high-purity magnesium alloys will have no significant influence in reducing the effects of galvanic corrosion.

Magnesium-to-Magnesium Assemblies. For all practical purposes, galvanic corrosion between magnesium alloys is negligible. However, good assembly practice dictates that the magnesium faying surfaces be given one or more coats of a chromate-pigmented primer.

Magnesium-to-Nonmetallic Assemblies. Although the joining of most nonmetallic materials, such as plastics and ceramics, to magnesium will not result in any potential corrosion hazard, there are some notable exceptions. Magnesium-to-wood assemblies present an unusual problem because of the water absorbency of wood and the tendency of the assemblies to leach out natural acids. To protect magnesium from attack, the wood should first be sealed with paint or varnish, and the faying surface of the magnesium treated as described previously for magnesium-to-magnesium assemblies. The joining of magnesium to carbon-fiber-reinforced plastics is another exception that, in the presence of a common electrolyte, could result in corrosion of the magnesium unless similar assembly precautions are observed.

Magnesium-to-Dissimilar-Metal Assemblies. Several techniques minimize or eliminate galvanic corrosion or, if breakdown occurs, reduce the effect in magnesium-to-dissimilar-metal couples. These include:

- Elimination of the common electrolyte
- Reduction of the relative area of dissimilar metal present
- Minimization of the potential difference of the dissimilar metal
- Protection of the dissimilar metal and the magnesium from the common electrolyte

Good design can play a vital role in reducing the threat of galvanic corrosion. Elimination of a common electrolyte may be possible by the provision of a simple drain hole or shield to prevent liquid entrapment at the dissimilar-metal junction. Alternatively, the location of screws or bolts on raised bosses may also help avoid common electrolyte contact, as would use of nylon washers, spacers, or similar moisture-impermeable gaskets. The use of studs in place of bolts will reduce the area of dissimilar metal exposed by up to 50%, provided the captive ends of the studs are located in blind holes.

The degree of attack resulting from galvanic corrosion is, among other things, proportional to the potential difference between the metals involved. Consequently, this should be reduced to a minimum by careful materials selection or the use of selected plating or coating of metals brought into contact with magnesium.

Dissimilar metals that are relatively compatible with magnesium are the aluminum-magnesium (5000-series) or aluminum-magnesium-silicon (6000-series) aluminum alloys, which should be used for washers, shims, fasteners (rivets and special bolts), and structural members, where possible. Other aluminum alloys, steels, titanium, copper, brass, and Monel will corrode magnesium when coupled with it under corrosive conditions, and protection is therefore required.

Aluminum, zinc, cadmium, and tin are used to coat steel or brass components to reduce the galvanic couple with magnesium. Reducing the potential difference or plating by using materials with high hydrogen overvoltages will help reduce galvanic corrosion under mildly corrosive environments but will have minimal effect in severe environments; additional precautions are required for corrosion protection.

Use of wet assembly techniques will eliminate galvanic corrosion in crevices. Caulking the metal junctions will increase the electrical resistance of the galvanic couple by lengthening the electrolytic path and thus reducing the degree of attack if it occurs. Vinyl tapes have also been used to separate magnesium from dissimilar metals or a common electrolyte and thus prevent galvanic attack. Finally, painting the magnesium and, more important, the dissimilar metal after assembly will effectively insulate the two materials externally from any common electrolyte.

3.17 Protective Coating Systems

Inorganic Surface Treatments. A full range of chemical and electrochemical cleaning and surface pretreatments before application of paint finishes is available for magnesium. Whichever pretreatment is selected, it must be applied to a clean metal surface. In the case of magnesium, this implies the removal of oil, dirt, or grease and, more important, a surface free of other contaminants. Heavy-metal contamination arising from blasting, brushing, tumbling, lapping, and other abrasive operations is particularly detrimental, as is contamination from graphite-containing die-forming lubricants. The use of abrasive materials compatible with magnesium, for example, high-purity alumina, silicon carbide, and glass, will help ensure that heavy-metal pickup is kept to a minimum.

Oil, dirt, and grease are removed by conventional solvent immersion or vapor degreasing techniques using chlorinated solvents. Alkali cleaning in high-pH cleaners is also suitable. Oxides, die-forming compounds, and other surface contaminants are removed by a range of acid pickling solutions. In addition, an electrochemical process known as fluoride anodizing will more effectively remove sand or heavy-metal contamination. This process is also applicable to finished work when dimensional losses cannot be tolerated.

The primary function of dip or anodic coatings on magnesium is to provide a suitable surface to promote the adhesion of subsequent organic coatings. Conversion coatings should not be regarded as protective treatments in their own right unless they are to be exposed only to noncorrosive environments. Under these conditions, they will delay the onset of natural surface oxidation and may provide a more visually attractive surface appearance.

The dip coatings (chrome pickle and dichromate treatments listed in Table 3.12) are very thin coatings used primarily for protection during shipment and storage and as primers for subsequent painting. These coatings should not be heated above 260 °C (500 °F). Exercise discretion when using chromate coatings without subsequent painting, because some chromate coatings are pyrophoric, that is, they spark when hit.

Several hard anodizing treatments are available for magnesium, but the most commonly used treatments are the No. 17 and HAE treatments (Table 3.13). Both may be applied as thin (0.005 mm, or 0.0002 in.) or thick (0.038 mm, or 0.0015 in.) coatings, with the thicker treatments imparting wear and abrasion resistance. These coatings are porous and provide excellent bases for subsequent painting. However, particularly for the thicker films, conventional painting may not completely seal the anodic pores. To prevent the risk of subsurface lateral corrosion spread from a point of damage, resin impregnation is used for maximum serviceability in aggressive corrosive environments. Inorganic chemical posttreatments are sometimes used to impregnate the anodic film with corrosion inhibitors, but these treatments can be detrimental to subsequently applied organic coatings and are not as effective as resin impregnation.

Selection of a suitable protective scheme depends on many factors, especially the expected operational environment, design life, inspection and maintenance costs, the component cost, and, of course, the cost of original surface protection. For new applications, it is advisable to err on the side of

Table 3.12 Examples of chemical conversion coating treatments for magnesium alloys

Name	Bath composition(a)	Procedure	Appearance	Typical metal removal	Comments
Chrome pickle (acid chromate)	180 g $Na_2Cr_2O_7$ $2H_2O$ (sodium dichromate), 187 mL 70% HNO_3, to 1 L H_2O	½ to 2 min immersion at room temperature; allow to drain for 5–30 s; rinse in cold water, then hot water to aid drying	Golden yellow, often with iridescence	Up to 0.015 mm (0.0006 in.)	Applicable to all alloys and forms; mainly applied to wrought and die castings; good paint base
Dichromate	50 g Nh_4F Hf, or 187 mL 60% HF, to 1 L H_2O	5 min immersion in activator at room temperature, except for AZ31 alloy, which should only be immersed for ½–1 min if HF activator is used; rinse thoroughly	Brassy to dark brown	Negligible	Applicable to most alloys and all forms; as-cast die-cast surfaces should be prepickled to remove skin segregation; excellent paint base
	180 g $Na_2Cr_2O_7$ $2H_2O$, 2.5 g CaF_2 or MgF_2 (calcium or magnesium fluoride), to 1 L H_2O	Immersion for 30 min in boiling solution (95 °C, or 205 °F, min) maintain pH 4.0–5.5; rinse and dry			

Source: Ref 3.8

Table 3.13 Details of two hard-anodizing treatments for magnesium alloys

Name	Bath composition(a)	Time, min	Temperature °C	Temperature °F	Coating appearance	Coating buildup mm	Coating buildup in.	Comments
HAE	135–165 g KOH, 34 g Al(OH)$_3$(aluminum hydroxide), 34 g KF (potassium fluoride), 34 g Na$_3$Po$_4$ (trisodium phosphate), 20 g K$_2$MnO$_4$ (potassium manganate), to 1 L H$_2$O	8 min at 200 A/m^2 (0.13 A/in.2) for thin coating (70 V)	15–30 max (cooling required)	60–85 max (cooling required)	Light tan (thin)	0.005	0.0002	Applicable to all alloys and forms. Thin coating provides excellent paint base. Thick coating provides excellent wear resistance, and if sealed with organic resins, provides superior corrosion protection as well. Process has good throwing power.
		60 min at 250 A/m^2 (0.16 A/in2) for thick coating (90 V) (ac anodize)			Dark brown (thick)	0.040	0.0016	
No. 17	240 g NH$_4$F[1] HF, 100 g Na$_2$Cr$_2$O$_7$ · H$_2$O, 90 mL 85% H$_3$PO$_4$, to 1 L H$_2$O	25 min at 200 A/m^2 (0.13 A/in.2) for thin coating (70 V)	70–80	160–175	Light green (thin)	0.006	0.0002	Applicable to all alloys and forms. Thin coating provides excellent paint base. Thick coating gives good wear resistance and, if sealed with organic resins provides superior corrosion protection. Process has excellent throwing power.
		25 min at 200 A/m^2 (0.13 A/in2) for thick coating (90 V) (ac anodize)			Dark green (thick)	0.030	0.0012	

Note: ac, alternating current. (a) Whenever water is specified, use deionized water. Source: Ref 3.8

overprotection until enough experience is gained to enable a more valued judgment to be made.

Corrosion-Protection Considerations

For indoor and similar noncorrosive environments, surface-protection requirements are minimal and may range from none to simple chromate or phosphate conversion coatings with primer only or a decorative paint finish. Even under apparently more corrosive conditions, other mitigating factors, such as the use of high-purity alloys, good ventilation, component warmth, good design, and oil films, will enable magnesium components to be used with little or no protection.

For mildly or moderately corrosive environments, a chromate pretreatment followed by one coat of suitable primer and one or more coats of compatible finish should be applied. Under these conditions, the effect of galvanic couples, if present, should be considered, and some precautions taken.

For moderately to severely corrosive environments, good-quality chromate conversion coatings or thin anodic pretreatments should be used. Chromate-inhibited epoxy primer systems, careful wet assembly procedures, and painting after assembly with primer and topcoat are recommended. Use of low-temperature baking paints is beneficial for improving humidity resistance (Fig. 3.19a).

For severely corrosive environments, for which maximum chemical, salt spray, and humidity resistance are required (Fig. 3.19b), specialized paints and coating techniques are used. Good-quality chromate or anodic pretreatments are required. Use of thick anodic coatings will also impart a measure of abrasion and damage resistance (Fig. 3.19c). Pretreatments should then be sealed with high-temperature baking organic resins. This is achieved by a process known as surface sealing (MIL-M-3171 section 3.9.3, MIL-M-46080), in which three coats of thinned resin are applied to a preheated component by spraying or, preferably, by dipping. Epoxy resin systems are preferred (MILC-46079), although the technique would be beneficial for other high-temperature baking resins, such as phenolics and epoxy silicones.

After this foundation treatment, full wet assembly procedures, including caulking of joints, should be performed before application of a cold

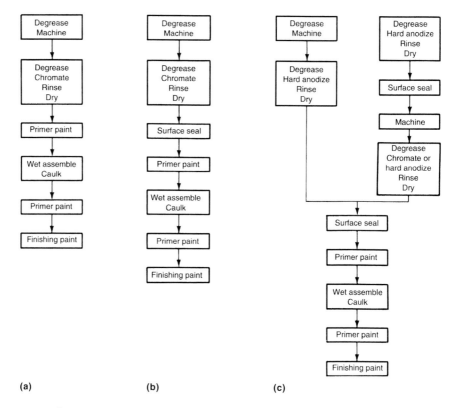

(a) (b) (c)

Fig. 3.19 Diagram of protection schemes for critical applications in corrosive environments. (a) For moderately corrosive environments. (b) For severely corrosive environments. (c) For severely corrosive environments with risk of abrasion or damage. Source: Ref 3.9

or low-temperature curing chromate-inhibited primer and a compatible topcoat system. For additional protection, a high-temperature baking paint system should be maintained throughout. The previous recommendations are necessarily very general in scope but represent the various protection schemes in worldwide use on magnesium alloys.

ACKNOWLEDGMENTS

Portions of the text for this chapter came from "Selection and Applications of Magnesium and Magnesium Alloys" and "Corrosion Resistance of Magnesium and Magnesium Alloys," both in *Metals Handbook Desk Edition,* 2nd ed., ASM International, 1998.

REFERENCES

3.1 P. Lyon, Electron 21 for Aerospace and Specialty Applications, *AeroMet Conference and Exposition,* June 7–10, 2004 (Seattle, WA)

3.2 H. Westengen and T.K. Aune, Casting Alloys, *Magnesium Technology,* Springer, 2006

3.3 J.D. Hanawalt, C.E. Nelson, and J.A. Peloubet, *Trans. AIME,* Vol 147, 1942, p 273

3.4 W. Unsworth, *Met. Mater.,* Feb 1988, p 82

3.5 Selection and Applications of Magnesium and Magnesium Alloys, *Metals Handbook Desk Edition,* 2nd ed., ASM International, 1998

3.6 K. Savage, Magnesium and Magnesium Alloys, *Casting,* Vol 15, *ASM Handbook,* ASM International, 2008

3.7 Microstructure of Magnesium Alloys, *Atlas of Microstructures of Industrial Alloys,* Vol 7, *ASM Metals Handbook,* 8th ed., American Society for Metals, 1972

3.8 Corrosion Resistance of Magnesium and Magnesium Alloys, *Metals Handbook Desk Edition,* 2nd ed., ASM International, 1998

3.9 B.A. Shaw and R.C. Wolfe, Corrosion of Magnesium and Magnesium-Base Alloys, *Corrosion: Materials,* Vol 13B, *ASM Handbook,* ASM International, 2005

SELECTED REFERENCES

• M. Avedesian and H. Baker, Ed., *ASM Specialty Handbook: Magnesium and Magnesium Alloys,* ASM International, 1999

• M.R. Bothwell, *The Corrosion of Light Metals,* H. Godard, W.B. Jepson, and M.R. Bothwell, Ed., John Wiley & Sons, 1967, p 259–311

• C.R. Brooks, Magnesium-Base Alloys, *Heat Treatment, Structure and Properties of Nonferrous Alloys,* American Society for Metals, 1982, p 253–274

- E.F. Emley, *Principles of Magnesium Technology,* Pergamon Press, 1966
- P.A. Fisher, Production, Properties, and Industrial Use of Magnesium and Its Alloys, *Int. Met. Rev.,* Vol 23 (No. 6), 1978, p 269
- T.K. Froats, D. Aune, W. Hawke, W. Unsworth, and J.E. Hillis, Corrosion of Magnesium and Magnesium Alloys, *Corrosion,* Vol 13, *ASM Handbook,* ASM International, 1987, p 740–754
- J.D. Hanawalt, C.E. Nelson, and J.A. Peloubet, *Trans. AIME,* Vol 147, 1942, p 273
- J.E. Hillis, Surface Engineering of Magnesium Alloys, *Surface Engineering,* Vol 5, *ASM Handbook,* ASM International, 1994
- J.E. Hillis, Magnesium, *Corrosion Tests and Standards: Application and Interpretation,* R. Baboian, Ed., ASTM, 1995, p 438–446
- S. Housh, B. Mikucki, and A. Stevenson, Formability: Selection and Application of Magnesium and Magnesium Alloys, *Properties and Selection: Nonferrous Alloys and Special-Purpose Materials,* Vol 2, *ASM Handbook,* ASM International, 1990
- S. Housh, B. Mikucki, and A. Stevenson, Machinability: Selection and Application of Magnesium and Magnesium Alloys, *Properties and Selection: Nonferrous Alloys and Special-Purpose Materials,* Vol 2, *ASM Handbook,* ASM International, 1990
- T.E. Leontis, Magnesium: Alloying, Fabrication, and Selection, *Encyclopedia of Materials Science and Engineering,* M.B. Bever, Ed., Pergamon Press and MIT Press, 1986, p 2624–2632
- L.F. Lockwood, G. Ansel, and P.O. Haddad, Magnesium and Magnesium Alloys, *Kirk-Othmer Encyclopedia of Chemical Technology,* 3rd ed., Wiley-Interscience, New York, 1981, p 570–615
- H.L. Logan, Magnesium Alloys, *The Stress Corrosion of Metals,* John Wiley & Sons, 1966, p 217–237
- W.K. Miller, Stress-Corrosion Cracking of Magnesium Alloys, *Stress-Corrosion Cracking: Materials Performance and Evaluation,* R.H. Jones, Ed., ASM International, 1992, p 251–263
- T.V. Padfield, Metallography and Microstructures of Magnesium and Its Alloys, *Metallography and Microstructure,* Vol 9, *ASM Handbook,* ASM International, 2004
- L.M. Pidgeon, Magnesium Production, *Encyclopedia of Materials Science and Engineering,* M.B. Bever, Ed., Pergamon Press and MIT Press, 1986, p 2635–2638
- I.J. Polmear, *Light Alloys: Metallurgy of the Light Metals,* 3rd ed., Halsted Press, 1995
- H. Proffitt, Magnesium Alloys, *Casting,* Vol 15, *ASM Handbook,* ASM International, 1988
- G.V. Raynor, *The Physical Metallurgy of Magnesium and Its Alloys,* Pergamon Press, 1959

- C.S. Roberts, *Magnesium and Its Alloys,* John Wiley & Sons, 1960
- W.F. Smith, *Structure and Properties of Engineering Alloys,* 2nd ed., McGraw-Hill Inc., 1993
- A. Stevenson, Heat Treating of Magnesium Alloys, *Heat Treating,* Vol 4, *ASM Handbook,* ASM International, 1991

CHAPTER **4**

Beryllium

BERYLLIUM is a metal with an unusual combination of physical and mechanical properties that make it particularly effective in optical components, precision instruments, and specialized aerospace applications. In each of these three general application areas, beryllium is selected because of its combination of low weight, high stiffness, and specific mechanical properties, such as a precise elastic limit. It is also useful because it is transparent to x-rays and other forms of high-energy electromagnetic radiation. The disadvantages of beryllium include high cost, somewhat difficult fabrication, and the toxicity of dust and airborne particulates (see the section "4.12 Health and Safety" in this chapter).

Unalloyed beryllium is readily joined by brazing. Fusion welding is not advisable in most situations, although beryllium can be fusion welded with aluminum filler metals with extreme care. Beryllium can be extruded into bar, rod, and tubing or rolled into sheet. The surface of beryllium can be polished to a very reflective mirror finish, and this finish is particularly effective at infrared wavelengths. Beryllium can be plated with nickel, silver, gold, and aluminum, and the surface can be anodized or chromate conversion coated to provide a measure of corrosion resistance.

Beryllium can be machined to extremely close tolerances. This attribute, in combination with its excellent dimensional stability, allows beryllium to be used for the manufacture of extraordinarily precise and stable components.

Almost all of the beryllium in use is a powder metallurgy (PM) product. Powder processing is required for a number of reasons. Castings of beryllium generally have porosity and other casting defects that make them unsuitable for use in critical applications. This stems from the high melting point, the high melt viscosity, and the narrow solid-liquid range of beryllium. The high melting point (1283 °C, or 2341 °F) promotes reaction of the molten metal with potential casting mold materials, and the high melt viscosity and the narrow solid-liquid range prevent the easy filling of complex castings. If high superheat temperatures are used to reduce viscosity, mold reaction limits the integrity of the component. In addition to the limitation on structural integrity imposed by casting, the grain size of as-cast beryllium is quite coarse (>50 μm, or >0.002 in.). The ductility and strength of beryllium depend primarily on grain size and require grain sizes of less than approximately 15 μm (600 μin.) in structural components. Powder metallurgy processing allows grain sizes as low as 1 to 10 μm when required. Consolidation by vacuum hot pressing, hot isostatic pressing, pressing and sintering, or other processes can produce parts with density values in excess of 99.5% of the theoretical value of 1.85 g/cm^3 (0.067 lb/in.^{3}).

The application of PM processing to beryllium does impose its own limits on the characteristics of the product. For example, the use of excessively fine powders increases the oxide content because of the increase in specific surface area. The oxide content of beryllium is one of the determinants of physical and mechanical properties; therefore, knowledge and control of the powder process are important to the proper fabrication of a beryllium component. Also, powder shape affects the anisotropy of mechanical properties.

4.1 Beryllium Properties and Applications

Of the mechanical and physical properties given for beryllium, the most noteworthy are its density, elastic modulus, mass absorption coefficient, and the other physical properties presented in Table 4.1. The specific modulus is of particular interest. The specific modulus for beryllium is substantially greater than that of other aerospace structural materials, such as steel, aluminum, titanium, or magnesium (Fig. 4.1). Also, beryllium has the highest heat capacity (1820 J/kg · °C, or 0.435 Btu/lb · °F) among metals and a thermal conductivity (210 W/m · K, or 121 Btu/ft · h · °F) comparable to that of aluminum (230 W/m · K, or 135 Btu/ft · h · °F). Beryllium is an efficient substrate material for conducting waste heat away from active solid-state electronic components, particularly in aerospace applications.

Table 4.1 Selected physical properties of beryllium

Property	Amount
Elastic modulus, GPa (10^6 psi)	303 (44)
Density, g/cm^3 (lb/in.^{3})	1.8477 (0.067)
Thermal conductivity, W/m · K (Btu/h · ft · °F)	210 (121)
Coefficient of thermal expansion, 10^{-6}/°C (10^{-6}/°F)	11.5 (6.4)
Specific heat at room temperature, kJ/kg · K (Btu/lb · °F)	2.17 (0.52)
Melting point, °C (°F)	1283 (2341)
Mass absorption coefficient (Cu K-alpha), cm^2/g	1.007
Specific modulus(a), m (in.)	16.7 × 10^6 (6.56 × 10^8)

(a) The specific modulus is defined (in inches) from the ratio of the elastic modulus (in psi) and the density (in lb/in.^{3}). Source: Ref 4.1

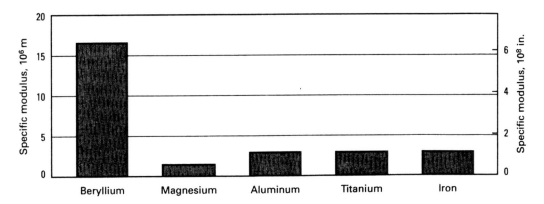

Fig. 4.1 Specific modulus comparison of lightweight materials. Source: Ref 4.1

The microyield strength of beryllium is important in guidance system components for ships, aircraft, and missiles. In a gyroscopic system, permanent errors can be introduced by even very small yielding of the guidance components. A permanent set of 1 part in 10^6 (microyield), as opposed to the common yield criterion of 2 parts in 10^4, is the yield strength value used in evaluating gyroscope materials. Instrument grades of beryllium thus have a microyield strength acceptance criterion.

The second important area of application for instrument-grade beryllium is in infrared optics. At long-wave infrared (LWIR) wavelengths, beryllium has a reflectivity in excess of 99% of the incident intensity. This enables beryllium to be used in LWIR surveillance and in deep-space observatories. The mirrors for satellites, such as the Infrared Astronomical Satellite, are often beryllium because the metal combines infrared reflectivity with light weight and high stiffness.

The low mass absorption coefficient of beryllium makes it practically transparent to x-rays and other high-energy electromagnetic radiation. It is therefore used as a window material in x-ray tubes and detectors such as those used in energy-dispersive analysis of x-ray equipment.

4.2 Beryllium Powder Production Operations

Powder metallurgy is required for a number of reasons, the primary one being that beryllium castings contain too much porosity and other casting defects to allow their use in critical applications. In addition to the casting defects, the grain size of cast beryllium is too coarse (>100 μm). Because the strength and ductility of beryllium depends on a fine grain size, grain sizes of less than 15 μm are required to obtain satisfactory mechanical properties. The effects of grain size on beryllium properties are shown in Fig. 4.2. Powder metallurgy techniques can produce grain sizes as fine as

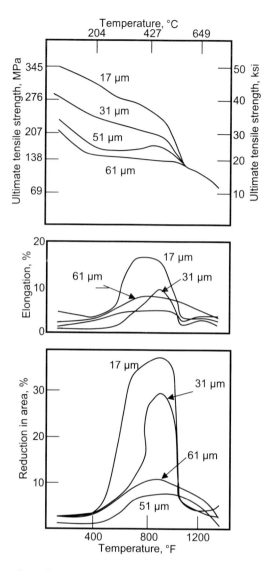

Fig. 4.2 Effect of grain size on beryllium mechanical properties. Source: Ref 4.2

1 to 10 µm when required. Beryllium powders are consolidated into near-net shapes by either vacuum hot pressing or hot isostatic pressing to obtain parts within 99.5% of theoretical density.

Although ingot castings are not useful as commercial products themselves because of the inherent limitations of cast beryllium, the ingot is quite acceptable from the chemical analysis point of view and provides a starting point for subsequent PM operations. The ingot is converted into chips using a lathe and a multihead cutting tool. The chips are then ground to powder using one of several mechanical methods: ball milling, attritioning, or impact grinding. These processes produce powders with varying characteristics, particularly with regard to particle shape.

Ball milling and attritioning are relatively slow processes, and they activate slip and fracture primarily upon the basal planes of the hexagonal close-packed beryllium crystal. This fracture mode gives rise to a flat-plate particle morphology. When the particles manufactured by this process are loaded into a die for consolidation, the flat surfaces align and give rise to areas of preferred orientation.

The impact-grinding process relies on a high-velocity gas stream to accelerate beryllium particles and drive them against a beryllium target. The impacts cause fracture on additional crystal planes other than the basal plane, resulting in a blocky particle. The blocky particles exhibit less tendency to align preferentially during powder loading, thereby reducing the tendency for preferred orientation in the final consolidated product. The reduction in preferred orientation leads to improved overall ductility, a property particularly sensitive to orientation, in all directions in the consolidated component. For this reason, impact grinding has largely replaced attritioning and ball milling as the major powder production technique.

In the impact-grinding process, shown in Fig. 4.3, coarse powder is fed from the feed hopper into a gas stream. As the gasborne powder is carried through a nozzle, it accelerates and impacts a beryllium target. The debris is then carried to the primary classifier, where the particles drop out and fines go to a secondary classifier and are discarded. This cycle continues until the desired particle size is achieved. Impact milling enables consis-

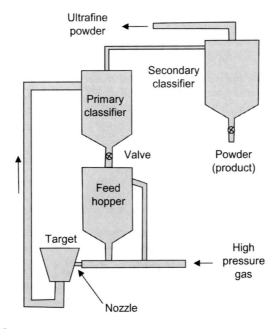

Fig. 4.3 Impact-grinding system. Source: Ref 4.2

tent control of powder composition by reducing impurity contamination and oxidation of powder particles. The process also yields improved powder configuration and morphology, resulting in improved isotropy and a cleaner microstructure of the final consolidated product.

Atomization is another powder production technique. Atomization has several potential advantages over mechanical comminution methods. Atomization typically produces a spherical particle, and the use of such a particle is logically the most effective means of eliminating property anisotropy resulting from the shape of powder particles. There are also economic factors that favor atomization. The inert gas atomization process is capable of very high production rates, and the effectiveness of such a process could conceivably reduce production costs significantly. The inert-gas atomization process has the capability of producing clean, fine-grained powders with excellent flow and packing characteristics. The properties of consolidated billets made from this material have been equal or superior to those made with comparable mechanically comminuted material.

4.3   Powder Consolidation Methods

Once powder is made, it must be consolidated. This step in the production of beryllium components has effects on the properties of the consolidated material that are as profound as the effects of the size and chemistry of the powder. Until the mid-1980s, the predominant consolidation method was vacuum hot pressing of powder into right circular cylinders. This remains the technique with the highest production volume, but net-shape technology is transforming the industry because of the cost advantages associated with greater material utilization.

Powder compaction is conducted either by vacuum hot pressing (VHP) or hot isostatic pressing (HIP), as shown schematically in Fig. 4.4. In VHP, the powder is loaded into a graphite die and vibrated, the die is placed into a vacuum hot press, a vacuum is established, and the powder is consolidated under 3.45 to 13.8 MPa (500 to 2000 psi) at temperatures between 1000 and 1100 °C (1830 and 2020 °F). Densities in excess of 99% of theoretical are obtained in diameters of 20 to 180 cm (8 to 72 in.). The HIP process is similar except that the powder must be enclosed in a mild steel can prior to consolidation. After canning, the can is degassed under vacuum at 593 to 704 °C (1100 to 1300 °F) to remove all air and gases absorbed on the particle surfaces. The can is then sealed and put through the HIP consolidation cycle at 103 MPa (15 ksi) and 760 to 1100 °C (1400 to 2010 °F). Although HIP is generally more expensive than VHP, it allows the best control of grain size because it allows greater latitude in temperature selection. Hot isostatic pressing is capable of attaining 100% of theoretical density. A comparison of the two methods on the properties of grade S-200 is shown in Fig. 4.5. The improvement in anisotropy for the

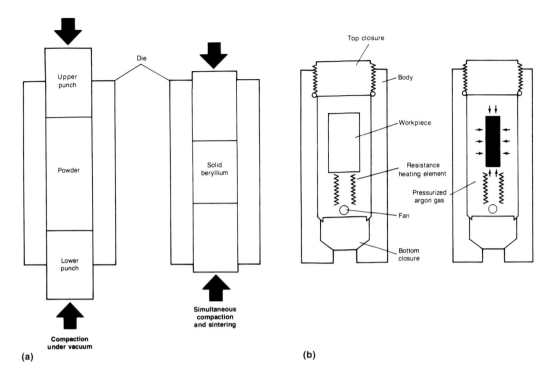

Fig. 4.4 Schematic diagrams of two powder consolidation methods. (a) Vacuum hot pressing. In this method, a column of loose beryllium powders is compacted under vacuum by the pressure of opposed upper and lower punches (left). The billet is then brought to final density by simultaneous compaction and sintering in the final stages of pressing (right). (b) Hot isostatic pressing. In this process, the powder is simultaneously compacted and sintered to full density inside a pressure vessel within a resistance-heated furnace (left). The powder is placed in a container, which collapses when pressure is exerted evenly on it by pressurized argon gas (right). Source: Ref 4.1

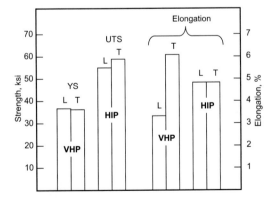

Fig. 4.5 Effects of compaction method on properties of S-200. VHP, vacuum hot press; HIP, hot isostatic press; YS, yield strength; UTS, ultimate tensile strength; L, longitudinal; T, transverse. Source: Ref 4.2

HIP-processed material is evident and is primarily a result of being able to use lower temperatures to keep the grain size fine and apply uniform and high pressures in all directions during consolidation.

Another net-shape process is cold isostatic pressing (CIP) followed by vacuum sintering. In this method, powder is loaded into an elastic (for example, neoprene or latex) bag and isostatically pressed by means of a liquid at room temperature to produce a green compact of approximately 80% theoretical density. The compact can then be vacuum sintered to achieve a density of 98% or more. At this density level, a can is not necessary for further densification by HIP because the porosity is isolated rather than interconnected. The expense of fabricating a complex disposable can is thus eliminated, and the bag that is used in place of the can has a relatively low cost and is reusable. The trade-off is that the high sintering temperature used in the bag method results in grain growth. However, this is offset somewhat by the elimination of residual porosity. Also, the CIP process permits the production of parts with complex, three-dimensional complexity that cannot be made by the uniaxial cold pressing route.

4.4 Beryllium Grades and Their Designations

Beryllium grades, as opposed to alloys, are nominally beryllium with beryllium oxide as the only other major component. Although oxide inclusions are regarded as undesirable contaminants in many systems, the oxide in beryllium is desirable as a grain-boundary pinning agent. The finer the powder, the greater the oxide content. When very fine grain sizes are required, the high oxide content of the powder acts to stabilize the grain size during consolidation. Beryllium oxide contents of commercial grades vary from an allowed maximum of 0.7% in I-70H to a minimum of 2.2% in I-220H. The chemical compositions of currently available commercial grades of beryllium are shown in Table 4.2, and minimum guaranteed mechanical properties are given in Table 4.3. Note that the mechanical properties in Table 4.3 are guaranteed minimums, and actual values may be substantially higher.

Table 4.2 Chemical composition for beryllium alloys

Grade	Be, % min	BeO	Al	C	Fe	Mg	Si	Other metallic impurities
					Chemical composition, % max			
I-70H	99.0	0.7	0.07	0.07	0.10	0.07	0.07	0.04
I-220H	98.0	2.2	0.10	0.15	0.15	0.08	0.08	0.04
O-30H	99.0	0.5	0.07	0.07	0.12	0.12	0.07	0.04
S-65	99.2	0.9	0.05	0.09	0.08	0.01	0.045	0.045(a)
S-200F	98.5	1.5	0.10	0.15	0.13	0.13	0.08	0.04
S-200FH	98.5	1.5	0.10	0.15	0.13	0.08	0.08	0.04
SR-200	98.0	2.0	0.16	0.15	0.18	0.08	0.08	0.04

(a) Grade also contains Cr (0.01 max), Ni, Cu, Ti, Zr (0.025 max), Zn, Mn, Ag, Co, Pb, Ca, Mo (0.005 max), and U (0.015 max). Source: Materion Brush Beryllium and Composites

Table 4.3 Guaranteed minimum mechanical properties for beryllium alloys

Grade	Tensile strength		Yield strength		Elongation, %	Microyield strength	
	MPa	ksi	MPa	ksi		kPa	ksi
I-70H	345	50	207	30	2.0	13.8	2.0
I-220H	448	65	345	50	1.0	55.2	8.0
O-30H	345	50	207	30	2.0	20.7	3.0
S-65	290	42	207	30	3.0	. . .	. . .
S-200F	324	47	241	35	2.0	. . .	. . .
S-200FH	414	60	296	43	3.0	. . .	. . .
SR-200	483	70	345	50	10.0	. . .	. . .

Source: Materion Brush Beryllium and Composites

Structural grades of beryllium are indicated by the prefix "S" in their designations. In general, these grades are produced to meet ductility and minimum strength requirements.

S-200F, an impact-ground powder grade, is the most commonly used grade of beryllium. This grade evolved from S-200E, which used attritioned powder as the input material. The properties of S-200F include a minimum ductility of 2% elongation in all directions within the vacuum hot-pressed billet, which is an increase of 1% over its predecessor, S-200E. In addition, S-200F has yield and ultimate strengths that are somewhat higher than those of S-200E, reflecting the general improvements in powder processing techniques and consolidation methods.

Among its applications, S-200F has been used successfully as an optical substrate and support bench in many astronomical telescopes, in fire control and forward-looking infrared systems, and in spacecraft and weather satellites. The material is most frequently used for parts machined from hot-pressed block. S-200F is a versatile material that is selected when weight and inertia factors exceed those of lower-cost aluminum. With its low mass, it can be driven through the scanning cycle much faster, with lower power requirements. S-200F contains a minimum 98.5% Be. In most applications, its optical surface is a hard, polished coating of electroless nickel, 24 to 150 µm thick. Electroless nickel is harder than bare beryllium and more easily polished to a fine surface finish.

Grade S-200FH has the suffix "H" to designate consolidation by HIP. S-200FH uses impact-ground powder that is consolidated in a sheet metal and can be formed into the shape of the final part. In production, the can is degassed, sealed, and hot isostatic pressed. This process conserves material and reduces the total machining time. S-200FH material is more isotropic and has higher density and higher mechanical properties than the traditional vacuum hot-pressed material. The material is useful for structural applications requiring low weight, high mechanical strength, and high fatigue endurance limit. A lightweight, high-stiffness material, S-200FH can significantly increase the operating speed of optical structures such as bar code readers, laser printers, and scanners.

S-200FC is the only beryllium grade currently available by CIP. This material uses impact-ground powders that are consolidated in a flexible rubber bag that approximate the final shape of the part. The powder is loaded into the bag and is degassed, sealed, and cold isostatic pressed at room temperature. The CIP process is useful for applications requiring lesser properties than those obtained by HIP. The CIP tooling is reusable, making it advantageous for parts required in the hundreds. Applications for S-200FC include optics for fire control systems (tanks and aircraft) and instrument applications (inertial measurement units).

SR-200 is a commercial grade of beryllium sheet produced by warm rolling billets of vacuum hot-pressed block. Cross-rolled beryllium sheet is selected by designers seeking high-performance heat sinks and structural supports in military electronic and avionics systems. SR-200 is also well suited in satellite structures as a hot-formed material. SR-200 has the best combination of strength and ductility.

Grade S-65 also is an impact-ground powder product. This grade was formulated to meet the damage tolerance requirements for use in the Space Shuttle. S-65 sacrifices some strength for improved ductility. The 3% minimum ductility requirement is achieved by using impact-ground powder in combination with tailored heat treatments. The heat treatments produce a desirable morphology of Fe-Al-Be-base precipitates. S-65H is a commercial beryllium grade that has the best high-temperature ductility of the conventional impact-ground beryllium grades. The material also has a lower metallic impurity level than all structural aluminum grades. The lower impurity content of S-65H makes it especially compatible with nuclear energy and fusion reactor applications.

Low levels of iron and aluminum are present in commercial grades of beryllium (Table 4.2). Although these elements cannot be eliminated economically, they can be balanced, and heat treatments can be applied to form discrete grain-boundary precipitates of $AlFeBe_4$. This minimizes the iron in solid solution or in the compound $FeBe_{11}$, either of which is embrittling. The precipitates also eliminate aluminum from the grain boundaries, thus precluding hot shortness at elevated temperatures.

At moderate temperatures, beryllium develops substantial ductility. At 800 °C (1470 °F), for example, the elongation of S-200F is in excess of 30%. Yield strength and ultimate strength decrease with increasing temperature, but usable strength and modulus are maintained up to approximately 600 to 650 °C (1100 to 1200 °F). The changes in strength and ductility of S-200F with temperature are shown in Fig. 4.6.

Instrument grades of beryllium, which are designated by the prefix "I," were developed to meet the specific needs of a variety of precision instruments. These instruments generally are used in inertial guidance systems where high geometrical precision and resistance to plastic deformation on a part per million scale are required. The resistance to deformation at this level is measured by microyield strength.

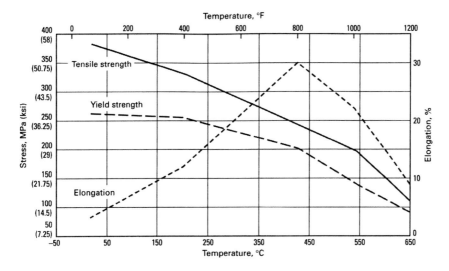

Fig. 4.6 Tensile properties of S-200F beryllium at elevated temperatures. Source: Ref 4.1

I-220H is an instrument-grade beryllium that has been developed for applications in which a high microyield is required. It is a high-strength, moderate-ductility grade. Consisting of a minimum 98% Be, I-220H is produced through the consolidation of beryllium powder by HIP.

I-220H is useful for structural, instrumental, and optical applications and those requiring high resistance to plastic deformation at low stress levels. I-220H is used as a telescope support on space exploration satellites. I-220H offers the best combination of high tensile strength, ductility, and microyield strength of any grade of beryllium. The microyield strength (amount of stress that produces one microinch of permanent strain) of I-220H is typically 97 MPa (14 ksi).

I-70H is an optical grade of beryllium that provides superior performance for bare polished beryllium optics. Containing 99% Be and 0.05% maximum beryllium oxide, it is the lowest-oxide beryllium ever marketed. I-70H is manufactured by controlling impact-ground powder in a hot isostatic process, yielding a more isotropic material than any previously available optical grade of beryllium.

O-30H is an optical grade of beryllium that is 99% Be and no more than 0.5% oxide. O-30H has the highest isotropy of thermal and mechanical properties of any grade of beryllium and is ideally suited for cryogenic applications. O-30H is produced through the consolidation of beryllium powder by HIP. O-30H is well suited for low-scatter optics, cryogenic optical substrates, high thermal isotropy, optical benches, and metering rods.

4.5　Wrought Products and Fabrication

Consolidated beryllium block can be rolled into plate, sheet, and foil, and it can be extruded into shapes or tubing at elevated temperatures. At present, working operations typically warm work the material, thereby avoiding recrystallization. The strength of warm-worked products increases significantly as the degree of working increases. The in-plane properties of sheet and plate (rolled from a PM material) increase as the gage decreases, as shown in Table 4.4. As with most hexagonal close-packed materials, beryllium develops substantial texture as a result of these working operations. The texture in sheet, for example, generally results in excellent in-plane strength and ductility, with almost no ductility in the short-transverse (out-of-plane) direction. Similar property trade-offs resulting from the anisotropy caused by warm working are inherent in extruded tube and rod. Sheet can be formed at moderate temperatures by standard forming methods. For some special applications, there has been interest in the improved formability of ingot-derived rolling stock as opposed to material rolled from a block of PM materials. The ingot-derived stock has improved weldability and formability, primarily because of its reduced oxide content.

In many instances, complex components must be built up from shapes made from sheet and foil. For example, beryllium honeycomb structures have been made by brazing formed sheet, as have complex satellite structural components. Beryllium tubing made by extrusion has also been used in satellite components. Space probes such as the Galileo Jupiter explorer have used extruded beryllium tubing to provide stiff, lightweight booms for precision antenna and solar array structures.

In the International Thermonuclear Experimental Reactor (ITER), power is released from the reaction between hydrogen and tritium (hydrogen isotope) atoms, resulting in the formation of a new atom—helium. This process gives an enormous thermal release; the plasma temperature in which nuclear fusion takes place is near to 100 °C million (by comparison, the sun's core temperature is 40 °C million). Establishing a controlled nuclear fusion reaction is the main task of the ITER project. For thermal shielding of the reactor core and withdrawing plasma heat power, ITER (Fig. 4.7) is equipped with internal shields, the so-called first wall. The blanket wall is a prefabricated structure consisting of water-cooling pan-

Table 4.4　In-plane properties of beryllium products rolled from a powder metallurgy source block

Thickness		0.2% yield strength		Ultimate strength		
mm	in.	MPa	ksi	MPa	ksi	Elongation, %
11–15	0.45–0.60	275	40	413	60	3
6–11	0.25–0.45	310	45	448	65	4

Source: Ref 4.1

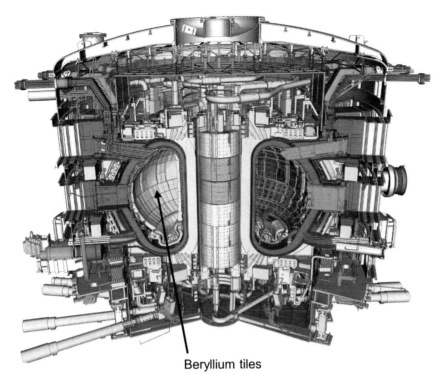

Beryllium tiles

Fig. 4.7 Beryllium is used in the blanket wall of the International Thermonuclear Experimental Reactor Organization fusion reactor. Source: Ref 4.3

els on fixed supporting stainless steel frames. These panels are three-layer composite structures of beryllium, a heat-removing interlayer made of special copper alloy, and a base water-cooled layer of special stainless steel.

Flat-Rolled Products. Sheet, plate, and foil are the most commonly produced wrought forms of beryllium. These flat-rolled products are produced by encasing beryllium rolling blocks in steel jackets, followed by cross rolling at a moderate temperature. Biaxial deformation is accomplished by alternating the direction of the rolling pack through the rolling mill. This procedure produces flat-rolled products with uniform biaxial properties. Typical tensile properties of beryllium sheet are shown in Table 4.5. The standard condition that sheet and plate are supplied in is hot rolled, stress relieved, ground, and pickled.

Extrusions. Input billets for extrusions are usually machined hot-pressed blocks. Unsintered pressed billets can also be extruded. The extrusion billets are jacketed in low-carbon steel cans with shaped nose plugs and are hot extruded through a steel die between 900 and 1065 °C (1650 and 1950 °F). Warm extrusions are produced at 400 to 450 °C (750 to 840 °F). Steel jackets are chemically removed from the beryllium. Beryllium powders

Table 4.5 Typical mechanical properties of different forms of beryllium alloys

Material	Test orientation (a)	Ultimate tensile strength		Yield strength(b)		Elongation (c), %
		MPa	ksi	MPa	ksi	
Block						
Normal-purity (hot-pressed) structural grade(d)	L	370	53	266	38	2.3
	T	390	56	273	39	2.6
Thermal or brake grade	L	294	42	196	28	2.7
	T	322	46	496	28	4.6
High-purity (isostatically pressed)	L	455	65	287	41	3.9
	T	455	65	287	41	4.4
High-oxide instrument grade (hot pressed)	L	476	68	406	58	1.5
	T	511	73	413	59	2.7
Fine grain size (isostatically pressed)	L	580	84	407	59	3.7
	T	587	85	407	59	4.2
Sheet (1–6.4 mm, or 0.040–0.250 in., thick)						
Normal-purity powder	. . .	531	77	372	54	16
Normal-purity ingot	. . .	352	51	172	25	7
Extrustions						
Normal-purity powder	. . .	655–690	95–100	345–518	50–75	8–13
High-purity powder	. . .	655–828	95–120	345–518	50–75	8–13
Forgings (no longer produced)						
Normal purity	. . .	483–600	70–87	435–600	63–87	0–4.5
Wire (0.05–0.64 mm, or 0.002–0.025 in., diameter) (no longer produced)						
High-purity ingot	. . .	966	140	793	115	3

(a) L, longitudinal; T, transverse. (b) 0.2% offset. (c) Elongation in 50 mm (2 in.), except 250 mm (10 in.) for wire. (d) Structural grade contains approximately 1.8% BeO, and thermal or brake grade approximately 0.9%. Source: Ref 4.2

can also be extruded. Warm extrusions are also produced bare using very low extrusion speeds, low reduction ratios, and special solid-film lubricants that strongly adhere to the billet. The main advantage is the ability to produce small, complex shapes with very good dimensional control.

Canning overcomes the strong tendency of beryllium to gall or stick to the extrusion tools. Furthermore, if not canned, abrasive oxides that score tooling would form at the elevated temperatures (above 600 °C, or 1110 °F). Also, because of the coarse-grained structure in the cast ingot and the differences in the crystallographic orientation of individual grains relative to the extrusion direction, a significant difference in deformation behavior may occur in different regions of the billet. Carbon steel is the preferred canning material.

Extrusion temperatures for beryllium range between 900 and 1065 °C (1650 and 1950 °F). Temperatures at the lower end of this range are preferred as higher strengths are then attainable, because the lower temperatures would produce finer grain sizes. Reduction ratios should be above 9:1 and not exceed 60:1. Low reduction ratios may not develop the required properties, while high reductions may lead to excessive alloying between the can and beryllium due to self-heating.

Large differences in mechanical properties are obtained between the directions perpendicular and parallel to the extrusion direction. Increased reduction ratios lead to increased degrees of preferred orientation of the basal plane being parallel to the extrusion direction. Although the longitudinal yield and ultimate strengths may increase significantly, they are only increased moderately or even may decrease in the transverse direction, depending on the degree of preferred orientation that is obtained. Elongations would increase in the longitudinal direction and decrease in the transverse direction.

4.6 Forming Beryllium

Beryllium has been successfully formed by most common sheet metal forming operations. The following are required:

- Equipment that can be controlled at slow speeds and that can withstand the use of heated dies
- Dies that can withstand the temperatures at which beryllium is commonly formed
- Facilities for preheating and controlling the temperature of dies and workpieces
- In some applications, facilities for stress relieving the work at 705 to 790 °C (1300 to 1450 °F)
- Special lubrication
- Safety precautions when grit blasting are required for cleaning after forming.

The formability of beryllium is low compared with that of most other metals. Beryllium has a hexagonal close-packed crystal structure; thus, there are relatively few slip planes, and plastic deformation is limited. For this reason, all beryllium products should be formed at elevated temperatures (generally 540 to 815 °C, or 1000 to 1500 °F) and at slow speeds.

Temperature, composition, strain rate, and previous fabrication history have marked effects on the results obtained in the forming of beryllium.

The effect of temperature on formability (in terms of bend angle at fracture) of two grades of powder sheet is shown in Fig. 4.8. Although these data show the effect of temperature on bendability, maximum strain on a 2*t* bend radius is not achieved at less than 90°. Therefore, it should not be assumed that the quantitative results shown in Fig. 4.8 can always be applied directly in practice. It should be noted that Fig. 4.8 was generated using beryllium sheet with a guaranteed elongation of only 5%. Current beryllium sheet products have guaranteed room-temperature elongations of 10%; typical values of 15 to 20% indicate that, if the test illustrated in Fig. 4.8 were repeated today, improvement in results would be significant. In one case, a 90° bend with a 2*t* radius was achieved in 0.5 mm (0.020 in.) thick beryllium sheet.

Effect of Composition. The oxide content of ingot and powder sheet has a significant effect on formability, as shown by the curves in Fig. 4.8. As the oxide content increases, yield strength increases and ductility decreases.

Effect of Strain Rate. Strain rate greatly influences the formability of beryllium. For instance, the stroke of a press brake is too fast for making sharp bends in hot beryllium. Slow bending, by means of equipment such as a hydraulic or air-operated press, is usually used. Minimum bend limits for the press-brake method and the slower-press method are compared in Fig. 4.9 for bending of cross-rolled powder sheet. In most forming applications, both the die and the workpiece must be preheated. Dies are specially constructed to permit heating; heat may be supplied by either electrical elements or gas burners. No specially prepared atmosphere is needed.

In the forming of thin sheet (less than ~1 mm, or 0.040 in., thick), cooling of the work between the furnace and the forming equipment is often a

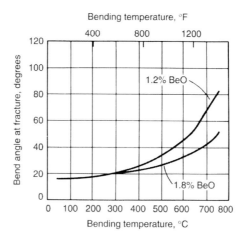

Fig. 4.8 Bend angle to fracture versus temperature of beryllium sheet using a 2*t* bend radius. Source: Ref 4.4

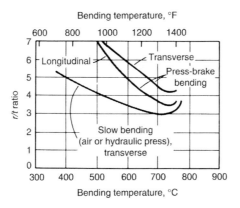

Fig. 4.9 Minimum bending limits for press-brake versus slower (hydraulic) bending of beryllium sheet in transverse and longitudinal directions. r, bend radius; t, sheet thickness. Source: Ref 4.4

problem. Overheating to compensate for this heat loss is not recommended. One satisfactory solution is to "sandwich" thin sheets of beryllium between two sheets of low-carbon steel. This sandwich is retained throughout heating and forming.

Stress relieving between stages of forming, or after forming is completed, is needed only in the forming of relatively thick sheet or in severe forming. For some finish-formed parts, stress relieving has proved an effective means of countering "oil canning" or excessive warpage. When stress relieving is used, regardless of whether it is an intermediate step or a final operation, holding at 705 to 760 °C (1300 to 1400 °F) for 30 min is recommended. No specially prepared atmosphere is needed.

If parts require cleaning after forming and if grit blasting is used, the wet method is recommended. Wet blasting minimizes the possibility that beryllium oxide dust will contaminate the surrounding atmosphere. Adequate ventilation must be provided if parts are processed by chemical etching after forming.

Deep Drawing. Lubrication is required to prevent galling between the beryllium workpiece and the die. A lubricant film must be maintained over that portion of the blank surface making contact with the drawing surfaces of the die throughout the entire draw. Because elevated temperatures (595 to 675 °C, or 1100 to 1250 °F, for the workpiece; 400 to 500 °C, or 750 to 930 °F, for the dies) are required to deep draw beryllium, conventional lubricants applied directly to the blank and die will burn off, causing galling between workpiece and die at high-pressure areas such as the draw ring. The solution to this problem is best achieved by using die materials that are self-lubricating, such as graphite or an overlay of colloidal suspension of graphite on an asbestos paper carrier. Blank development for deep drawing of beryllium generally follows the same rules as for other metals.

Blanks too thin to support themselves during the early stages of drawing will buckle or wrinkle. A restraining force is required to prevent this.

There are numerous factors involved in determining whether blank restraint is required during any drawing operation. The two most important are the ratio of blank diameter, d, to blank thickness, t, and the percentage of reduction from one draw to the next. The relationship between reduction, R, and d/t is shown in Fig. 4.10 for cylindrical parts, whether they are flat-bottomed or hemispherical cups. The areas under the curves were determined experimentally, with the curves themselves being the normal limit of formability for a given reduction at a given d/t ratio. The curves in Fig. 4.10 describe formability limits; therefore, some consideration should be given during design to avoid borderline cases. Reductions of more than 50% are possible but will require partial drawing followed by several anneals, usually with a high failure rate. A more practical approach involves several stages of tooling that require smaller reductions.

Strain rates during deep drawing of beryllium may vary widely, depending on the severity of the draw. For deep drawing simple hemispherical shells, punch speeds of 760 to 1270 mm/min (30 to 50 in./min) are commonly used. With optimal die clearance and lubrication, strain rates in excess of 2500 mm/min (100 in./min) have been observed in successful deep draws.

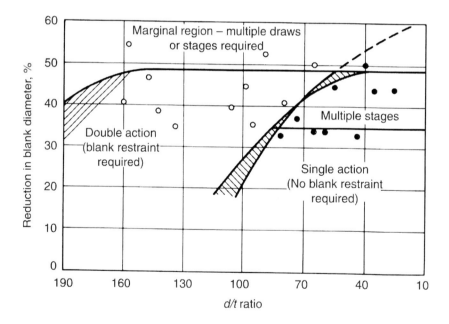

Fig. 4.10 Percent reduction in deep drawing versus diameter-to-thickness (d/t) ratio for deep drawing of cylindrical beryllium shells. Data points are experimental observations (double action or single action) used to derive the curve limits; d, blank diameter; t, blank thickness; shaded areas, marginal. Source: Ref 4.4

Numerous shapes have been deep drawn from beryllium. Considerable material savings may be achieved by deep drawing rather than machining thin-walled parts. The process lends itself to cup-shaped parts that have a slightly thicker wall at the equator than at the pole because of thickening in this area during forming.

Three-roll bending is a process for shaping smoothly contoured, large-radius parts by applying three-point bending forces progressively along the part surface. Usually, one or more of the forming rolls is driven. The process has been used to form curved panel sections and full cylinders from beryllium. As in all forming operations for beryllium, it is necessary to heat the blank to achieve the necessary ductility to avoid cracking.

Three-roll bending has been used to form precision beryllium cylinders. The cylinders were joined by an electron beam fusion weld and are round within 0.5 mm (0.02 in.) total indicator reading on the diameter.

Panels for the Agena spacecraft also have been formed to a 762 mm (30 in.) radius of curvature.

There were two sizes of panels formed, 635 by 635 mm (25 by 25 in.) and 559 by 355 mm (22 by 14 in.), at thicknesses of 1.4 and 1.88 mm (0.055 and 0.074 in.), respectively, from cross-rolled beryllium powder sheet. The flat beryllium sheet was heated to approximately 427 °C (800 °F), placed on a stainless steel sheet somewhat longer than the beryllium, and manually rolled to contour. The stainless steel sheet was used to "lead in" the beryllium and reduce the flat end inherent to roll forming. The rolled panels were stress relieved at 732 °C (1350 °F) for 20 min.

Stretch Forming. When stretch forming beryllium, the wrapping operation usually takes place quite slowly. Two commonly used types of tooling to stretch form beryllium are generally described as open-die and closed-die tooling. Open die, the most common, consists of a male die with the desired contour and some means of forcing the blank to assume that contour. Tension is not normally required for beryllium because the high modulus resists buckling and wrinkling. Closed-die tooling has male and female counterparts. The male die is used to force the blank into the female die, thereby causing the blank to assume the contour of the male die. This type of tooling lends itself well to beryllium forming because both portions of the die may be heated to facilitate maintenance of the heat necessary in the blank to avoid cracking. Friction forces on the female die can help to restrain the part and cause stretching.

Spinning. Beryllium sheets up to 5.1 mm (0.200 in.) thick have been successfully formed by spinning. For sheets less than approximately 1 mm (0.040 in.) thick, a common practice is to sandwich the beryllium between two 1.5 mm (0.060 in.) sheets of low-carbon steel and heat the sandwich to 620 °C (1150 °F) for spinning. The steel sheets not only help to maintain temperature but also help to prevent buckling. Beryllium sheets more than approximately 1 mm (0.040 in.) thick usually are not sandwiched between steel sheets for spinning and are heated to 730 to 815 °C (1350 to 1500 °F).

Hemispherical shapes have been spun in as many as nine stages with no adverse effect on the properties of the beryllium. The part and mandrel often are torch heated during spinning. Lubrication is especially important in spinning. Colloidal graphite or glass is usually used. Wet blasting is the recommended means of cleaning the workpiece after spinning.

Figure 4.11 plots combinations of conditions under which parts of a variety of shapes have been successfully produced by spinning cross-rolled beryllium powder sheet. The points plotted, however, represent only limited data, and many more points would have to be established before it would be safe to designate dimensional limitations for spinning specific shapes.

4.7 Beryllium Machining

Because of its high cost and toxicity potential, machining operations should be minimized where possible by the use of near-net shapes. For near-net shapes produced by PM, as little as 2.5 mm (0.10 in.) of material is all that must be removed by machining. Because the chips from machining operations are normally recycled, the majority of machining operations are conducted dry. In addition, due to the toxic nature of beryllium dust, the chips are collected by vacuum as close to the cutting edge as possible. Generally, the exhaust system is located within 13 to 19 mm (0.50 to 0.75 in.) of the cutting tool. Beryllium, being extremely brittle, must be securely clamped during machining operations to minimize any vibration or chatter. Carbide tools are most often used in machining beryllium. Finishing cuts are usually 50 to 130 μm (0.002 to 0.005 in.) in depth to minimize surface damage. Although most machining operations are performed without coolant to avoid chip contamination, the use of coolant can reduce the depth of damage and give longer tool life. The drilling of sheet can lead to

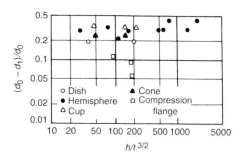

Fig. 4.11 Dimensional combinations for the successful spinning of beryllium sheet. d_0, blank diameter; d_1, diameter of spun part; h, height of spun part; t, blank thickness. Source: Ref 4.4

delamination and breakout unless the drill head is of the controlled torque type and the drills are carbide burr type.

Machining beryllium results in surface damage that extends from 25 to 500 μm (0.001 to 0.020 in.) deep, depending on the machining process. This surface damage consists of a dense network of intersecting twins and microcracks. Because surface damage adversely affects the mechanical properties, the machining process must be closely controlled; however, even with controlled machining, some surface damage will occur. To restore the mechanical properties, the surface damage must either be removed by chemical etching or "healed" by heat treatment. Chemical milling of beryllium usually removes approximately 50 to 100 μm (0.002 to 0.004 in.) from the surface, which is about the typical thickness of the damaged surface. Chemical milling can be conducted using sulfuric acid, nitric/hydrofluoric acid, and ammonium bifluoride. One disadvantage of chemical milling is that the removal is sometimes uneven and can affect precision tolerances. Vacuum heat treating at 800 °C (1470 °F) for 2 h will remove most of the machining damage and restore almost all of the mechanical properties. Heat treating removes the thermally induced twins and does not impact part tolerances. However, heat treating will not remove microcracks.

4.8 Beryllium Joining

Parts may be joined mechanically by riveting, but only by careful squeeze riveting to avoid damage to the beryllium, by bolting, threading, or by using press fittings specifically designed to avoid damage. Parts can also be joined by brazing, soldering, braze welding, adhesive bonding, and diffusion bonding. Fusion welding is not recommended. Brazing can be accomplished with zinc, aluminum-silicon, or silver-base filler metals. Many elements, including copper, can cause embrittlement when used as brazing filler metals. However, specific manufacturing techniques have been developed by various beryllium fabricators to use many of the common braze materials.

4.9 Aluminum-Beryllium Alloys

In addition to the beryllium alloys, there are a limited number of aluminum-beryllium alloys that are attractive in stiffness-driven designs. An example is the alloy AlBeMet AM 162, which contains 62% Be and 38% Al by weight. The alloy is made by PM followed by CIP to approximately 80% of its final density. The material is then rolled or extruded to its final shape and density. A flow chart of the processing is given in Fig. 4.12.

The mechanical properties are compared with high-strength aluminum and titanium in Table 4.6. With its low density and high elastic modulus,

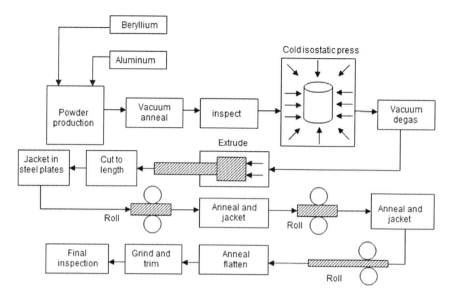

Fig. 4.12 AlBeMet AM 162 manufacturing flow. Source: Ref 4.2

Table 4.6 Property comparison for AlBeMet AM 162

Property	AlBeMet AM 162	7075-T73 Al	Ti-6Al-4V
Density, g/cm^3 (lb/in.^{3})	2.10 (0.076)	2.70 (0.098)	4.41 (0.160)
Ultimate tensile strength, MPa (ksi)	338 (49)	386 (56)	896 (130)
Yield strength, MPa (ksi)	283 (41)	303 (44)	827 (120)
Elongation, %	10	7	10
Modulus of elasticity, MPa (10^6 psi)	200 (29)	68.9 (10)	110 (16)
Fatigue endurance limit, MPa (ksi)	165 (24)	159 (23)	552 (80)
Fracture toughness, MPa · m$^{1/2}$ (ksi · in.$^{1/2}$)	24 (22)	47 (42)	669 (97)
Coefficient of thermal expansion, μm/m · °C (μin./in. · °F)	13.9 (7.7)	23.4 (13)	8.82 (4.9)
Thermal conductivity, W/m · K (Btu/h · ft · °F)	241.9 (139.0)	163.2 (93.8)	7.31 (4.2)

Note: Extrusions; transverse properties. Source: Ref 4.5

AlBeMet AM 162 would be 3.9 times stiffer than 7075-T73 Al and Ti-6Al-4V in a tension tie application. The disadvantages of AlBeMet AM 162 are the limitations on extrusion and rolled sheet sizes available and high cost (approximately 400 times the cost of aluminum). While this alloy is somewhat more difficult to machine than aluminum (20% lower speeds and feeds), it is easier to machine than titanium. Like all beryllium-containing alloys, the machining process must be controlled due to the hazardous nature of beryllium.

4.10 Beryllium Corrosion

The corrosion of beryllium is not a major problem. Most applications of beryllium subject the part to rather benign environments, such as the near vacuum of outer space for gyro systems, mirror reentry vehicle structures, and space platform structural parts.

Corrosion of Beryllium in Air. Beryllium is commonly referred to as being self-protective against atmospheric oxidation; it resembles aluminum and titanium in this respect. The protective nature of its oxide is partially due to the fact that temperature poses no reported corrosion problems for beryllium. On the other hand, air that contains water at a level at which condensation can occur with temperature cycling can cause beryllium to corrode under certain conditions. There are two primary situations in which beryllium has been found to be susceptible to corrosion in a water-bearing air atmosphere. One cause of corrosion in air is when the beryllium contains beryllium carbide (Be_2C) particles exposed at its surface by a machining operation. The water available from the surrounding air will slowly react with the carbides to form a flowery white buildup of beryllium oxide (BeO). This type of corrosion is commonly manifested by the bursting out of the beryllium at the edges of the carbide particles or by the formation of blisters as a result of the localized volume expansion during reaction. White corrosion products and blisters were observed to form on the surface of extruded beryllium that had been exposed for approximately six months to the local humid air. The corrosion products were caused by carbide inclusions that were parallel to the extrusion axis. Similar corrosion attack was discovered during the shelf storage of finished precision-machined components. The components were scrapped because of the disruption of the finely machined, unetched surfaces.

The second major cause of corrosion in a humid air environment occurs when the beryllium surface contains contaminants in the form of chlorides and sulfates that remain on the beryllium after a final drying operation or are introduced after proper rinsing and drying. The use of tap water or a chloride-contaminated water bath without adequate cascade rinsing in deionized water prior to drying has been reported as a cause of corrosion during subsequent processing or storage. The residual surface contaminants become concentrated upon drying and require very little water or condensation derived from temperature and humidity changes to provide a condition for corrosion at active sites.

Fingerprinting of clean, dried beryllium components is one of the most commonly encountered sources of surface contamination that can result in subsequent etching and corrosion of beryllium. The corrosive nature of fingerprints is apparent from the need to abrade and renew the affected surface mechanically before most chemical treatments. All producers and users of beryllium strongly specify the avoidance of hand contact with semifinished and finished beryllium components.

The final cleaning treatment can be very important, particularly in view of the present-day trend of avoiding the chlorinated solvents that had often been used to remove oily surface contamination. Solvent and/or chemical cleaning is extremely important when the final machined surface will not be given a final acid etching treatment to remove mechanically produced surface imperfections.

An acid etching treatment is commonly used on machined beryllium components to remove machining-damaged surface material. The material removed by etching may contain microcracks, which are open defects into which contaminants can penetrate and cause corrosion. Precision components that cannot be acid etched but are lapped to final size are susceptible to corrosion as a result of the unremoved microcracks. In the removal of only a few tenths of a mil of stock by lapping with an improper lapping agent, the microcracks can be smeared over with adjacent material. This results in entrapment of contaminants and corrosion of the component at a later time. A free-cutting diamond paste is recommended as the lapping agent.

Contamination from the Atmosphere. Rainwater has been recognized as a potential source of corrosive agents that can be inadvertently transported into a protective environment and cause deterioration of susceptible materials. Atmospheric moisture in silos and launch control centers of missile systems has been cited as the cause of most of the general corrosion problems. Despite the fact that corrosion problems have been encountered in missile silos, the maintenance of a humidity-controlled environment has generally been effective in protecting most components from corrosion. For example, one manufacturer has not found a single case of corrosion of beryllium components in rocket thrusters during periodic examination of all operational units through years of continued service. These components were maintained under humidity-controlled conditions.

The open atmosphere above the ocean provides a site of unusually high salt content. Storms at sea and the mist generated by a moving vessel have been found to produce severe corrosion of beryllium brake shoe pads on military carrier bombers. The brake shoes were typical of those that have performed well for years on the C5A military transport in less corrosive environments. The salt-laden environment of the aircraft carrier caused the brake pads to pit and to become covered with corrosion products.

Aqueous Corrosion of Beryllium. Beryllium that is clean and free of surface impurities has very good resistance to attack in low-temperature high-purity water. The corrosion rate in good-quality water is typically less than 0.025 mm/yr (1 mil/yr). Beryllium has been reported to perform without problems for 10 years in slightly acidified demineralized water in a nuclear test reactor. This high-purity water environment also produced no evidence of accelerated corrosion when the beryllium was galvanically coupled with stainless steel or aluminum. However, beryllium of normal commercial purity is susceptible to attack, primarily in the form of localized pitting, when exposed to impure water. Chloride and sulfate ions (SO^{2-}_4) are the most critical contaminants in aqueous corrosion. Because these contaminants are present in tap water, most processing specifications warn against its use. In cases where tap water can be used, the final rinses are done in deionized water to ensure that the Cl^- or SO^{2-}_4 impurities are not retained on the subsequently dried metal surface. Electron microprobe analysis showed that the localized pitting occurred at sites rich in either silicon or aluminum.

On the basis of the examination of pitted specimen surfaces by microprobe analysis, the following general conclusions were drawn:

- Pitting corrosion is determined by the distribution of alloy impurities.
- Sites that have high concentrations of iron, aluminum, or silicon (probably in solid solution) tend to form corrosion pits.
- The corrosion sites are often characterized by the presence of alloy-rich particles, which are probably beryllides; however, the particles themselves do not appear to be corrosion-active.
- Pitting density is believed to be related to the number of segregated regions in the matrix containing concentrations of iron, aluminum, or silicon higher than concentrations present in the surrounding areas.

Segregated impurities, such as Be_2C particles, intermetallic compounds, or alloy-rich zones, can contribute to localized attack when present in an exposed beryllium surface.

Stress corrosion has been reported on cross-rolled sheet when synthetic seawater was used as the test medium. Studies of time to failure versus applied stress revealed that the time decreased from 2340 to 40 h as the applied stress was increased from 8.4 to 276 MPa (1220 to 40,000 psi, or 70% of yield strength). Failure appeared to be closely associated with random pitting attack. Certain pits appeared to remain active, which promoted severe localized attack.

In-Process Corrosion Problems. Many of the specified practices used by manufacturers and users result from knowledge of the problems experienced in processing and handling of components. One example of this is the common specification to avoid contact of beryllium by tap water. Early processors encountered serious problems of pitting or surface corrosion in beryllium that was allowed to stand in tap water or was rinsed off with tap water and, after drying, was allowed to stand in a damp atmosphere where water condensation could easily occur on the contaminated surfaces. Improper handling, inadequate control of chemical and rinse baths, and improper storage or packaging procedures are the primary causes of in-process corrosion problems.

Chloride ions from sources other than tap water can also lead to pitting attack in aqueous baths. Chlorinated solvents used to remove oils and greases have been found to be sources of Cl⁻ ions when the solvent is carried into a cleaning bath by an as-machined surface.

Fingerprinting of in-process parts has been a commonly cited problem. Under certain atmospheric conditions, the salt in a fingerprint can seriously etch a beryllium surface. Such etching of a finished, critical-sized component can cause rejection of the component. Etching of a semifinished surface, if not properly conducted, can lead to localized corrosion of a chemically treated surface.

Dyes in penetrant solutions for nondestructive testing, if merely wiped off a surface, can later seep out of crevices and cause localized attack. A second example is when localized galvanic attack is found on beryllium. Contact with apparently dry materials has also been found to be the cause of corrosion of beryllium. In one case, beryllium became etched when it was placed on an inspection bench on which a rubberized mesh was used to prevent nicking and scratching of finished parts. In another instance, a molded Styrofoam (Dow Chemical Company) support used for shipment of a complex satellite boom caused severe localized etching of the beryllium at points of pressure contact. The beryllium part was not shipped in direct contact with the Styrofoam, but the part had been laid on the foam after being removed from its polyethylene bag for inspection. Attack was caused by chemical compounds contained in the contacting materials.

Handling and Storage Procedures. In handling and packaging beryllium components for shipment or storage, it is important to maintain an initially clean surface and to avoid unduly moist environments. The various practices employed by industry are discussed as follows.

Handling of beryllium parts with bare hands is often done and is not harmful when the part is in a rough or semifinished stage. In general, a finish-machined surface, even if it is to be acid etched to remove machine-damaged surface material, should not be handled with bare hands. A fingerprint left on the surface for some time, during which it may etch the beryllium, may retard the beneficial aspects of the acid etching treatment or at least may provide a questionable artifact after etching. Fingerprints that have etched the beryllium adversely affect conversion or passivation coatings, disrupting the continuity of the coating and thus its effectiveness. The common practice in handling finished and semifinished parts (or when handprinting is unacceptable) is to use protective gloves. White cotton, nylon, polyethylene, or rubber gloves are usually used.

Packaging for Shipment. Polyethylene bags are extensively used by industry as protective containment and/or barriers to contact between beryllium and its surroundings. A permeable container with a desiccant is often used inside the polyethylene bag to remove undesirable moisture. An argon gas purge of the interior of the bag has been used as a means of displacing the moisture initially present. Because of the desire to minimize the availability of moisture to the beryllium, an attempt is generally made to provide an effective seal at the bag opening. An integrally molded interlocking seal or the practice of overlapping and taping provides an effective seal against free moisture access to the interior.

Molded foam, formed Styrofoam, and contoured fiber-type insulation materials are typical means of supporting and protecting large beryllium structures or components against mechanical damage inside a shipping container. In some cases, when support is desired in all directions, a molded upper closure is used to encase the structure totally with supporting material. In almost every case, however, a polyethylene bag provides an effective separator between the beryllium and the contacting support material.

Storage of beryllium components while awaiting assembly or use can be safely accomplished under the conditions used for shipment, that is, in a compatible container made of metal or polyethylene with a desiccant to maintain a dry storage atmosphere. Heating a cabinet with a lightbulb to keep the interior approximately 50 °C (120 °F) has been found to be effective for long-term storage of gyro components. A room or chamber with humidity control, as well as a humidity-controlled missile silo, can be effective in preventing corrosion of beryllium rocket motor components during storage periods of several years.

The following points should be considered in safely storing components for an extended period of time:

- Ensure that the component entering storage is free of corrosion-causing contaminants.
- Maintain a moderately dry and noncondensing environmental atmosphere.
- Ensure against contact with materials of nonproven compatibility.

4.11 Corrosion-Protection Surface Treatments and Coatings

In addition to the treatment designed to provide normal storage and handling protection, many of the coatings and treatments developed for more severe applications are useful for less demanding applications. For example, those treatments designed to provide protection against seawater are effective in protecting against corrosion of beryllium during processing or use in air environments.

Chromate-type coatings are applied by a simple dip treatment. The coatings thus produced provide reasonable protection for beryllium during handling and storage and against attack by salt-containing environments for moderately long time periods. A passivation treatment consisting of a 30 min dip in a solution consisting of 25% H_3PO_4, 25% saturated solution of potassium chromate (K_2CrO_4), and 50% deionized water reduces the susceptibility to white spot formation in moist air. Commercial manufacturers of beryllium components are often required by drawing specifications to provide passivation-type coatings on finished parts. These coatings are effective in providing protection during shipment and short periods of storage under less-than-ideal conditions.

Fluoride coatings are produced by treating beryllium in fluorine above 520 °C (970 °F), which produces a glassy-appearing water-insoluble coating. This type of coating has been shown to be very effective in resisting corrosion in chloride-containing water and in distilled water. Coatings 0.2 and 1.2 μm (0.008 and 0.05 mil) thick were unchanged after 3000 h in distilled water at room temperature. A 0.2 μm (0.008 mil) thick coating provided effective protection in water containing a 150 ppm concentration of Cl^- ion.

Anodic coatings that are similar in character to the well-known anodizing on aluminum have been shown to improve the resistance of beryllium to corrosion in normally corrosive aqueous anodizing solution compositions that provide differing qualities and results in various test environments. Nitric acid (HNO_3), H_2CrO_4, sodium dichromate ($Na_2Cr_2O_7$), and sodium hydroxide (NaOH) are common ingredients in many anodizing bath formations.

Anodized films vary in thickness from 2.5 μm (0.1 mil) to several mils, depending on solution type, applied voltage, current density, bath temperature, and time of treatment. The film qualities vary in accordance with descriptive terms, such as color, thickness, adherence uniformity, density, and electrical characteristics. It was concluded from a survey of many anodizing procedures that the following simple formulations produced the most uniform and adherent coatings:

- Solution of 50% HNO_3 with a current density of 2.15 A/m^2 (0.20 A/ft^2) for 5 min
- Solution of 7.5% NaOH with a current density of 108 A/m^2 (10 A/ft^2) for 20 min

Anodic coatings were found to protect beryllium effectively for the following indicated times in the following environments:

- 2400 h in a humidity cabinet
- 3 months at 40 °C (100 °F) in tap water
- 2000 h in ASTM International salt spray test

Organic paint coatings are used when an electrically insulating barrier is required between the beryllium component and another metallic structure. Coatings used for this purpose provide protection against galvanic attack in the event of an electrolyte gaining access to the joint. An epoxy primer has been found to be effective for general protection. The primer is used on all exposed surfaces. The primer is applied immediately after a flash etching of the surfaces to be coated and is cured by baking.

4.12 Health and Safety

The main concern associated with the handling of beryllium is the effect on the lungs when excessive amounts of respirable beryllium powder or dust are inhaled. Two forms of lung disease are associated with beryllium: acute berylliosis and chronic berylliosis. The acute form, which can have an abrupt onset, resembles pneumonia or bronchitis. Acute berylliosis is now rare because of the improved protective measures that have been enacted to reduce exposure levels.

Chronic berylliosis has a very slow onset. It still occurs in industry and seems to result from the allergic reaction of an individual to beryllium. At present, there is no way of predetermining those who may be hypersensitive. Sensitive individuals exposed to airborne beryllium may develop the lung condition associated with chronic berylliosis.

The current Occupational Safety and Health Administration (OSHA) permissible exposure limits for beryllium are 2 $\mu g/m^3$ as an 8 h time-weighted average, 5 $\mu g/m^3$ as a ceiling not to be exceeded for more than 30 min at a time, and 25 $\mu g/m^3$ as a peak exposure never to be exceeded. The American Conference of Governmental Industrial Hygienists current threshold limit value for beryllium is 0.05 $\mu g/m^3$ averaged over an 8 h work shift. Facilities and personnel working with beryllium should comply with the guidelines specified in the most recent OSHA safety and health bulletin.

ACKNOWLEDGMENTS

Portions of the text for this chapter came from "Beryllium" by A.J. Stonehouse and J.M. Marder in *Properties and Selection: Nonferrous Alloys and Special-Purpose Materials,* Vol 2, *ASM Handbook,* ASM International, 1990; "Powder Metallurgy Beryllium" by J.M. Marder in *Powder Metal Technologies and Applications,* Vol 7, *ASM Handbook,* ASM International, 1998; "Forming of Beryllium" in *Metalworking: Sheet Forming,* Vol 14B, *ASM Handbook,* ASM International, 2006; and "Corrosion of Beryllium" by J.J. Mueller and D.R. Adolphson in *Corrosion,* Vol 13, *ASM Handbook,* ASM International, 1987.

REFERENCES

4.1 A.J. Stonehouse and J.M. Marder, Beryllium, *Properties and Selection: Nonferrous Alloys and Special-Purpose Materials,* Vol 2, *ASM Handbook,* ASM International, 1990

4.2 J.M. Marder, Powder Metallurgy Beryllium, *Powder Metal Technologies and Applications,* Vol 7, *ASM Handbook,* ASM International, 1998

4.3 International Thermonuclear Experimental Reactor Organization, http://www.iter.org/

4.4 Forming of Beryllium, *Metalworking: Sheet Forming,* Vol 14B, *ASM Handbook,* ASM International, 2006

4.5 W. Speer and O.S. Es-Said, Applications of an Aluminum-Beryllium Composite for Structural Aerospace Components, *Eng. Fail. Anal.,* Vol 11, 2004, p 895–902

SELECTED REFERENCES

- R.N. Caron and J.T. Staley, Beryllium: Effects of Composition, Processing, and Structure on Properties of Nonferrous Alloys, *Materials Selection and Design,* Vol 20, *ASM Handbook,* ASM International, 1997

- D.V. Gallagher and R.E. Hardesty, Machining of Beryllium, *Machining,* Vol 16, *ASM Handbook,* ASM International, 1989

- A. Goldberg, "Beryllium Manufacturing Processes," UCRL-TR-222539, Lawrence Livermore National Laboratory, 2006

- L. Grant, Forming of Beryllium, *Forming and Forging,* Vol 14, *ASM Handbook,* ASM International, 1988

- J.J. Mueller and D.R. Adolphson, Corrosion of Beryllium, *Corrosion,* Vol 13, *ASM Handbook,* ASM International, 1987

CHAPTER **5**

Titanium Alloys

TITANIUM is a lightweight metal (approximately 60% of the density of iron) that can be highly strengthened by alloying and deformation processing. Titanium is nonmagnetic and has good heat-transfer properties. Its coefficient of thermal expansion is somewhat lower than that of steels and less than half that of aluminum. Titanium and its alloys have melting points higher than those of steels, but maximum useful temperatures for structural applications generally range from 425 to 595 °C (800 to 1100 °F). Titanium has the ability to passivate and thereby exhibit a high degree of immunity to attack by most mineral acids and chlorides. Titanium is nontoxic and generally biologically compatible with human tissues and bones. The combination of high strength, stiffness, good toughness, low density, and good corrosion resistance provided by various titanium alloys at very low-to-elevated temperatures allows weight savings in aerospace structures and other high-performance applications. The excellent corrosion resistance and biocompatibility coupled with good strength make titanium and its alloys useful in chemical and petrochemical applications, marine environments, and biomaterial applications.

The primary reasons for using titanium alloys include:

- *Weight savings.* The high strength-to-weight ratio of titanium alloys allows them to replace steel in many applications requiring high strength and fracture toughness. With a density of 4.4 g/cm^3 (0.16 lb/in.^{3}), titanium alloys are only approximately half as heavy as steel and nickel-base superalloys, yielding excellent strength-to-weight ratios.
- *Fatigue strength.* Titanium alloys have much better fatigue strength than aluminum alloys and are frequently used for highly loaded bulkheads and frames in fighter aircraft.
- *Operating temperature capability.* When the operating temperature exceeds approximately 130 °C (270 °F), aluminum alloys lose too much strength, and titanium alloys are often required.

- *Corrosion resistance.* The corrosion resistance of titanium alloys is superior to both aluminum and steel alloys.
- *Space savings.* Titanium alloys are used for landing gear components on commercial aircraft where the size of aluminum components would not fit within the landing gear space envelope.

Due to their outstanding resistance to fatigue, high-temperature capability, and resistance to corrosion, titanium alloys comprise approximately 42% of the structural weight of the F-22 fighter aircraft, over 4080 kg (9000 lb) in all. In commercial aircraft, the older Boeing 747-100 contained only about 2.6% Ti, while the newer Boeing 777 contains approximately 8.3%. In commercial passenger aircraft engines, the fan, the low-pressure compressor, and approximately $^2/_3$ of the high-pressure compressor are made from titanium alloys.

5.1 Titanium Applications

Titanium and its alloys are used primarily in two areas where the unique characteristics of these metals justify their selection: corrosion-resistant service and strength-efficient structures. For these two diverse areas, selection criteria differ markedly. Corrosion applications normally use low-strength unalloyed titanium mill products fabricated into tanks, heat exchangers, or reactor vessels for chemical processing, desalination, or power generation plants. In contrast, high-performance applications typically use high-strength titanium alloys in a very selective manner, depending on factors such as thermal environment, loading parameters, available product forms, fabrication characteristics, and inspection and/ or reliability requirements. For example, higher-strength alloys such as Ti-6Al-4V and Ti-3Al-8V-6Cr-4Mo-4Zr are used for offshore drilling applications and geothermal piping. As a result of their specialized usage, alloys for high-performance applications normally are processed to more stringent and costly requirements than unalloyed titanium for corrosion service.

Historically, titanium alloys have been used instead of iron or nickel alloys in aerospace applications because titanium saves weight in highly loaded components that operate at low to moderately elevated temperatures. Many titanium alloys have been custom designed to have optimal tensile, compressive, and/or creep strength at selected temperatures and, at the same time, to have sufficient workability to be fabricated into mill products suitable for specific applications. During the life of the titanium industry, various compositions have had transient usage, but one alloy, Ti-6Al-4V, has been consistently responsible for approximately 45% of industry application. Ti-6Al-4V is unique in that it combines attractive properties with inherent workability, which allows it to be produced in all types of mill products, in both large and small sizes; good shop fabricability, which allows the mill products to be made into complex hardware; and the production experience and commercial availability that lead to reliable and economic usage. Thus, Ti-6Al-4V has become the standard alloy against which other alloys must be compared when selecting a titanium alloy for a specific application. Ti-6Al-4V also is the standard alloy selected for castings that must exhibit superior strength. For elevated-temperature applications, the most commonly used alloy is Ti-6Al-2Sn-4Zr-2Mo + Si. This alloy is primarily used for turbine components and in sheet form for afterburner structures and various hot airframe applications.

Gas Turbine Engine Components. Rotating components in gas turbine engines require titanium alloys that maximize strength efficiency and metallurgical stability at elevated temperatures. These alloys also must exhibit low creep rates along with predictable behavior in stress rupture and low-cycle fatigue. To reproducibly provide these properties, stringent user requirements are specified to ensure controlled homogeneous microstructures and total freedom from melting imperfections such as alpha segregation, high- or low-density tramp inclusions, and unhealed ingot porosity or pipe. Currently, Ti-6Al-4V, Ti-6Al-2Sn-4Zr-2Mo, Ti-5Al-2Sn-2Zr-4Mo-4Cr, and Ti-6Al-2Sn-4Zr-6Mo are some alloys used in gas turbine engine applications.

Aerospace pressure vessels similarly require optimized strength efficiency in addition to auxiliary properties such as weldability and predictable fracture toughness at cryogenic to moderately elevated temperatures. To provide this combination of properties, stringent user specifications require controlled microstructures and freedom from melting imperfections. For cryogenic applications, the interstitial elements oxygen, nitrogen, and carbon are carefully controlled (extra low interstitial, or ELI) to improve ductility and fracture toughness. For these applications, the basic titanium alloy Ti-6Al-4V (or Ti-6Al-4V-ELI), processed in the annealed or the solution treated and aged (STA) condition, is widely used. Ti-5Al-2.5Sn-ELI is an attractive alternative.

Applications Requiring Corrosion Resistance and High Strength. Aircraft structural applications as well as high-performance automotive and marine applications require corrosion resistance and high-strength efficiency. The latter normally is achieved by judicious alloy selection combined with close control of mill processing; corrosion resistance is primarily a function of composition. Ti-6Al-4V and Ti-6Al-2Sn-4Zr-2Mo are two of the alloys currently used for high-performance applications. Beta-titanium alloys are used extensively for spring applications. Titanium alloys are ideal spring materials because of the lower modulus and lower density relative to steels. They can provide weight savings of up to 70% and volume reductions of 50%.

Titanium alloys have been used in submarine hulls and submersible research vehicles. Because of favorable heat-transfer behavior, thermal stability, and corrosion resistance, titanium is used in heat exchangers for power generation (Fig. 5.1). Corrosion resistance also is a factor in the selection of titanium for use in outer containers for high-level solidified nuclear waste.

Economic considerations normally determine whether titanium alloys can be used for corrosion service. Capital expenditures for titanium equipment generally are higher than for equipment fabricated from competing materials, such as stainless steel, brass, bronze, copper-nickel, or carbon steel. As a result, titanium equipment must yield lower operating costs, longer life, or reduced maintenance to justify selection based on lower total life-cycle cost. Commercially pure titanium satisfies the basic requirements for corrosion service.

Optic-system support structures are a little-known but very important structural application for titanium. Complex castings are used in surveillance and guidance systems for aircraft and missiles to support the optics where wide temperature variations are encountered in service. The chief reason for selecting titanium for this application is that the thermal expansion coefficient of titanium most closely matches that of the optics.

Prosthetic Devices. Because of its unique corrosion behavior, titanium is used extensively in prosthetic devices such as artificial heart pumps,

Fig. 5.1 Large titanium heat exchanger. Source: Ref 5.1

pacemaker cases, heart valve parts, and load-bearing bone or hip joint replacements or bone splints. In general, body fluids are chloride brines that have pH values from 7.4 into the acidic range and also contain a variety of organic acids and other components—media to which titanium is totally immune.

Consumer goods have become another significant portion of the titanium market. Titanium golf club heads for drivers occupy the major share of this market.

5.2 Titanium Physical Metallurgy

Titanium exists in two crystalline forms, as shown in Fig. 5.2: the one stable at room temperature is called alpha (α) and has a hexagonal close-packed (hcp) structure, while the one that is stable at elevated temperature is body-centered cubic (bcc) and is called beta (β). In pure titanium, the alpha phase is stable up to 880 °C (1620 °F), where it transforms to the beta phase; the transition temperature is known as the beta transus. The beta phase is then stable from 880 °C (1620 °F) to the melting point of 1720 °C (3130 °F).

At room temperature, commercially pure titanium is composed primarily of the alpha phase. As alloying elements are added to titanium, they tend to change the amount of each phase present and the beta tran-

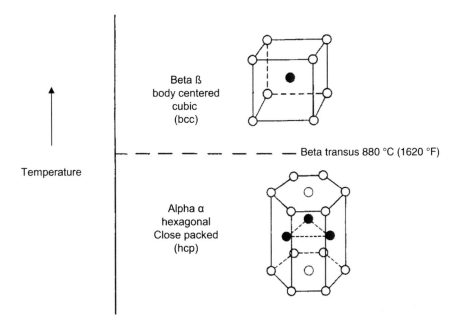

Fig. 5.2 Crystal structures of pure titanium

sus temperature in the manner shown in Fig. 5.3. Those elements that increase the beta transus temperature through stabilizing the alpha phase are called alpha stabilizers and include aluminum, oxygen, nitrogen, and carbon. Those elements that decrease the beta transus temperature are called beta stabilizers. The beta stabilizers are further subdivided into beta isomorphous elements, which have a high solubility in titanium, and beta eutectoid elements, which have only limited solubility and tend to form intermetallic compounds. The beta isomorphous elements are molybdenum, vanadium, niobium, and tantalum, while beta eutectoid elements include manganese, chromium, silicon, iron, cobalt, nickel, and copper. Finally, tin and zirconium are considered neutral because they neither raise nor lower the beta transus temperature, but since both contribute to solid-solution strengthening, they are frequently used as alloy additions.

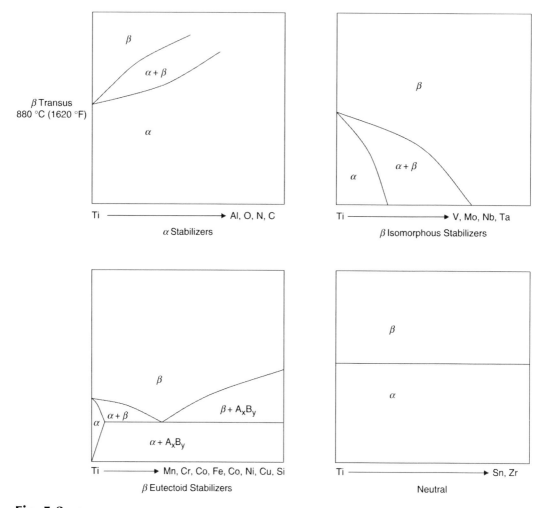

Fig. 5.3 Phase diagrams for binary titanium alloys. Source: Ref 5.2

Titanium alloys are classified according to the amount of alpha and beta retained in their structures at room temperature. Classifications include commercially pure, alpha and near-alpha, alpha-beta, and metastable beta. The commercially pure and alpha alloys have essentially all-alpha microstructures. Beta alloys have largely all-beta microstructures after air cooling from the solution-treating temperature above the beta transus. Alpha-beta alloys contain a mixture of alpha and beta phases at room temperature. Within the alpha-beta class, an alloy that contains much more alpha than beta is often called a near-alpha alloy. The names superalpha and lean-beta alpha are also used for this type of alpha-beta alloy. While these classifications are useful, many of them are actually very close to each other in the total amount of beta stabilizer present, as illustrated in the Fig. 5.4 phase diagram. For example, Ti-6Al-4V is classified as an alpha-

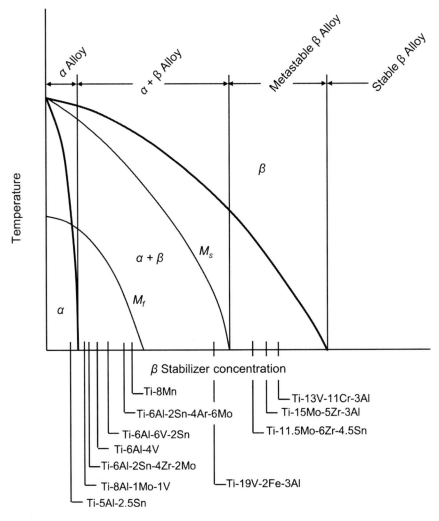

Fig. 5.4 Pseudo-binary titanium phase diagram. Source: Ref 5.3

beta alloy, and Ti-6Al-2Sn-4Zr-2Mo is classified as a near-alpha alloy, yet they differ little in the total amount of beta stabilizer concentration.

Alpha and near-alpha alloys usually contain 5 to 6% Al as the main alloying element, with additions of the neutral elements tin and zirconium and some beta stabilizers. Because these alloys retain their strength at elevated temperatures and have the best creep resistance of the titanium alloys, they are often specified for high-temperature applications. The addition of silicon, in particular, improves the creep resistance by precipitating fine silicides, which hinders dislocation climb. Aluminum also contributes to oxidation resistance. These alloys also perform well in cryogenic applications. Due to the limitation of slip systems in the hcp structure, the alpha phase is less ductile and more difficult to deform than the bcc beta phase. Because these are single-phase alloys containing only alpha, they cannot be strengthened by heat treatment.

The alpha-beta alloys have the best balance of mechanical properties and are the most widely used. In fact, one alpha-beta alloy, Ti-6Al-4V, is by far the most widely used alloy. In this alloy, aluminum stabilizes the alpha phase while vanadium stabilizes the beta phase; therefore, alpha-beta alloys contain both alpha and beta phases at room temperature. In contrast to the alpha and near-alpha alloys, the alpha-beta alloys can be heat treated to higher strength levels, although their heat treat response is not as great as that for the beta alloys. In general, the alpha-beta alloys have good strength at room temperature and for short times at elevated temperatures, although they are not noted for their creep resistance. The weldability of many of these alloys is poor due to their two-phase microstructures.

The metastable beta alloys contain a sufficient amount of beta stabilizing elements that the beta phase is retained to room temperature. Because the bcc beta phase exhibits much more deformation capability than the hcp alpha phase, these alloys exhibit much better formability than the alpha or alpha-beta alloys. Where the alpha and alpha-beta alloys would require hot forming operations, some of the beta alloys can be formed at room temperature. The beta alloys can be STA to higher strength levels than the alpha-beta alloys while still retaining sufficient toughness. The biggest drawbacks of the beta alloys are increased densities due to alloying elements such as molybdenum, vanadium, and niobium; reduced ductility when heat treated to peak strength levels; and some have limited weldability.

Titanium has a great affinity for interstitial elements, such as oxygen and nitrogen, and readily absorbs them at elevated temperatures, which increases strength and reduces ductility (Fig. 5.5). Oxygen tends to increase the strength and decrease the ductility. As the amount of oxygen and nitrogen increases, the yield and ultimate strengths increase and the ductility and fracture toughness decrease. Titanium absorbs oxygen at temperatures above 704 °C (1300 °F), which complicates the processing and

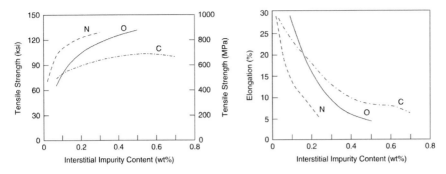

Fig. 5.5 Effects of interstitial content on strength and ductility of titanium. Source: Ref 5.4

increases the cost, because many hot working operations are conducted at temperatures exceeding 704 °C (1300 °F). However, oxygen, in controlled amounts, is actually used to strengthen the commercially pure (CP) grades. Some alloys are available in an ELI grade that is used for applications requiring maximum ductility and fracture toughness. Hydrogen is always minimized in titanium alloys because it causes hydrogen embrittlement by the precipitation of hydrides, so the maximum limit allowed is approximately 0.015 wt% (~100 ppm). When the solubility limit of hydrogen in titanium (~100 to 150 ppm for CP titanium) is exceeded, hydrides begin to precipitate. Absorption of several hundred ppm of hydrogen results in embrittlement (Fig. 5.6) and the possibility of stress cracking. Note that the addition of 20 ppm does not cause embrittlement, but when the hydrogen content goes up to 250 ppm, the reduction in area is seriously impaired.

Titanium alloys derive their strength from the fine microstructures produced by the transformation from beta to alpha. If alpha-beta or beta alloys are solution heat treated and aged, titanium martensite can form during the quenching operation; however, the martensite formed in titanium alloys is not like the extremely hard and strong martensite formed during the heat treatment of steels. For example, the tensile strength of Ti-6Al-4V only increases from 896 to 1172 MPa (130 to 170 ksi) on STA, while the tensile strength of 4340 steel can be increased from 758 to 1930 MPa (110 to 280 ksi) by heat treatment. While the grain size does not normally affect the ultimate tensile strength, the finer the grain size, the higher the yield strength, ductility, and fatigue strength. However, to resist grain-boundary sliding and rotation, larger grain sizes are actually preferred for some applications requiring creep resistance.

Although the melting point of titanium is higher than 1649 °C (3000 °F), it is not possible to create titanium alloys that operate at temperatures approaching their melting point. Titanium alloys are usually restricted to maximum temperatures of 316 to 593 °C (600 to 1100 °F), depending on alloy composition.

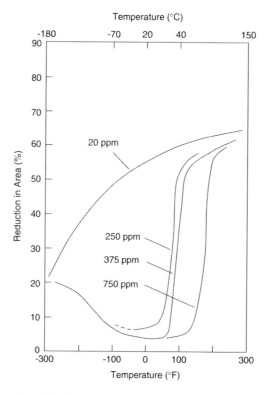

Fig. 5.6 Effect of hydrogen content (ppm) on ductility of alpha titanium. Source: Ref 5.4

Titanium Alloy Systems

There are several grades of unalloyed titanium. The primary difference between grades is oxygen and iron content. Grades of higher purity (lower interstitial content) are lower in strength, hardness, and transformation temperature than those higher in interstitial content, and have greater formability. The high solubility of the interstitial elements oxygen and nitrogen makes titanium unique among metals and also creates problems not of concern in most other metals. For example, heating titanium in air at high temperature results not only in oxidation but also in solid-solution hardening of the surface as a result of inward diffusion of oxygen (and nitrogen). A surface-hardened zone of alpha case or air contamination layer is formed. Because this hard, embrittled surface layer degrades fatigue strength and ductility, it is removed by machining, chemical milling, or other mechanical means prior to placing a part in service.

Table 5.1 lists some commercial and semicommercial titanium grades and alloys currently available, which are subdivided into four groups: unalloyed (CP) grades, alpha and near-alpha alloys, alpha-beta alloys, and beta alloys. Ti-6Al-4V is the most widely used titanium alloy, accounting for approximately 45% of total titanium production. Unalloyed grades comprise approximately 30% of production, and all other alloys combined comprise the remaining 25%. Selection of an unalloyed grade of titanium, an alpha or near-alpha alloy, an alpha-beta alloy, or a beta alloy depends on desired mechanical properties, service requirements, cost considerations, and the other factors that enter into any materials selection process.

Unalloyed Titanium. Unalloyed titanium or CP grades are available with yield strengths ranging from 182 to 463 MPa (25 to 70 ksi), with the

Table 5.1 Summary of commercial and semicommercial grades and alloys of titanium

| Designation | Tensile strength, min | | 0.2% yield strength, min | | Impurity limits, wt%, max | | | | | Nominal composition, wt% | | | | |
	MPa	ksi	MPa	ksi	N	C	H	Fe	O	Al	Sn	Zr	Mo	Others
Unalloyed grades														
ASMT grade 1	240	35	170	25	0.03	0.08	0.015	0.20	0.18	...	...	...	...	...
ASMT grade 2	340	50	280	40	0.03	0.08	0.015	0.30	0.25	...	...	...	...	...
ASMT grade 3	450	65	380	55	0.05	0.08	0.015	0.30	0.35	...	...	...	...	...
ASMT grade 4	550	80	480	70	0.05	0.08	0.015	0.50	0.40	...	...	...	...	...
ASMT grade 7	340	50	280	40	0.03	0.08	0.015	0.30	0.25	...	...	...	...	0.2Pd
ASMT grade 11	240	35	170	25	0.03	0.08	0.015	0.20	0.18	...	...	...	...	0.2Pd
α and near-α alloys														
Ti-0.3Mo-0.8Ni	480	70	380	55	0.03	0.10	0.015	0.30	0.25	...	...	...	0.3	0.8Ni
Ti-5Al-2.5Sn	790	115	760	110	0.05	0.08	0.02	0.50	0.20	5	2.5	...	...	...
Ti-5Al-2.5Sn-ELI	690	100	620	90	0.07	0.08	0.0125	0.25	0.12	5	2.5	...	...	...
Ti-8Al-1Mo-1V	900	130	830	120	0.05	0.08	0.015	0.30	0.12	8	...	...	1	1V
Ti-6Al-2Sn-4Z-2Mo	900	130	830	120	0.05	0.05	0.0125	0.25	0.15	6	2	4	2	0.08Si
Ti-6Al-2Nb-1Ta-0.8Mo	790	115	690	100	0.02	0.03	0.0125	0.12	0.10	6	...	...	1	2Nb, 1Ta
Ti-2.25Al-11Sn-5Zr-1Mo	1000	145	900	130	0.04	0.04	0.008	0.12	0.17	2.25	11	5	1	0.2Si
Ti-5.8Al-4Sn-3.5Zr-0.7Nb-0.5Mo-0.35Si	1030	149	910	132	0.03	0.08	0.006	0.05	0.15	5.8	4	3.5	0.5	0.7Nb, 0.35Si
α–β alloys														
Ti-6Al-4V(a)	900	130	830	120	0.05	0.10	0.0125	0.30	0.20	6	...	...	...	4V
Ti-6Al-4V-ELI(a)	830	120	760	110	0.05	0.08	0.0125	0.25	0.13	6	...	...	...	4V
Ti-6Al-6V-2Sn(a)	1030	150	970	140	0.04	0.05	0.015	1.0	0.20	6	2	...	...	0.75Cu, 6V
Ti-6Al-2Sn-4Zr-6Mo(b)	1170	170	1100	160	0.04	0.04	0.0125	0.15	0.15	6	2	4	6	...
Ti-5Al-2Sn-2Zr-4Mo-4Cr(b)	1125	163	1055	153	0.04	0.05	0.0125	0.30	0.13	5	2	2	4	4Cr
Ti-6Al-2Sn-2Mo-2Cr(b)	1030	150	970	140	0.03	0.05	0.0125	0.25	0.14	5.7	2	2	2	2Cr, 0.25Si
Ti-3Al-2.5V(c)	620	90	520	75	0.015	0.05	0.015	0.30	0.12	3	...	...	...	2.5V
Ti-4Al-4Mo-2Sn-0.5Si	1100	160	960	139	(d)	0.02	0.0125	0.20	(d)	4	2	...	4	0.5Si
β alloys														
Ti-10V-2Fe-3Al	1170	170	1100	160	0.05	0.05	0.015	2.5	0.16	3	...	...	...	10V
Ti-3Al-8V-6Cr-4Mo-4Zr	900	130	830	120	0.03	0.05	0.020	0.25	0.12	3	...	4	4	6Cr, 8V
Ti-15V-3Cr-3Al-3Sn	1000(b)	145(b)	965(b)	140(b)	0.05	0.05	0.015	0.25	0.13	3	3	...	...	15V, 3Cr
	1241(e)	180(e)	1172(e)	170(e)										
Ti-15Mo-3Al-2.7Nb-0.2Si	862	125	793	115	0.05	0.05	0.015	0.25	0.13	3	...	...	15	2.7Nb, 0.2Si

ELI, extra low interstitial. (a) Mechanical properties given for the annealed condition; may be solution treated and aged to increase strength. (b) Mechanical properties given for the solution-treated-and-aged condition; alloy not normally applied in annealed condition. (c) Primarily a tubing alloy; may be cold drawn to increase strength. (d) Combined $O_2 + 2N_2 = 0.27\%$. (e) Also solution treated and aged using an alternative aging temperature (480 °C, or 900 °F). Source: Ref 5.5

higher-strength grades containing more oxygen. The CP grades have good formability, are readily weldable, and have excellent corrosion resistance. The CP grades are supplied in the mill-annealed condition, which permits extensive forming at room temperature, while severe forming operations can be conducted at 149 to 482 °C (300 to 900 °F). However, property degradation can be experienced after severe forming if the material is not stress relieved. One of the largest usages of CP alloys is for corrosion-resistant tubing, tanks, and fittings in the chemical processing industry. Palladium can be added to further enhance corrosion resistance.

Alpha and Near-Alpha Alloys. The alpha and near-alpha alloys are not heat treatable, have medium formability, are weldable, and have medium strength, good notch toughness, and good creep resistance in the range of 316 to 593 °C (600 to 1100 °F). Aluminum is the most important alloying element in titanium, because it is a potent strengthener and also reduces density. However, the aluminum content is usually restricted to approximately 6% because higher contents run the risk of forming the brittle intermetallic compound Ti_3Al. The only true alpha alloy that is commercially available is the alloy Ti-5Al-2.5Sn. Ti-5-2.5 is used in cryogenic applications, because it retains its ductility and fracture toughness down to cryogenic temperatures. The remainder of the commercially available alloys in this class are classified as near-alpha alloys.

Unlike alpha-beta and beta alloys, alpha alloys cannot be significantly strengthened by heat treatment. Generally, alpha alloys are annealed or recrystallized to remove residual stresses induced by cold working. Alpha alloys have good weldability because they are insensitive to heat treatment. They generally have poorer forgeability and narrower forging temperature ranges than alpha-beta or beta alloys, particularly at temperatures below the beta transus. This poorer forgeability is manifested by a greater tendency for strain-induced porosity or surface cracks to occur, which means that small reduction steps and frequent reheats must be incorporated in forging schedules.

Near-alpha alloys are those that contain some beta phase dispersed in an otherwise all-alpha matrix. The near-alpha alloys generally contain 5 to 8% Al, some zirconium and tin, along with some beta stabilizer elements. Because these alloys retain their properties at elevated temperature and possess good creep strength, they are often specified for elevated-temperature applications. Silicon in the range of 0.10 to 0.25% enhances the creep strength. High-temperature near-alpha alloys include Ti-6242S (Ti-6Al-2Sn-4Zr-2Mo-0.25Si) and IMI 829 (Ti-5.5Al-3.5Sn-3Zr-1Nb-0.3Si), which can be used to 540 °C (1000 °F), and IMI 834 (Ti-5.8Al-4Sn-3.5Zr-0.7Nb-0.5Mo-0.35Si) and Ti-1100 (Ti-6Al-2.8Sn-4Zr-0.4Mo-0.4Si), a modification to Ti-6242S, which can be used to 593 °C (1100 °F). Ti-8Al-1Mo-1V is an older alloy that has rather poor stress-corrosion resistance (due to its extremely high aluminum content) but has a low density and a higher modulus than other titanium alloys.

Alpha alloys that contain small additions of beta stabilizers (Ti-8Al-1Mo-1V or Ti-6Al-2Nb-1Ta-0.8Mo, for example) sometimes have been classed as superalpha or near-alpha alloys. Although they contain some retained beta phase, these alloys consist primarily of alpha and behave more like conventional alpha alloys than alpha-beta alloys.

Alpha-Beta Alloys. The alpha-beta alloys are heat treatable to moderate strength levels but do not have as good elevated-temperature properties as the near-alpha alloys. Their weldability is not as good as the near-alpha alloys, but their formability is better. The alpha-beta alloys, which include Ti-6Al-4V, Ti-6Al-6V-2Sn, and Ti-6Al-2Sn-4Zr-6Mo, are capable of higher strengths than the near-alpha alloys. They have a good combination of mechanical properties, rather wide processing windows, and can be used in the range of approximately 315 to 399 °C (600 to 750 °F). They can be strengthened by STA, but the strength obtainable decreases with section thickness. The lean alloys, such as Ti-6-4, are weldable. Alpha-beta alloys contain elements, in particular aluminum, to strengthen the alpha phase and beta stabilizers to provide solid-solution strengthening and response to heat treatment. As the percentage of beta stabilizing elements increases, the hardenability increases and the weldability decreases. Ti-6-4 accounts for approximately 60% of the titanium used in aerospace and up to 80 to 90% for airframes. There are four different heat treatments that are often used for wrought Ti-6-4:

- *Mill anneal.* The most common heat treatment. It produces a tensile strength of approximately 896 MPa (130 ksi), good fatigue properties, moderate fracture toughness ($K_{Ic} = 67$ MPa $\sqrt{m}$, or 60 ksi $\sqrt{in.}$), and reasonable fatigue crack growth rate.
- *Recrystallization anneal.* Can be used for parts requiring increased damage tolerance, because the K_{Ic} value for Ti-6Al-4V goes from approximately 67 to 77 MPa $\sqrt{m}$ (60 to 70 ksi $\sqrt{in.}$). This heat treatment produces slightly lower strength and fatigue properties and improved fracture toughness and slower fatigue crack growth rates.
- *Beta anneal.* Used where it is important to maximize fracture toughness (minimum $K_{Ic} \sim 89$ MPa $\sqrt{m}$, or 80 ksi $\sqrt{in.}$) and minimize fatigue crack grow rate. However, the fatigue strength is significantly degraded. Alpha-beta alloys for fracture-critical applications are often beta annealed to develop a transformed beta structure. A transformed beta structure produces tortuous crack paths with secondary cracking, which enhances fracture toughness and slows fatigue crack growth; however, the equiaxed alpha-beta microstructure has twice the ductility and better fatigue life than the transformed beta microstructure.

- *Solution treated and aged.* Provides maximum strength, but full hardenability is Ti-6-4 limited to sections 25 mm (1 in.) thick or less. This heat treatment is used for mechanical fasteners with a minimum tensile strength of 1103 MPa (160 ksi). The STA treatment is not normally used for structural components due to its limited hardenability and warping problems during heat treating.

Ti-6Al-4V-ELI, with a maximum oxygen content of 0.13%, is used for fracture-critical structures and for cryogenic applications. Oxygen is a potent strengthening element in Ti-6Al-4V and must be held to low limits to develop high fracture toughness. The commercial grade of Ti-6Al-4V has an oxygen content of 0.16 to 0.18%, while the ELI grade is limited to 0.10 to 0.13%. With higher oxygen content, the commercial grade has higher strength and slightly lower ductility, while the ELI grade has approximately a 25% higher fracture toughness. A comparison of the properties of commercial and ELI Ti-6Al-4V is given in Fig. 5.7. The beta transus temperature is also influenced by the oxygen content, with the beta transus being in the range of 1010 to 1021 °C (1850 to 1870 °F) for the commercial grade and in the range of 960 to 982 °C (1760 to 1800 °F) for the ELI grade.

Ti-6-22-22S (Ti-6Al-2Sn-2Zr-2Mo-2Cr-0.25Si) is similar to Ti-6Al-4V but is a higher-strength alloy with a tensile strength of 1034 MPa (150 ksi) in the mill-annealed condition and 1172 MPa (170 ksi) in the STA condition. Ti-6-22-22S was originally developed as a deep-hardening, high-strength alloy for moderate-service temperatures. Due to its higher alloying content, it has better room- and elevated-temperature static and fatigue strength than Ti-6Al-4V. It can be given a triplex heat treatment

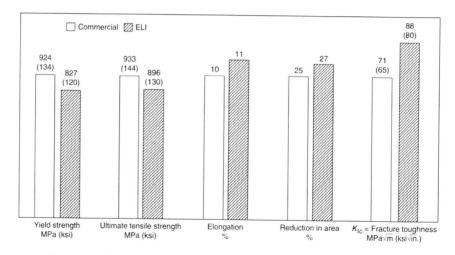

Fig. 5.7 Room-temperature properties of commercial versus extra low interstitial (ELI) Ti-6Al-4V. YS, yield strength; UTS, ultimate tensile strength; El, elongation; RA, reduction in area; K_{Ic}, fracture toughness. Source: Ref 5.6

to maximize damage-tolerant properties. Typical properties are a tensile strength of 1034 MPa (150 ksi) with a minimum fracture toughness of $K_{Ic} = 78$ MPa $\sqrt{m}$ (70 ksi $\sqrt{in}$.). The heat treatment consists of solution treating above the beta transus followed by a solution treatment below the beta transus and then a lower-temperature aging treatment. This alloy also exhibits excellent superplastic forming characteristics; it can be formed at temperatures lower than that for Ti-6Al-4V and yet have higher strengths.

SP-700 (Ti-4.5Al-3V-2Mo-2Fe) was developed as a lower-temperature superplastic forming alloy that can be formed at temperatures as low as 704 °C (1300 °F), compared to approximately 899 °C (1650 °F) for Ti-6Al-4V. This reduces energy costs, lengthens die life, and reduces the amount of alpha case that must be removed after part fabrication. Ti-3Al-2.5V is a lean version of Ti-6Al-4V that is often used for aircraft hydraulic tubing in sizes ranging from 6.4 to 38 mm (0.25 to 1.5 in.) diameter and for honeycomb core. Ti-6246 (Ti-6Al-2Sn-4Zr-6Mo) and Ti-17 (Ti-5Al-2Sn-2Zr-4Mo-4Cr) are engine alloys that are used up to 316 and 399 °C (600 and 750 °F), respectively, where higher strengths than Ti-6Al-4V are needed. Where higher-temperature creep resistance is required, the near-alpha alloy Ti-6242S is normally specified.

The microstructure of alpha-beta alloys can take different forms, ranging from equiaxed to acicular, or some combination of both. Equiaxed structures are formed by working the alloy in the alpha-beta range and annealing at lower temperatures. Acicular structures (Fig. 5.8c) are formed by working or heat treating above the beta transus and rapid cooling. Rapid cooling from temperatures high in the alpha-beta range (Fig. 5.8d, e) will result in equiaxed primary (prior) alpha and acicular alpha from the transformation of beta structures. Generally, there are property advantages and disadvantages for each type of structure. Equiaxed structures have higher ductility and formability, higher threshold stresses for hot salt stress corrosion, higher strength for an equivalent heat treatment, and better low-cycle fatigue (crack-initiation) properties. The advantages of the acicular structures are better creep properties, higher fracture toughness, superior stress-corrosion resistance, and lower fatigue crack propagation rates.

In the alpha-beta alloys, the presence of nonequilibrium phases, such as titanium martensite or metastable beta, results in substantial increases in tensile and yield strengths following the aging treatment. No response to aging occurs on furnace cooling from solution temperatures. Only a slight response occurs on air cooling that produces the Fig. 5.8(b) and (d) microstructures. The greatest response is experienced with water quenching from the solution temperature, typical of microstructures shown in Fig. 5.8(c) and (e). A good response to aging takes place on water quenching from the beta field, as in Fig. 5.8(c); however, ductilities are quite low. The best combination of properties can be produced by solution treating and rapidly quenching from close to but below the beta transus temperature (Fig. 5.8e), followed by an aging treatment.

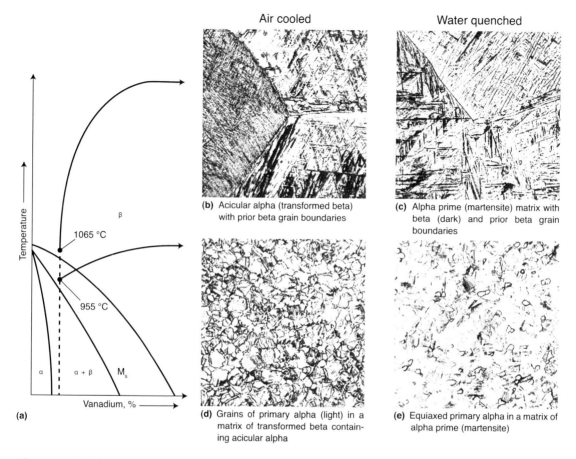

Fig. 5.8 Ti-6Al-4V microstructures produced by cooling from different temperatures. Original magnification: 250×. Source: Ref 5.1

Alpha-beta alloys contain one or more alpha stabilizers or alpha-soluble elements plus one or more beta stabilizers. These alloys retain more beta phase after solution treatment than do near-alpha alloys, the specific amount depending on the quantity of beta stabilizers present and on heat treatment.

Solution treating usually is done at a temperature high in the two-phase alpha-beta field and is followed by quenching in water, oil, or other suitable quenchant. As a result of quenching, the beta phase present at the solution-treating temperature may be retained or may be partly, or fully, transformed during cooling by either martensitic transformation or nucleation and growth. The specific response depends on alloy composition, solution-treating temperature (beta-phase composition at the solution temperature), cooling rate, and section size. Solution treatment is followed by aging, normally at 480 to 650 °C (900 to 1200 °F), to precipitate alpha and produce a fine mixture of alpha and beta in the retained or transformed beta phase. Transformation kinetics, transformation products, and specific response of a given alloy can be quite complex; a detailed review of the subject is beyond the scope of this chapter.

Solution treating and aging can increase the strength of alpha-beta alloys 30 to 50%, or more, over the annealed or overaged condition. Response to solution treating and aging depends on section size; alloys relatively low in beta stabilizers (e.g., Ti-6Al-4V) have poor hardenability and must be quenched rapidly to achieve significant strengthening. For Ti-6Al-4V, the cooling rate of a water quench is not rapid enough to significantly harden sections thicker than approximately 25 mm (1 in.). As the content of beta stabilizers increases, hardenability increases; for example, Ti-5Al-2Sn-2Zr-4Mo-4Cr can be through hardened with relatively uniform response throughout sections up to 150 mm (6 in.) thick. For some alloys of intermediate beta stabilizer content, the surface of a relatively thick section can be strengthened, but the core may be 10 to 20% lower in hardness and strength. The strength that can be achieved by heat treatment is also a function of the volume fraction of beta phase present at the solution-treating temperature. Alloy composition, solution temperature, and aging conditions must be carefully selected and balanced to produce the desired mechanical properties in the final product.

Beta alloys are richer in beta stabilizers and leaner in alpha stabilizers than alpha-beta alloys. They are characterized by high hardenability, with beta phase completely retained on air cooling of thin sections or water quenching of thick sections. Beta alloys have excellent forgeability, cold rolling capabilities, and, in sheet form, can be cold brake formed more readily than high-strength alpha-beta or alpha alloys. It is more difficult to perform more complex triaxial strain-type forming operations with beta alloys because they exhibit almost no work hardening, and necking occurs early. After solution treating, beta alloys are aged at temperatures of 450 to 650 °C (850 to 1200 °F) to partially transform the beta phase to alpha. The alpha forms as finely dispersed particles in the retained beta, and strength levels comparable or superior to those of aged alpha-beta alloys can be attained. The chief disadvantages of beta alloys in comparison with alpha-beta alloys are higher density, lower creep strength, and lower tensile ductility in the aged condition. Although tensile ductility is lower, the fracture toughness of an aged beta alloy generally is higher than that of an aged alpha-beta alloy of comparable yield strength.

In the solution-treated condition (100% retained beta), beta alloys have good ductility and toughness, relatively low strength, and excellent formability. Solution-treated beta alloys begin to precipitate alpha phase at slightly elevated temperatures and thus are unsuitable for elevated-temperature service without prior stabilization or overaging treatment.

Beta alloys (at least commercial beta alloys), despite the name, actually are metastable, because cold work at ambient temperature can induce a martensitic transformation, or heating to a slightly elevated temperature can cause partial transformation to alpha or other transformation products. The principal advantages of beta alloys are that they have high hardenability, excellent forgeability, and good cold formability in the solution-treated condition.

Beta alloys contain high percentages of the bcc beta phase that greatly increases their response to heat treatment, provides higher ductility in the annealed condition, and provides much better formability than the alpha or alpha-beta alloys. In general, they exhibit good weldability, high fracture toughness, and a good fatigue crack growth rate; however, they are limited to approximately 370 °C (700 °F) due to creep.

The beta alloys, including Ti-10-2-3 (Ti-10V-2Fe-3Al), Ti-15-3 (Ti-15V-3Al-3Cr-3Sn), and Beta 21S (Ti-15Mo-3Al-2.7Nb-0.25Si), are high-strength alloys that can be heat treated to tensile strength levels approaching 1379 MPa (200 ksi). In general, they are highly resistant to stress-corrosion cracking. The fatigue strength of these alloys depends on the specific alloy; for example, the fatigue strength of Ti-10-2-3 is very good, but that of Ti-15-3 is not so good. The beta alloys possess fair cold formability, which can eliminate some of the hot forming operations normally required for the alpha-beta alloys.

Ti-10-2-3 is the most widely used of the beta alloys in airframes. Ti-10-2-3 is a popular forging alloy because it can be forged at relatively low temperatures, offering flexibility in die materials and forging advantages for some shapes. It is used extensively in the main landing gear of the Boeing 777. It is used at three different tensile strength levels: 965, 1103, and 1172 MPa (140, 160, and 170 ksi). At the 1172 MPa (170 ksi) level, it has a fairly narrow processing window to meet the strength, ductility, and toughness requirements. The lower strength levels have wider processing windows and improved toughness. It exhibits excellent fatigue strength but only moderate fatigue crack growth rates. Although Ti-10-2-3 is not as deep hardening or as fatigue resistant as Ti-6-22-22S, it is capable of being heat treated to higher strengths than Ti-6-22-22S. A Russian near-beta alloy, Ti-5-5-5-3 (Ti-5Al-5V-5Mo-3Cr), is receiving quite a bit of consideration for new applications. Forged bar that has been STA heat treated has a tensile strength of 1206 MPa (175 ksi), a yield strength of 1117 MPa (162 ksi), and an elongation of 10%. Although the fracture toughness is somewhat lower than comparatively processed Ti-10-2-3, it is not a concern for the applications being considered. Although primarily a wrought alloy, it has also been evaluated for investment castings with promising results.

Heat treated Beta C (Ti-3Al-8V-6Cr-4Mo-4Zr) is often used for springs. Ti-15-3 can also be used for springs when it is heat treated to an ultimate tensile strength (UTS) of 1034 MPa (150 ksi). An advantage of Ti-15-3 is the ability to cold form the material in thin gages and then STA to high strengths. Ti-15-3 can also be used for castings where higher strength (UTS = 1138 MPa, or 165 ksi) than Ti-6-4 (UTS = 837 MPa, or 120 ksi) is required. Beta 21S (Ti-15Mo-3Al-2.7Nb-0.25Si) was originally developed as an oxidation-resistant alloy for high-temperature metal-matrix composites for the National Aerospace Plane. Even though it is a beta alloy, it has reasonable creep properties, better than Ti-6-4. It can be heat treated to a tensile strength of 862 MPa (125 ksi) for temperature uses between 482 and 566 °C (900 and 1050 °F) or to higher strength levels (UTS = 1034 MPa, or 150 ksi) for lower-temperature usage. One of the distinguishing features of Beta 21S is its resistance to hydraulic fluids, presumably due to the synergistic effects of molybdenum and niobium.

5.3 Titanium Properties

Properties are a function of composition and processing. Typically, tensile elastic moduli lie in the range from 100 to 120 GPa (14.7 to 17×10^6 psi) and decrease with temperature. Texture control and heat treatment can cause changes in moduli. Physical properties vary with alloy composition, and some may vary with processing as it influences microstructure. Tensile properties cover a wide range of values; moderate- to high-strength alloys typically show tensile strength values in the range from 895 to 1065 MPa (130 to 155 ksi).

Creep properties of the highest-tensile-strength alloys usually drop off markedly with temperature. A comparison of tensile strengths and 150 h creep strengths of a few titanium alloys are shown in Fig. 5.9 and 5.10.

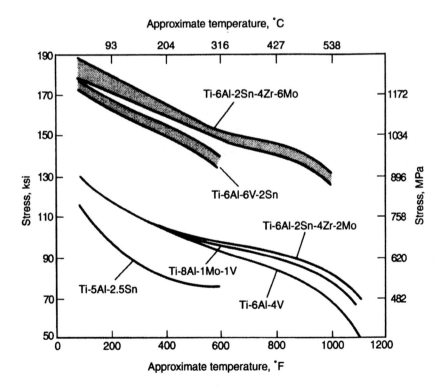

Fig. 5.9 Comparison of typical tensile strengths reported for various titanium alloys. Source: Ref 5.7

The fatigue strength of titanium alloys is of interest because it does not show a marked drop with temperature until temperatures in excess of 315 to 425 °C (600 to 800 °F) are reached. A related design property, fracture toughness (K_{Ic}), is of interest, particularly in applications of high-strength titanium alloys. Table 5.2 gives typical fracture toughness values for three high-strength alloys as a function of microstructure. Toughness is influenced by texture and is a function of test direction (Table 5.3). Fatigue crack growth rate (da/dN) shows considerable variation from lot-to-lot, as

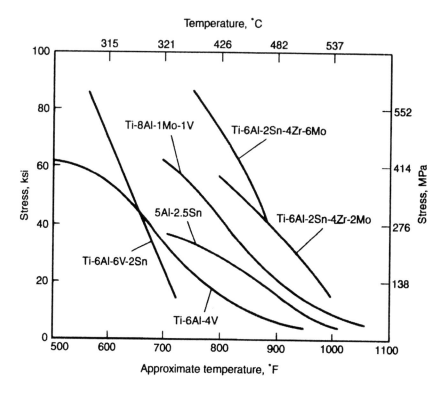

Fig. 5.10 Comparison of typical 150 h, 0.1% creep strengths reported for various titanium alloys. Source: Ref 5.7

Table 5.2 Typical fracture toughness values of high-strength titanium alloys

Alloy	Alpha morphology	Yield strength		Fracture toughness (K_{Ic})	
		MPa	ksi	MPa·m$^{1/2}$	ksi·in.$^{1/2}$
Ti-6Al-4V	Equiaxed	910	130	44–66	40–60
	Transformed	875	125	88–110	80–100
Ti-6Al-6V-2Sn	Equiaxed	1085	155	33–55	30–50
	Transformed	980	140	55–77	50–70
Ti-6Al-2Sn-4Zr-6Mo	Equiaxed	1155	165	22–23	20–30
	Transformed	1120	160	33–55	30–50

Source: Ref 5.7

illustrated in Fig. 5.11. This property is strongly microstructure (heat treat) dependent. Beta annealing can reduce *da/dN* by a factor of 10.

5.4 Processing of Titanium and Titanium Alloys

Titanium metal passes through three major steps during processing from ore to finished product:

1. Reduction of titanium ore to a porous form of titanium metal called sponge
2. Melting of sponge and scrap to form ingot
3. Remelting and casting into finished shape, or primary fabrication, in which ingots are converted into general mill products followed by secondary fabrication of finished shapes from mill products

Table 5.3 Effect of test direction on mechanical properties of textured Ti-6Al-2Sn-4Zr-6Mo plate

| Test direction(a) | Tensile strength | | Yield strength | | Elongation, % | Reduction in area, % | Elastic modulus | | K_{Ic} | | K_{Ic} specimen orientation |
	MPa	ksi	MPa	ksi			GPa	10⁶ psi	MPa·m^{1/2}	ksi·in.^{1/2}	
L	1027	149	952	138	11.5	18.0	107	15.5	75	68	L-T
T	1358	197	1200	174	11.3	13.5	134	19.4	91	83	L-T
S	938	136	924	134	6.5	26.0	104	15.1	49	45	S-T

(a) High basal-pole intensities reported in the transverse direction, 90° from normal, and also intensity nodes in positions 45° from the longitudinal (rolling) direction and approximately 40° from the plate normal. L, longitudinal; T, transverse; S, short. Source: Ref 5.7

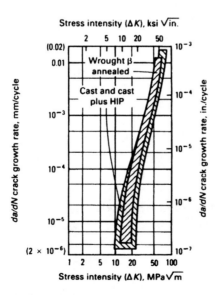

Fig. 5.11 Scatter-band comparison of fatigue crack growth rate behavior of wrought β-annealed Ti-6Al-4V to cast and cast plus hot isostatic pressed HIP Ti-6Al-4V data. Source: Ref 5.7

Machining of cast structures or mill products is required, and welding invariably is needed in casting repair or buildup of structures from castings or mill products. Powder metallurgy processing is used for some applications. At each of these steps, the mechanical and physical properties of the titanium in the finished shape may be affected by any one of several factors, or by a combination of factors. Among the most important are the amounts of specific alloying elements and impurities, melting process used to make ingot, casting process and volume of cast article plus use of densification techniques such as hot isostatic pressing to reduce casting porosity, method for mechanically working ingots into mill products, and the final step employed in working, fabrication, or heat treatment.

Titanium Melting. Control of raw materials is extremely important in producing titanium and its alloys because there are many elements of which small amounts can have major effects on the properties of these metals in finished form. The raw materials used in producing titanium are titanium in the form of sponge metal, alloying elements (in the form of a master alloy), and reclaimed titanium scrap (usually called revert).

Titanium sponge must meet stringent specifications for control of ingot composition. Most important, sponge must not contain hard, brittle, and refractory titanium oxide, titanium nitride, or complex titanium oxynitride particles that, if retained through subsequent melting operations, could act as crack-initiation sites in the final product. Carbon, nitrogen, oxygen, silicon, and iron commonly are found as residual elements in sponge. These elements must be held to acceptably low levels because they raise the strength and lower the ductility of the final product. Alloying element purity is as important as purity of sponge and must be controlled with the same degree of care to avoid undesirable residual elements, especially those that can form refractory or high-density inclusions in the titanium matrix.

Basically, oxygen and iron contents determine strength levels of CP titanium (ASTM and ASME grades 1, 2, 3, and 4) and the differences in mechanical properties between ELI grades and standard grades of titanium alloys. This effect is illustrated in Table 5.4. In higher-strength grades, oxygen and iron are intentionally added to the residual amounts already in the sponge to provide extra strength. On the other hand, carbon and nitrogen usually are held to minimum residual levels to avoid embrittlement.

Use of reclaimed scrap makes production of ingot titanium more economical than production solely from sponge. If properly controlled, addition of scrap is fully acceptable, even in materials for critical structural applications, such as rotating components for jet engines. All forms of scrap can be remelted: machining chips, cut sheet, trim stock, and chunks. To use properly, scrap must be thoroughly cleaned and carefully sorted by alloy and by purity before being remelted. During cleaning, surface scale must be removed because adding titanium scale to the melt could

Table 5.4 Tensile properties of annealed titanium sheet as influenced by oxygen and iron contents

Material	Maximum content, %		Minimum tensile strength		Minimum yield strength(a)	
	O	Fe	MPa	ksi	MPa	ksi
Unalloyed titanium, grade 1	0.18	0.2	240	35	170	2
Unalloyed titanium, grade 2	0.25	0.30	345	50	275	40
Unalloyed titanium, grade 3	0.35	0.30	450	65	380	55
Unalloyed titanium, grade 4	0.40	0.50	655	95	485	70
Ti-6Al-4V	0.20	0.30	925	134	870	126
Ti-6Al-4V-ELI	0.13	0.25	900	130	830	120
Ti-5Al-2.5Sn	0.20	0.50	830	120	780	113
Ti-5Al-2.5Sn-ELI	0.12	0.25	690	100	655	95

ELI, extra low interstitial. (a) At 0.2% offset. Source: Ref 5.4

produce refractory inclusions or excessive porosity in the ingot. Machining chips from fabricators who use carbide tools are acceptable for remelting only if all carbide particles adhering to the chips are removed; otherwise, hard high-density inclusions could result. Improper control of alloy revert would produce off-composition alloys and could degrade the properties of the resulting metal. Cold hearth melting was developed to eliminate the occurrence of high-density inclusions and minimize the incidence of low-density inclusions for premium-quality product, and as a more economical means of melting certain types of scrap.

Titanium Ingot. Melting practice for most titanium and titanium alloy ingot is to melt twice in an electric arc furnace under vacuum, a procedure known as the double consumable electrode vacuum melting process. As shown in Fig. 5.12, during melting, the electrode is the anode and the water-cooled copper crucible is the cathode; that is, the consumable titanium electrode is arc remelted in a water-cooled copper crucible in a vacuum arc melt furnace. In this two-stage process, titanium sponge, revert, and alloy additions are initially mechanically consolidated, welded together to form an electrode, and then melted to produce an ingot. Ingots from the first melt are used as the consumable electrodes for second-stage melting. Processes other than consumable electrode arc melting are used in some instances for first-stage melting of ingot for noncritical applications. Usually, all melting is done under vacuum, but in any event, the final stage of melting must be done by the consumable electrode vacuum arc process.

Double melting is considered necessary for all applications to ensure an acceptable degree of homogeneity in the resulting product. Triple melting is used to achieve better uniformity. Triple melting also reduces oxygen-rich or nitrogen-rich inclusions in the microstructure to a very low level by providing an additional melting operation to dissolve them. Engine manufacturers usually require triple melting for rotating-grade material.

Segregation and other compositional variations directly affect the final properties of mill products. Melting in a vacuum reduces the hydrogen

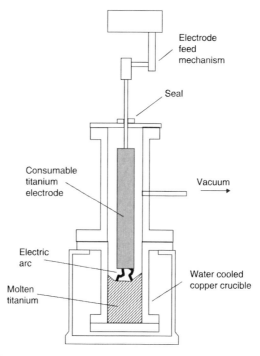

Fig. 5.12 Vacuum arc melting of titanium ingots. Source: Ref 5.2

content of titanium and essentially removes other volatiles. This tends to result in high purity in the cast ingot. However, anomalous operating factors such as air leaks, water leaks, arc-outs, or even large variations in power level affect both the soundness and the homogeneity of the final product. Still another factor is ingot size. Normally, ingots are 650 to 900 mm (26 to 36 in.) in diameter and weigh 3600 to 6800 kg (8000 to 15,000 lb). Larger ingots are economically advantageous to use and are important in obtaining refined macrostructures and microstructures in very large sections, such as billets with diameters of 400 mm (16 in.) or greater. Ingots up to 1000 mm (40 in.) in diameter and weighing more than 9000 kg (20,000 lb) have been melted successfully, but there appear to be limitations on the improvements that can be achieved by producing large ingots due to increasing tendency for segregation with increasing ingot size.

Segregation in titanium ingot must be controlled because it leads to several different types of imperfections that cannot be readily eliminated either by homogenizing heat treatments or by combinations of heat treatment and primary mill processing. Type I imperfections, usually called high interstitial defects, are regions of interstitially stabilized alpha phase that have substantially higher hardness and lower ductility than the surrounding material, and that also exhibit a higher beta transus temperature. They arise from very high nitrogen or oxygen concentrations in sponge, master alloy, or revert. Type I imperfections frequently, but not always, are associated with voids or cracks. Although type I imperfections sometimes are referred to as low-density inclusions, they often are of higher density than is normal for the alloy.

Type II imperfections, sometimes called high-aluminum defects, are abnormally stabilized alpha-phase areas that may extend across several beta grains. Type II imperfections are caused by segregation of metallic alpha stabilizers, such as aluminum, and contain an excessively high proportion of primary alpha having a microhardness only slightly higher than that of the adjacent matrix. Type II imperfections sometimes are accompanied by adjacent stringers of beta, areas low in both aluminum content and hardness. This condition is generally associated with closed solidification pipe into which alloy constituents of high vapor pressure migrate, only to be incorporated into the microstructure during primary mill fabrication. Stringers normally occur in the top portions of ingots and can be detected by macroetching or anodized blue etching. Material containing stringers usually must undergo metallographic review to ensure that the indications revealed by etching are not artifacts.

Beta flecks, another type of imperfection, are small regions of stabilized beta in material that has been alpha-beta processed and heat treated. In size, they may be less than 1 μm, or they may encompass several prior beta grains. Beta flecks are either devoid of primary alpha or contain less than some specified minimum level of primary alpha. They are caused by localized regions either abnormally high in beta stabilizer content or abnormally low in alpha stabilizer content. Beta flecks are attributed to microsegregation during solidification of ingots of alloys that contain strong beta stabilizers. They are most often found in products made from large-diameter ingots. Beta flecks also may be found in beta-lean alloys such as Ti-6Al-4V that have been heated to a temperature near the beta transus during processing.

Types I and II imperfections are not acceptable in aircraft-grade titanium because they degrade critical design properties. Beta flecks are not considered harmful in alloys lean in beta stabilizers if they are to be used in the annealed condition. On the other hand, they constitute regions that incompletely respond to heat treatment, and for this reason microstructural standards have been established for allowable limits on beta flecks in various alpha-beta alloys. Beta flecks are more objectionable in beta-rich alpha-beta alloys than in leaner alloys and are not acceptable in beta alloys.

5.5 Titanium Primary Fabrication

The product flow for various titanium product forms is shown in Fig. 5.13. Titanium alloys are available in most mill product forms: billet, bar, plate, sheet, strip, foil, extrusions, wire, and tubing; however, not all alloys are available in all product forms. Primary fabrication includes the operations

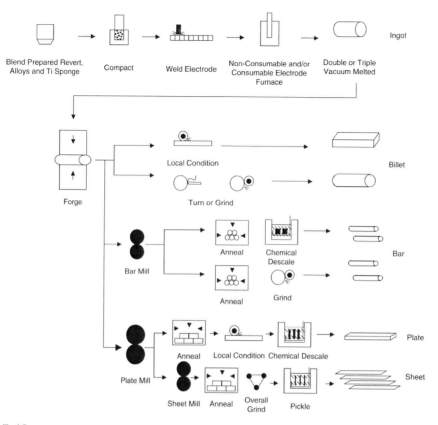

Fig. 5.13 Titanium product flow. Source: Ref 5.2

performed at the mill to convert ingot into products. Besides producing these shapes, primary fabrication hot working is used to refine the grain size, produce a uniform microstructure, and reduce segregation. It has long been recognized that these initial hot working operations will significantly affect the properties of the final product.

Prior to thermomechanical processing, the as-cast ingot is conditioned by grinding to remove surface defects. The first step in deforming as-cast ingots is a series of slow-speed steps including upset forging, side pressing, and press cogging to help homogenize the structure and break up the transformed beta structure. Cogging is a simple open-die forging process between flat dies conducted in slow-speed machines such as a hydraulic press. The ingot is fed through the press in a series of short bites that reduces the cross-sectional area. Electrically heated air furnaces are used to preheat the ingot to 704 to 760 °C (1300 to 1400 °F), and then it is forged at 927 to 1177 °C (1700 to 2150 °F) in large presses so that the deformation can be applied slowly to avoid cracking. These operations are initially done above the beta transus, but significant amounts of deformation are also done below the beta transus, but high in the alpha + beta field, to further refine the microstructure to a fine, equiaxed alpha-beta structure while avoiding surface rupturing. Working as the temperature falls through the beta transus is also an effective way of eliminating grain-boundary alpha, which has an adverse effect on fatigue strength. Final hot working must be carried out in the alpha + beta field to develop a microstructure that has better ductility and fatigue properties than if all of the hot working were conducted above the beta transus. In alpha-beta alloys, slow cooling from above the beta transus must be avoided or alpha will precipitate at the prior beta grain boundaries, leading to a decrease in strength and ductility.

To obtain an equiaxed structure in near-alpha and alpha-beta alloys, the structure is sufficiently worked to break up the lamellar structure and then annealed to cause recrystallization of the deformed structure into an equiaxed structure. The equiaxed structure obtained is a function of the prior microstructure, the temperature of deformation, the type of deformation, the extent of deformation, the rate of deformation, and the annealing temperature and time. The most important variable is to obtain sufficient deformation to cause recrystallization. In general, the finer the initial microstructure and the lower the deformation temperature (i.e., greater percent of cold work), the more efficient is the deformation in causing recrystallization. After forging, billets and bars are straightened, annealed, finished by turning or surface grinding to remove surface defects and alpha case, and ultrasonically inspected.

Slabs from the forging operations are hot rolled into plate and sheet products using two- and three-high mills. For thin sheet, pack rolling is often used to maintain the temperatures required for rolling. Four or five sheets are coated with parting agent and sandwiched together during rolling. Cross-rolling can be used to reduce the texture effect in plate and sheet. Specific hot rolling procedures for bar, plate, and sheet are proprietary to the individual producers; however, typical hot rolling temperatures for Ti-6Al-4V are 954 to 1010 °C (1750 to 1850 °F) for bar, 927 to 982 °C (1700 to 1800 °F) for plate, and 899 to 927 °C (1650 to 1700 °F) for sheet. Typical finishing operations for hot-rolled material are annealing, descaling in a hot caustic bath, straightening, grinding, pickling, and ultrasonic inspection.

Bars up to approximately 100 mm (4 in.) in diameter are unidirectionally rolled, and their properties commonly reflect total reduction in the alpha-beta range. For example, a round bar 50 mm (2 in.) in diameter rolled from a Ti-6Al-4V billet 100 mm (4 in.) square typically is 140 to 170 MPa (20 to 25 ksi) lower in tensile strength than rod 7.8 mm ($^5/_{16}$ in.) in diameter rolled on a rod mill from a billet of the same size at the same rolling temperatures. For bars approximately 50 to 100 mm (2 to 4 in.) in diameter, strength may not decrease with section size, but transverse ductility and notched stress-rupture strength at room temperature do become lower. In diameters greater than approximately 75 to 100 mm (3 to 4 in.), annealed Ti-6Al-4V bars may not meet prescribed limits for stress rupture at room temperature (1170 MPa, or 170 ksi, minimum to cause rupture of a notched specimen in 5 h) unless the material is given a special duplex anneal. Transverse ductility is lower in bars approximately 65 to 100 mm (2.5 to 4 in.) in diameter because it is not possible to obtain the preferred texture throughout bars of this size.

Plate and sheet commonly exhibit the same tensile properties in both the transverse and longitudinal directions relative to the final rolling direction. These characteristics favorably affect tensile properties of Ti-6Al-4V sheet in various gages. Other properties, such as fatigue resistance, also are improved by this type of rolling. Directionality in properties is observed only as a slight drop in transverse ductility of plate greater than 25 mm (1 in.) thick. For forming applications, some customers specify a maximum allowable difference between tensile strengths in the transverse and longitudinal directions.

5.6 Titanium Secondary Fabrication

Secondary fabrication refers to manufacturing processes such as die forging, extrusion, hot and cold forming, machining, chemical milling, and joining, all of which are used to produce finished parts from mill products and, in the case of machining, chemical milling, and joining, for producing finished parts from castings. Each of these processes may strongly influence properties of titanium and its alloys, either alone or by interacting with effects of processes to which the metal has previously been subjected.

Titanium Forging. Although not as difficult as superalloys and refractory alloys, titanium alloys are difficult to forge. Titanium alloy forgings can be produced by a wide variety of forging processes including both open- and closed-die methods. Quite large titanium forgings have been produced, as shown in the Fig. 5.14 closed-die forging. In addition to providing a structural shape, the forging process improves the mechanical properties. Tensile strength, fatigue strength, fracture toughness, and creep resistance can all be improved by forging. The forging process selected depends on the shape to be produced, the cost, and the desired mechanical properties. Often, two or more forging processes are used to produce the final part. For example, open-die forging may be conducted prior to closed-die forging to conserve material and produce more grain flow. Because titanium alloys are quite a bit more difficult to forge than either aluminum or steel, the final forging will normally have more stock that will have to be removed by machining.

The forging conditions for titanium alloys have a greater effect on the final microstructure and mechanical properties than for other metals, such as steel or aluminum. While forging of many metals is primarily a shap-

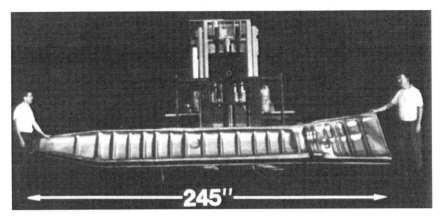

Fig. 5.14 Largest closed-die titanium alloy forging ever manufactured, a Boeing 747 main landing gear beam. Area, 4 m^2 (6200 in.^{2}); weight, 1630 kg (3600 lb). Part was produced on a 450 MN (50,000 tonf) hydraulic press. Dimensions given in inches. Source: Ref 5.8

ing process, forging of titanium alloys is used to produce specific shapes but is also a method used to control the microstructure. To help control microstructure and minimize contamination, titanium is forged at only about 60 to 70% of its melting point, while steels are normally forged at 80 to 90% of their melting points, and nickel alloys are forged at 85 to 95% of their melting points. Whether the part is forged above or below the beta transus temperature will have a pronounced effect on both the forging process and the resultant microstructure and mechanical properties. The properties of the alpha-beta alloys can be varied significantly depending on whether they are forged above or below the beta transus, as shown in Table 5.5. It should be noted that while beta forging can be used to increase the fracture toughness and fatigue crack growth rate, there is a significant penalty on fatigue strength, as much as a 50% reduction compared to alpha + beta-processed material. Alpha-beta forging is used to develop optimal combinations of strength and ductility and optimal fatigue strength. Forging in the two-phase alpha + beta field is a process in which all or most of the deformation is conducted below the beta transus. The resultant microstructure will contain either deformed or equiaxed alpha and transformed beta. For alpha-beta forging, the starting forging temperature is usually 28 °C (50 °F) lower than the beta transus. This is not as simple as it may seem, because small compositional changes will affect the actual beta transus temperature. For example, if Ti-6Al-4V-ELI material is forged at the same temperature as the standard-grade Ti-6Al-4V, microstructural defects can occur, because the beta transus can be as much as 28 to 61 °C (50 to 110 °F) lower for the ELI grade. Because the deformation history and forging parameters have a significant effect on both the microstructure and mechanical properties of the finished part, close control over the

Table 5.5 Effect of beta forging on Ti-6Al-4V

Property	Beta forging
Strength	Lower
Ductility	Lower
Fracture toughness	Higher
Fatigue life	Lower
Fatigue crack growth rate	Lower
Creep strength	Higher
Aqueous stress-corrosion cracking resistance	Higher
Hot salt stress-corrosion cracking resistance	Lower

Source: Ref 5.9

forging process and any subsequent heat treatment is required to obtain a satisfactory product.

Hot-worked billet and bar are the primary product forms used as starting stock for forging operations. Titanium alloys, and in particular the beta and near-beta alloys, are highly strain-rate sensitive in deformation processes such as forging. Therefore, relatively slow strain rates are used to reduce the resistance to deformation. Open-die forging is often used when the number of parts does not warrant the investment in closed dies; however, the majority of titanium parts are produced as closed-die forgings. Closed-die forging includes blocker forging with a single die set, conventional forging with two or more die sets, high-definition forging with two or more die sets, and precision forging with two or more die sets under isothermal conditions. Conventional closed-die forgings are more expensive than blocker forgings but yield better properties and reduce machining costs.

The dies are heated to reduce the pressure required for forging and to reduce surface chilling that can lead to inadequate die filling and surface cracking. Conventional alpha-beta forging is conducted with die temperatures in the range of 204 to 260 °C (400 to 500 °F) for hammer forging and up to 482 °C (900 °F) for press forging. Typical die materials are H11 and H12 die steels. The billets are normally coated with a glass to provide surface-contamination protection and to act as lubricant. Lubricants are also applied to the dies to prevent sticking and galling. In conventional forging, the temperature is allowed to fall during the forging process, while in isothermal forging, the dies and part are kept at a constant forging temperature. The pressure requirements for forging are dependent on the specific alloy composition, the forging process being used, the forging temperature range, and the strain rate of the deformation. In general, the pressures are somewhat higher than those required for steel.

During alpha-beta forging, defects such as wedge cracks and cavities, known as strain-induced porosity, can develop as a result of the coarse beta structure previously developed during ingot breakdown. Wedge cracks are microcracks that develop at the beta grain-boundary triple points, while cavitation is microvoiding that forms along the grain boundaries. Slow strain rates and high temperatures help to reduce these defects. The presence of grain-boundary alpha appears to promote these defects. Nonuniform deformation, which can result from die chilling and/or excessive friction, can lead to bulging during upsetting/pancake forging of cylindrical preforms or ingots. Shear banding and cracking have been observed in both conventional and isothermal forgings.

The key to successful forging and heat treatment is the beta transus temperature. The possible locations for the temperature of forging and/or heat treatment of a typical alpha-beta alloy, such as Ti-6Al-4V, are shown in Fig. 5.15. The higher the processing temperature in the alpha + beta region, the more beta is available to transform on cooling. On quenching from above the beta transus, a completely transformed, acicular structure arises. The exact form of the globular (equiaxed) alpha and the transformed beta structures produced by processing depends on the forging temperature

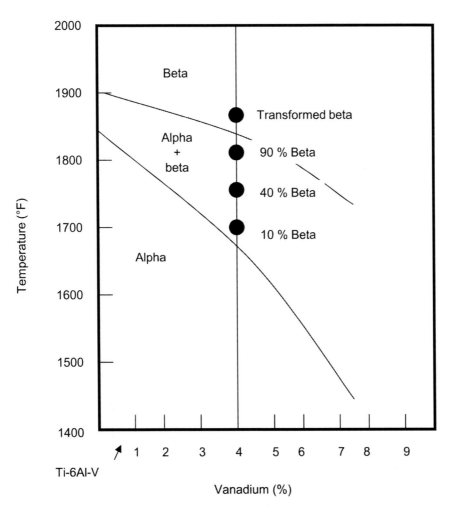

Fig. 5.15 Phase transformations on heating through alpha + beta region. Source: Ref 5.3

relative to the beta transus, which varies from heat to heat, and the degree and nature of deformation produced. Section size is important, and the number of working operations can be significant. Conventional forging may require two or three operations, whereas isothermal forging may require only one. A schematic of a conventional forging and subsequent heat treatment sequence is shown in Fig. 5.16. The solution heat treatment offers a chance to modify or tune the as-forged microstructure, while the aging cycle modifies the transformed beta structures to an optimal dispersion and increases strength.

Hot die and isothermal forging are near-net shape processes in which the dies are maintained at significantly higher temperatures than in conventional forging. This reduces die chill and increases metal flow. Isothermal forging, as shown in Fig. 5.17, is done at a constant temperature where the flow stress is constant, resulting in more uniform microstructures and less property variation. Isothermal forging requires less die pressure and helps ensure that the dies are filled during forging. However, expensive high-temperature die materials, such as molybdenum alloys, and vacuum or inert atmospheres are required to prevent excessive die oxidation. While conventional die forging can require two or three separate operations, isothermal forging can often be accomplished in a single operation. Like so many other advanced manufacturing processes, forging has greatly benefited from automated process control and computer modeling.

Microstructural control is basic to successful processing of titanium alloys. Undesirable structures (grain-boundary alpha, beta fleck,

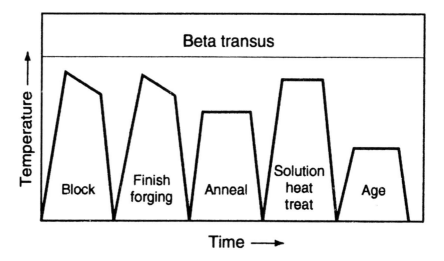

Fig. 5.16 Schematic diagram of a conventional forging and subsequent heat treatment sequence for producing an α-β structure. Typical temperatures during processing would be 955 °C (1750 °F) for the forging and solution treatment, 730 °C (1350 °F) for annealing, and 540 °C (1000 °F) for aging. Typical times during processing would be 30 min to 2 h for both annealing and solution treatment, and 8 h for aging. Source: Ref 5.4

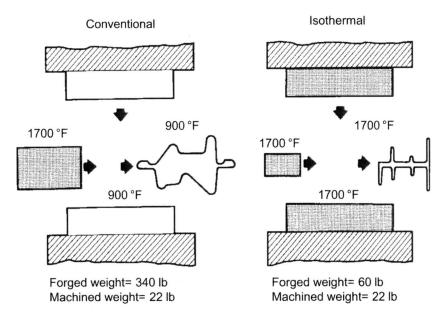

Fig. 5.17 Comparison of conventional and isothermal forging. Source: Ref 5.10

"spaghetti," or elongated alpha) can interfere with optimal property development. Titanium ingot structures can carry over to affect the forged product. Beta processing, despite its adverse effects on some mechanical properties, can reduce forging costs, while isothermal forging offers a means of reducing forging pressures and/or improving die fill and part detail and provides better microstructural control. Isothermal beta forging is finding use in the production of more creep-resistant components of titanium alloys.

Typical microstructures representative of those most commonly found in alpha-beta alloys are shown in Fig. 5.18. Proceeding from Fig. 5.18(a) to (d) will generally result in progressively decreased tensile and fatigue strengths, with increasing improvements in damage-tolerance-type properties. The difference in microstructure between Fig. 5.18(a) and (b) is caused by the differences in working history. The temperature during sheet rolling decreases as rolling proceeds, and the final rolling temperature is significantly lower than the final forging temperature. Thus, there is less retained beta at the final working temperature for the sheet, and a predominantly globular-type alpha microstructure results (the black features in Fig. 5.18a are retained or transformed beta). The final forging temperature is significantly higher, with more retained beta, accounting for the higher amount of lamellar transformed beta microstructure. The slow cooling of the recrystallize-annealed structure (Fig. 5.18c) permits the primary alpha to grow during cooling, consuming most of the beta. Retained beta is observed at some alpha-alpha boundaries and triple points. Solution

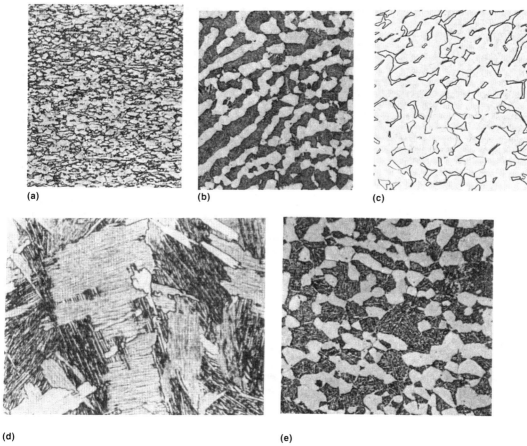

Fig. 5.18 Effects of thermomechanical processing on the microstructures of Ti-6Al-4V. (a) Sheet, rolled starting at 925 °C (1700 °F), annealed for 8 h at 730 °C (1350 °F), and furnace cooled. Structure consists of slightly elongated grains of α (light) and intergranular β (gray). Original magnification: 250×. (b) As-forged at 955 °C (1750 °F), below the β transus. Elongated α (light), caused by low reduction (20%) of a billet that had coarse, platelike α in a matrix of transformed β containing acicular α. Original magnification: 250×. (c) Plate, recrystallize annealed at 925 °C (1700 °F) 1 h, cooled to 760 °C (1400 °F) at 50 to 55 °C/h (90 to 100 °F/h), then air cooled. Equiaxed α with intergranular β. Original magnification: 500×. (d) Forging, β annealed 2 h at 705 °C (1300 °F) exhibiting 92% basketweave structure. (e) Forging, solution treated 1 h at 955 °C (1750 °F), water quenched, and annealed 2 h at 705 °C (1300 °F). Equiaxed α grains (light) in transformed β matrix (dark) containing fine acicular α. Original magnification: 500×. See text for an explanation of how these processing/microstructure combinations affect properties. Source: Ref 5.4

treating and aging is not commonly used for Ti-6Al-4V (Fig. 5.18e) but is the standard heat treatment for aerospace fasteners.

Figure 5.19 summarizes the results of an extensive study of alpha-beta forging versus beta forging for several titanium alloys. Although yield strength after beta forging was not always as high as that after alpha-beta forging, values of notched tensile strength and fracture toughness were consistently higher for beta-forged material. The property combinations obtained with beta forging would be similar to those from alpha-beta forging and beta annealing, which is a more common practice.

The beta-forged alloys tend to show a transformed beta or acicular microstructure (Fig. 5.18d), whereas alpha-beta-forged alloys show a more equiaxed structure (Fig. 5.18b). Trade-offs are required for each structural type (acicular versus equiaxed) because each structure has unique capabilities. The relative advantages of equiaxed and acicular microstructures are given in Table 5.6.

Titanium Extrusion. Extrusion is used to make long lengths of constant cross-section products, as an alternative mill process to rolling. Properties can be affected by processing, as with other product forms. The important variables are extrusion temperature and extrusion ratio. The overall properties differ somewhat from other wrought material because extrusions are normally produced above the beta transus temperature, resulting in a fully transformed beta structure.

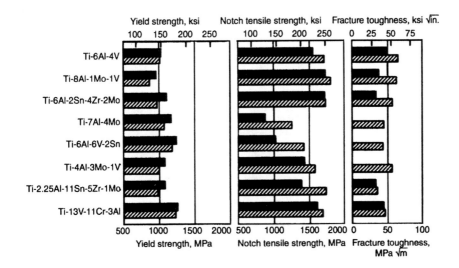

Fig. 5.19 Comparison of mechanical properties of α-β-forged and β-forged titanium alloys. Shaded bars represent α-β-forged material; striped bars represent β-forged material. Source: Ref 5.4

Table 5.6 Relative advantages of equiaxed and acicular microstructures

Equiaxed

Higher ductility and formability
Higher threshold stress for hot salt stress corrosion
Higher strength (for equivalent heat treatment)
Better hydrogen tolerance
Better low- and high-cycle fatigue (initiation) properties

Acicular

Superior creep properties
Higher fracture toughness values
Higher fatigue crack growth resistance

Source: Ref 5.4

5.7 Titanium Powder Metallurgy

Powder metallurgy (PM) products having properties similar to those of other forms are now being manufactured. The room-temperature properties of titanium and several titanium alloys in wrought, cast, and powder forms are compared in Table 5.7. The processes for manufacture of titanium powders are slow and costly, however, and this has resulted in slow growth of PM as a means of manufacturing titanium parts.

The most important considerations in producing titanium PM components are oxygen content, purity, and contaminants. Oxygen has the same undesirable effects on PM parts as it has on those of wrought products. Powders, especially very fine powders, must be handled very carefully because they have a high affinity for oxygen and can be highly pyrophoric. Purity is important for all powder products and is critical for the consolidation of blended elemental fines where the void content must be minimal; the residual chloride content is a major factor in the amount of residual porosity of the final product. Contaminants are key in terms of the final

Table 5.7 Comparison of typical room-temperature properties of wrought, cast, and powder metallurgy (PM) titanium products

Product and condition	Tensile strength MPa	Tensile strength ksi	Yield strength MPa	Yield strength ksi	Elongation, %	Reduction in area, %	Charpy impact strength J	Charpy impact strength ft·lbf	Fracture toughness MPa·m$^{1/2}$	Fracture toughness ksi·in.$^{1/2}$
Unalloyed titanium										
Wrought bar, annealed	550	80	480	70	18	33	35(a)	26(a)	...	...
Cast bar, as-cast	635	92	510	74	20	31	26(a)	19(a)	...	...
PM compact, annealed(b)	480	70	370	54	18	22	...	...	...	...
Ti-5Al-2.5Sn-ELI										
Wrought bar, annealed	815	118	710	103	19	34	...	...	...	...
Cast bar, as-cast	795	115	725	105	10	17	...	...	...	...
PM compact, annealed and forged(c)	795	115	715	104	16	27	...	...	...	...
Ti-6Al-4V										
Wrought bar, mill annealed	965	140	875	127	13	25	...	...	...	...
Wrought bar, recrystallize annealed	970	141	875	127	16	27	27	20	52	47
Wrought bar, beta annealed	955	139	860	125	9	21	...	...	91	83
Cast bar, as-cast	1000	145	895	130	8	16	...	...	107	97
Cast bar, annealed	930	135	825	120	12	22	17.5	24	103	94
Cast bar, annealed(d)	895–930	130–135	825–855	120–124	6–10	10–15	...	...	...	...
Cast bar, STA(e)	935–970	136–141	855–900	124–130	5–8	6–14	...	...	...	...
Cast bar, STA(f)	965–1025	140–149	860–925	125–134	5–8	10–14	...	...	...	...
Cast bar, HIP	1000	145	870	126	8	16	...	...	109	99
PM compact, annealed(b)	825–855	120–124	740–785	107–114	5–8	8–14	...	...	...	...
PM compact, annealed and forged(c)	925	134	840	122	12	27	...	...	...	...
PM compact, STA	965	140	895	130	4	6	...	...	...	...

ELI, extra low interstitial; HIP, hot isostatic pressed. (a) Charpy values at –40 °C (–40 °F). (b) Approximately 94% dense. (c) Almost 100% dense. (d) Annealed at 730 or 845 °C (1345 or 1555 °F). (e) Alpha-beta solution treated and aged (STA): 955 °C, 1 h, cool + 620 °C, 2 h (1750 °F, 1 h, cool + 1150 °F, 2 h). (f) Beta solution treated and aged (STA): 1025 °C, 1 h, cool + 620 °C, 2 h (1875 °F, 1 h, cool + 1150 °F, 2 h). Source: Ref 5.4

fatigue properties that will be obtained (they are detrimental to the fatigue performance).

5.8 Directed Metal Deposition

Directed metal deposition, also known as laser powder deposition, laser direct manufacturing, and electron beam, free-form fabrication, is a rather recent development that can help to reduce the cost of titanium parts. A focused laser beam, shown in Fig. 5.20, is used to melt titanium powder and deposit the melt in a predetermined path on a titanium substrate plate. The metal-deposited preform is then machined to the final part shape. This near-net process leads to savings in materials, machining costs, and cycle times over conventional forged or machined parts.

Directed metal deposition is conducted in a chamber that is constantly being purged with high-purity argon or contains a vacuum atmosphere to prevent atmospheric contamination. Computer-aided design files are used to generate the desired trajectory paths. The trajectory paths are then transmitted as machine instructions to the system, which contains a high-power heating source. The laser or electron beam traces out the desired part by moving the titanium substrate plate beneath the beam in the appropriate x-y trajectories. Titanium prealloyed powder is introduced into the molten metal puddle and provides for buildup of the desired shape, as the laser is traversed over the target plate. A three-dimensional preform is fabricated by repeating the pattern, layer-by-layer, over the desired geometry and indexing the focal point up one layer for each repeat pattern. Preforms generally are deposited with 0.75 to 5.1 mm (0.030 to 0.20 in.) of excess material, which is removed by final machining. A digital model and completed part are shown in Fig. 5.21. The process is capable of depositing between 0.9 and 4.5 kg (2 and 10 lb) of material per hour, depending on part complexity. Preforms have been laser formed from a number of titanium alloys, including Ti-6Al-4V, Ti-5-2.5, Ti-6242S, Ti-6-22-22S, and CP Ti. Depending on the exact process used, the static properties will be 0 to 34 MPa (0 to 5 ksi) lower than equivalent wrought properties. The fatigue properties are essentially equivalent to wrought material; however, some processes require hot isostatic pressing after forming to obtain equivalence by closing internal pores.

Potential savings with this technology include much better material utilization. While the buy-to-fly ratio for forgings is in the range of 5:1 for simple shapes and 20:1 for complex shapes, the ratio for laser forming is approximately 3:1, which includes the substrate that forms $^2/_3$ of the as-deposited part. Machining times are reduced by as much as 30%. In addition, there are no expensive dies or tools required. Finally, the part lead times are much shorter: 12 to 18 months for forgings versus only several months for laser-formed preforms. This process can also be used to conduct local repairs on damaged or worn parts.

(a)

(b)

Fig. 5.20 Laser-additive manufacturing process. (a) Powder feed stock is added by gas jets to the melt pool formed by laser. (b) Direct metal deposition of titanium foil shape. Courtesy of S. Kelly, Pennsylvania State University. Source: Ref 5.11

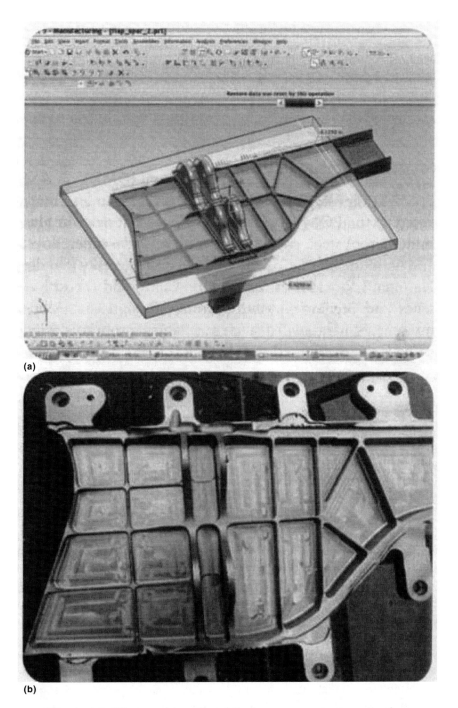

(a)

(b)

Fig. 5.21 Direct metal deposition of titanium aerospace structure using electron beam technology. (a) Digital representation. (b) Fabricated and machined component. Courtesy of Sciaky Inc. Source: Ref 5.11

5.9   Titanium Forming

The yield-strength-to-modulus ratio in titanium results in a significant amount of springback, as much as 15 to 25° after cold forming. To compensate for springback, titanium is normally overformed or hot sized after cold forming. Besides springback, cold forming takes more force, requires stress relieving between forming operations, and must be done at slow forming speeds to prevent cracking. Strain hardening also increases the yield and tensile strengths while causing a slight drop in ductility.

In all forming operations, titanium and its alloys are susceptible to the Bauschinger effect, where a plastic deformation in one direction causes a reduction in yield strength when stress is applied in the opposite direction. The Bauschinger effect is most pronounced at room temperature; plastic deformation (1 to 5% tensile elongation) at room temperature always introduces a significant loss in compressive yield strength, regardless of the initial heat treatment or strength of the alloys. At 2% tensile strain, for instance, the compressive yield strengths of Ti-4Al-3Mo-1V and Ti-6Al-4V drop to less than half the values for solution-treated material. Increasing the deformation temperature reduces the Bauschinger effect; subsequent full thermal stress relieving completely removes it. The effect for Ti-6Al-4V compared to the stress-relieved condition is shown in Fig. 5.22. The Bauschinger effect then dictates that all cold- and warm-formed parts that are structural must be annealed. Temperatures as low as the aging temperature remove most of the Bauschinger effect in solution-treated beta titanium alloys; the aging temperature is not sufficient for alpha and alpha-beta alloys. Heating or plastic deformation at temperatures above the normal aging temperature for solution-treated Ti-6Al-4V causes overaging to occur and, as a result, all mechanical properties decrease.

Except for thin gages, near-alpha and alpha-beta alloys are usually hot formed. Beta alloys, which contain the bcc structure, are very amenable to cold forming, and one of them (Ti-15-3), along with the CP grades, can be successfully formed at room temperature in sheet form. In general, the bends must be of a larger radii than in hot forming. Titanium can also be stretch formed at room temperature at slow speeds with dies heated to approximately 149 °C (300 °F). Cold forming is usually followed by hot sizing and stress relieving to reduce residual stresses, restore the compressive yield strength, improve dimensional accuracy, and make the part more resistant to delayed cracking. During hot sizing, the part is held in fixtures or dies to prevent distortion. All sheet products that are going to be formed should be free of scratches and gouges, and burrs and sharp edges should be filed smooth to prevent edge cracking.

Hot forming, conducted at the temperatures shown in Table 5.8, greatly improves formability, reduces springback, and eliminates the need for stress relieving. However, the requirement to hot form increases the cost of titanium structures by having to heat the material; the need for more

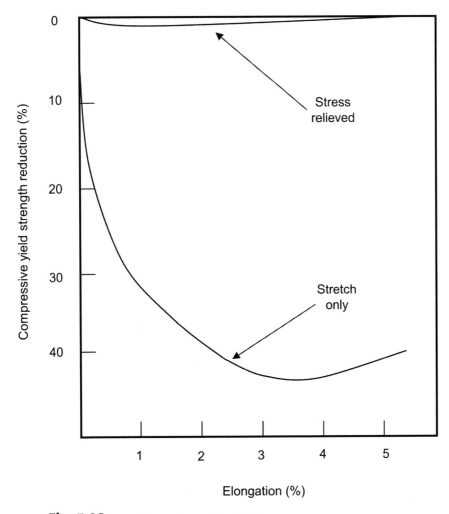

Fig. 5.22 Bauschinger effect on Ti-6Al-4V

expensive tools capable of withstanding the temperature; and the requirement to remove surface contamination (i.e., alpha case) after forming. In addition, the material must be free of all grease and any residue that could cause stress-corrosion cracking. Severe forming operations are done in hot dies with preheated stock. Due to the tendency of titanium to gall, forming lubricants containing graphite or molybdenum sulfide will reduce both part and tool damage, especially for severe forming operations such as drawing. Because forming is usually done above 649 °C (1200 °F), alpha case will form on the surface of unprotected titanium and must be removed by either machining or chemical milling. Although forming can be done in a vacuum, under a protective atmosphere, or with oxidation-resistant coatings, the normal practice is to conduct the forming in air and then remove the alpha case by chemical milling.

Table 5.8 Temperatures for the hot forming and stress relieving of titanium alloys

Alloy	Stress-relief temperature		
	°C	°F	Time, min
Commercially pure titanium (all grades)	480–595	900–1000	15–240
Alpha alloys			
5Al-2.5Sn	540–650	1000–1200	15–360
5Al-2.5Sn (ELI)(a)	540–650	1000–1200	15–360
6Al-2Cb-1Ta-0.8Mo	540–650	1000–1200	15–360
8Al-1Mo-1V	595–760	1100–1400	15–75
11Sn-5Zr-2Al-1Mo	480–540	900–1000	120–480
Alpha-beta alloys			
3Al-2.5V	370–595	700–1100	15–240
6Al-4V	480–650	900–1200	60–240
6Al-4V (ELI)(a)	480–650	900–1200	60–240
6Al-6V-2Sn	480–650	900–1200	60–240
6Al-2Sn-4Zr-2Mo	480–650	900–1200	60–240
6Al-2Sn-2Zr-4Mo-4Cr	480–650	900–1200	60–240
6Al-2Sn-2Zr-2Mo-2Cr-0.25Si	480–650	900–1200	60–240
Metastable beta alloys			
13V-11Cr-3Al	705–730	1300–1350	30–60
3Al-8V-6Cr-4Mo-4Zr	705–760	1300–1400	30–60
15V-3Al-3Cr-3Sn	790–815	1450–1500	30–60
10V-2Fe-3Al	675–705	1250–1300	30–60

(a) ELI, extra low interstitial. Source: Ref 5.12

Vacuum, or creep, forming can also be used to form sheet and plate to mild contours. In this process, the plate is placed on a ceramic tool that contains integral heaters. Insulation is then placed over the sheet followed by a silicone rubber vacuum bag. The part is slowly heated to temperature while a vacuum is applied to the bag. The part slowly creeps to shape under the combined influence of heat and pressure.

5.10 Superplastic Forming

The advantages of superplastic forming (SPF) include the ability to make part shapes not possible with conventional forming, reduced forming stresses, improved formability with essentially no springback, and reduced machining costs. In addition, as shown in Fig. 5.23, there is a reduction in the number of detailed parts and fasteners, leading to a weight savings and reduced manufacturing costs. In general, titanium alloys exhibit much higher superplastic elongations than aluminum alloys, and there are a much wider variety of titanium alloys that exhibit superplasticity. In addition, for titanium, SPF can be combined with diffusion bonding (DB) to make large, one-piece, unitized structures. Because superplasticity depends on microstructure, the fine-grained, equiaxed, two-phase alpha-beta alloys exhibit inherent grain stability and are therefore resistant to grain growth at elevated temperature. Although the optimal structure depends on the specific alloy, the optimal volume fraction of beta is approximately 20% in

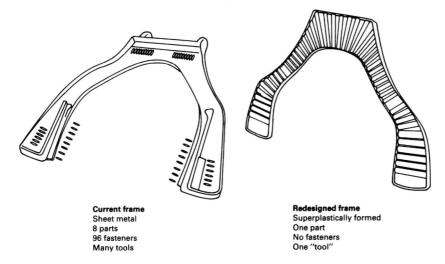

Current frame
Sheet metal
8 parts
96 fasteners
Many tools

Redesigned frame
Superplastically formed
One part
No fasteners
One "tool"

Fig. 5.23 Ti-6Al-4V aircraft nacelle frame that was redesigned from a conventional configuration to one suitable for superplastic forming having fewer parts and fasteners. The redesigned version of this B-1B aircraft component, having 0.161 m² (250 in.²) plan view area, resulted in a 33% weight savings and a 55% cost savings over a conventional multiple-piece assembly. Source: Ref 5.13

Ti-6Al-4V. While internal void formation by cavitation is a concern with aluminum and some other alloys, it has not been a problem in titanium alloys; therefore, the additional complication of back pressure to suppress cavitation is not required. Superplastic titanium alloys include Ti-6-4, Ti-6242S, Ti-6-22-22S, Ti-3-2.5, Ti-8-1-1, Ti-1100, Timetal 550, and SP700.

In the single-sheet SPF process, illustrated in Fig. 5.24, a single sheet of metal is sealed around its periphery between an upper and lower die. The lower die is either machined to the desired part shape, or a die inset is placed in the lower die box. The dies and sheet are heated to the SPF temperature, and argon gas pressure is used to slowly form the sheet down over the tool. The lower cavity is maintained under vacuum to prevent atmospheric contamination. After the sheet is heated to its superplastic temperature range, argon gas is injected through inlets in the upper die. This pressurizes the cavity above the metal sheet, forcing it to superplastically form to the shape of the lower die. Gas pressurization is slowly applied so that the strains in the sheet are maintained in the superplastic range. Typical forming cycles for Ti-6Al-4V are 0.69 to 1.38 MPa (100 to 200 psi) at 899 to 954 °C (1650 to 1750 °F) for 30 min to 4 h.

During the forming operation, the metal sheet is reduced uniformly in thickness; however, where the sheet makes contact with the die, it sticks and no longer thins out. Therefore, if the sheet is formed over a male die, it will touch the top of the die first, and this area will be the thickest. The thickness tapers down along the sides of the die to its thinnest point in the bottom corners, which are formed last. For the same reason, when a sheet

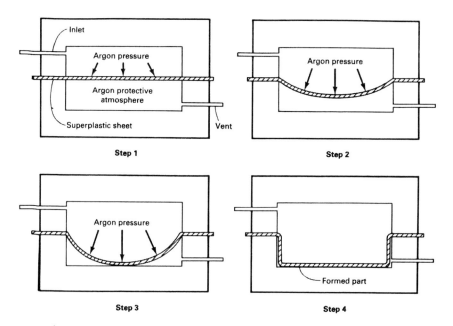

Fig. 5.24 Schematic for single-sheet superplastic forming. Source: Ref 5.13

is formed into a female cavity, the first areas that make contact are the center of the bottom and the top of the sides. These areas are the thickest. The thickness tapers down the sides to the thinnest point in the bottom corners, which again form last. To reduce these variations in thickness, overlay forming can be used.

In overlay forming, the sheet that will become the final part is cut smaller than the tool periphery. A sacrificial overlay sheet is then placed on top of it and clamped to the tool periphery. As gas is injected into the upper die cavity, the overlay sheet forms down over the lower die, forming the part blank simultaneously. While overlay forming does help to minimize thickness variations, it requires a sacrificial sheet for each run that is discarded. Dies for titanium SPF are high-temperature steels such as ESCO 49C (Fe-22Cr-4Ni-9Mn-5Co) lubricated with either boron nitride or yttria.

For titanium alloys, SPF can be combined with DB, a process known as superplastic forming/diffusion bonding (SPF/DB), to form a one-piece unitized structure. Titanium is very amenable to DB because the thin protective oxide layer (TiO_2) dissolves into the titanium above 620 °C (1150 °F), leaving a clean surface. Several processes have been developed, including two-, three-, and four-sheet processes.

In the two-sheet process, shown in Fig. 5.25, two sheets are welded around the periphery to form a closed envelope. The sheets can be welded by either resistance seam welding or by laser welding; what is important is that the weld joints are vacuumtight and capable of resisting up to 1.38

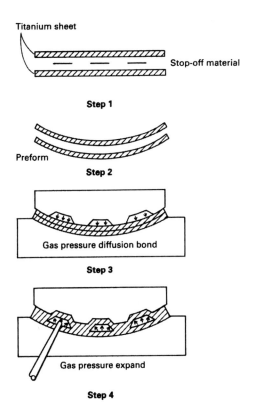

Fig. 5.25 Two-sheet superplastic forming/diffusion bonding process. Source: Ref 5.13

MPa (200 psi) gas pressure. To prevent the two sheets from bonding within the hat stiffeners, stop-off agents such as boron nitride or yttria suspended in an acrylic binder are used. The welded pack is placed in the die and heated to the forming temperature, where the two sheets diffusion bond together. Then, argon gas pressure is used to form the portions of the upper sheet coated with stop-off agent into the corrugated die cavity.

The three-sheet process is shown in Fig. 5.26. The three sheets are welded around the periphery. In this process, selected areas of the center sheet are masked with a stop-off material to prevent bonding. The pack is then placed between the dies, and the sheets are diffusion bonded together except at the locations that have been masked. After DB, gas pressure is applied to each side of the center sheet to expand the substructure. The areas of the center sheet that were masked stretch between the top and bottom sheets to form the stiffening ribs.

The four-sheet process, shown in Fig. 5.27, uses two sheets that form the skin (skin pack) and two sheets that form the substructure (core pack). The two core pack skins are first selectively welded together in a pattern that will form the substructure. The two skins for the skin pack are

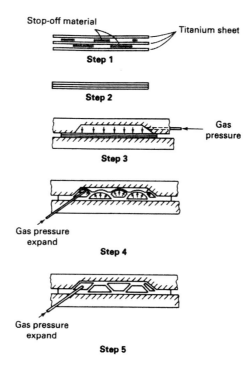

Fig. 5.26 Three-sheet superplastic forming/diffusion bonding process. Source: Ref 5.13

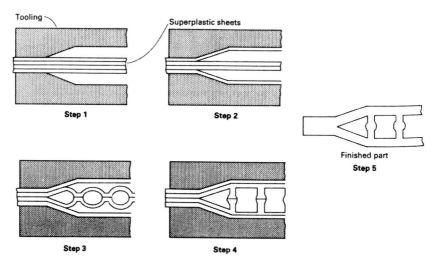

Fig. 5.27 Example of a four-sheet superplastic forming/diffusion bonding process in which the outer sheets are formed first, and the center sheets are then formed and bonded to the outer two sheets. Source: Ref 5.13

then welded to the periphery of the core pack. Although either resistance seam welding or laser welding can be used, one advantage of laser welding is that an automated laser welder can be programmed to make weld patterns and resulting core geometries other than rectangular, possibly resulting in more structurally efficient substructure designs. The total pack is then placed in the die, heated to the SPF temperature, the face sheets are expanded against the tool to form the skins, and then the core pack is expanded to form the substructure. The process is not limited to four sheets; experimental structures with five or more sheets have been designed and fabricated. A distinct advantage of both the three- and four-sheet processes is that the tooling is much simpler; instead of having cavities for substructure as in the two-sheet process, the tools have the more gentle curvature of the inner and outer part moldlines. The completed parts are pickled to remove any alpha case that may have formed at elevated temperature.

Superplastic forming and SPF/DB of titanium has been successfully used on a number of aircraft programs. An example is the F-15E, in which these processes allowed the elimination of over 726 detail parts and over 10,000 fasteners.

5.11   Heat Treating

Heat treatments for titanium alloys include stress relieving, annealing, and STA. All titanium alloys can be stress relieved and annealed, but only the alpha-beta and beta alloys can be STA to increase their strength. Alpha and near-alpha alloys do not undergo a phase change during processing and cannot be strengthened by STA. Because the response to STA is determined by the amount of beta phase, the beta alloys, with higher percentages of beta phase, are heat treatable to higher strength levels than the alpha-beta alloys. In addition, the beta alloys can be through-hardened to thicker sections than the alpha-beta alloys.

One disadvantage of heat treating titanium alloys is the lack of a nondestructive test method to determine the actual response to heat treatment. While hardness may be used with steel alloys and a combination of hardness and conductivity for aluminum alloys, there are no equivalent methods for titanium. Therefore, if verification is required, actual mechanical property tests must be conducted to determine heat treat response.

Stress relief is used to remove residual stresses that result from mechanical working, welding, cooling of castings, machining, and heat treatment. Stress-relief cycles may be omitted if the part is going to be annealed or solution treated. All titanium alloys can be stress relieved without affecting their strength or ductility. Like most thermal processes, combinations of time and temperature may be used, with higher temperatures requiring shorter times and lower temperatures requiring longer times. Thicker sections require longer times to ensure uniform temperatures. Either furnace or air cooling is usually acceptable. While the cooling rate is not critical, uniformity of cooling is important, especially through the 316 to 482 °C (600 to 900 °F) range. The stress-relief temperature for the near-alpha and alpha-beta alloys is in the range of 454 to 816 °C (850 to 1500 °F). Ti-6Al-4V is normally stress relieved at 538 to 649 °C (1000 to 1200 °F), with the time being dependent on the temperature used and the percent stress relief desired. At 8 h, 538, 593, and 649 °C (1000, 1100, and 1200 °F) will result in 55, 75, and 100% stress relief, respectively. During stress relieving of STA parts, care must be taken to prevent overaging to lower-than-desired strength levels.

Annealing is similar to stress relief but is done at higher temperatures to remove almost all residual stresses and the effects of cold work. Common types of annealing operations for titanium alloys include the following:

- *Mill annealing.* As the name implies, mill annealing is conducted at the mill. It is not a full anneal and may leave traces of cold or warm working in the microstructure of heavily worked products, particularly sheet. For near-alpha and alpha-beta alloys, this is the heat treatment normally supplied by the manufacturer. Mill annealing of Ti-6-4 can be achieved by heating to 704 to 782 °C (1300 to 1440 °F) and holding for a minimum of 1 h. Beta alloys are not supplied in the mill-annealed condition because this condition is not stable at elevated temperatures and can lead to the precipitation of embrittling phases.
- *Duplex annealing.* Can be used to provide better creep resistance for high-temperature alloys such as Ti-6242S. It is a two-stage annealing process that starts with an anneal high in the alpha + beta field, followed by air cooling. The second anneal is conducted at a lower temperature to provide thermal stability, again followed by air cooling.
- *Recrystallization annealing.* Used to improve the fracture toughness. The part is heated into the upper range of the alpha + beta field, held for a period of time, and then slowly cooled.
- *Beta annealing.* Conducted by annealing at temperatures above the beta transus, followed by air cooling or water quenching to avoid the formation of grain-boundary alpha. This treatment maximizes fracture toughness at the expense of a substantial decrease in fatigue strength.

A summary of these different annealing procedures and their effects on properties is given in Table 5.9. With the appropriate use of constraint fixtures, operations such as straightening, sizing, and flattening can be combined with annealing. The elevated-temperature stability of alpha-beta alloys is improved by annealing because the beta phase is stabilized.

Solution Treating and Aging. The purpose of the solution treatment is to transform a portion of the alpha phase into beta and then to cool rapidly enough to retain the beta phase at room temperature. During aging, alpha

Table 5.9 Effects of different anneal cycles on titanium properties

Property	Mill anneal(a)	Recrystallization anneal(b)	Duplex anneal(c)	Beta anneal(d)
Ultimate tensile strength	High	Low	Low	Low
Ductility	High	High	High	Lower
Fatigue strength	Intermediate	Intermediate	Lower	Lower
Fracture toughness	Lowest	High	Intermediate	Highest
Fatigue crack growth rate	Lowest	Intermediate	Intermediate	Highest
Creep resistance	Lowest	Lowest	Intermediate	Highest

(a) Roughly 165 to 250 °C (300 to 450 °F) below beta transus, air cool. (b) Roughly 28 to 56 °C (50 to 100 °F) below beta transus, slow cool. (c) Roughly 28 to 56 °C (50 to 100 °F) below beta transus, air cool followed by mill anneal. (d) Usually 28 to 56 °C (50 to 100 °F) above beta transus, air cool.

precipitates from the retained beta. Solution treating and aging is used with both alpha-beta and beta alloys to achieve higher strength levels than can be obtained by annealing. Solution treating consists of heating the part to high in the two-phase alpha + beta field, followed by quenching. Solution treatments for alpha-beta alloys are conducted by heating to slightly below the beta transus. To obtain the maximum strength with adequate ductility, it is necessary to solution treat within approximately 28 to 83 °C (50 to 150 °F) of the beta transus. When an alpha-beta alloy is solution treated, the ratio of beta phase to alpha phase increases and is maintained during quenching. The effect of solution-treating temperature on the strength and ductility of Ti-6Al-4V sheet is shown in Fig. 5.28. During aging, the unstable retained beta transforms into fine alpha phase, which increases the strength.

The cooling rate after solution heat treating has an important effect on the strength of alpha-beta alloys. For most alpha-beta alloys, quenching in water or an equivalent quenchant is required to develop the desired strength levels. The time between removing from the furnace and the initiation of the quench is usually about 7 s for alpha-beta alloys and as long as 20 s for beta alloys. For alloys with appreciable beta-stabilizing elements and moderate section thickness, air or fan cooling is usually adequate. Essentially, the amount and type of beta stabilizers in the alloy will determine the depth of hardening. Unless an alloy contains appreciable amounts of beta stabilizers, it will not harden through thick sections and will exhibit lower properties in the center where the cooling rates are lower.

Aging consists of reheating the solution-treated part in the range of 427 to 649 °C (800 to 1200 °F). A typical STA cycle for Ti-6Al-4V would be to solution treat at 904 to 927 °C (1660 to 1700 °F), followed by water quenching. Aging would then be conducted at 538 °C (1000 °F) for 4 h, followed by air cooling. Ti-6Al-4V is sometimes solution treated and overaged to achieve modest decreases in strength while obtaining improved fracture toughness and good dimensional stability.

The solution treatment for beta alloys is carried out above the beta transus. Commercial beta alloys are usually supplied in the solution-treated

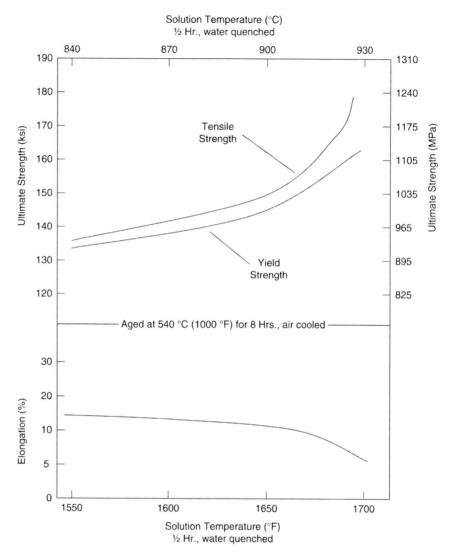

Fig. 5.28 Effect of solution heat treat temperature on Ti-6Al-4V sheet. Source: Ref 5.2

condition with a 100% beta structure to provide maximum formability and only need to be aged to achieve high strength levels. After forming, the part is aged to provide maximum strength. Beta-processed alloys have improved fracture toughness, better creep strength, and more resistance to stress-corrosion cracking; however, there is a considerable loss in ductility and fatigue strength. Beta alloys are usually air cooled from the solution-treating temperature.

Although there are coatings that can be used to protect titanium alloys during heat treatment from oxygen, hydrogen, and nitrogen, the best practice is to conduct the heat treatments in a vacuum furnace. Prior to heat treating, it is important that the surfaces are clean and free of all organic contaminants, including fingerprints. After heat treatment, any alpha case must be removed from the surface, either by machining or chemical milling.

5.12 Titanium Casting

All titanium castings have compositions based on those of the common wrought alloys. There are no commercial titanium alloys developed strictly for casting applications. This is unusual because in other metallic systems, alloys have been developed specifically as casting alloys, often to overcome certain problems such as poor castability of a wrought alloy composition. No unusual problems regarding castability or fluidity have been encountered in any of the titanium metals cast to date.

The major reason for selecting a titanium casting instead of a wrought titanium product is cost. This cost advantage may be attained through increased design flexibility, better use of available metal, or reduction in the cost of machining or forming parts. Generally, the more complex the part, the better the economics of using a casting. An example of the complexity that can be achieved with investment casting is shown in Fig. 5.29.

Titanium castings are unlike castings of other metals in that they may be equal or nearly equal in tensile and creep-rupture strengths to their wrought counterparts. Strength guarantees in most specifications for titanium castings are the same as for wrought forms. Typical ductilities of cast products, as measured by elongation and reduction in area, are lower than typical values for wrought products of the same alloys. Fracture toughness and crack propagation resistance may exceed those of corresponding wrought material. However, the fatigue strength of cast titanium is inferior to that of wrought titanium. Fortunately, the fatigue strength of cast titanium can be enhanced by further processing and heat treatment.

Fig. 5.29 Complex investment-cast titanium components used for gas turbine applications. Source: Ref 5.14

Titanium castings, similar to wrought titanium products, are used primarily in four areas of application: aerospace products, marine service, industrial (corrosion) service, and sporting goods. Commercially pure titanium is used for the vast majority of corrosion applications, whereas Ti-6Al-4V is the dominant alloy for aerospace and marine applications. Ti-6Al-2Sn-4Zr-2Mo-Si is selected with increasing frequency for elevated-temperature service. Castings also have been supplied in alloys Ti-5Al-2.5Sn, Ti-15V-3Cr-3Al-3Sn, and Ti-6Al-6V-2Sn, as well as in several European alloys.

Titanium castings now are used extensively in the aerospace industry and to lesser but increasing measure in the chemical process, marine, and other industries. Sporting goods, primarily investment-cast golf club heads, are now the second leading application area for titanium (second only to the commercial aerospace industry). Titanium castings still represent a small portion of the titanium industry, only about 1 to 2% of the total weight shipped.

Titanium castings have been cast in machined graphite molds, rammed graphite molds, and proprietary investments used for precision investment casting. Significant design complexity, tolerance, and surface finish control have been achieved, and large parts can be cast. Porosity continues to be a potential problem. However, by use of hot isostatic pressing (HIP), internal soundness of titanium castings can be improved to the point that no porosity or small voids can be detected.

In the investment casting process, shown in Fig. 5.30, a pattern of the part is produced from wax. The waxes are formulated to give smooth, defect-free surfaces, be stable, maintain tolerances, and have a relatively long shelf life. The wax patterns are then robotically dipped in a fine ceramic slurry that contains refractories such as silica or alumina. The coated patterns are then stuccoed with dry coarser particles of the same material to make the slurry dry faster and ensure adhesion between the layers. The dipping and stuccoing process is repeated until the desired thickness is obtained, usually six to eight times. After the mold is completely dry, it is placed in an oven and the wax is melted out. The ceramic mold is then fired at approximately 982 °C (1800 °F).

The titanium melt is produced by vacuum arc remelting titanium in a water-cooled copper crucible before pouring it into the mold. Sufficient preheat of the melt and preheating of the molds is used to maximize flow to achieve complete mold filling. As-cast titanium has a microstructure typical of titanium alloys worked in the beta field, which has lower ductility and fatigue strength than equiaxed structures. Due to the slow cooling rate from the HIP temperature, titanium castings are often heat treated to refine the microstructure and eliminate grain-boundary alpha, large alpha plate colonies, and individual alpha plates.

One of the problems with investment castings has been shell inclusions (Fig. 5.31), which are small pieces of the ceramic shell that flake off dur-

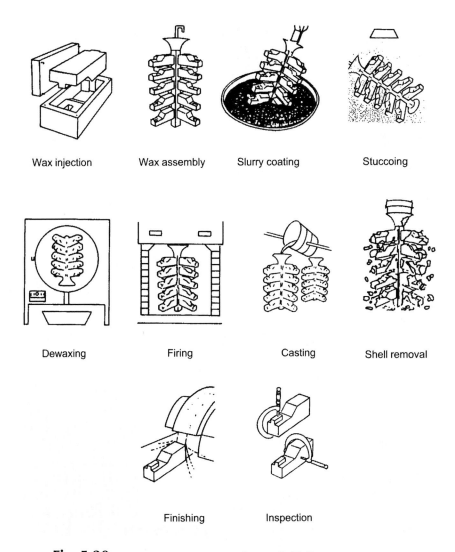

Wax injection   Wax assembly   Slurry coating   Stuccoing

Dewaxing   Firing   Casting   Shell removal

Finishing   Inspection

Fig. 5.30 Investment casting process. Source: Ref 5.15

ing casting and can cause contamination that adversely affects fatigue strength. Very extensive and expensive nondestructive inspection procedures have been developed to detect this defect and others in castings. To eliminate surface contamination, castings are usually chemical milled after casting. Unfortunately, titanium investment castings have a tendency not to completely fill the mold during casting, particularly for large and/or complex castings. Weld repair of surface defects must be carefully done to avoid oxygen and hydrogen pickup. Repairs are usually done using gas tungsten arc welding with filler wire. Extralow interstitial filler wire is often used to help minimize the potential for oxygen contamination. Weld-repaired castings must be stress relieved after welding. Fortunately, test

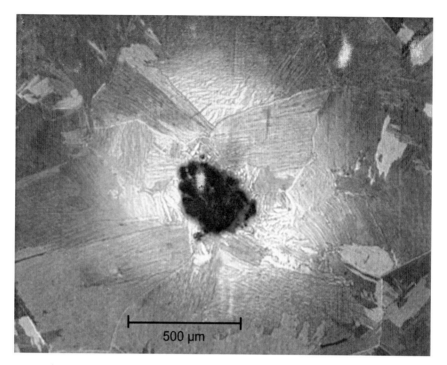

Fig. 5.31 Shell inclusion cross section

programs have shown that the fatigue properties of castings with properly conducted weld repairs are not degraded.

To improve the fatigue properties, all aerospace-grade titanium castings are processed by HIP to close off any internal porosity. Typical HIP processing is done under argon pressure of 103 MPa (15 ksi) at 954 °C (1750 °F) for 2 h. It should be noted that HIP will collapse internal porosity but will not close off surface-connected porosity, hence the need to repair those areas by welding prior to HIP. The microstructure improvement of a Ti-6Al-4V investment casting as a result of HIP is shown in Fig. 5.32.

Initially, HIP was used with excellent results to salvage parts that had been rejected after radiographic inspection. The effectiveness of the technique gave rise to plans to use HIP for routine parts, but high costs made such plans economically questionable (virtually all aerospace castings are HIPed). However, for certain casting configurations, adequate feeding by use of conventional risering is virtually impossible, and, to meet aerospace nondestructive inspection standards, shrinkage voids are closed by welding. From a technical viewpoint, HIP is a heat treatment, although some studies have claimed that HIP alone does little, if anything, to enhance mechanical properties of Ti-6Al-4V castings. Properties of HIPed alloys are a function of the HIP temperature relative to the beta transus and the post-HIP heat treatment. With castings of marginal to substandard quality, HIP raises the lower limit of data scatter and raises the degree of confidence in the reliability of cast products.

Hot isostatic pressing is considered by many to be a process that simplifies the problem of defining a standard for internal casting quality. At the same time, use of HIP ensures that subsurface microporosity will be healed and therefore will not become exposed on a subsequently machined or polished surface to mar the finish or to act as a possible site for fatigue crack propagation.

Cast titanium alloys are equal or nearly equal in strength to wrought alloys of the same compositions. However, typical ductilities are below the typical values for comparable wrought alloys but still above the guaranteed minimum values for the wrought metals. Because castings of Ti-6Al-4V have been used in aerospace applications, the most extensive data have been developed for this alloy. Typical room-temperature tensile properties are given in Table 5.10 for cast CP titanium and for eight cast titanium alloys. Figure 5.33 compares plane-strain fracture toughness values for

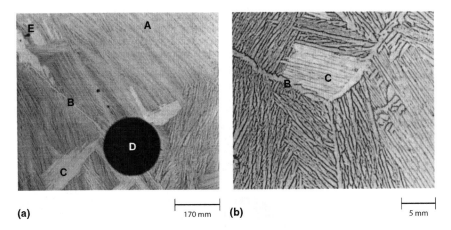

(a) 170 mm **(b)** 5 mm

Fig. 5.32 Comparison of the microstructures of (a) as-cast versus (b) cast + hot isostatic pressed Ti-6Al-4V alloys, illustrating lack of porosity in (b). Grain-boundary α (area B) and α plate colonies (area C) are common to both alloys; β grains (area A), gas (area D), and shrinkage voids (area E) are present only in the as-cast alloy. Source: Ref 5.16

Table 5.10 Typical room-temperature tensile properties of titanium alloy castings (bars machined from castings)

Alloy(a)(b)	Ultimate strength		Yield strength		Elongation, %	Reduction of area, %
	MPa	ksi	MPa	ksi		
Commercially pure (grade 2)	552	80	448	65	18	32
Ti-6Al-4V, annealed	930	135	855	124	12	20
Ti-6Al-4V-ELI	827	120	758	110	13	22
Ti-6Al-2.75Sn-4Zr-0.4Mo-0.45Si, β-STA(c)	938	136	848	123	11	20
Ti-6Al-2Sn-4Zr-2Mo, annealed	1006	146	910	132	10	21
Ti-5.8Al-4.0Sn-3.5Zr-0.5Mo-0.7Nb-0.35Si, β-STA(c)	1069	155	952	138	5	8
Ti-6Al-2Sn-4Zr-6Mo, β-STA(c)	1345	195	1269	184	1	1
Ti-3Al-8V-6Cr-4Zr-4Mo, β-STA(c)	1330	193	1241	180	7	12
Ti-15V-3Al-3Cr-3Sn, β-STA(c)	1275	185	1200	174	6	12

Note: Specification minimums are less than these typical properties. (a) Solution treated and aged (STA) heat treatments may be varied to produce alternate properties. (b) ELI, extra low interstitial. (c) Beta STA, solution treatment within the β phase field followed by aging. Source: Ref 5.4

Ti-6Al-4V castings with values for Ti-6Al-4V plate and other wrought titanium alloys. The high toughness is due to the beta-processed-type microstructure inherent in castings.

Generally, an improvement in fatigue properties and a reduction in the scatter of fatigue data is achieved through HIP (Fig. 5.34). Castings of

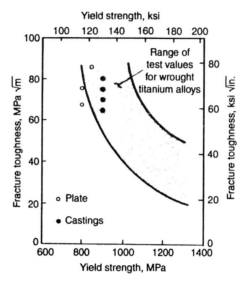

Fig. 5.33 Fracture toughness of Ti-6Al-4V castings compared with that of Ti-6Al-4V plate and other titanium alloys. Source: Ref 5.4

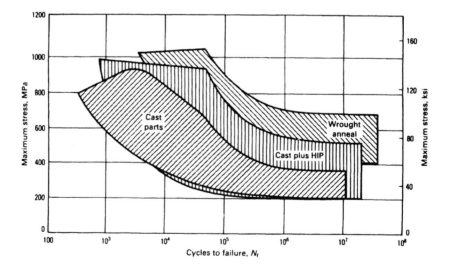

Fig. 5.34 Effect of hot isostatic pressing (HIP) on fatigue properties of Ti-6Al-4V investment castings. Room-temperature smooth bar; tension-tension fatigue; $R = +0.1$. Source: Ref 5.4

alloys such as Ti-6Al-4V will generally have static and fatigue properties lower than wrought products. Fracture properties, such as fracture toughness, fatigue crack growth rate, and stress-corrosion resistance, are superior to those of mill-annealed wrought Ti-6Al-4V.

5.13 Machining

Use low cutting speeds . . . maintain high feed rates . . . use copious amounts of cutting fluid . . . use sharp tools and replace them at the first signs of wear . . . moving parts of the machine tool should be free from backlash or torsional vibrations . . .

"Titanium Machining Techniques," Timet Titanium Engineering Bulletin No. 7, 1960s.

Unfortunately, not much has changed in machining titanium since the mid-1960s. Titanium was a difficult-to-machine metal then and remains a difficult-to-machine metal. Titanium is difficult to machine for several reasons:

- Titanium is very reactive, and the chips tend to weld to the tool tip, leading to premature tool failure due to edge chipping. Almost all tool materials tend to react chemically with titanium when the temperature exceeds 510 °C (950 °F).
- The low thermal conductivity of titanium causes heat to build up at the tool-workpiece interface. High temperatures at the cutting edge are the principal reason for rapid tool wear. When machining Ti-6-4, approximately 80% of the heat generated is conducted into the tool due to the low thermal conductivity of titanium. The thermal conductivity of titanium is approximately $\frac{1}{6}$ that of steel. This should be contrasted with high-speed machining of aluminum, in which almost all of the heat of machining is ejected with the chip.
- The relatively low modulus of titanium causes excessive workpiece deflection when machining thin walls; that is, there is a bouncing action as the cutting edge enters the cut. The low modulus of titanium is a principal cause of chatter during machining operations.
- Titanium maintains its strength and hardness at elevated temperatures, contributing to cutting tool wear. Very high mechanical stresses occur in the immediate vicinity of the cutting edge when machining titanium. This is another difference between titanium and the high-speed machining of aluminum; aluminum becomes very soft under high-speed machining conditions.

Table 5.11 gives machinability comparisons of several titanium alloys with other materials (higher numbers indicate improved/lower-cost machinability). *Improper machining procedures, especially grinding*

Table 5.11 Machinability comparisons of several titanium alloys with other materials

Alloy	Condition(a)	Machinability rating(b)
Aluminum alloy 2017	STA	300
B1112 resulfurized steel	HR	100
1020 carbon steel	CD	70
4340 alloy steel	A	45
Commercially pure titanium	A	40
302 stainless steel	A	35
Ti-5Al-2.5Sn	A	30
Ti-6Al-4V	A	22
Ti-6Al-6V-2Sn	A	20
Ti-6Al-4V	STA	18
HS25 (Co-base)	A	10
Rene' 41 (Ni-base)	STA	6

(a) STA, solution treated and aged; HR, hot rolled; CD, cold drawn; A, annealed. (b) Based on a rating of 100 for B1112 steel. Source: Ref 5.4

operations, can cause surface damage to the workpiece that will dramatically reduce fatigue life.

The following guidelines are well established for the successful machining of titanium:

- Use slow cutting speeds. A slow cutting speed minimizes tool edge temperature and prolongs tool life. Tool life is extremely short at high cutting speeds. As speed is reduced, tool life increases.
- Maintain high feed rates. The depth of cut should be greater than the work-hardened layer resulting from the previous cut.
- Use generous quantities of cutting fluid. Coolant helps in heat transfer, reduces cutting forces, and helps to wash chips away.
- Maintain sharp tools. As the tool wears, metal builds up on the cutting edge, resulting in a poor surface finish and excessive workpiece deflection.
- Never stop feeding while the tool and workpiece are in moving contact. Tool dwell causes rapid work hardening and promotes smearing, galling, and seizing.
- Use rigid setups. Rigidity ensures a controlled depth of cut and minimizes part deflection.

Rigid machine tools are required to machine titanium. Sufficient horsepower must be available to ensure that the desired speed can be maintained for a given feed rate and depth of cut. Titanium requires approximately 0.8 hp/in.^{3} of material removed per minute. The base and frame should be massive enough to resist deflections, and the shafts, gears, bearings, and other moving parts should run smoothly with no backlash, unbalance, or torsional vibrations. Rigid spindles with larger taper holders are recommended; a number 50 taper or equivalent provides stability and the mass to counter the axial and radial loads encountered when machining titanium.

Cutting tools used for machining titanium include cobalt-containing high-speed tool steels, such as M33, M40, and M42, and the straight tungsten carbide grade C-2 (ISO K20). While carbides are more susceptible to chipping during interrupted cutting operations, they can achieve approximately a 60% improvement in metal-removal rates compared to high-speed steels (HSS). Ceramic cutting tools have not made inroads in titanium machining due to their reactivity with titanium, low fracture toughness, and poor thermal conductivity. It should be noted that although improvements in cutting tool materials and coatings have resulted in tremendous productivity improvements in machining for a number of materials (e.g., steels), none of these improvements have been successful with titanium.

Cutting fluids are required to achieve adequate cutting tool life in most machining operations. Flood cooling is recommended to help remove heat and act as a lubricant to reduce the cutting forces between the tool and workpiece. A dilute solution of rust inhibitor and/or water-soluble oil at 5 to 10% concentration can be used for higher-speed cutting operations, while chlorinated or sulfurized oils can be used for slower speeds and heavier cuts to minimize frictional forces that cause galling and seizing. The use of chlorinated oils requires careful cleaning after machining to remove the possibility of stress-corrosion cracking.

For the production of airframe parts, end milling and drilling are the two most important machining processes, while turning and drilling are the most important for jet engine components. In turning operations, carbide tools are recommended for continuous cuts to increase productivity, but for heavy interrupted cutting operations, high-speed tool steel tools are needed to resist edge chipping. Tools must be kept sharp and should be replaced at a wear land of approximately 0.38 mm (0.015 in.) for carbide and 0.76 mm (0.030 in.) for HSS. Tool geometry is important, especially the rake angle. Negative rake angles should be used for rough turning with carbide, while positive rakes are best for semifinishing and finishing cuts and for all operations using HSS.

When milling titanium, climb milling (Fig. 5.35), rather than conventional milling, is recommended to minimize tool chipping, the predominant failure mode in interrupted cutting. In climb milling, the tooth cuts a minimum thickness of chip, minimizing the tendency of the chip to adhere to the tool as it leaves the workpiece.

Slow speeds and uniform positive feeds help to reduce tool temperature and wear. Tools should not be allowed to dwell in the cut or rub across the workpiece. This will result in rapid work hardening of the titanium, making it even more difficult to cut. Both carbide and HSS cutting tools can be used; however, carbide tools are more susceptible to chipping and may not perform as well in heavy interrupted cutting operations. Increased

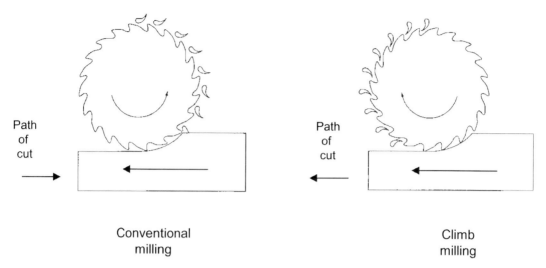

Path
of
cut

**Conventional
milling**

Path
of
cut

**Climb
milling**

Fig. 5.35 Conventional and climb milling

relief angles increase tool life by reducing pressure and deflection. Rigid machine tools and setups are necessary, along with flood cooling.

Many of the same guidelines for turning and milling also apply to drilling titanium. During drilling, positive power-feed equipment is preferred over hand drilling to reduce operator fatigue and provide more consistent hole quality. Blunt point angles (140°) work best for small-diameter holes (6.3 mm, or $\frac{1}{4}$ in., and less), while sharp points (90°) work better for larger-diameter holes where higher pressures are needed to feed the drill. Relief angles also affect tool life; too small a relief angle will cause smearing and galling, while too large a relief angle results in edge chipping. When drilling holes deeper than one hole diameter, peck drilling will produce more consistent and higher-quality holes. During peck drilling, the drill motor periodically extracts the bit from the hole to clear chips from the flutes and hole. Because it is not practical to use flood coolant during many assembly drilling operations, drills equipped with air blast cooling have been found to be effective in thin-gage sheet.

In all machining operations, and especially grinding, the surface of titanium alloys can be damaged to the extent that it adversely affects fatigue life. Damage, including microcracks, built-up edges, plastic deformation, heat-affected zones, and tensile residual stresses, can result from improper machining operations. As shown in Fig. 5.36, machining operations that induce residual compressive stresses on the surface, such as gentle grinding, milling, and turning, are much less susceptible to fatigue life reductions. Note the devastating results for improper grinding procedures. Postmachining processes, such as grit blasting or shot peening, can also

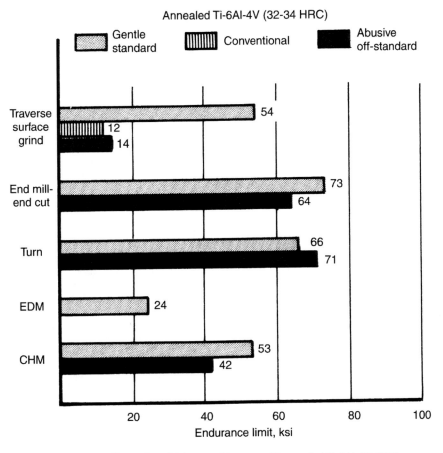

Fig. 5.36 Effects of machining conditions on mill-annealed Ti-6Al-4V. EDM, electrical discharge machining. Source: Ref 5.3

be used to induce residual compressive stresses at the surface and improve fatigue life, as shown in Fig. 5.37.

Titanium is an extremely reactive metal, and fine particles of titanium can ignite and burn; however, the use of flood coolant in most machining operations eliminates this danger to a large extent. Chips should be collected, placed in covered steel containers for recycling, and preferably stored outside of the building. In the event of a titanium fire, water should never be applied directly to the fire; it will immediately turn to steam and possibly cause an explosion. Instead, special extinguishers containing dry salt powders developed specifically for metal fires should be used.

Chemical milling is often used to machine pockets in skins in lower-stressed areas to save weight. The use of maskants allows multiple step-cuts, and tolerances as tight as 0.025 mm (0.001 in.) are possible. Chemical milling is conducted by masking the areas in which no milling is desired and then etching in a solution of nitric-hydrofluoric acid. The

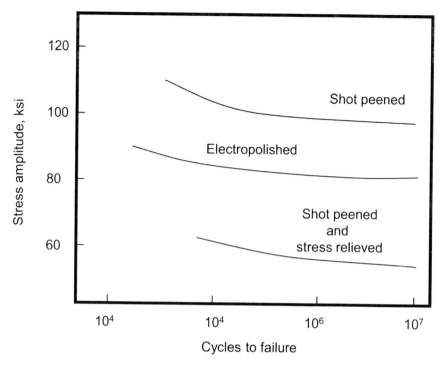

Fig. 5.37 Effects of surface residual stress on fatigue strength. Source: Ref 5.17

hydrofluoric acid removes the titanium by etching, while the nitric acid limits the pickup of hydrogen. After milling, the maskant is stripped.

5.14 Titanium Joining

Adhesive bonding, mechanical fastening, metallurgical bonding (e.g., diffusion bonding), and welding are used routinely and successfully to join titanium and its alloys. The first three processes do not affect the properties of these metals as long as joints are properly designed. Metallurgical bonding includes all solid-state joining processes in which diffusion or deformation plays the major role in bonding the members together. Because these processes are performed below but close to the beta transus, metallurgical effects, either normally caused by heating at that temperature or resulting from contamination, should be anticipated. Properly processed joints have similar properties to the base metal and, because diffusion bonding is carried out at a temperature high in the alpha-beta field, material properties appear similar to those resulting from high-temperature annealing. With most alloys, a final low-temperature anneal produces properties characteristic of typical annealed material and provides thermal stability.

Welding has the greatest potential for affecting material properties. In all types of welds, contamination by interstitial impurities such as oxygen and nitrogen must be closely controlled to maintain useful ductility in the weldment. Alloy composition, welding procedure, and subsequent heat treatment are very important in determining the final properties of welded joints.

Mechanical properties for representative alloys and types of welds can be summarized as follows:

- Welding generally increases strength and hardness.
- Welding generally decreases tensile and bend ductility.
- Welds in unalloyed titanium grades 1, 2, and 3 do not require postweld treatment unless the material will be highly stressed in a strongly reducing atmosphere. In such event, stress relieving or annealing may prove useful. In addition, if the components operate at elevated temperatures, a stress relief is recommended. The elevated temperature could accelerate hydrogen migration to the high residual stresses, resulting in hydrogen embrittlement.
- Welds in beta-rich alloys such as Ti-6Al-6V-2Sn may have a high tendency to fracture with little or no plastic straining. Weld ductility can be improved by postweld heat treatment consisting of slow cooling from a high annealing temperature.
- Rich beta-stabilized alloys can be welded, and such welds can be produced with good ductility. Satisfactory properties are extremely difficult to produce in some beta alloy welds, however.

The CP grades, near-alpha alloys, and some of the alpha-beta alloys exhibit excellent weldability. The weldability of some of the higher-strength alpha-beta alloys, such as Ti-6-6-2 and Ti-6246, are not as good as Ti-6Al-4V because they have a tendency to hot crack. Some of the beta-stabilizing alloying elements added to these alloys lower the melting point and extend the freezing range. Thermal stresses introduced during cooling can cause the partially solidified weld to separate, a defect known as liquation cracking. Some of the highly beta-alloyed systems also have low ductility in both the fusion and heat-affected zones. While most titanium alloys require a postweld stress relief, especially if the part is going to be subjected to fatigue loading, CP grades 1, 2, and 3 do not require postweld stress relief unless the part will be highly stressed in a reducing atmosphere.

Gas tungsten arc welding (GTAW), gas metal arc welding (GMAW), plasma arc welding (PAW), and electron beam (EB) welding are the well-established fusion welding methods for titanium, and laser welding is a rapidly emerging technology. The relative merits of GTAW, GMAW, PAW, and EB welding are given in Table 5.12. Many common welding processes are unsuitable for titanium because titanium tends to react with the fluxes and gases used, including gas welding, shielded metal arc, flux core, and submerged arc welding. In addition, titanium cannot be welded to many

Table 5.12 Relative comparison of titanium welding methods

	Gas tungsten arc welding	Gas metal arc welding	Plasma arc welding	Electron beam welding
Common thickness range, mm (in.)	0.8–6.4 (1/32–$\frac{1}{4}$)	6.4–76+ ($\frac{1}{4}$–3+)	3.2–9.5 ($\frac{1}{8}$–$\frac{3}{8}$)	Foil–76 (3)
Ease of welding	Good	Fair	Good	Excellent
Grooved joint required	Often	Always	No	No
Automatic or manual	Both	Automatic	Both	Automatic
Mechanical properties	Fair	Fair	Good	Excellent
Quality of joint	Good	Fair–good	Excellent	Excellent
Equipment cost	Low	High	Moderate	Very high
Wire required	Often	Always	Sometimes	No
Distortion	Very high	High	Moderate	Very low

dissimilar metals due to the formation of intermetallic compounds. However, successful welds can be made to zirconium, tantalum, and niobium.

Because fusion welding results in melting and resolidification, there will exist a microstructural gradient from the as-cast nugget through the heat-affected zone to the base metal. Preheating the pieces to be welded helps to control residual-stress formation on cooling. It is also a common practice to use weld start and run-off tabs to improve weld quality. Fusion welding usually increases the strength and hardness of the joint material while decreasing its ductility.

Attention to cleanliness and the use of inert gas shielding, or vacuum, are critical to obtaining good fusion welds. Molten titanium weld metal must be totally protected from contamination by air. Because oxygen and nitrogen from the atmosphere will embrittle the joint, all fusion welding must be conducted using either a protective atmosphere (i.e., argon or helium) or in a vacuum chamber. Argon, helium, or a mixture of the two is used for shielding. Helium gases operate at higher temperatures than argon, allowing greater weld penetrations and faster speeds, but the hotter helium-shielded arc is less stable, requiring better joint fit-up and more operator skill. Therefore, argon is usually the preferred shielding gas. All shielding gases should be free of water vapor; a dewpoint of –46 °C (–50 °F) is recommended. The hot heat-affected zones and root side of the welds must also be protected until the temperature drops below 425 °C (800 °F). If fusion welding is conducted in an open environment, local trailing and backing shields must be used to prevent weld contamination. The color of the welded joint is a fairly good way to assess atmospheric contamination. Welds that appear bright silver to straw indicate no to minimal contamination, while light blue to dark blue indicates unacceptable contamination.

Cleanliness is important because titanium readily reacts with moisture, grease, refractories, and most other metals to form brittle intermetallic compounds. Prior to welding, all grease and oil must be removed with a nonchlorinated solvent such as toluene or methyl ethyl ketone. Surface oxide layers can be removed by pickling or stainless steel wire brushing of the joint area. If the oxide layer is heavy, grit blasting or chemical descaling should be conducted prior to pickling. After pickling, the cleaned material should be wrapped in wax-free kraft paper and handled with clean white cotton gloves. Filler wire should be wrapped and stored in a clean dry location when not in use.

Gas tungsten arc welding is the most common fusion welding method for titanium. In GTAW, the welding heat is provided by an arc maintained between a nonconsumable tungsten electrode and the workpiece. In GTAW, as shown in Fig. 5.38, the power supply is direct current with a negative electrode. The negative electrode is cooler than the positive weld joint, enabling a small electrode to carry a large current, resulting in a deep weld penetration with a narrow weld bead. The weld puddle and adjacent heat-affected zone (HAZ) on the weld face are protected by the nozzle gas; trailing shields are used to protect the hot solidified metal and the HAZ behind the weld puddle; and back-up shielding protects the root of the weld and its adjacent HAZ. GTAW can be accomplished either manually or automatically in sheet up to approximately 3.2 mm (0.125 in.) in thickness without special joint preparation or filler wire. For thicker gages, grooved joints and filler wire additions are required. If high joint ductility is needed, unalloyed filler wire can be used at some sacrifice in joint strength. In manual welding, it is important the operator make sure the tungsten electrode does not make contact with the molten weld bead, because tungsten contamination can occur. Electrodes with ceria (2% cerium oxide) or lanthana (1 to 2% lanthanum oxide) are recommended because they produce better weld stability, superior arc starting characteristics, and operate cooler for a given current density than pure tungsten electrodes. Conventional GTAW equipment can be used but requires the addition of appropriate argon or helium shielding gases. Protection can be provided by either rigid chambers or collapsible plastic tents that have been

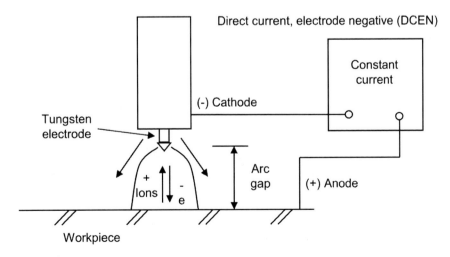

Fig. 5.38 Gas tungsten arc welding schematic

thoroughly purged with argon. Other methods of local shielding have also been successfully used.

Plasma arc welding is similar to GTAW except that the plasma arc is constricted by a nozzle which increases the energy density and welding temperature. The higher energy density allows greater penetration than GTAW and faster welding speeds.

Gas metal arc welding uses a consumable electrode rather than a non-consumable electrode that is used in the GTAW process. In GMAW, as shown in Fig. 5.39, the power supply is direct current with a positive electrode. The positive electrode is hotter than the negative weld joint, ensuring complete fusion of the wire in the weld joint. GMAW has the advantage of more weld metal deposit per unit time and unit of power consumption. For plates 12.7 mm (0.5 in.) and thicker, it is a more cost-effective process than GTAW. However, poor arc stability can cause appreciable spatter during welding, which reduces its efficiency.

Electron beam welding uses a focused beam of high-energy electrons, resulting in a high depth of penetration and the ability to weld sections up to 76 mm (3 in.) thick. Other advantages of EB are a very narrow HAZ, low distortion, and clean welds, because the welding is conducted in a vacuum chamber. The biggest disadvantage of EB is the high capital equipment cost, as shown for the EB welding chamber in Fig. 5.40. Electron beam welding has been successfully used on two major fighter programs to make large unitized structures. In the late 1960s, Grumman Aerospace used EB welding on the F-14 aircraft to weld the folding wingbox. Because computer controls were not available at that time, almost every weld was

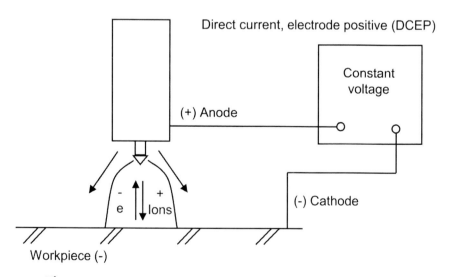

Fig. 5.39 Gas metal arc welding schematic

made through constant-thickness material with the beam perpendicular to the surface. In addition, a separate setup was required for almost every weld. Nevertheless, Grumman successfully delivered over 700 production units. More recently, Boeing used EB welding extensively on the aft fuselage of the F-22 aircraft. The four major assemblies welded were the aft and forward booms (two each), which were fabricated from beta-annealed Ti-6Al-4V. This amounts to over 76 linear meters (3000 linear in.) of weldment, which required many less setups due to improvements in computer controls. Weld thickness ranged from 6.4 mm (0.25 in.) to slightly more than 25 mm (1 in.). Gun-to-work distances varied from 25 to 64 cm (10 to 25 in.), with beam swings of more than 130°. Both of these applications avoided the costly and error-prone process of installing thousands of mechanical fasteners.

Laser beam (LB) welding uses a high-intensity coherent beam of light, which, similar to EB, results in a very narrow HAZ. However, LB welding is limited to sheet and plate up to approximately 13 mm (0.50 in.). One big advantage of LB is that it can be conducted in the open atmosphere with appropriate shielding, while EB requires a vacuum chamber.

Fig. 5.40 Low-voltage electron beam welding unit consisting of a 3505 × 2845 × 2690 mm (138 × 112 × 106 in.) chamber. Courtesy of Sciaky Bros., Inc. Source: Ref 5.18

Resistance Welding. Titanium alloys can be readily resistance welded using both spot and seam welding. As with all titanium welding, cleanliness of the material to be welded is mandatory. Due to the rapid thermal cycles experienced with resistance welding, inert gas shielding is not required.

Diffusion bonding is a solid-state joining process that relies on the simultaneous application of heat and pressure to facilitate a bond that can be as strong as the parent metal. Diffusion bonding, as shown in Fig. 5.41, occurs in four steps:

1. Development of intimate physical contact through the deformation of surface roughness through creep at high temperature and low pressure
2. Formation of a metallic bond
3. Diffusion across the faying surfaces
4. Grain growth across the original interface.

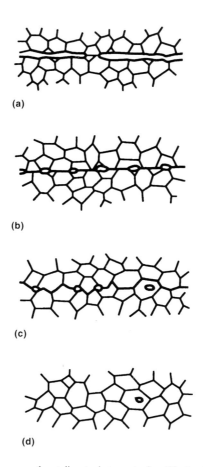

(a)

(b)

(c)

(d)

Fig. 5.41 Sequence of metallurgical stages in the diffusion bonding process. (a) Deformation of surfaces at high temperature and low pressure. (b) Formation of a metallic bond. (c) Diffusion across the faying surfaces. (d) Grain growth across interface. Source: Ref 5.19

Clean and smooth faying surfaces are necessary to create high-quality bonds. One advantage of titanium is that at temperatures exceeding 621 °C (1150 °F), it dissolves its own surface oxide (TiO_2), leaving a surface that is very amenable to diffusion bonding. Another advantage is that many titanium alloys display superplasticity at the bonding temperatures (899 to 954 °C, or 1650 to 1750 °F), which greatly facilitates intimate contact. Typical diffusion-bonding parameters for Ti-6-4 are 899 to 949 °C (1650 to 1740 °F) at 1.4 to 13.8 MPa (200 to 2000 psi) for 1 to 6 h.

Brazing is often used to fabricate sandwich assemblies that contain titanium skins and welded titanium honeycomb core. Titanium brazing must be conducted in either a vacuum or an inert gas atmosphere using either induction, furnace, or vacuum chambers. Fixture material selection is important; for example, nickel will react with titanium to form low-melting-point eutectics, causing the part and fixture to fuse; coated graphite and carbon steel fixtures work well. Brazing can be achieved using several different classes of alloys: aluminum alloys, silver-base alloys, Ti-Cu or Ti-Cu-Ni/Zr alloys, Cu-Ni or Cu-Ni-Ti alloys, and Ti-Zr alloys. Braze materials for use in the 870 to 925 °C (1600 to 1700 °F) range include Ag-10Pd, Ti-15Ni-15Cu, Ti-20Ni-20Cu, and 48Ti-48Zr-4Be, while lower-temperature materials, such as 3003 Al and other aluminum compositions, have been used when the use temperature is in the range of 315 °C (600 °F). When corrosion resistance is a priority, the 48Ti-48Zr-4Be and 43Ti-43Zr-12Ni-2Be materials should be considered. Another alloy, Ag-9Pd-9Ga, which flows at 899 to 913 °C (1650 to 1675 °F), offers good filling when large gaps are encountered. The disadvantages of silver- and aluminum-base alloys is that they form brittle intermetallics, are corrosion prone, and are limited to lower-temperature (<316 °C, or 600 °F) usage.

Diffusion brazing, also known as liquid interface diffusion bonding and transient liquid-phase (TLP) bonding, is a process similar yet somewhat different than conventional brazing. Like brazing, a separate interfacial material is used that melts at a lower temperature than the base metals and forms a liquid at the interface. Interlayers in the form of foils or electroplated, evaporated, or sputtered coatings are used in the joint. In TLP bonding, the interlayer is heated to the bonding temperature and melted. Liquid film formation depends on the formation of a low-melting-point eutectic or peritectic at the joint interface. During an isothermal hold at the bonding temperature, diffusion changes the composition of the joint, and the joint filler layer solidifies. Additional time at temperature allows time for further diffusion, resulting in a dilution of the interlayer elements with minimal effect on the parent metal. All stages of TLP bonding depend on solute diffusion from the joint region into the base material. Structural titanium honeycomb panels have been bonded using a nickel-titanium eutectic alloy.

5.15 Corrosion Resistance of Titanium and Titanium Alloys

Unalloyed titanium is highly resistant to corrosion by many natural environments, including seawater, body fluids, and fruit and vegetable juices. Titanium is used extensively for handling salt solutions (including chlorides, hypochlorides, sulfates, and sulfides), wet chlorine gas, and nitric acid solutions. On the other hand, hot concentrated low-pH chloride salts (such as boiling 30% $AlCl_3$ and boiling 70% $CaCl_2$) corrode titanium. Warm or concentrated solutions of HCl, H_2SO_4, H_3PO_4, and oxalic acid also are damaging. In general, all acidic solutions that are reducing in nature corrode titanium unless they contain inhibitors. Strong oxidizers, including anhydrous red fuming nitric acid and 90% H_2O_2, also cause attack. Ionizable fluoride compounds, such as NaF and HF, activate the surface and can cause rapid corrosion; dry chlorine gas is especially harmful.

Titanium has limited oxidation resistance in air at temperatures above approximately 540 °C (1000 °F), and extended exposure above this temperature for structural applications is not recommended, although some alloys have recently been developed that can be used at higher temperatures. Exposure to liquid or gaseous oxygen, nitrogen tetroxide, or red fuming nitric acid can cause titanium to react violently under impact loading or loading above the yield point where the protective surface ruptures, exposing an activated surface.

Titanium has been used to contain liquid or supercritical hydrogen at cryogenic temperatures, but above –100 °C (–150 °F), hydrogen may severely embrittle titanium. The potential for embrittlement is enhanced where hydrogen flow rates are high or where coatings on the titanium become damaged.

In unalloyed titanium and many titanium alloys, weld zones are just as resistant to corrosion as the base metal. Other fabrication processes (such as bending, forming, and machining) also appear to have no influence on basic corrosion resistance.

Galvanic Corrosion. Coupling of titanium to dissimilar metals usually does not accelerate corrosion of the titanium except in reducing environments, where titanium does not become passivated. Under reducing conditions, it has a galvanic potential similar to that of aluminum and undergoes accelerated corrosion when coupled to more noble metals.

In most environments, titanium is the cathodic member of any galvanic couple. It may accelerate corrosion of the other member but, in most instances, is itself unaffected. If the surface area of the titanium exposed to the environment is small in relation to the exposed surface area of the other metal, the effect of the titanium on the corrosion rate of the other metal is negligible. However, if the exposed area of titanium greatly exceeds that of the other metal, severe corrosion of the other metal can result.

Because titanium is nearly always the cathodic member of any galvanic couple, hydrogen may be evolved at its surface in an amount proportional to the galvanic current flow. This can result in formation of surface hydride films that generally are stable and cause no problems. At temperatures above 75 °C (170 °F), however, the hydrogen may diffuse into the titanium metal, causing embrittlement. In some environments, titanium hydride is unstable and decomposes or reacts, with a resultant loss of metal.

Anodic control of the corrosion reaction predominates when titanium is exposed to a reducing acid such as hydrochloric or sulfuric. Alloying with elements that reduce anodic activity therefore should improve corrosion resistance. This can be accomplished by using alloying elements that:

- Shift the corrosion potential of the alloy in the positive direction (cathodic alloying)
- Increase the thermodynamic stability of the alloy and thus reduce the ability of the titanium to dissolve anodically
- Increase the tendency of titanium to passivate

The first group includes noble metals such as platinum, palladium, and rhodium. The second includes nickel, molybdenum, and tungsten. The third group includes zirconium, tantalum, chromium, and possibly molybdenum. Considerable work has been done on the use of noble metals as alloying additions in titanium. An outgrowth of this work has been the development of Ti-0.2Pd, which has considerably greater resistance to corrosion in reducing environments than does unalloyed titanium (ruthenium additions have been found to have the same effect). In addition, work on alloying for thermodynamic stability has resulted in Ti-2Ni, which was developed for service in hot brine environments where crevice corrosion is sometimes a problem.

Various studies have shown that crevice corrosion resistance of titanium is improved by addition of molybdenum. The commercial alloy Ti Code 12, which contains 0.3% Mo and 0.8% Ni, combines some of the favorable properties of nickel and molybdenum additions while avoiding the negative aspects. This alloy has excellent resistance to pitting and crevice corrosion in high-temperature brines, which sometimes attack CP titanium, and also has better resistance to oxidizing environments such as nitric acid.

Crevice Corrosion. Titanium is subject to crevice corrosion in brine solutions containing oxidizers. Although crevice corrosion of titanium is observed most often in hot chloride solutions, it also occurs in iodide, bromide, and sulfate solutions. Susceptibility increases with increasing temperature, increasing concentration of chloride ions, decreasing concentration of dissolved oxygen, and decreasing pH. In solutions with neutral pH, crevice corrosion of titanium has not been observed at temperatures below 120 °C (250 °F). At lower pH values, crevice corrosion sometimes is encountered at temperatures below 120 °C (250 °F).

Erosion-Corrosion and Cavitation. For most materials there are critical velocities beyond which protective films are swept away and accelerated corrosion attack occurs. This accelerated attack is known as erosion-corrosion. The critical velocity differs greatly from one material to another and may be as low as 0.6 to 0.9 m/s (2 to 3 ft/s). For titanium, the critical velocity in seawater is more than 27 m/s (90 ft/s). Numerous erosion-corrosion tests have shown titanium to have outstanding resistance to this form of attack. Erosion-corrosion can be greatly aggravated by the presence of abrasive particles (such as sand) in a flowing fluid. Titanium exhibited superior resistance to this type of attack in seawater containing fine sand that flowed through conventional titanium condenser tubes at the rate of 1.8 m/s (6 ft/s).

Cavitation resistance tests have proved titanium to be one of the metals most resistant to cavitation damage.

Stress-Corrosion Cracking. Unalloyed titanium generally is immune to stress-corrosion cracking (SCC) unless it has a high oxygen content (0.3% or more). For this reason, SCC is of little concern in the chemical process industries, where unalloyed titanium is most commonly used. On the other hand, certain alloys of titanium used principally in the aerospace industry are subject to SCC. Generally, stress corrosion of titanium and its alloys does not occur unless there is a preexisting crack or sharp cracklike defect.

One of the first reported instances of SCC of unalloyed titanium occurred in red fuming nitric acid. It was also found that a pyrophoric surface deposit was formed on exposure to this acid. There was no evidence that these two phenomena are related, but addition of 1.5 to 2.0% water to the acid completely inhibited both reactions. Since then, SCC has been demonstrated in hot, dry sodium chloride, methanol, hydrochloric acid solutions, seawater, chlorinated solvents, nitrogen tetroxide, mercury, and cadmium.

One of the important variables affecting susceptibility to SCC is alloy composition. Aluminum additions increase susceptibility to SCC; alloys containing more than 6% Al generally are susceptible to stress corrosion. Additions of tin, manganese, and cobalt are detrimental, whereas zirconium appears to be neutral. Beta stabilizers such as molybdenum, vanadium, and niobium are beneficial. Susceptibility of titanium alloys to SCC also can be affected by heat treatment.

Accelerated Crack Propagation in Seawater. Titanium is known to be highly resistant to corrosion by seawater. However, for certain alloys, components containing very sharp notches or cracks exhibit accelerated crack propagation and thus lose resistance to fracture when exposed to seawater. Exposure to seawater does not appear to diminish the service life of titanium alloys, such as Ti-6Al-4V and Ti-5Al-2.5Sn, that exhibit this phenomenon in laboratory testing. These two alloys have been used successfully in aircraft for many years without reported failures. Apparently, the conditions leading to accelerated crack propagation (primarily, the existence of a crack) have not been encountered in service.

Accelerated crack propagation in seawater can be avoided by proper alloy selection. Alloys containing more than 6% Al are particularly susceptible. Additions of tin, manganese, cobalt, and oxygen are detrimental, whereas beta stabilizers such as molybdenum, niobium, and vanadium tend to reduce or eliminate susceptibility to this phenomenon. Unalloyed titanium is not susceptible unless it contains more than approximately 0.3% O_2.

Hot Salt Corrosion. Titanium and titanium alloys can be damaged by halogenated compounds at temperatures above 260 °C (500 °F). Chloride salts, especially sodium chloride, can be detrimental. Residual salts cause surface pitting or even cracking of certain alloys under high tensile loads. Although rarely encountered in service, cracking of titanium parts because of hot salt corrosion was encountered by fabricators during stress-relieving operations. Responsibility was traced to vapors of chlorinated cleaning fluids that were not completely removed prior to thermal processing, chloride traces from other process fluids (including tap water), and even salt residues from fingerprints.

The extent of damage by salts is directly related to temperature, exposure time, and tensile stress level. Processing history, alloy composition, salt composition, and other environmental conditions also have important effects. Susceptibility to hot salt corrosion appears to be influenced considerably by processing and alloy additions. Therefore, control of these factors should make it possible to avoid this phenomenon in service.

Solid and Liquid Metal Embrittlement. Some titanium alloys crack under tensile stress when in contact with liquid cadmium, mercury, or silver-base brazing alloys. This type of embrittlement differs from SCC, although there are some similarities. Liquid metal embrittlement appears to result from diffusion along grain boundaries and formation of brittle phases, which in turn produce the loss of ductility.

Titanium also can be embrittled by contact with certain solid metals (cadmium and silver, for example) when the titanium is under tensile stress. Service failures have occurred in cadmium-plated titanium alloys at ambient temperatures and in silver-brazed titanium parts at temperatures above 315 °C (600 °F). Silver-plated components should not be used in contact with titanium under stress at temperatures above 230 °C (450 °F). Cadmium-plated parts such as interference-fit fasteners or press-fit bushings should not be used in contact with titanium at any temperature. As a general rule, titanium is not used in contact with cadmium.

ACKNOWLEDGMENTS

Portions of the text for this chapter came from "Introduction and Overview of Titanium and Titanium Alloys," "Properties, Compositions, and Applications of Selected Titanium Alloys," "Processing of Titanium and Titanium Alloys," and "Corrosion Resistance of Titanium and Titanium Alloys," all by R.R. Boyer in *Metals Handbook Desk Edition,* 2nd ed., ASM International, 1998.

REFERENCES

5.1 S. Lampman, Wrought Titanium and Titanium Alloys, *Properties and Selection: Nonferrous Alloys and Special-Purpose Materials,* Vol 2, *ASM Handbook,* ASM International, 1990

5.2 F.C. Campbell, *Elements of Metallurgy and Engineering Alloys,* ASM International, 2008

5.3 M.J. Donachie, *Titanium: A Technical Guide,* 2nd ed., ASM International, 2000

5.4 R.R. Boyer, Processing of Titanium and Titanium Alloys, *Metals Handbook Desk Edition,* 2nd ed., ASM International, 1998

5.5 R.R. Boyer, Introduction and Overview of Titanium and Titanium Alloys, *Metals Handbook Desk Edition,* 2nd ed., ASM International, 1998

5.6 Y.V.R.K. Prasad et al., Titanium Alloy Processing, *Adv. Mater. Process.,* June 2000, p 85–89

5.7 R.R. Boyer, Properties, Compositions, and Applications of Selected Titanium Alloys, *Metals Handbook Desk Edition,* 2nd ed., ASM International, 1998

5.8 G.W. Kuhlman, Forging of Titanium Alloys, *Metalworking: Bulk Forming,* Vol 14A, *ASM Handbook,* ASM International, 2005

5.9 J.C. Williams, Titanium Alloys: Production, Behavior and Application, *High Performance Materials in Aerospace,* Chapman & Hall, 1995, p 85–134

5.10 F.C. Campbell, *Manufacturing Technology for Aerospace Structural Materials,* Elsevier Scientific, 2006

5.11 S. Copley, R. Martukanitz, W. Frazier, and M. Rigdon, Mechanical Properties of Parts Formed by Laser Additive Manufacturing, *Adv. Mater. Process.,* Vol 169 (No. 9), Sept 2011, p 26–29

5.12 J.D. Beal, R. Boyer, and D. Sanders, Forming of Titanium and Titanium Alloys, *Metalworking: Sheet Forming,* Vol 14B, *ASM Handbook,* ASM International, 2006

5.13 C.H. Hamilton and A.K. Ghosh, Superplastic Sheet Forming, *Forming and Forging,* Vol 14, *ASM Handbook,* ASM International, 1988

5.14 J.R. Newman, D. Eylon, and J.K. Thome, Titanium and Titanium Alloys, *Casting,* Vol 15, *ASM Handbook,* ASM International, 1988

5.15 *Metals Handbook Desk Edition,* 2nd ed., ASM International, 1998, p 744

5.16 M. Guclu, Titanium and Titanium Alloy Castings, *Casting,* Vol 15, *ASM Handbook,* ASM International, 2008

5.17 S. Yue and S. Durham, Titanium, *Handbook of Materials for Product Design,* McGraw-Hill, 2001, p 3.1–3.61

5.18 Electron-Beam Welding, *Welding, Brazing, and Soldering,* Vol 6, *ASM Handbook,* ASM International, 1993

5.19 Solid-State Welding Processes, *Metals Handbook Desk Edition,* 2nd ed., ASM International, 1998

SELECTED REFERENCES

- P. Allen, Titanium Alloy Development, *Adv. Mater. Process.,* Oct 1996, p 35–37
- F.G. Arcella, D.H. Abbott, and M.A. House, "Titanium Alloy Structures for Airframe Application by the Laser Forming Process," AIAA-2000-1465, American Institute of Aeronautics and Astronautics, Inc., 2000
- F.G. Arcella and F.H. Froes, Producing Titanium Aerospace Components from Powder Using Laser Forming, *JOM,* May 2000, p 28–30
- R.R. Boyer, An Overview of the Use of Titanium in the Aerospace Industry, *Mater. Sci. Eng. A,* Vol 213, 1996, p 103–114
- W.D. Brewer, R.K. Bird, and T.A. Wallace, Titanium Alloys and Processing for High Speed Aircraft, *Mater. Sci. Eng. A,* Vol 243, 1998, p 299–304
- J.D. Cotton, L.P. Clark, and H.R. Phelps, Titanium Alloys on the F-22 Fighter Aircraft, *Adv. Mater. Process.,* May 2002, p 25–28
- M.J. Donachie, Selection of Titanium Alloys for Design, *Handbook of Materials Selection,* John Wiley & Sons, Inc., 2002, p 201–234
- E.O. Ezugwu and Z.M. Wang, Titanium Alloys and Their Machinability—A Review, *J. Mater. Process. Technol.,* Vol 68, 1997, p 262–274
- S.J. Gerdemann, Titanium Process Technologies, *Adv. Mater. Process.,* July 2001, p 41–43
- S.J. Hatakeyama, "SPF/DB Wing Structures Development for the High-Speed Civil Transport," SAE 942158, Society of Automotive Engineers, 1994
- X. Huang and N.L. Richards, Activated Diffusion Brazing Technology for Manufacture of Titanium Honeycomb Structures—A Statistical Study, *Weld. J.,* March 2004, p 73–81
- R. Irving, EB Welding Joins the Titanium Fuselage of Boeing's F-22 Fighter, *Weld. J.,* Dec 1994, p 31–36
- J.H. Lee, "Applications and Development of EB Welding on the F-22," SAE 972201, Society of Automotive Engineers, 1997

- A. Mitchell, Melting, Casting and Forging Problems in Titanium Alloys, *Mater. Sci. Eng. A,* Vol 243, 1998, p 257–262
- Y.V.R.K. Prasad et al., A Study of Beta Processing of Ti-6Al-4V: Is It Trivial?, *J. Eng. Mater. Technol.,* Vol 123, 2002, p 355–360
- S.L. Semiatin, V. Seetharaman, and I. Weiss, Hot Workability of Titanium and Titanium Aluminide—An Overview, *Mater. Sci. Eng. A,* Vol 243, 1998, p 1–24
- S.L. Semiatin et al., Plastic Flow and Microstructure Evolution during Thermomechanical Processing of Laser-Deposited Ti-6Al-4V Preforms, *Metall. Mater. Trans. A,* Vol 32, 2002, p 1801–1811
- G. Shen and D. Furrer, Manufacturing of Aerospace Forgings, *J. Mater. Process. Technol.,* Vol 98, 2000, p 189–195
- "Titanium Machining Techniques," Titanium Engineering Bulletin No. 7, Titanium Metals Corporation, 1960s
- "Titanium Welding Techniques," Titanium Engineering Bulletin No. 6, Titanium Metals Corporation, 1964
- S. Veeck, D. Lee, and T. Tom, Titanium Investment Castings, *Adv. Mater. Process.,* Jan 2002, p 59–62
- J. Williams, Thermo-Mechanical Processing of High-Performance Ti Alloys: Recent Progress and Future Needs, *J. Mater. Process. Technol.,* Vol 117, 2001, p 370–373
- J.C. Williams, Titanium Alloys: Production, Behavior and Application, *High Performance Materials in Aerospace,* Chapman & Hall, 1995, p 85–134
- J.C. Williams and E.A. Starke, Progress in Structural Materials for Aerospace Systems, *Acta Mater.,* Vol 51, 2003, p 5775–5799
- Y. Zhou, W.F. Gale, and T.H. North, Modelling of Transient Liquid Phase Bonding, *Int. Mater. Rev.,* Vol 40 (No. 5), 1995, p 181–196

CHAPTER **6**

Titanium Aluminide Intermetallics

ALLOYS based on ordered intermetallic compounds constitute a unique class of metallic material that form long-range ordered crystal structures (Fig. 6.1) below a critical temperature, generally referred to as the critical ordering temperature (T_c). These ordered intermetallics usually exist in relatively narrow compositional ranges around simple stoichiometric ratios.

The search for new high-temperature structural materials has stimulated much interest in ordered intermetallics. Recent interest has focused on nickel aluminides based on Ni_3Al and $NiAl$, iron aluminides based on Fe_3Al and $FeAl$, and titanium aluminides based on Ti_3Al and $TiAl$. These aluminides possess many attributes that make them attractive for

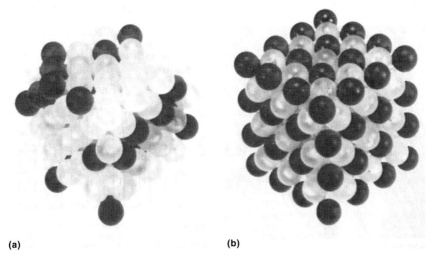

(a) (b)

Fig. 6.1 Atomic arrangements of conventional alloys and ordered intermetallic compounds. (a) Disordered crystal structure of a conventional alloy. (b) Long-range ordered crystal structure of an ordered intermetallic compound. Source: Ref 6.1

high-temperature structural applications. They contain enough aluminum to form, in oxidizing environments, thin films of alumina (Al_2O_3) that are compact and protective. They have low densities, relatively high melting points, and good high-temperature strength properties. Properties of nickel, iron, and titanium aluminides are given in Table 6.1, and the attributes and upper-use temperatures are shown in Table 6.2.

Nickel, iron, and titanium aluminides, like other ordered intermetallics, exhibit brittle fracture and low ductility at ambient temperatures. It has also been found that quite a number of ordered intermetallics, such as iron aluminides, exhibit environmental embrittlement at ambient temperatures. The embrittlement involves the reaction of water vapor in air with reactive elements (aluminum, for example) in intermetallics to form atomic hydrogen, which drives into the metal and causes premature fracture. Thus, the poor fracture resistance and limited fabricability have restricted the use of aluminides as engineering materials in most cases. However, in recent years, alloying and processing techniques have been used to overcome the brittleness problem of ordered intermetallics. Success in this work has inspired parallel efforts aimed at improving strength properties. The results have led to the development of a number of attractive intermetallic alloys having useful ductility and strength.

Table 6.1 Properties of nickel, iron, and titanium aluminides

Alloy	Crystal structure(a)	Critical ordering temperature (T_c)		Melting point (T_m)		Material density, g/cm³	Young's modulus	
		°C	°F	°C	°F		GPa	10⁶ psi
Ni_3Al	$L1_2$ (ordered fcc)	1390	2535	1390	2535	7.50	179	25.9
NiAl	B2 (ordered bcc)	1640	2985	1640	2985	5.86	294	42.7
Fe_3Al	DO_3 (ordered bcc)	540	1000	1540	2805	6.72	141	20.4
	B2 (ordered bcc)	760	1400	1540	2805	...	...	...
FeAl	B2 (ordered bcc)	1250	2280	1250	2280	5.56	261	37.8
Ti_3Al	DO_{19} (ordered hcp)	1100	2010	1600	2910	4.2	145	21.0
TiAl	$L1_0$ (ordered tetragonal)	1460	2660	1460	2660	3.91	176	25.5
$TiAl_3$	DO_{22} (ordered tetragonal)	1350	2460	1350	2460	3.4	...	...

(a) fcc, face-centered cubic; bcc, body-centered cubic; hcp, hexagonal close-packed. Source: Ref 6.1

Table 6.2 Attributes and upper-use temperature limits for nickel, iron, and titanium aluminides

Alloy	Attributes	Maximum-use temperature, °C (°F)	
		Strength limit	Corrosion limit
Ni_3Al	Oxidation, carburization, and nitridation resistance; high-temperature strength	1000 (1830)	1150 (2100)
NiAl	High melting point; high thermal conductivity; oxidation, carburization, and nitridation resistance	1200 (2190)	1400 (2550)
Fe_3Al	Oxidation and sulfidation resistance	600 (1110)	1100 (2010)
FeAl	Oxidation, sulfidation, molten salt, and carburization resistance	800 (1470)	1200 (2190)
Ti_3Al	Low density; good specific strength	760 (1400)	650 (1200)
TiAl	Low density; good specific strength; wear resistance	1000 (1830)	900 (1650)

Source: Ref 6.1

The crystal structures showing the ordered arrangements of atoms in several of these aluminides are shown in Fig. 6.2. For most of the aluminides listed in Table 6.1, the critical ordering temperature is equal to the melting temperature. Others disorder at somewhat lower temperatures, and Fe_3Al passes through two ordered structures (DO_3 and $B2$) before becoming disordered. Many of the aluminides exist over a range of compositions, but the degree of order decreases as the deviation from stoichiometry increases. Additional elements can be incorporated with-

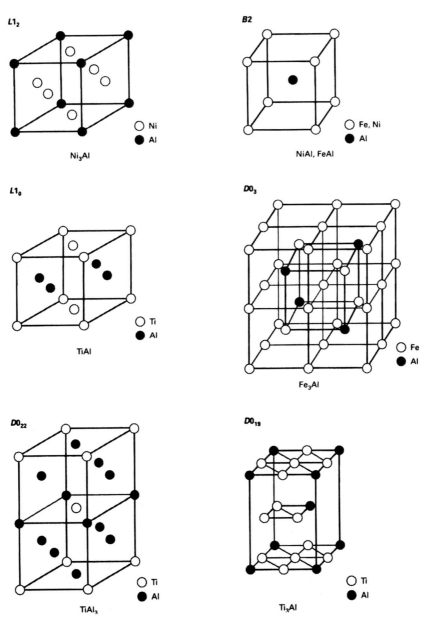

Fig. 6.2 Crystal structures of nickel, iron, and titanium aluminides. Source: Ref 6.1

out losing the ordered structure. For example, in Ni_3Al, silicon atoms are located in aluminum sites, cobalt atoms on nickel sites, and iron atoms on either. In many instances, the so-called intermetallic compounds can be used as bases for alloy development to improve or optimize properties for specific applications.

The remainder of this chapter deals with the lower-density titanium aluminides. The main barriers to wide-scale use include high processing costs, limited ductility, and environmental stability.

6.1 Titanium Aluminides

Because of their low density, titanium aluminides based on Ti_3Al and $TiAl$ are attractive candidates for applications in advanced aerospace engine components (latter stages of the compressor or turbine sections), airframe components, and automotive valves and turbochargers. The characteristics of titanium aluminides alongside those of other aluminides are shown in Table 6.2, and the creep behavior of titanium aluminides is compared with that of conventional titanium alloys in Fig. 6.3. Despite a lack of fracture resistance (low ductility, fracture toughness, and fatigue crack growth rate), the titanium aluminides Ti_3Al (α-2) and $TiAl$ (γ) have great potential for enhanced performance. Table 6.3 compares properties of these aluminides with those of conventional titanium alloys and superalloys. Because they have slower diffusion rates than conventional titanium alloys, the titanium aluminides feature enhanced high-temperature properties such as strength retention, creep and stress rupture, and fatigue resistance.

In addition to their low ductility at ambient temperatures, another negative feature of titanium aluminides is their oxidation resistance, which is lower than desirable at elevated temperatures. The titanium aluminides are characterized by a strong tendency to form TiO_2, rather than the protec-

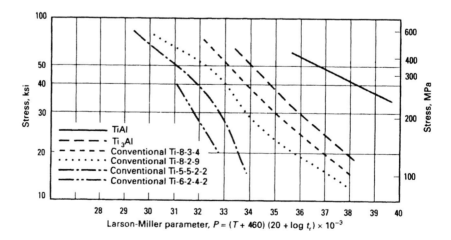

Fig. 6.3 Comparison of the creep behavior of conventional titanium alloys and titanium aluminide intermetallics. Source: Ref 6.1

Table 6.3 Properties of titanium aluminides, titanium-base conventional alloys, and nickel-base superalloys

Property	Conventional titanium alloys	Ti₃Al	TiAl	Nickel-base superalloys
Density, g/cm³	4.5	4.1–4.7	3.7–3.9	8.3
Modulus, GPa (10⁶ psi)	96–100 (14–14.5)	100–145 (14.5–21)	160–176 (23.2–25.5)	206 (30)
Yield strength(a), MPa (ksi)	380–1150 (55–167)	700–990 (101–144)	400–650 (58–94)	...
Tensile strength(a), MPa (ksi)	480–1200 (70–174)	800–1140 (116–165)	450–800 (65–116)	...
Creep limit, °C (°F)	600 (1110)	760 (1400)	1000 (1830)	1090 (1995)
Oxidation limit, °C (°F)	600 (1110)	650 (1200)	900 (1650)	1090 (1995)
Ductility at room temperature, %	20	2–10	1–4	3–5
Ductility at high temperature, %	High	10–20	10–60	10–20
Structure(b)	hcp/bcc	DO_{19}	Ll_0	fcc/L_2

(a) At room temperature. (b) hcp, hexagonal close-packed; bcc, body-centered cubic; fcc, face-centered cubic. Source: Ref 6.1

tive Al_2O_3, at high temperatures. Because of this tendency, a key factor in increasing the maximum-use temperatures of these aluminides is enhancing their oxidation resistance while maintaining adequate levels of creep and strength retention at elevated temperatures.

6.2 Alpha-2 Alloys

As shown on the titanium-aluminum phase diagram (Fig. 6.4), Ti_3Al has a wide range of composition stability, with aluminum contents of 22 to 39 at.%. The compound is congruently disordered at a temperature of 1180 °C (2155 °F) and an aluminum content of 32 at.%. The stoichiometric composition, Ti-25Al, is stable up to approximately 1090 °C (1995 °F).

The semicommercial and experimental α-2 alloys developed are two phase (α-2 + β/B2), with contents of 23 to 25 at.% Al and 11 to 18 at.% Nb. Alloy compositions with current engineering significance are Ti-24Al-11Nb, Ti-25Al-10Nb-3V-1Mo, Ti-25Al-17Nb-1Mo, and modified alloy compositions such as Ti-24.5Al-6Nb-6(Ta,Mo,Cr,V). Increasing the niobium content generally enhances most material properties, although excessive niobium can degrade creep performance. Niobium can be replaced by specific elements for improved strength (molybdenum, tantalum, or chromium), creep resistance (molybdenum), and oxidation resistance (tantalum, molybdenum). However, for full optimization of mechanical properties, control of the microstructure must be maintained, particularly for tensile, fatigue, and creep performance. Microstructural features, such as primary α-2 grain size and volume fraction and secondary α-2 plate morphology and thickness, are varied by thermomechanical processing.

Typical mechanical properties for a number of α-2 alloys are listed in Table 6.4. Production of two-phase alloys by alloying Ti_3Al with β-stabilizing elements results in up to a doubling of strength. Interface strengthening of the two-phase mixture appears to be predominantly responsible for the increased strength, but other strengthening factors, such as long-range order, solid solution, and texture effects, also contribute.

A fine Widmanstätten microstructure with a small amount of primary α-2 grains exhibits better ductility than microstructures with a coarse

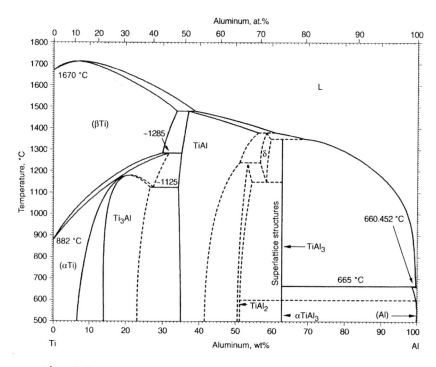

Fig. 6.4 Titanium-aluminum binary phase diagram. Source: Ref 6.1

Table 6.4 Properties of α-2 Ti₃Al alloys with various microstructures

| Alloy | Microstructure(a) | Yield strength | | Ultimate tensile strength | | Elongation, % | Plane-strain fracture toughness (K_{Ic}) | | Creep rupture(b) |
		MPa	ksi	MPa	ksi		MPa √m	ksi √in.	
Ti-25Al	E	538	78	538	78	0.3	...	...	...
Ti-24Al-11Nb	W	787	114	824	119	0.7	...	...	44.7
	FW	761	110	967	140	4.8	...	...	...
Ti-24Al-14Nb	W	831	120	977	142	2.1	...	...	59.5
Ti-25Al-10Nb-3V-1Mo	W	825	119	1042	151	2.2	13.5	12.3	>360
	FW	823	119	950	138	0.8	...	...	...
	C + P	745	108	907	132	1.1	...	...	...
	W + P	759	110	963	140	2.6	...	...	...
	FW + P	942	137	1097	159	2.7	...	...	...
Ti-24.5Al-17Nb	W	952	138	1010	146	5.8	28.3	25.7	62
	W + P	705	102	940	136	10.0	...	...	...
Ti-25Al-17Nb-1Mo	FW	989	143	1133	164	3.4	20.9	19.0	476

(a) E, equiaxed α-2; W, Widmanstätten; FW, fine Widmanstätten; C, colony structure; P, primary α-2 grains. (b) Time to rupture, h, at 650 °C (1200 °F) and 380 MPa (55 ksi). Source: Ref 6.1

Widmanstätten microstructure or an aligned acicular α-2 morphology. A detailed investigation into the effect of microstructure on creep behavior in Ti-25Al-10Nb-3V-1Mo has shown that the colony-type microstructure shows better creep resistance than other microstructures. Creep resistance of Ti-25-10-3-1 is raised by a factor of 10 in the steady-state regime over that of conventional alloy Ti-1100 (Ti-6Al-3Sn-4Zr-0.4Mo-0.45Si) and 2 orders of magnitude over that of Ti-6Al-2Sn-4Zr-2Mo-0.1Si. However, 0.4% creep strain in Ti-25-10-3-1 is reached within 2 h.

Additions of silicon and zirconium appear to improve creep resistance, but the most significant improvement is attained by increasing the aluminum content to 25 at.% and limiting β-stabilizing elements to approximately 12 at.%. However, the Ti-24.5Al-17Nb-1Mo alloy exhibits a rupture life superior to that of other α-2 alloys.

6.3　Orthorhombic Alloys

At higher niobium levels, the α-2 phase evolves to a new ordered orthorhombic structure that is based on the composition Ti_2AlNb (O phase). This has been observed in titanium aluminides with compositions near Ti-(21-25)Al-(21-27)Nb (at.%).

Although the orthorhombic alloys have a lower use temperature than the γ aluminides (approximately 650 °C, or 1200 °F), they offer much higher absolute strengths. Room-temperature tensile strengths on the order of 1380 MPa (200 ksi) with close to 5% elongation have been reported. In addition, 0.2% yield strengths in the range of 590 to 690 MPa (86 to 100 ksi) at 700 °C (1290 °F) have also been reported for Ti-21Al-25Nb (at.%) alloy. This material had a room-temperature yield strength of approximately 1070 MPa (155 ksi) with 3.5% elongation.

The ordered orthorhombic alloys with the best combination of tensile, creep, and fracture toughness properties are two-phase O + β alloys such as Ti-22Al-27Nb (at.%). The elevated-temperature tensile properties of a two-phase O + β Ti-22Al-27Nb at.% alloy are shown in Table 6.5.

Table 6.5　Tensile properties of a two-phase (O + β) alloy (Ti-22Al-27Nb at.%)

Test temperature		Aging treatment	Tensile yield strength		Ultimate tensile strength		Elongation, %
°C	°F		MPa	ksi	MPa	ksi	
22	72	None	1056	153	1152	167	3.4
		None	1028	149	1083	157	2.2
		540 °C (1000 °F), 100 h	1083	157	1166	169	3.3
		540 °C (1000 °F), 100 h	1090	158	1159	168	2.8
		650 °C (1200 °F), 100 h	1090	158	1145	166	2.6
		650 °C (1200 °F), 100 h	1076	156	1145	166	2.5
		760 °C (1400 °F), 100 h	987	143	1076	156	5.2
		760 °C (1400 °F), 100 h	966	140	1083	157	5.0
540	1000	None	849	123	1007	146	14.3
		None	856	124	1049	152	14.3
		540 °C (1000 °F), 100 h	876	127	1049	152	17.9
		540 °C (1000 °F), 100 h	890	129	1070	155	16.1
650	1200	None	794	115	938	136	14.3
		None	807	117	945	137	12.5
		650 °C (1200 °F), 100 h	794	115	938	136	10.7
		650 °C (1200 °F), 100 h	807	117	952	138	10.7
760	1400	None	559	81	787	114	10.7
		None	593	86	766	114	14.3
		760 °C (1400 °F), 100 h	462	67	649	94	21.4
		760 °C (1400 °F), 100 h	552	80	731	106	16.1

Source: Ref 6.1

6.4 Gamma Alloys

The γ-TiAl phase has an $L1_0$ ordered face-centered tetragonal structure (Fig. 6.2), which has a wide range (49 to 66 at.% Al) of temperature-dependent stability (Fig. 6.4). The γ-TiAl phase apparently remains ordered up to its melting point of approximately 1450 °C (2640 °F).

The strong titanium-aluminum bond leads to a high activation energy for diffusion. This high-energy barrier helps to retain strength and resist creep to high temperatures when diffusion becomes the rate-controlling process. It also results in high stiffness over a wide temperature range, as shown in Fig. 6.5, making possible its use for static parts that need only to sustain elastic deflection. However, the rigid bond structure restricts the ability to accommodate plastic deformation. As with most other intermetallic com-

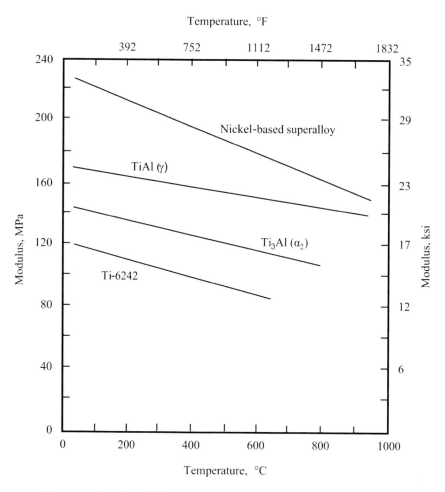

Fig. 6.5   Modulus of TiAl as a function of temperature in comparison to other materials. Source: Ref 6.2

pounds, γ, without modification, lacks the ductility and toughness that are necessary for structural applications.

One effective way to improve the deformability of γ is through microstructural control. Three distinctly different types of microstructures can be obtained by annealing. These are shown in Fig. 6.6 and consist of equiaxed, duplex, and lamellar.

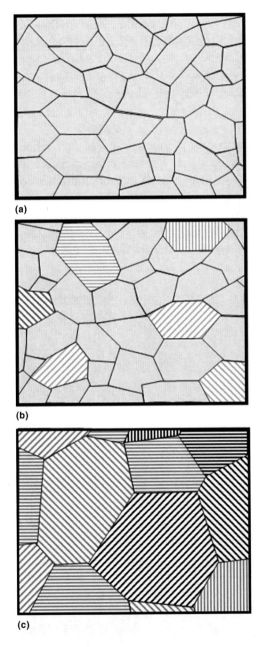

(a)

(b)

(c)

Fig. 6.6 Idealized wrought microstructures of (a) single-phase γ, (b) duplex of γ and lamellar, and (c) fully lamellar

Equiaxed. Alloys above 52 at.% Al generally lie in the single-phase γ field during heat treatment and are single-phase γ after cooling to room temperature. The grains are equiaxed and approximately 50 μm in diameter.

Duplex. For alloys between 46 and 50 at.% Al, heat treatment in the α + γ phase field results in a two-phase structure upon cooling. This structure consists of γ grains and grains of a lamellar structure. The lamellar grains contain alternating α-2 and γ plates, which form as a result of transformation from the primary α during cooling to room temperature. The grains are typically 10 to 35 μm in diameter, and the lamellar plates are 0.1 to 1 μm thick.

Lamellar. Alloys below 48 at.% Al that are heat treated in the single-phase α field can form the fully lamellar structure. The grains are typically greater than 500 μm in diameter, which reflects the rapid coarsening rate of the disordered α.

Among the three microstructures in Fig. 6.7, the duplex structure is significantly more ductile at room temperature than the others. The ductile duplex structure exists in a narrow range of aluminum concentration. Only the duplex alloys containing 45 to 50 at.% Al show appreciable plasticity. The ductility peak occurs at 48 at.% Al, particularly when the alloy is heat treated near the center of the α + γ phase field where the two phases have equal volume fractions. Alloys outside this composition range tend

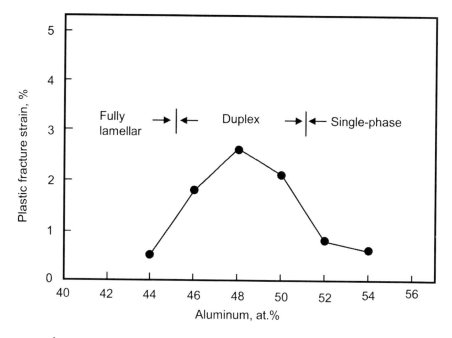

Fig. 6.7 Room-temperature tensile ductility of wrought binary γ-based alloys as a function of aluminum concentration, showing the effects of microstructures. Source: Ref 6.2

to have either the single-phase γ or the lamellar structure, and they are relatively brittle. Because the low-temperature ductility is a major concern for structural applications, the 45 to 50 at.% Al range is usually the basis for further alloying.

Generally, the duplex structure is desirable for ductility but poor for toughness and creep resistance. The reverse is true for the fully lamellar structure. The lamellar structure is not only tougher (Fig. 6.8) but more resistant to creep than the γ and duplex structures. It significantly reduces the transient creep and the rate of the steady-state creep (Fig. 6.9). As a result, the creep time to 0.2% strain can be increased by nearly 2 orders over that of the duplex structure.

The γ alloys of engineering importance contain approximately 45 to 48 at.% Al and 1 to 10 at.% M, with M being at least one of the following: vanadium, chromium, manganese, niobium, tantalum, and tungsten. These alloys can be divided into two categories: single-phase (γ) alloys and two-phase (γ + α-2) materials. The (α-2 + γ)/ γ phase boundary at 1000 °C (1830 °F) occurs at an aluminum content of approximately 49 at.%, depending on the type and level of solute M. Single-phase γ alloys contain third alloying elements such as niobium or tantalum that promote strengthening and further enhance oxidation resistance. Third alloying elements in two-phase alloys can raise ductility (vanadium, chromium,

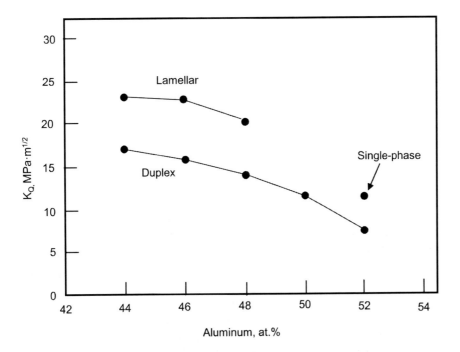

Fig. 6.8 Fracture toughness as a function of microstructure and aluminum concentration of wrought binary γ-based alloys. Source: Ref 6.2

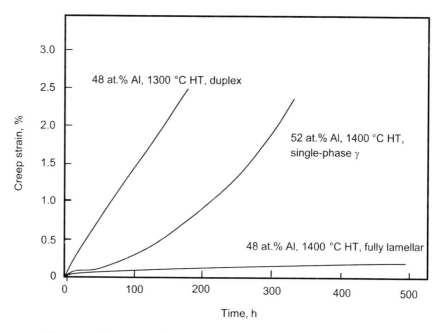

Fig. 6.9 Comparison of creep curves of the three typical microstructures in γ-based alloys at 800 °C (1470 °F) and 69 GPa (10 × 10⁶ psi). HT, heat treatment. Source: Ref 6.2

and manganese), increase oxidation resistance (niobium and tantalum), or enhance combined properties.

Gamma alloys are processed by conventional methods, including casting, ingot metallurgy, and powder metallurgy. Important alloying/melting processes include induction skull melting, vacuum arc melting, and plasma melting. Other methods under study include mechanical alloying, spray forming, shock reactive synthesis, physical vapor deposition, and hot pressing and rolling of elemental sheet into multilayer composite sheets. By appropriate thermomechanical processing (TMP), the morphology of the phases in nominally γ alloys can be adjusted to produce either lamellar or equiaxed morphologies, or a mixture of the two.

Table 6.6 lists the tensile properties and fracture toughness of γ alloys as functions of processing/microstructure and temperature. Ternary alloys of Ti-48Al with approximately 1 to 3% V, Mn, or Cr exhibit enhanced ductility, but Ti-48Al alloys with approximately 1 to 3% Nb, Zr, Hf, Ta, or W show lower ductility than binary Ti-48Al. The brittle-ductile transition (BDT) occurs at 700 °C (1290 °F) in Ti-56Al and at lower temperatures with decreasing aluminum levels. Increased room-temperature ductility generally results in a reduced BDT temperature. Above the BDT temperature, ductility increases rapidly with temperature, approaching 100% at 1000 °C (1830 °F) for the most ductile γ alloy compositions. The trend bands for variations in yield strength and tensile ductility with test tem-

Table 6.6 Tensile properties and fracture toughness values of gamma titanium aluminides tested in air

Alloy designation and composition, at. %	Processing and microstructure(a)	Temperature(b) °C	°F	Yield strength MPa	ksi	Tensile strength MPa	ksi	Elongation, %	Fracture toughness(c), K_q MPa √m	ksi √in.
48-1-(0.3C)/Ti-48Al-1V-0.3C-0.2O	Forging + HT/duplex	RT	RT	392	57	406	59	1.4	12.3	11.2
		437	819	. . .	. . .	. . .	. . .	. . .	22.8	20.7
		760	1400	320	46	470	68	10.8	. . .	. . .
	Casting/duplex	RT	RT	490	71	. . .	. . .	. . .	24.3	22.1
48-1(0.2C)/Ti-48Al-1V-0.2C-0.14O	Forging + HT/duplex + NL	RT	RT	480	70	530	77	1.5	. . .	. . .
		815	1500	360	52	450	65	. . .	. . .	. . .
48-2-2/Ti-48Al-2Cr-2Nb	Casting + HIP + HT/duplex	RT	RT	331	48	413	60	2.3	20–30	18–27
		760	1400	310	45	430	62	. . .	. . .	. . .
	Extrusion + HT/duplex	RT	RT	480	70		. . .	3.1	. . .	. . .
		760	1400	403	58	. . .	. . .	40	. . .	. . .
		870	1600	330	48	. . .	. . .	53	. . .	. . .
	Extrusion + HT/FL	RT	RT	454	66	. . .	. . .	0.5	. . .	. . .
		760	1400	405	59	. . .	. . .	3.0	. . .	. . .
		870	1600	350	51	. . .	. . .	19	. . .	. . .
	PM extrusion + HT/NL	RT	RT	510	74	597	87	2.9	. . .	. . .
		700	1290	421	61	581	84	5.2	. . .	. . .
G1/Ti-47Al-1Cr-1V-2.6Nb	Forging + HT/duplex	RT	RT	480	70	548	79	2.3	12	10.8
		600	1110	383	56	507	74	3.1	16	14.6
		800	1470	324	47	492	71	55	. . .	. . .
	Forging + HT/FL	RT	RT	330	48	383	56	0.8	30–36	27–33
		800	1470	290	42	378	55	1.5	40–70	36–64
Sumitomo/Ti-45Al-1.6Mn	Reactive sintering/NL	RT	RT	465	67	566	82	1.4	. . .	. . .
		800	1470	370	54	540	78	14	. . .	. . .
ABB alloy/Ti-47Al-2W-0.5Si	Casting + HT/duplex	RT	RT	425	62	520	75	1.0	22	20
		760	1400	350	51	460	67	2.5	. . .	. . .
47XD/Ti-47Al-2Mn-2Nb-0.8TiB₂	Casting + HIP + HT/NL + TiB₂	RT	RT	402	58	482	70	1.5	15–16	13.6–14.6
		760	1400	344	50	458	66	. . .	. . .	. . .
45XD/Ti-45Al-2Mn-2Nb-0.8TiB₂	Casting + HIP + HT	RT	RT	570	83	695	101	1.5	15–19	13.6–17.3
		600	1110	440	64	650	94	. . .	. . .	. . .
		760	1400	415	60	510	74	19	. . .	. . .
GE alloy 204b/Ti-46.2Al-x Cr-y (Ta,Nb)	Casting + HIP + HT/NL	RT	RT	442	64	575	83	1.5	34.5	31.4
		760	1400	382	55.4	560	81	12.4	. . .	. . .
		840	1545	381	55.2	549	80	12.2	. . .	. . .
Ti-47Al-2Nb-2Cr-1Ta	Casting + HIP + HT/duplex	RT	RT	430	62	515	75	1.0	. . .	. . .
		800	1470	363	53	495	72	23.3	. . .	. . .
		870	1600	334	48	403	58	14.6	. . .	. . .
Ti-47Al-2Nb-1.75Cr	Permanent mold casting + HIP + HT	RT	RT	429	62	516	75	1.4	. . .	. . .
		760	1400	286	41	428	62	13.3	. . .	. . .
		815	1500	368	53	531	77	23.3	. . .	. . .
Alloy 7/Ti-46Al-4Nb-1W	Extrusion + HT/NL	RT	RT	648	94	717	104	1.6	. . .	. . .
		760	1400	517	75	692	100	. . .	. . .	. . .
Alloy K5/Ti-46.5Al-2Cr-3Nb-0.2W	Forging + HT/duplex	RT	RT	462	67	579	84	2.8	11	10
		800	1470	345	50	468	68	40	. . .	. . .
	Forging + HT/RFL	RT	RT	473	69	557	81	1.2	20–22	18–20
		800	1470	375	54	502	73	3.2	. . .	. . .
		870	1600	362	53	485	70	12.0	. . .	. . .

(a) HT, heat treated; NL, nearly lamellar; HIP, hot isostatically pressed; FL, fully lamellar; PM, powder metallurgy; RFL, refined fully lamellar. (b) RT, room temperature. (c) K_q = provisional K_{Ic} value. Source: Ref 6.1

perature are shown in Fig. 6.10. The elastic moduli of γ alloys range from 160 to 176 GPa (23 to 25.5×10^6 psi) and decrease slowly with temperature.

Low-cycle fatigue experiments suggest that fine grain sizes increase fatigue life at temperatures below 800 °C (1470 °F). Fatigue crack growth rates for γ alloys are more rapid than those for superalloys, even when den-

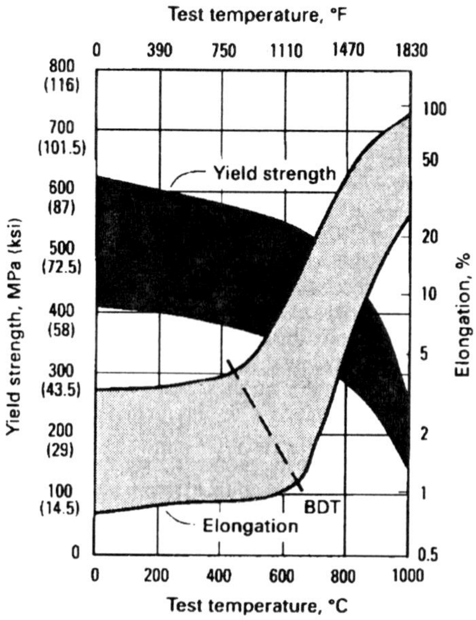

Fig. 6.10 Ranges of yield strength and tensile elongation as functions of test temperature for γ-TiAl alloys. BDT, brittle-ductile transition. Source: Ref 6.1

sity is normalized. Both fracture toughness and impact resistance are low at ambient temperatures, but fracture toughness increases with temperature; for example, the plane-strain fracture toughness (K_{Ic}) for Ti-48Al-1V-0.1C is 24 MPa $\sqrt{m}$ (21.8 ksi $\sqrt{in.}$) at room temperature. Fracture toughness is strongly dependent on the volume fraction of the lamellar phase. In a two-phase quaternary γ alloy, a fracture toughness of 12 MPa $\sqrt{m}$ (10.9 ksi $\sqrt{in.}$) is observed for a fine structure that is almost entirely γ; K_{Ic} is greater than 20 MPa $\sqrt{m}$ (18.2 ksi $\sqrt{in.}$) when a large volume fraction of lamellar grains are present.

Creep properties of γ alloys, when normalized by density, are better than those of superalloys, but they are strongly influenced by alloy chemistry and TMP. Increased aluminum content and additions of tungsten or carbon increase creep resistance. Increasing the volume fraction of the lamellar structure enhances creep properties but lowers ductility. The level of creep strain from elongation upon initial loading and primary creep is of concern because it can exceed projected design levels for maximum creep strain in the part.

Gamma titanium aluminides also serve as the matrix for titanium-matrix composites. Investment-cast Ti-47Al-2Nb-2Mn (at.%) + 0.8 vol% TiB_2 and Ti-45Al-2Nb-2Mn (at.%) + 0.8 vol% TiB_2 XD (exothermic dispersion) alloys have been developed. These alloys form in situ titanium diborides that cause grain refinement and subsequent enhancement of mechanical and physical properties. The XD type of in situ reinforcements offer better thermal stability than conventional metal-matrix composites and the opportunity to introduce reinforcements in the micrometer scale range (at 0.8 vol%, the TiB_2 is simply serving as a grain refiner). The XD aluminides offer excellent fatigue properties, with endurance limits in the range of 95 to 115% of the yield strength (non-XD γ alloys exhibit similar behavior). The high fatigue strength is attributed to the microstructural refinement mentioned and a strain aging effect that occurs during loading. The XD titanium aluminide composites have been used for applications such as missile fins. Ti-47Al-2V + 7 vol% TiB_2 has a higher strength than 17-4PH steel above 600 °C (1110 °F) and a higher modulus at approximately 750 °C (1380 °F). These improvements extend the operating temperature of the missile wing, which led to a redesign using the XD composite. The tensile properties of XD composites are given in Table 6.6.

Although the γ class of titanium aluminides offers oxidation and interstitial (oxygen, nitrogen) embrittlement resistance superior to that of the α-2 and orthorhombic (Ti_2AlNb) classes of titanium aluminides, environmental durability is still a concern, especially at temperatures of approximately 750 to 800 °C (1380 to 1470 °F) in air. This is because all titanium aluminides are characterized by a strong tendency to form TiO_2, rather than the protective Al_2O_3 scale, at elevated temperature. To improve the oxidation resistance of γ aluminides, two areas are being explored: alloy development and the development of improved protective coatings.

Ternary and higher-order alloying additions can reduce the rate of oxidation of γ alloys. Of particular benefit are small (1 to 4%) ternary additions of tungsten, niobium, and tantalum. When combined with quaternary additions of 1 to 2% Cr or Mn, further improvement in oxidation resistance is gained. However, it is important to stress that these small alloying additions do not result in continuous Al_2O_3 scale formation. Rather, a complex intermixed Al_2O_3/TiO_2 scale is still formed, but the rate of growth of this scale is reduced.

Further improvements in oxidation resistance can be obtained through the use of protective coatings. Three general coating alloy approaches have been taken for protecting titanium aluminides: MCrAlY (M = Ni, Fe, Co), aluminizing, and silicides/ceramics. Protection of titanium aluminides under oxidizing conditions has been achieved with all three approaches; however, the fatigue life of coated material is often reduced to below that of uncoated material.

The degradation in the fatigue life of titanium aluminides by coatings results from three main factors: the formation of brittle coating/substrate reaction zones (chemical incompatibility), the brittleness of the coating alloy, and the differences in the coefficient of thermal expansion (CTE) between the coating and the substrate. The MCrAlY coatings, which are successfully used to protect nickel-, iron-, and cobalt-base superalloys, are not chemically compatible with titanium aluminides and form brittle coating/substrate reaction zones at 800 °C (1470 °F). Aluminizing treatments result in the surface formation of the $TiAl_3$ and $TiAl_2$ phases, which are brittle and exhibit CTE mismatches with α-2 orthorhombic and γ titanium aluminides. Silicide and ceramic coatings are also generally too brittle to survive fatigue conditions.

The ideal oxidation-resistant coating for γ alloys would be titanium-aluminum base for optimal chemical and mechanical compatibility with γ substrates, be capable of forming a continuous Al_2O_3 scale for protection from both oxidation and interstitial oxygen/nitrogen embrittlement, and possess reasonable mechanical properties to survive high-cycle fatigue. No ideal combination of these properties exists at present. However, reasonable compromises have been achieved with coating alloys based in the Ti-Al-Cr system. These coatings have been applied by sputtering (Ti-44Al-28Cr on Ti-47Al-2Cr-2Ta), hot isostatic pressing (Ti-44Al-28Cr and Ti-50Al-20Cr coatings), and low-pressure plasma spraying. Figure 6.11 shows the results of interrupted weight gain oxidation data for Ti-48Al-2Cr-2Nb coated with Ti-51Al-12Cr. These tests showed that the coating successfully protected the substrate at 800 and 1000 °C (1470 and 1830 °F) in air.

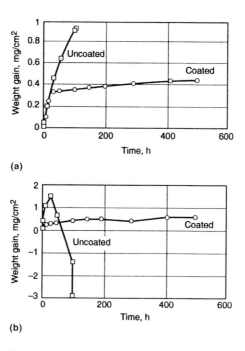

(a)

(b)

Fig. 6.11 Effect of low-pressure plasma-sprayed Ti-51Al-12Cr coating on (a) 800 °C (1470 °F) and (b) 1000 °C (1830 °F) interrupted oxidation behavior of Ti-48Al-2Cr-2Nb γ-TiAl alloy in air. Source: Ref 6.1

6.5 Gamma Titanium Aluminide Applications

General Electric has certified and implemented TiAl in the new GEnx-1B engine that powers the Boeing 787 Dreamliner that entered service in 2011. The last two stages of the GEnx-1B low-pressure turbine are made from cast gamma TiAl blades. The blades are cast from alloy 48Al-2Cr-2Nb that has a nominal room-temperature elongation of approximately 2% and environmental resistance so that it does not require coatings for applications up to approximately 800 °C (1470 °F).

TiAl is also being considered for both intake and exhaust valve applications for automotive engines. Intake valves can reach temperatures of up to approximately 600 °C (1110 °F), while exhaust valves can reach temperatures in excess of 800 °C (1470 °F). Thus, while intake valves could be made from titanium alloy, TiAl has a real opportunity to replace the conventional exhaust valve steels and the wrought nickel-base superalloys that are used in high-performance/heavy-duty engines. The speed of the engine valve-train is limited by the mass of the intake valves, which are larger than the exhaust valves. Thus, TiAl can lead to improved engine performance when used as an intake valve material, even though titanium alloys are also suitable from a property-envelope standpoint. Compared to steel valves, TiAl offers a mass reduction of approximately 49%. Although ceramic valves made of silicon nitride offer a 57% reduction in mass, there are concerns about high manufacturing costs and its intrinsic brittle behavior.

The second main area of interest for the application of TiAl within the automotive sector has been as a turbocharger wheel material in diesel engines. At present, the nickel-base superalloy Inconel 713C is most commonly used to manufacture diesel engine turbocharger wheels. The implementation of TiAl as an alternative material would result in a reduction of particulates within the exhaust gas and reduced turbo lag resulting from reduced inertia. A shorter response time between pressing down on the car accelerator and the car starting to accelerate is thus possible. The better mileage and reduced emissions would lead to a more environmentally friendly car. Another benefit includes reduced noise and vibration resulting from the resonant frequencies of the rotor being shifted to higher levels.

6.6 Processing Near-Gamma Alloys

Because of the improved ability to control microstructure in two-phase materials as well as their more attractive properties, near-gamma alloys, which contain a small amount of second-phase alpha-two (Ti_3Al) or ordered beta, are the most common materials in this class. The near-gamma alloys typically contain between 45 and 48 at.% Al as well as 0.1

to 6 at.% of secondary alloying elements such as niobium, chromium, manganese, vanadium, tantalum, and tungsten. Gamma titanium aluminide alloys can be processed by ingot metallurgy, powder metallurgy, or by casting to final shape (Fig. 6.12).

The typical ingot metallurgy approach for processing near-gamma titanium aluminide alloys usually comprises ingot production, ingot breakdown with or without intermediate and final heat treatment, and secondary processing.

Ingot Production and Ingot Structure. Three principal methods have been used successfully to melt near-gamma titanium aluminide ingots; these are induction skull melting, vacuum arc melting, and plasma (cold hearth) melting. The first of these, induction skull melting, has been used primarily to produce small-diameter (~75 to 125 mm, or 3 to 5 in.) ingots for laboratory research. Larger ingots (up to ~650 mm, or 26 in., diameter) have been made by the other two techniques. Thermal stresses developed by nonuniform temperature fields during arc or plasma melting and casting may be quite large, especially for larger-diameter ingots, and thus give rise to cracking of the low-ductility gamma titanium aluminide alloys. One method of alleviating the thermal cracking tendency to some extent involves a modified consumable arc melting process in which the elec-

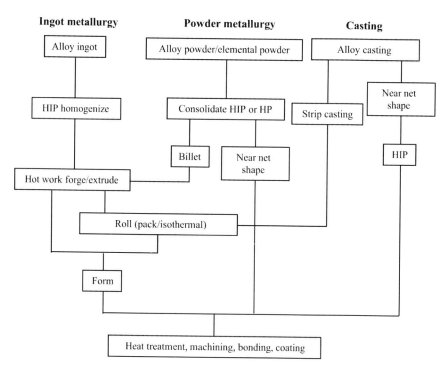

Fig. 6.12 Processing methods and routes for gamma titanium aluminides. HIP, hot isostatic pressing; HP, hot pressing. Source: Ref 6.3

trode is melted using high-power input to keep the entire charge molten, after which the heat is poured into a metal mold. Alternatively, in the plasma-melting process, the ingot is solidified in a continuous manner, thereby allowing special techniques for reducing the development of thermal stresses.

Ingot Breakdown Techniques. Three processes have been successfully used to break down the cast structure of gamma titanium aluminide alloys: isothermal forging, conventional (canned) forging, conventional (canned) extrusion, and equal-channel angular extrusion (Table 6.7). For each approach, the ingot is usually hot isostatically pressed or given a homogenization heat treatment prior to working. Hot isostatic pressing is usually done at 1260 °C (2300 °F) and a pressure of approximately 175 MPa (25 ksi). Near-gamma alloys with an aluminum content less than approximately 46 at.% may be hot isostatically pressed at a slightly lower temperature to avoid an incursion into the single-phase alpha field and a large amount of alpha grain growth during the long thermal exposure typical of the process. Isothermal forging to break down the coarse ingot structure typically consists of pancaking cylindrical preforms to reductions between 4:1 and 6:1 at temperatures between 1065 and 1175 °C (1950 and 2145 °F) and nominal strain rates between 10^{-3} and 10^{-2} s^{-1}.

Conventional hot pancake forging of cast plus hot isostatically pressed near-gamma titanium aluminide alloys, such as Ti-45.5Al-2Cr-2Nb and Ti-48Al-2Cr, has also been successfully demonstrated. In this process, the dies are usually at ambient or slightly higher (~200 °C, or 390 °F)

Table 6.7 Bulk forming alternatives for breakdown of near-gamma titanium aluminide alloys

Method	Advantages	Disadvantages
Isothermal pancake forging	Modest workability requirements Large experience base	Slow speed/long cycle time Process parameters dictated by die material (e.g., TZM) characteristics Product yield losses Large multistep reductions required to break down cast microstructure
Canned hot pancake forging	High rate process Conventional steel tooling Wide working temperature range Refined microstructure	Workability-limited process parameter selection Can/decan costs; can design Metal flow (can/workpiece) control Product yield losses
Canned hot extrusion	High rate process Conventional steel tooling Wide working temperature range Refined microstructure High product yield Large experience base	Workability-limited process parameter selection Can/decan costs; can design Preform/extrusion diameter trade-off
Equal-channel angular extrusion	High rate process Conventional steel tooling Refined microstructure High product yield No change in cross section	Workability limitations (e.g., shear localization) Can/decan costs; can design Tooling design; tool life

Source: Ref 6.4

temperatures. To minimize die chilling and thus the tendency for fracture, strain rates typical of conventional hot working processes (i.e., ~1 s^{-1}) are used. However, even with these strain rates, the workpiece must be canned to produce a sound forging. Because of can-workpiece flow-stress differences and heat-transfer effects, uniform flow of typical can materials (e.g., type 304 stainless steel) and gamma titanium aluminide preforms can be difficult to achieve. To remedy this problem, finite-element modeling techniques have been used to design cans and select process variables. It was shown that moderately uniform gamma pancake thicknesses can be achieved through a judicious choice of can geometry and can-workpiece insulation.

Because of the higher strain rates involved in conventional hot forging, as compared to those in isothermal forging, more hot work is imparted by the conventional process conducted at the same nominal workpiece temperature and to the same level of reduction. Thus, the as-forged microstructure from conventional hot forging of near-gamma alloys is typically finer, more uniform, and contains very little if any remnant lamellar colonies. In addition, with optimal can design and insulation, temperature and deformation nonuniformities within the gamma preform can be minimized during conventional forging, and relatively uniform macrostructure and microstructure throughout wrought pancakes are obtained.

Considerable effort has been expended to develop conventional (canned) hot extrusion techniques for the breakdown of a variety of near-gamma ingot materials. As with conventional hot forging, the selection of can materials and geometry, can-workpiece insulation, and process variables is extremely important with regard to obtaining sound wrought products with attractive microstructures. Typical process variables for conventional hot extrusion to break down the cast structure of gamma titanium aluminide alloys include ram speeds of 15 to 50 mm/s (0.6 to 2 in./s), reductions between 4:1 and 12:1, and preheat temperatures ranging from 1050 to 1450 °C (1920 to 2640 °F).

Dies with streamline or conical geometry have been used with equal success in round-to-round extrusion. Streamline dies have also been employed in producing round-to-rectangle extrusions to make sheet bar having a width-to-thickness ratio as large as 6:1. Can materials for conventional hot extrusion are usually type 304 stainless steel and sometimes carbon steels (for preheat temperatures of 1250 °C, or 2280 °F, or lower) or either Ti-6Al-4V or commercial-purity titanium (for preheat temperatures higher than 1250 °C, or 2280 °F). Even with canning, however, substantial temperature (and hence microstructural) nonuniformities may develop during extrusion due to the complex interaction of heat transfer and deformation heating effects. The temperature nonuniformities are most marked for the extrusion of billets of small diameter (i.e., of the order of 75 mm, or 3 in.) (Fig. 6.13). These temperature nonuniformities can be decreased, but not eliminated, by the use of insulation between the billet and can. One of the best materials for reducing heat losses has been found to be woven silica fabric, although other materials such as various foil alloys are also effective. Nevertheless, even with such measures, the temperature gradients are sufficiently large to produce noticeable radial microstructure variations in the extrudate. For example, it was found that the gamma grain size varied from 6 to 14 mm (0.2 to 0.6 in.) from the surface to the center of a Ti-49.5Al-2.5Nb-1.1Mn workpiece extruded at 1050 °C (1920 °F) to a 6:1 reduction. A similar effect is seen in TMP extrusion of near-gamma alloys to obtain fully lamellar microstructures. This extrusion technique involves billet preheating at or just below the alpha transus temperature. Deformation heating raises the workpiece temperature well into the alpha phase field, thereby promoting recrystallization of single-phase alpha that then transforms to the lamellar structure during cooldown. A typical variation in alpha grain size from the surface to the center of a Ti-45Al-2Cr-2Nb extrusion hot worked by the TMP extrusion technique is shown in Fig. 6.14.

Secondary Processing. The development of uniform, fine microstructures during breakdown of ingots of gamma titanium aluminide alloys leads to improved workability with regard to both fracture resistance and reduced flow stresses. These improvements are useful in secondary processes such as sheet rolling, superplastic forming of sheet, and isothermal, closed-die forging.

Two major techniques for rolling of sheet are conventional hot pack rolling and bare isothermal rolling. For the near-gamma titanium aluminide alloys, rolling is usually most easily conducted in the alpha-plus-gamma phase field at temperatures approximately 40 to 150 °C (70 to 270 °F)

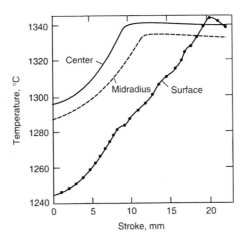

Fig. 6.13 Finite-element-predicted temperature-versus-time curves at the center, midradius, and outer diameter of a Ti-45Al-2Cr-2Nb billet encapsulated in a Ti-6Al-4V can, preheated at 1300 °C (2370 °F), and extruded to a 6:1 reduction. Source: Ref 6.5

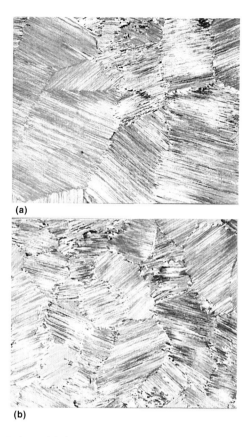

Fig. 6.14 Polarized light optical microstructures developed in a canned Ti-45Al-2Cr-2Nb billet preheated at the alpha transus temperature and extruded to a 6:1 reduction. (a) Center of extrudate. (b) Outer diameter of extrudate. Source: Ref 6.5

below the alpha transus using reductions per pass of 10 to 15% and rolling speeds that produce effective strain rates of the order of 1 s^{-1}. Using these parameters, sheets as large as 400 by 700 mm (16 by 28 in.) and ranging in thickness from 0.2 to 2.0 mm (0.008 to 0.08 in.) have been produced. In addition, a variety of microstructures have been developed in rolled sheet products, some with gamma grain sizes as small as 2 mm (0.08 in.).

An alternate sheet fabrication technique, involving rolling of uncanned gamma titanium aluminide preforms under isothermal conditions, has been developed and demonstrated on laboratory scale. The rolling equipment includes a mill with 300 mm (12 in.) wide, 60 mm (2.4 in.) diameter, ceramic work rolls and 150 mm (6 in.) diameter TZM molybdenum backup rolls. The rolls and gamma workpiece are enclosed in a vacuum chamber and heated under an argon atmosphere. Using this equipment, Ti-46Al and Ti-51Al binary alloys have been rolled to 0.75 to 1.0 mm (0.03 to 0.04 in.) thick, 150 mm (6 in.) wide sheet from 3.0 mm (0.12 in.) thick preforms.

Typical processing parameters include a preform/roll temperature between 1000 and 1100 °C (1830 and 2010 °F), rolling speed between 2 and 6 mm/min (0.08 and 0.24 in./min), and reduction per pass between 5 and 15%. For the reduction per pass and rolling speed used, the effective strain rate of the preform as it is rolled is approximately 10^{-3} s^{-1}, or a rate at which the workability of gamma alloys is good in both as-cast and wrought forms. Unfortunately, these very low strain rates lead to relatively long processing times. However, the microstructures produced by isothermal rolling are similar to those produced by conventional, hot pack rolling conducted under higher-temperature/higher-strain-rate conditions.

The fact that fine-grained microstructures can be developed in two-phase (alpha-two + gamma) or three-phase (alpha-two + gamma + beta) near-gamma titanium aluminide alloys during ingot breakdown and/or rolling suggests that these materials may be prime candidates for superplastic forming. Materials were tested over a wide range of temperatures and strain rates. For most of the test conditions, the strain rate sensitivity (m) values generally ranged from 0.4 to 0.8, or conditions under which tensile ductilities in the range of approximately 800 to 8000% may be expected. In the vast majority of the cases, however, the observed tensile ductilities were much less (i.e., ~200 to 500%), suggesting the influence of fracture processes such as cavitation in controlling formability. A noteworthy exception to this general trend was the achievement of an elongation of approximately 1000% at 1200 °C (2190 °F) and a nominal strain rate of 10^{-3} s^{-1} in rolled, fine-gamma grain samples of Ti-45.5Al-2Cr-2Nb. Under these test conditions, the m value was estimated to be between 0.4 and 0.5.

In isothermal, closed-die forging, the high m values developed in fine-grained gamma titanium aluminide alloys deformed at low strain rates have been used to make parts such as jet engine blades. Forging trials have shown excellent die-filling capability even in complex areas such as blade root sections. The enhanced metal flow of the fine-grained titanium aluminide alloys has also spurred efforts to develop higher-rate, conventional forging processes (using unheated tooling) for parts such as automotive engine valves.

The fine, equiaxed gamma grain microstructures that enable the forming of intricate parts during superplastic sheet forming or isothermal closed-die forging provide good ductility and strength but inferior fracture toughness and creep resistance. On the other hand, near-gamma titanium aluminide alloys containing a fully lamellar microstructure with a moderately small (~50 to 200 mm, or 2 to 8 in.) alpha grain size have been found to provide a better property mix. Several processing techniques have been developed to obtain such microstructures. These include the supertransus heat treatment of alloys containing grain-growth-inhibiting elements, such as boron, in solid solution or in the form of precipitates or controlled transient heating into the alpha phase field, a method suitable for parts of thin cross section.

Powder Metallurgy Processing of Near-Gamma Titanium Aluminide Alloys. A number of rapid solidification techniques have been used over the past 25 years to synthesize powders of various near-gamma and single-phase gamma titanium aluminide alloys. Perhaps the first process to be used, the Pratt and Whitney rapid solidification rate method involved pouring a molten stream of liquid onto a rapidly spinning disk, thereby producing fine droplets that solidified in flight. More recently, the plasma rotating-electrode process, developed by Nuclear Metals, and gas atomization, developed by Pratt and Whitney and Crucible Materials Corporation, have been used to make powders of these materials.

Powder consolidation has been performed primarily by hot isostatic pressing (in metal or ceramic cans) or canned hot extrusion. Hot isostatic pressing has been used to make finished parts as well as preforms for subsequent sheet rolling and forging. Hot isostatic pressing at 1050 to 1150 °C (1920 to 2100 °F) has been found to be capable of transforming the dendritic solidification structure of as-produced powders into a fine, equiaxed gamma structure while bringing about consolidation to full or nearly full theoretical density. As-hot isostatically pressed compacts or hot isostatically pressed plus rolled/forged powder products have exhibited mechanical properties comparable to or exceeding those in wrought ingot metallurgy near-gamma titanium aluminide products. Other powder metallurgy techniques for gamma titanium aluminide alloys that have been investigated in less detail include powder production via reaction of elemental powders, mechanical alloying, and magnetron sputtering, and the manufacture of components with microstructure/property gradients.

ACKNOWLEDGMENTS

Portions of the text for this chapter came from "Structural Intermetallics" in *Metals Handbook Desk Edition*, 2nd ed., ASM International, 1998, and "Bulk Forming of Intermetallic Alloys" by S.L. Semiatin in *Metal Working: Bulk Forming*, Vol 14A, *ASM Handbook*, ASM International, 2005.

REFERENCES

6.1 Structural Intermetallics, *Metals Handbook Desk Edition*, 2nd ed., ASM International, 1998

6.2 S.C. Huang and J.C. Chesnutt, Gamma TiAl and Its Alloys, *Structural Applications of Intermetallic Compounds*, John Wiley & Sons, 2000

6.3 Y.-W. Kim, Ordered Intermetallic Alloys, Part III: Gamma Titanium Aluminides, *JOM*, July 1994, p 30–39

6.4 S.L. Semiatin, Bulk Forming of Intermetallic Alloys, *Metal Working: Bulk Forming*, Vol 14A, *ASM Handbook*, ASM International, 2005

6.5 R.L. Goetz, S.L. Semiatin, and S.-C. Huang, unpublished research, Wright Laboratory Materials Directorate, Wright-Patterson AFB, 1994

SELECTED REFERENCES

- F. Appel, J.D. Heaton, P. Oehring, and M. Oehring, *Gamma Titanium Aluminide Alloys,* Wiley-VCH, 2011
- P.J. Bania, An Advanced Alloy for Elevated Temperatures, *J. Met.,* Vol 40 (No. 3), 1988, p 20–22
- P.A. Bartolotta and D.L. Krause, "Titanium Aluminide Applications in the High Speed Civil Transport," NASA/TM—1999-209071, 1999
- M.J. Blackburn and M.P. Smith, "Research to Conduct an Exploratory Experimental and Analytical Investigation of Alloys," Technical Report AFWAL-TR-80-4175, U.S. Air Force Wright Aeronautical Laboratories, 1980
- M.J. Blackburn and M.P. Smith, "R&D on Composition and Processing of Titanium Aluminide Alloys for Turbine Engine," Technical Report AFWAL-TR-82-4086, U.S. Air Force Wright Aeronautical Laboratories, 1982
- E.F. Bradley, "The Potential Structural Use of Aluminides in Jet Engines," Paper presented at the Gorham Advanced Materials Institute Conference on Investment, Licensing and Strategic Partnering Opportunities, Emerging Technology, Applications, and Markets for Aluminides, Iron, Nickel and Titanium, Nov 1990 (Monterrey, CA)
- N.S. Choudhury, H.C. Graham, and J.W. Hinze, in *Properties of High Temperature Alloys With Emphasis on Environmental Effects,* Electrochemical Society, 1976, p 668–680
- F.H. Froes, *Mater. Edge,* No. 5, May 1988
- T.J. Jewett et al., *High-Temperature Ordered Intermetallic Alloys III, Materials Research Society Symposia Proceedings,* Vol 133, C.T. Liu, A.I. Taub, N.S. Stoloff, and C.C. Koch, Ed., Materials Research Society, 1989, p 69–74
- C. Leyens and M. Peters, *Titanium and Titanium Alloys—Fundamentals and Applications,* Wiley-VCH Verlag GmbH & Co., 2003
- H.A. Lipsitt, in *Advanced High Temperature Alloys: Processing and Properties,* S.S. Allen, R.M. Pellous, and R. Widmer, Ed., American Society for Metals, 1986
- H.H. Lipsitt, in *High-Temperature Ordered Intermetallic Alloys, Materials Research Society Symposia Proceedings,* Vol 39, C.C. Koch, C.T. Liu, and N.S. Stoloff, Ed., Materials Research Society, 1985, p 351–364
- G. Lutjering and J.C. Williams, *Titanium,* Springer-Verlag, 2003
- R.G. Rowe, *High Temperature Aluminides and Intermetallics,* S.H. Whang, C.T. Liu, and D. Pope, Ed., TMS, 1990

CHAPTER **7**

Engineering Plastics

AN ENGINEERING PLASTIC can be defined as a synthetic polymer capable of being formed into load-bearing shapes and possessing properties that enable it to be used in the same manner as traditional materials, and that has high-performance properties that permit it to be used in the same manner as metals and ceramics. This definition includes thermoplastic and thermoset resins. It also includes continuous fiber-reinforced resins (composite materials) as well as the short fiber- and/or particulate-reinforced plastics commonly used for parts and components. Because composites are a subset of engineering plastics, they are addressed in much greater detail in Chapter 8, "Polymer-Matrix Composites," in this book.

A plastic is a polymeric material composed of molecules made up of many (poly-) repeats of some simpler unit, the mer. All polymers are chemically constructed of repeats of the basic mer unit, which is chemically bonded to others of its kind to form one-, two-, or three-dimensional molecules. The mer units of polymers are bonded to one another with strong covalent bonds. Most polymers contain mainly carbon in their backbone structures because of the unique ability of carbon to form extensive, stable covalent bonds. While covalent bonds are stronger than metallic bonds, they are highly directional in character. In thermoplastics, the strong covalent bonds occur within the polymer molecules but not between them. Only weaker, secondary bonds occur between polymer molecules. Thus, unless the polymer is the type that is bonded together in three dimensions (as in a thermoset), it may be relatively easy to disrupt its structure with moderate heat, even though it is difficult to rupture the covalent bonds within the polymer molecule itself.

7.1 General Characteristics

Engineering plastics offer some unique product benefits. These are usually physical properties, or combinations of physical properties, that allow vastly improved product performance, such as:

- *Electrical insulation.* Most combinations of plastic resin and filler are natural electrical insulators.
- *Thermal insulation.* All common resin-filler combinations are excellent thermal insulators, compared to metals and most other nonmetals.
- *Chemical resistance.* Some plastics offer extreme resistance to chemical reagents and solvents.
- *Magnetic inertness.* Most plastics will not respond to an electromagnetic field.
- *Extremely lightweight.* Some resin-reinforcement combinations offer extremely favorable strength-to-weight and stiffness-to-weight ratios.
- *Transparency.* Several plastics offer transparency and a high degree of optical clarity.
- *Toughness.* Many plastics offer toughness, most commonly regarded as freedom from the tendency to fracture.
- *Colorability.* All plastics are colorable to some extent. Many are capable of a nearly unlimited color spectrum while maintaining high visual quality.

Plastics have the following limitations:

- Strength and stiffness is low relative to metals and ceramics.
- Service temperatures are usually limited to only a few hundred degrees because of the softening of thermoplastic plastics or degradation of thermosetting plastics.
- Some plastics degrade when subjected to sunlight, other forms of radiation, and some chemicals.
- Thermoplastics exhibit viscoelastic properties, which means they creep under relatively low stress levels and can be a distinct limitation in load-bearing applications.

An engineer should immediately disqualify engineering plastics if the application requires maximum efficiency of heat transfer, electrical conductivity, or nonflammable properties. Although some plastics can be formulated to retard flames, their organic nature makes them inherently flammable. Applications under constant stress for which a close tolerance must be held need to be scrutinized in terms of the effects of creep at the application temperature. The engineer also must be cognizant of the poor ultraviolet (UV) resistance of most, though not all, engineering plastics, and the resulting requirement of a UV absorber.

7.2 Thermosets and Thermoplastics

Polymeric plastics are classified as either thermosets or thermoplastics (Fig. 7.1). Thermosets are low-molecular-weight, low-viscosity monomers ($\approx$2000 cps) that are converted during curing into three-dimensional cross-linked structures that are infusible and insoluble. Cross-linking results from chemical reactions that are driven by heat generated either by chemical reactions themselves (i.e., exothermic heat of reaction) or by externally supplied heat. Prior to cure, the resin is a relatively low-molecular-weight semisolid that melts and flows during the initial part of the cure process. As the molecular weight builds during cure, reactions accelerate and the available volume within the molecular arrangement decreases, resulting in less mobility of the molecules and an increase in viscosity. The viscosity increases until the resin gels (solidifies), and then strong covalent bond cross links form during cure. After the resin gels and forms a rubbery solid, it cannot be remelted. Further heating causes additional cross-linking until the resin is fully cured. This progression through cure is shown in Fig. 7.2. Because cure is a thermally-driven event requiring chemical reactions, thermosets are characterized as having rather long processing times. Due to the high cross-link densities obtained for high-performance thermoset systems, they are inherently brittle unless steps are taken to enhance toughness.

Thermoset polymers cure by either addition or condensation reactions. A comparison of these two cure mechanisms is shown in Fig. 7.3. In the addition reaction shown for an epoxy reacting with an amine curing agent, the epoxy ring opens and reacts with the amine curing agent to form a cross link. The amine shown in this example is what is known as an aliphatic amine and would produce a cross-linked structure with only moderate temperature capability. Higher-temperature capabilities can be produced by curing with what are known as aromatic amines. Aromatic amines contain the large and bulky benzene ring, which helps to restrict chain movement when the network is heated. A typical curing agent, diamino diphenyl sulfone, and a typical epoxy, tetraglycidyl methylene dianiline (TGMDA), are shown in Fig. 7.4. Note that both the curing agent and

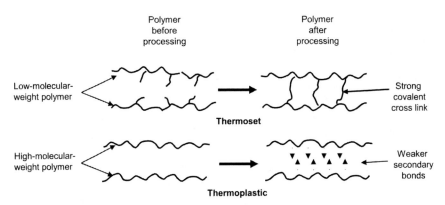

Fig. 7.1 Comparison of thermoset and thermoplastic polymer structures

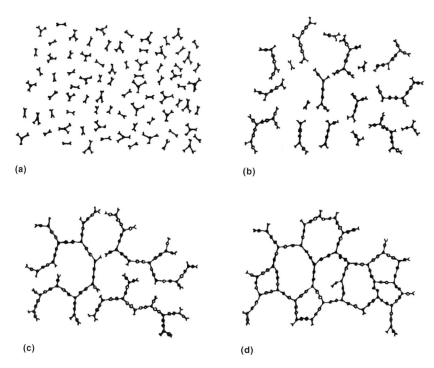

Fig. 7.2 Stages of cure for thermoset resin. (a) Polymer and curing agent prior to reaction. (b) Curing initiated, with size of molecules increasing. (c) Gelation with full network formed. (d) Fully cured and cross-linked. Source: Ref 7.1

the epoxy contain benzene rings. Also note that the TGMDA has four epoxy rings, each capable of reacting. Hence, this epoxy has what is known as a functionality of four. Because it has four active epoxide rings, it is capable of forming high-density cross-linked structures. This particular curing agent and epoxy are used in many of the composite-matrix resin formulations.

In the lower portion of Fig. 7.3, two phenol molecules are shown to react with a formaldehyde molecule to form a phenolic linkage in what is known as a condensation reaction. The significant feature of this reaction is the evolution of a water molecule each time the reaction occurs. When thermosets such as phenolics and polyimides that cure by condensation reactions are used for molded parts, they often contain high porosity levels, due to the water or alcohol vapors created by the condensation reactions. Addition-curing thermosets, such as epoxies, polyesters, vinyl esters, and bismaleimides, are generally much easier to process with low void levels.

An important consideration in selecting any plastic (thermoset or thermoplastic) is the glass transition temperature. The glass transition temperature (T_g) is a good indicator of the temperature capability of the matrix. The

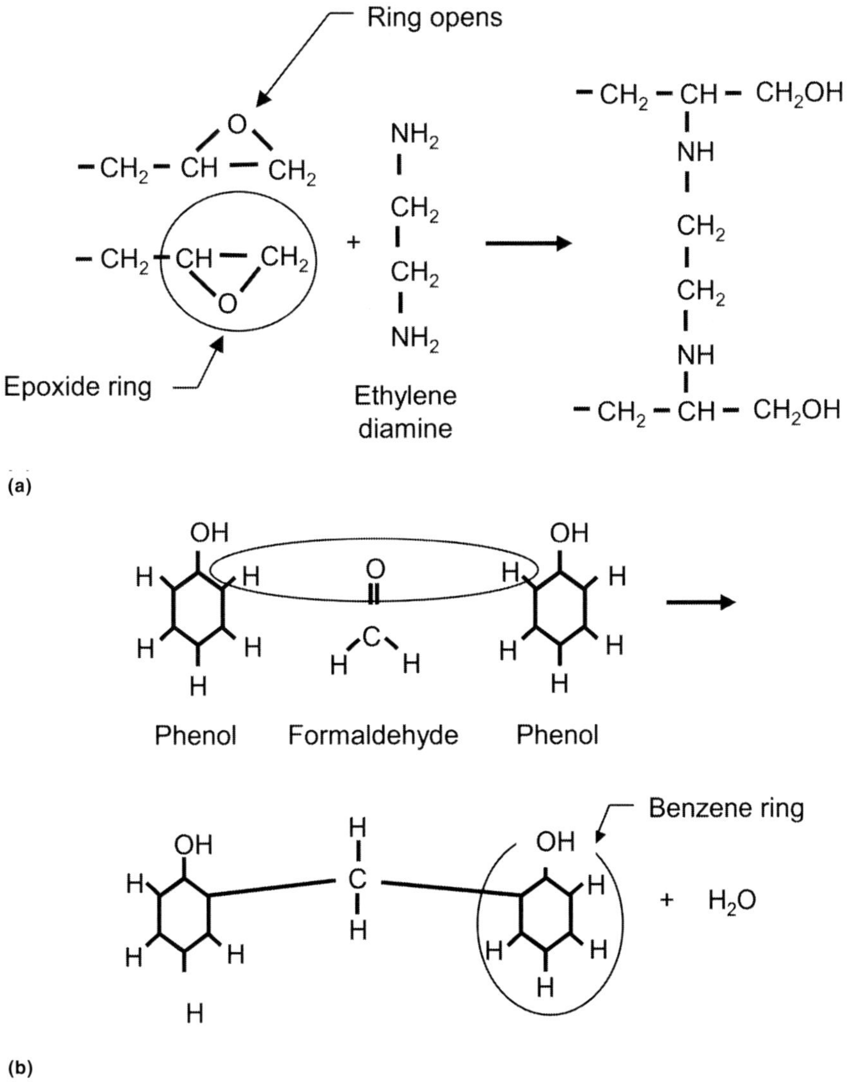

(a)

(b)

Fig. 7.3 Thermoset cure reactions. (a) Addition reaction to form cross link. (b) Condensation reaction to form cross link

glass transition temperature of a polymeric material is the temperature at which it changes from a rigid glassy solid into a softer, semiflexible material. At this point, the polymer structure is still intact, but the cross links are no longer locked in position. A resin should never be used above its T_g unless the service life is very short.

In contrast to thermosets, thermoplastics are high-molecular-weight resins that are fully reacted prior to processing. They melt and flow during processing but do not form cross-linking reactions. Their main

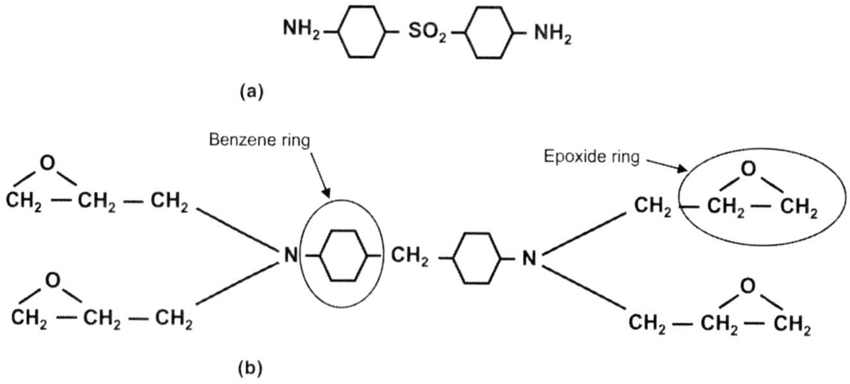

Fig. 7.4 Composite-matrix resins. (a) Diamino diphenyl sulfone (DDS) curing agent. (b) Tetraglycidyl methylene dianiline (TGMDA) epoxy

chains are held together by relatively weak secondary bonds. When heated to a sufficiently high temperature, these secondary bonds break down, and the thermoplastic eventually reaches the fluid state. On cooling, the secondary bonds reform and the thermoplastic becomes a solid again. Being high-molecular-weight resins, the viscosities of thermoplastics during processing are orders of magnitude higher than that of thermosets (e.g., 10^4 to 10^7 poise for thermoplastics versus 10 poise for thermosets). Because thermoplastics do not cross link during processing, they can be reprocessed; for example, they can be thermoformed into structural shapes by simply reheating to the processing temperature. Thermosets, due to their highly cross-linked structures, cannot be reprocessed and will thermally degrade and eventually char if heated to high-enough temperatures.

Many thermoplastics are polymerized by what is called chain polymerization, as shown in Fig. 7.5 for the simple thermoplastic polyethylene. Chain polymerization consists of three steps: initiation, propagation, and termination. During initiation, an active polymer capable of propagation is formed by the reaction between an initiator species and a monomer unit. In the figure, "R" represents the active initiator that contains an unpaired electron (·). Propagation involves the linear growth of the molecule as monomer units become attached to each other in succession, to produce a long chain molecule. Chain growth, or propagation, is very rapid, with the period required to grow a molecule consisting of 1000 repeat units on the order of 100 to 1000 s. Propagation can terminate in one of two ways. In the first, the active ends of two propagating chains react together to form a nonreactive molecule. The second method of termination occurs when an active chain end reacts with an initiator. Polyethylene can normally have anywhere from 3500 to 25,000 of these repeat units.

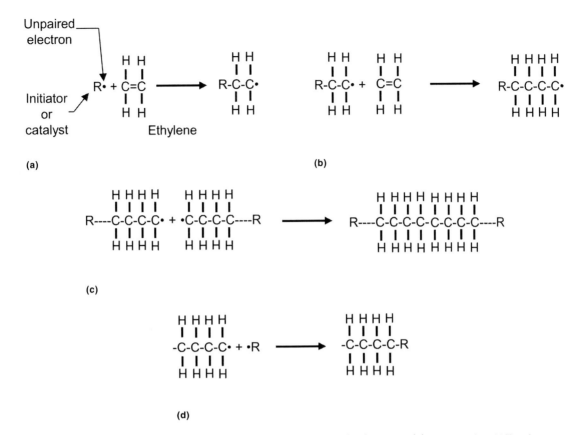

Fig. 7.5 Chain polymerization of polyethylene. (a) Initiation. (b) Chain growth by propagation. (c) Termination by reaction of active end groups. (d) Termination by reaction with initiator

Thermoplastics are further classified as being either amorphous or semicrystalline. The differences between an amorphous and a semicrystalline thermoplastic are shown in Fig. 7.6. An amorphous thermoplastic contains a massive random array of entangled molecular chains. The chains themselves are held together by strong covalent bonds, while the bonds between the chains are much weaker secondary bonds. When the material is heated to its processing temperature, it is these weak secondary bonds that break down and allow the chains to move and slide past one another. Amorphous thermoplastics exhibit good elongation, toughness, and impact resistance. As the chains get longer, the molecular weight increases, resulting in higher viscosities, higher melting points, and greater chain entanglement, all leading to higher mechanical properties.

Semicrystalline thermoplastics contain areas of tightly folded chains (crystallites) that are connected together with amorphous regions. Amorphous thermoplastics exhibit a gradual softening on heating, while semicrystalline thermoplastics exhibit a sharp melting point when the crystalline regions start dissolving. As the polymer approaches its melting point, the crystalline lattice breaks down and the molecules are free

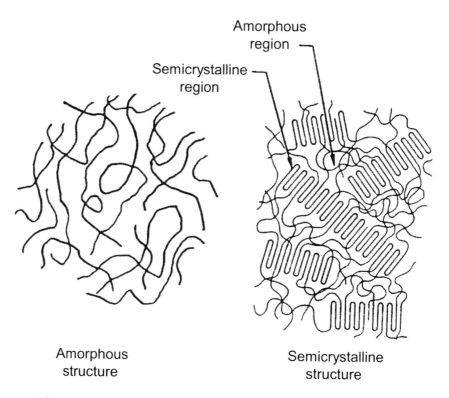

Fig. 7.6 Comparison of amorphous and semicrystalline thermoplastic structures

to rotate and translate, while noncrystalline amorphous thermoplastics exhibit a more gradual transition from a solid to a liquid. In general, the melting point (T_m) increases with increasing chain length, greater attractive forces between the chains, greater chain stiffness, and increasing crystallinity, while the glass transition temperature (T_g) increases with lower free volume, greater attractive forces between the molecules, decreasing chain mobility, increasing chain stiffness, increasing chain length, and, for thermosets, increasing cross-link density. Crystallinity provides the following attributes to a thermoplastic resin:

- Crystalline regions are held together by amorphous regions. The maximum crystallinity obtainable is approximately 90%, whereas metallic structures are usually 100% crystalline and exhibit much more ordered structures.
- Crystallinity increases density. The density increase helps to explain the improved solvent resistance, because it becomes more difficult for the solvent molecules to penetrate the tightly packed crystallites.
- Crystallinity increases strength, stiffness, creep resistance, and temperature resistance but usually decreases toughness. The tightly packed crystalline structure behaves somewhat like cross-linking in thermosets by decreasing and restricting chain mobility.

- Crystalline polymers are either opaque or translucent, while transparent polymers are always amorphous.
- Crystallinity can be increased by mechanical stretching.

In general, most thermoplastics contain approximately 20 to 35% crystallinity. It should also be noted that all thermoset resins are amorphous but are cross-linked to provide strength, stiffness, and temperature stability. As a general class of polymers, thermoplastics are much more widely used than thermosets, accounting for approximately 80% of the polymers produced. A number of industrially important plastics are listed in Table 7.1, along with their chemical names, and whether they are amorphous or semicrystalline.

7.3   Metals versus Plastics

Designing with plastics is often a more complex task than designing with metals. Engineering plastics, which are viscoelastic, do not respond to mechanical stress in the linear, elastic manner for which most designers have developed an intuitive feel. In many applications, engineering plastics exhibit a more complex property mix than do metals. Because

Table 7.1   Abbreviations, chemical names, and structures of select plastics

Material	Chemical name	Structure
ABS	Acrylonitrile-butadiene-styrene	Amorphous
ABS-PC	Acrylonitrile-butadiene-styrene-polycarbonate	Amorphous
DAP	Diallyl phthalate	Amorphous
POM	Polyoxymethylene (acetal)	Semicrystalline
PMMA	Polymethyl methacrylate (acrylic)	Amorphous
PAR	Polyarylate	Amorphous
LCP	Liquid crystal polymer	Semicrystalline
MF	Melamine formaldehyde	Amorphous
Nylon 6	Polyamide	Semicrystalline
Nylon 6/6	Polyamide	Semicrystalline
Nylon 12	Polyamide	Semicrystalline
PAE	Polyaryl ether	Amorphous
PBT	Polybutylene terephthalate	Semicrystalline
PC	Polycarbonate	Amorphous
PBT-PC	Polybutylene terephthalate-polycarbonate	Amorphous
PEEK	Polyetheretherketone	Semicrystalline
PEI	Polyether-imide	Amorphous
PESV	Polyether sulfone	Amorphous
PET	Polyethylene terephthalate	Semicrystalline
PF	Phenol formaldehyde (phenolic)	Amorphous
PPO (mod)	Polyphenylene oxide	Amorphous
PPS	Polyphenylene sulfide	Semicrystalline
PSU	Polysulfone	Amorphous
SMA	Styrene-maleic anhydride	Amorphous
UP	Unsaturated polyester	Amorphous

engineering plastics, as a family, are much younger than engineering metals, their database is not yet complete. In addition, their rapid evolution makes the materials selection process more difficult.

Engineering plastics have a broad range of mechanical properties, and interpreting those listed on a data sheet requires an understanding of the test methods and the method of reporting. Mechanical properties for a variety of plastics are summarized in Table 7.2. Because many of the properties shown in Table 7.2 can change dramatically depending on processing parameters, the engineer must allow for any such changes when selecting materials.

Mechanical properties are often the most important properties in the design and selection of engineering plastics, but the mechanical behavior of plastics is quite different from that of other engineering materials in several respects. Plastics are often modified by fillers, plasticizers, flame retardants, stabilizers, and impact modifiers, all of which substantially change mechanical properties. For example, Table 7.3 shows the range of three mechanical properties for various commercial grades of acrylonitrile-butadiene-styrene. Considerable variation exists among different grades, even though they have similar chemical structures.

Table 7.2 Mechanical properties of selected plastics

Material	Tensile strength		Tensile modules		Flexural strength		Impact strength		Hardness, Rockwell	Flame rating, UL 94
	MPa	ksi	GPa	10⁶ psi	MPa	ksi	J/m	Ft lbf/in.		
ABS	41	6.0	2.3	0.33	72.4	10.5	3.47	6.5	R103	HB
ABS-PC	59	8.5	2.6	0.38	89.6	13.0	560	10.5	R117	HB
DAP	48	7.0	10.3	1.50	117	17.0	37	0.7	E80	HB
POM	69.0	10.0	3.2	0.47	98.6	14.3	133	2.5	R120	HB
PMMA	72.4	10.5	3.0	0.43	110	16.0	21	0.4	M68	HB
PAR	68	9.9	2.1	0.30	82.7	12.0	288	5.4	R122	HB
LCP	110	16.0	11.0	1.60	124	18.0	101	1.9	R80	V0
MF	52	7.5	9.65	1.40	93.1	13.5	16	0.3	M120	V0
Nylon 6	81.4	11.8	2.76	0.40	113	16.4	59	1.1	R119	V2
Nylon 6/6	82.7	12.0	2.83	0.41	110	16.0	53	1.0	R121	V2
Nylon 12	81.4	11.8	2.3	0.34	113	16.4	64	1.2	R122	V2
PAE	121	17.6	8.96	1.30	138	20.0	64	1.2	M85	V0
PBT	52	7.5	2.3	0.34	82.7	12.0	53	1.0	R117	HB
PC	69.0	10.0	2.3	0.34	96.5	14.0	694	13.0	R118	V2
PBT-PC	55	8.0	2.2	0.32	86.2	12.5	800	15.0	R115	HB
PEEK	93.8	13.6	3.5	0.51	110	16.0	59	1.1	...	V0
PEI	105	15.2	3.0	0.43	152	22.0	53	1.0	M109	V0
PESV	84.1	12.2	2.6	0.38	129	18.7	75	1.4	M88	V0
PET	159	23.0	8.96	1.30	245	35.5	101	1.9	R120	HB
PF	41	6.0	5.9	0.85	62	9.0	21	0.4	M105	HB
PPO (mod)	54	7.8	2.5	0.36	88.3	12.8	267	5.0	R115	V0
PPS	138	20.0	11.7	1.70	179	26.0	69	1.3	R123	V0
PSU	73.8	10.7	2.5	0.36	106	15.4	64	1.2	M69	HB
SMA	31	4.5	1.9	0.27	55	8.0	133	2.5	R95	HB
UP	41	6.0	5.5	0.80	82.7	12.0	32	0.6	M88	HB

Note: Refer to Table 7.1 for chemical name. Source: Ref 7.2

Table 7.3   Range of mechanical properties of commercial acrylonitrile-butadiene-styrene

Grade	Tensile strength		Elongation at break, %	Notched impact strength	
	MPa	ksi		J/m	ft lbf/in.
Injection molding	20–76	2.9–11	1–46	50–100	1–2
Injection molding and extrusion	29–49	4.2–7.1	2–46	100–400	2–8
Extrusion	2–59	0.3–8.5	2–40	100–700	2–14
Glass fiber filled	66–100	9.6–16	1–4	50 to >100	1 to >2

Source: Ref 7.3

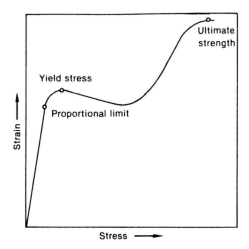

Fig. 7.7   Typical stress-strain curve for an engineering metal. Note high degree of linearity below the proportional limit. Source: Ref 7.4

The mechanical behavior of polymers is time dependent, or viscoelastic. Therefore, design data based on short-term tests have the possibility of misrepresenting that polymer in a design application that involves long-term loading. The magnitude of the time dependence of polymers is very temperature-dependent. Well below their glass transition temperature, glassy or semicrystalline polymers are only very weakly viscoelastic. For these polymers, test data based on a time-independent analysis will probably be adequate. As the temperature is increased, either by the environment or by heat given off during deformation, the time dependence of the mechanical response will increase. The characterization of such materials must consider viscoelasticity.

A typical stress-strain curve for an engineering metal is shown in Fig. 7.7. The salient features of metal behavior are that the slope of the stress-strain curve is a constant up to the proportional limit and is known as the elastic modulus, and that the elastic modulus remains a constant over a wide range of strain rates and temperatures. A large number of design equations have been derived for use in the design of structural members such as beams, plates, and columns. All of these equations use the elastic modulus as a fundamental measure of the response of the material to stress. However, if the designer attempts to use these equations for plastics part design using a modulus value taken from a plastics property specifications sheet, serious design errors will be made in many cases. The implicit assumption behind the design equations is that the material behaves in a linear, elastic manner. As already noted, this assumption is not true for plastics.

When comparing the mechanical behavior of metals to that of plastics, as shown in Fig. 7.7 and 7.8, it can be seen that plastic, unlike metal, has no true proportional limit and that its stress-strain curve is not linear. Furthermore, the shape of the plastic stress-strain curve is greatly affected by the rate at which stress is applied, and stress-strain behavior is greatly affected by temperature. Clearly, the mechanical response of plastic is more complex than that of metal. Because a plastic behaves in a manner that seems to combine elastic solid and highly viscous liquid behaviors, it is said to be viscoelastic.

Another important consideration when considering metals versus plastics for a particular application is fatigue behavior. The design approach for plastics is the same as that for metals exposed to conditions that are expected to produce fatigue. Curves of stress versus number of stress cycles, or *S-N* curves, are used to determine the allowable stress for an application. However, the application of fatigue data to plastics design

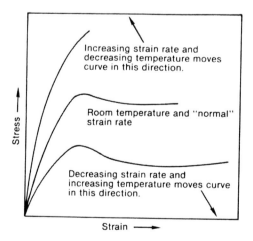

Fig. 7.8 Typical stress-strain curve for an engineering plastic. Note that there is no true proportional limit. Source: Ref 7.4

requires greater caution than is the case for metals. Changes in strain rate and temperature have a much more profound impact on the mechanical properties of plastics than of metals. As a further complication, changes in the mode of deformation have a much greater effect on fatigue results for plastics than for metals; for example, tests under constant fatigue stress may not agree with tests under constant fatigue strain. When designing for fatigue resistance, it is necessary to ascertain that the candidate plastic fatigue data were measured using strain rates, temperatures, and deformation modes appropriate for the application.

When designing with engineering plastics, it is generally best to choose the properties region in which the response of the material is still elastic. This is called the proportional region in a stress-strain curve, which obeys Hookean laws, and is the region where conventional strength of materials design calculations is likely a reasonable approximation. Otherwise, the time-dependent factors of viscoelasticity must be considered.

The chemical nature of engineering plastics is so profoundly different from that of metals that these two material classes respond to the environment in very different ways. For example, metals often rust or tarnish when exposed to atmospheric moisture, while plastics do not. Metals are electrical conductors, while plastics tend to be insulators. Most metals do not burn under normal conditions, while many plastics burn readily.

Plastics are generally not as strong as metals. They are also prone to dimensional change, especially under load at high temperatures. However, because plastics are generally less dense than metals, their strength-to-weight ratios are highly favorable. Table 7.4 shows that glass-filled plastics have strength-to-weight ratios that are twice those of steel and cast aluminum alloys.

In theory, any combination of part design and material should compete in cost and function with an infinite variety of other approaches. In

Table 7.4 Range of mechanical properties for common engineering materials

Material	Elastic modulus		Tensile strength		Maximum strength/density		Elongation at break, %
	GPa	10^6 psi	MPa	ksi	(km/s)^2	(kft/s)^2	
Ductile steel	200	30	350–800	50–120	0.1	1	0.2–0.5
Cast aluminum alloys	65–72	9–10	130–300	19–45	0.1	1	0.01–0.14
Polymers	0.1–21	0.02–30	5–190	0.7–28	0.05	0.5	0–0.8
Glasses	40–140	6–20	10–140	1.5–21	0.05	0.5	0
Copper alloys	100–117	15–18	300–1400	45–200	0.17	1.8	0.02–0.65
Moldable glass-filled polymers	11–17	1.6–2.5	55–440	8–64	0.2	2	0.003–0.015

Source: Ref 7.5

practice, the competition is narrowed by some easily recognized trends. The competitive position of various metals to plastics is compared in Table 7.5.

7.4 Effect of Fillers and Reinforcements

As already noted, most engineering plastics are available in neat or reinforced grades. Although the effects of fibers or fillers differ, depending on the matrix plastic, some generalizations can be made. Glass fibers are the most common reinforcement used in engineering plastics. The type of glass and the surface treatment on the fibers, which enhances adhesion to the matrix, are crucial to property improvement. Generally, the introduction of glass fibers increases tensile strength, modulus, and impact strength. The effect of glass fillers on tensile strength and modulus are shown in Fig. 7.9 and 7.10. However, as shown in Fig. 7.11, while the addition of glass increases the impact strength of nylon 6/6 and polybutylene terephthalate, small additions actually reduce the impact strength of polycarbonate.

Mold shrinkage is reduced with the addition of glass fibers. Although a single shrinkage value is usually listed, the addition of glass fibers typically induces anisotropic flow in the material. The result is a significant

Table 7.5   Common plastic/metal competitors

Material	Processes		Generalized plastics competitive position
	Primary	Secondary	
Sheet steel	Stamped	Welded fabrication	Sheet molding compound (SMC) or thermoplastic. SMCs offer 30% weight savings at similar room-temperature performance. Cost comparison depends on steel complexity.
Sheet stainless steel	Stamped	Welded fabrication	As above, except SMCs offer a vast cost saving.
Sheet aluminum	Stamped	Riveted skin and stronger construction	Integrated SMC offers equal performance at equal weight with substantial cost saving.
Cast aluminum	Sand permanent or die cast	Machined	Plastics save 20 to 30% weight at equal performance. Cost saving only if metal requires heavy machining and plastic does not.
Forged aluminum	Forge	Machined	Plastics cannot match structure in equal sections. Super-premium thermoplastics compete on cost basis.
Brass or bronze	Cast	Machined	Usually, for saltwater resistance. Several thermoplastics and thermoset materials compete successfully.
Magnesium	Cast	Machined	For light weight requirement. Several plastics compete successfully.
Zinc	Die cast	Machined	Very low cost castings to finished shape. Nylons compete successfully.
Iron	Sand cast	Machined	Plastics more expensive per cubic inch for equal performance. Plastics usually cannot compete if iron is used correctly.
Titanium and exotics	Cast or forged	Machined	If exotic metal is used for extreme corrosion resistance, plastics compete well.

Source: Ref 7.4

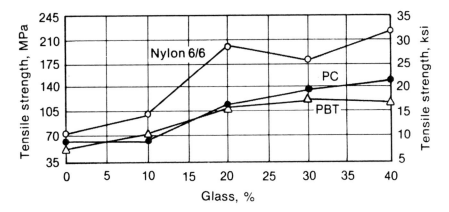

Fig. 7.9 Effect of glass addition on tensile strength. PC, polycarbonate; PBT, polybutylene terephthalate. Source: Ref 7.2

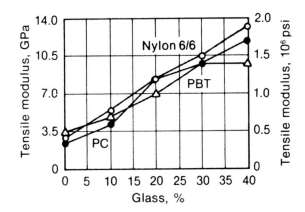

Fig. 7.10 Effect of glass addition on tensile modulus. PC, polycarbonate; PBT, polybutylene terephthalate. Source: Ref 7.2

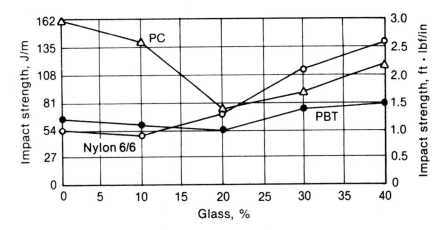

Fig. 7.11 Effect of glass addition on impact strength. PC, polycarbonate; PBT, polybutylene terephthalate. Source: Ref 7.2

difference in strength and shrinkage in many molded parts, depending on whether the portion is parallel or perpendicular to the flow direction. If it is parallel to the flow, there would be an increase in tensile and impact strength values and a reduction in shrinkage, which would coincide with the orientation of the fibers.

Fillers are used in engineering plastics for a variety of reasons. They generally reduce the cost of the material as well as the mold shrinkage. They improve the thermal conductivity and reduce the coefficient of thermal expansion. Their effect on mechanical properties depends on the amount and type of filler, but they generally increase tensile strength and modulus, as shown in Fig. 7.12.

Typical fillers, such as talc or calcium carbonate, have a very low aspect ratio and do not contribute to anisotropic behavior. They can be used to modify properties and reduce shrinkage without causing property changes that are dependent on flow direction. The amount and type of reinforcement affect other properties. Glass fibers are very abrasive and tend to increase the wear and coefficient of friction of engineering plastics. When a reinforcement is necessary and increased wear is undesirable, additives such as molybdenum disulfide, graphite, or synthetic fluorine-containing resin are commonly used to combat wear.

Flammability can be affected by the addition of fibers or fillers, especially in the case of crystalline materials with sharp melting points. Unfilled polyamide (nylon) typically is self-extinguishing on some flammability tests because the portion that burns also melts, and the flaming portion drops away, leaving an extinguished remainder. However, with the addition of glass fiber, the ignited sample is held together by the fiber and continues to burn. In such cases, a flame-retardant additive is required to extinguish the flame.

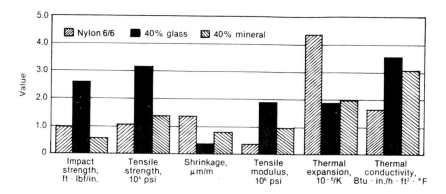

Fig. 7.12 Comparison of filler and reinforcements. Source: Ref 7.2

7.5   Temperature and Molecular Structure

Temperature affects mechanical properties by means of its influence on intermolecular interactions. The effect of temperature on mechanical properties can be understood by examining the relationship between tensile modulus and temperature. A typical modulus-temperature curve is shown in Fig. 7.13. At temperatures below the glass transition temperature (T_g), most plastic materials have a tensile modulus of approximately 2 GPa (0.3×10^6 psi). If the material is crystalline, a small drop in modulus is generally observed at the T_g, while a large drop is seen at the melting temperature, T_m. The T_g is primarily associated with amorphous, rather than semicrystalline, resins. Resins that are partially crystalline have at least a 50% amorphous region, which is the region that has a T_g. If a material is amorphous, a single decrease is usually seen at temperatures near the T_g. At even higher temperatures, there is another similar drop in modulus, and the plastic flows easily as a high viscosity liquid. At this condition, the plastic can be processed by extrusion or molding.

Mechanical properties are also affected by molecular weight. Most material manufacturers provide grades with different molecular weights. High-molecular-weight materials have high melt viscosities and low melt indexes. When the molecular weight is low, the applied mechanical stress

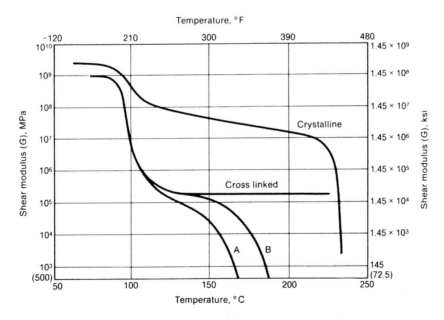

Fig. 7.13   Shear modulus versus temperature for crystalline polystyrene (PS), two linear PS materials (A and B) with different molecular weights, and lightly cross-linked PS. Source: Ref 7.6

tends to slide molecules over each other and separate them. The solid, with very little mechanical strength, has negligible structural value. With a continuing increase in molecular weight, the molecules become entangled, the attractive force between them becomes greater, and mechanical strength begins to improve. Both elongation at break and impact strength increase with increasing molecular weight.

It is generally desirable for material manufacturers to make plastics with sufficiently high molecular weights to obtain good mechanical properties. For polystyrene the molecular weight is 100,000, and for polyethylene the value is 20,000. It is not desirable to increase molecular weight further because melt viscosity will increase rapidly, although there are occasional exceptions to this rule. The yield strength of polypropylene decreases when molecular weight increases. Studies of morphology indicate that high molecular weight and branching reduce crystallinity. Polymers with high intermolecular interactions, such as hydrogen bonding, do not require high molecular weight to achieve good mechanical properties. With low molecular weight, viscosity is very low, which is commonly observed for polyamides (nylons).

Plastics with narrow molecular weight are preferred for low warpage in thin-wall injection molding, film extrusion, and rotational molding. Moderately low-molecular-weight plastics are suitable for high-speed processing, such as high-draw-down-rate extrusion, high-speed calendering, and injection molding. Most processing conditions require materials with high molecular weights. This is especially true for extrusion and blow molding, which require sufficient melt strength for the extrudate to support itself as it exits from the die.

Copolymerization is also frequently used to change the properties of plastics. For example, copolymerization with vinyl acetate increases the processability and thermal stability of polyvinyl chloride. Copolymerization of acrylonitrile with styrene increases the T_g.

Plasticizers are low-molecular-weight compounds that are often compounded into high-molecular-weight polymers to improve processability, impact strength, and elongation. Plasticizers lower melt viscosity and processing temperature. Fundamentally, they function by broadening the molecular weight distribution and increasing the low-molecular-weight fraction of the total composition. Plasticizers are essentially nonvolatile solvents.

7.6 Effect of Crystallinity

In semicrystalline polymers, the mobility of polymer segments is reduced considerably. The effect is an increase in rigidity, modulus, and hardness and a decrease in solvent solubility. Impact strength sometimes increases with crystallinity at temperatures above the T_g because the crystals act as cross links. However, at a very high degree of crystallinity, the impact strength usually decreases. The tensile strength of semicrystalline materials generally shows a small decrease when the temperature increases above the T_g. Tensile properties decrease only at temperatures near the crystalline melt point. Materials with an excessive crystalline fraction become brittle at temperatures below the T_g.

During crystallization, the crystalline polymer packs all of the low-molecular-weight components and impure species into the interstices between the spherulites, leaving these as contaminated boundaries of lower strength and modulus. Shrinkage during crystallization may leave stresses and voids in these interstices, weakening them even more. The surface between spherulites and amorphous interstices is the weak interface at which cracking is most likely to begin.

The amount of crystalline fraction and the size of crystalline spherulites can be affected by the addition of nucleating agents, or seed particles, which can be small, inorganic particles. Plastics with seeds contain a higher crystalline fraction with small domains. Crystallinity is also affected by the temperature gradient in processing. For example, a high melt pressure in molding reduces dwell time in the barrel. This reduces the temperature gradient, which tends to decrease crystallization. Conversely, low mold temperature increases the crystallization rate.

The cooling temperature rate also affects the amount of crystallinity. Generally, the maximum crystallization rate is observed at approximately 0.9 of the melting point, measured in absolute temperature. For a material cooled at approximately this temperature, sufficient crystallinity will develop. If the material has a high T_g and the cooling process takes place below it, an amorphous structure can result.

For semicrystalline material, control of crystallinity is generally more important than control of molecular weight in changing mechanical properties. For these materials, the property can be correlated with density, which in turn is related to crystallinity. One primary example is polyethylene, which in the commercial market is classified according to density. Hydraulic stress during injection molding flow and calendering aligns the polymer molecules parallel to each other and favors crystallization. In these cases, tensile strength in the machine direction is generally higher. During tension loading, elongation can reach several times the original length if necking occurs. In the necking region, the unoriented polymer chains are transformed into thin, oriented chains, resulting in a single, sharp moving neck. Liquid crystal polymers are one extreme of such aligned polymers. Because of rigid molecules, these materials tend to align themselves in melts or solutions. By properly aligning them with stress during the solidifying stage, high tensile strength in one direction can be obtained.

7.7   Effect of Viscoelasticity

Because of the viscoelastic nature of plastics, their deformation response to an applied stress turns out to be very dependent on temperature as well as time. The time dependence can be broken down in terms of the rate at which a stress is applied, the duration of the applied stress, and the overall stress history. The exact temperature dependence will, in turn, be influenced by the thermal properties of the plastic itself, which are categorically different for amorphous thermoplastics, semicrystalline thermoplastics, and thermosets. For some conditions, the time and temperature dependence of the relationship between stress and strain may be superposable and therefore essentially equivalent. Time- and temperature-dependent viscoelastic properties can be examined experimentally by creep- and stress-relaxation tests and by dynamic mechanical testing methods.

Creep tests measure the time-dependent deformation of a material while it is held under a constant applied load at a given temperature, and stress-relaxation tests measure the decay of an applied load within a material while it is held at a constant deformation and at a given temperature. An objective of a creep test may be to define a time-, temperature-, and stress-dependent function for the overall tensile or shear strain (ε or γ) of a particular plastic material. For a stress-relaxation test, the objective may be to obtain the time, temperature, and strain dependence of a function describing the measured stress (σ or τ). In either case, at rather low levels of strain, a time-dependent function for the modulus or compliance may be specifically obtained as well.

7.8   Thermal Properties

The thermal characteristics that are important in the application of engineering plastics are listed in Table 7.6. Long-term temperature resistance is the temperature at which the part must perform for the life of the device. One of the most common measures of long-term temperature resistance is the thermal index determined by the Underwriters' Laboratories (UL). In this test, standard test specimens are exposed to different temperatures and tested at varying intervals. Failure is said to occur when property values drop to 50% of their initial value. The property criterion for determining the long-term use temperature depends on the application.

The most common change that takes place during high-temperature exposure is an oxidation reaction, which decreases the molecular weight of the polymer. This reduction in molecular weight often is first evidenced by a reduction in physical strength and, most frequently, in impact strength, as the plastic embrittles. As the degradation reaction continues, other physical properties drop off, and eventually the electrical properties are affected. For this reason, the long-term temperature resistance is often

Table 7.6 Thermal properties of selected plastics

Material	Heat deflection temperature at 1.82 MPa (0.264 ksi)		UL Index		Thermal conductivity		Coefficient of thermal expansion, 10^{-5}/K
	°C	°F	°C	°F	W/m K	Btu in./h ft³ °F	
Acrylonitrile-butadiene-styrene	99	210	60	140	0.27	1.9	5.3
ABS-polycarbonate (ABS-PC) alloy	115	240	60	140	0.25	1.7	3.5
Diallyl phthalate (DAP)	285	545	130	265	0.36	2.5	2.7
Polyoxymethylene (POM)	136	275	85	185	0.37	2.6	3.7
Polymethyl methacrylate (PMMA)	92	200	90	195	0.19	1.3	3.4
Polyarylate (PAR)	155	310	...	...	0.22	1.5	3.1
Liquid crystal polymer (LCP)	311	590	220	430	...	...	0.5
Melamine-formaldehyde (MF)	183	360	130	265	0.42	2.9	2.2
Nylon 6	65	150	75	165	0.23	1.6	2.5
Nylon 6/6	90	195	75	165	0.25	...	4.0
Amorphous nylon 12	140	285	65	150	0.25	...	7.0
Polyarylether (PAE)	160	320	160	320	...	...	3.0
Polybutylene terephthalate (PBT)	...	...	120	250	...	...	4.5
PC	129	265	115	240	0.20	...	3.8
PBT-PC	129	265	105	220	...	...	2.8
PEEK	...	...	250	480	0.25	1.7	2.6
Polyether-imide (PEI)	210	410	170	340	0.22	1.5	3.1
Polyether sulfone (PESV)	203	395	70	340	...	...	5.5
PET	224	435	140	285	0.17	...	1.5
Phenol-formaldehyde (PF)	163	325	150	300	0.25	1.7	1.6
Unsaturated polyester (UP)	279	535	130	265	0.12	0.8	1.6
Modified polyphenylene oxide alloy (PPO)	100	212	80	175	...	...	3.8
Polyphenylene sulfide (PPS)	260	500	200	390	0.17	...	3.0
Polysufone (PSU)	174	345	140	285	0.26	1.8	3.1
Styrene-maleic anhydride terpolymer (SMA)	103	215	80	175	...	...	...

Note: UL, Underwriters' Laboratory. Source: Ref 7.4

rated differently for different applications, such as those requiring impact or other mechanical properties, as opposed to those requiring only electrical insulation.

Heat deflection temperature (also known as deflection temperature under load) can give an insight into the temperature at which a part would begin to deflect under load. In the standard ASTM International test (D 648), the heat deflection temperature is the temperature at which a 125 mm (5 in.) bar deflects 0.25 mm (0.010 in.) when a load is placed in the center. It is typically reported at both 0.44 and 1.82 MPa (0.064 and 0.264 ksi) stresses.

Because it is a measure of the rigidity of a material, the heat deflection temperature can be influenced by the addition of glass fibers. Softening or relaxation is also a function of the crystallinity of the plastic, that is, the uniform compactness of the molecular chains forming it. As shown in Table 7.7, glass-fiber reinforcements increase the heat deflection temperature of a crystalline material, such as polyamide or polybutylene terephthalate, to a greater extent than they do in an amorphous material, such as polycarbonate.

Table 7.7 Heat deflection temperature versus glass content for selected plastics

Material(a)	Glass content							
	0%		15%		30%		40%	
	°C	°F	°C	°F	°C	°F	°C	°F
PBT, crystalline	54	130	190	375	207	405	204	400
PA, crystalline	90	195	243	470	249	480	249	480
PC, amorphous	129	265	146	295	146	295	146	295

(a) PBT, polybutylene terephthalate; PA, polyamide; PC, polycarbonate. Source: Ref 7.4

Thermal conductivity is often very important to the function of a device. When internal parts generate heat, as in computers or other electrical devices, it is necessary for the housing materials to conduct heat to the ambient environment or to mounting hardware. In an industrial computer, heat generated internally by electrical components flows through a plastic material into a metal heat sink. If the heat were not removed, the life of the electrical components would be reduced. Thermal conductivity does not differ significantly among plastics. The organic plastics are basically very good insulators. Consequently, to improve thermal conductivity, plastics filled with mineral or conductive materials must be used.

The coefficient of thermal expansion also can be an important factor in any application involving two different materials. Because plastics have wide variations in thermal expansion, stresses are created whenever one material is connected to or encapsulated in another material. These stresses, which can be very large, can be calculated from:

$$\sigma = E(\delta_1 - \delta_2)\Delta T \qquad \text{(Eq 7.1)}$$

where σ is the stress, E is the modulus of elasticity, δ_1 is the coefficient of thermal expansion for material 1, δ_2 is the coefficient of thermal expansion for material 2, and ΔT is the change in temperature. This expansion varies from material to material and is affected by the amount and type of fillers or reinforcements. The coefficient of thermal expansion varies, depending on polymer structure, and is generally anisotropic in nature. Parts molded with oriented molecules expand differentially, as long as the glass transition temperature, T_g, is not reached. Above the T_g, the polymer tends to expand isotropically, and hysteresis is noted in the expansion curve upon cooling.

7.9 Electrical Properties

Plastics have insulating and dielectric properties that make them ideal for many electrical and electronic applications. The environment for these types of applications may include such factors as high voltage differences, electrical ruptures due to high current arcing, and heat due to overloaded circuits.

Plastics are not perfect insulators. Under some conditions, they may pass significant leakage current, either through the bulk of the material or across the surface of the plastic part. The terms used to describe these behaviors are bulk and surface resistivity. As dielectrics, plastics interact with and change the characteristics of nearby electric fields. The terms used to describe this behavior include dielectric strength and dielectric constant.

Electrical applications tend to fall into two categories, depending on whether the insulating or the dielectric properties are of primary importance. Applications in which the insulating properties are of primary importance include electrical housings, connectors, and circuit support elements. The dielectric properties are of primary importance in housings for active electrical components and coverings for communication wiring. In both cases, it is often necessary to include flammability properties as critical design constraints. As with mechanical properties, the electrical properties of plastics are highly dependent on temperature. They also depend on electrical frequency, environmental exposure, and heat aging. The requisite properties for specific electrical applications are often well defined by regulatory agencies and applicable codes.

7.10   Environmental Factors

The effects of environmental factors on plastics are unique and sometimes crucial, because plastics are generally more sensitive to environmental factors than other engineering materials. Ultraviolet light causes degradation through a breakdown of molecular weight. To overcome this problem, UV absorbers, such as carbon black or aromatic ketones, are added to plastics.

Plastics are also sensitive to chemicals. Alkalis and acids affect them through hydrolysis of polymer chains. The effects of solvents on stress cracking are more difficult to predict, and experimental results should be consulted. Solvent-induced cracking occurs when a plastic is subjected to a moderate mechanical stress and immersed in a weak solvent. The cause appears to be a lowering of the cohesive bond energies when the normal polymer-polymer bonds, which sustain the mechanical stress, are replaced by the new polymer-solvent bonds, which cannot contribute to the overall strength of the material. Resistance to stress cracking increases with molecular weight and density. In a crystalline material, resistance to environmental stress cracking increases with crystallinity for short-term peak loads and decreases with increasing crystallinity for long-term constant loads.

To select the proper material, the engineer must understand the type of environment in which the plastic part will function, and then determine four critical parameters: operating temperature, stress level, chemical exposure, and adjoining materials.

Operating Temperature. The operating temperature is important not only because of its effect on properties but also because of its effect on chemical reactions. For example, raising the temperature 10 °C (18 °F) approximately doubles the reaction rate. The temperature of interest is that of the material when it is exposed to the chemical environment, or the temperature of the environment itself. There may be situations in which this temperature may be different from the long-term operating temperature for the part. For example, the temperature in an oven may exceed 200 °C (390 °F), but the material may be exposed to detergents only at temperatures in the range of 80 to 90 °C (175 to 195 °F), and therefore, the chemical resistance to detergents should be evaluated at the 90 °C (195 °F) level.

Stress Level. A part can encounter two types of stresses. Dynamic stresses are those that result from the function of the part. Because stress occurs only when the device is used, the stress level is zero when the device is not in use. The design engineer must understand the function of the part to calculate dynamic stresses. Residual, or molded-in, stresses are those that are initiated during the molding operation. They can be due to orientation of the polymer molecules or to alignment of the fibrous reinforcements commonly used with engineering plastics. Many factors influence molecular or fibrous orientation, such as design of the part, design of the mold, and processing conditions. The most common areas of high stress are points of attachment, sharp corners, gate areas, sections behind bosses, and openings in the part that may contribute to knit-line formation.

It is always good design practice to minimize the amount of molded-in stresses in a plastic part. Radiusing all corners (especially inside corners), maintaining uniform wall thicknesses, and other good practices can contribute to this end result. However, most designs retain some level of molded-in stresses. If these must be limited to ensure proper part function, a quality-control test must be specified, for example, impact, flexural, or tensile, with limits. This is significant because it tends to increase part cost.

Chemical Exposure. The chemical resistance of many engineering plastics is excellent. Plastics are used routinely in environments in which metals would rapidly fail, such as in equipment used for handling corrosive liquids. However, most plastics have specific weaknesses in terms of chemical attack. A particular plastic with a high resistance to strong acids and bases, for example, may be rapidly attacked by chlorinated hydrocarbon solvents. Various engineering plastic families vary widely in the degree of chemical resistance they exhibit. If aggressive substances are to be encountered in an application, the resistance of the candidate plastic must be carefully considered.

Plastic parts fail because of chemical exposure in several characteristic ways, which can be classified as plasticization, chemical reaction, and environmental stress cracking:

- *Plasticization.* May occur if a part is exposed to a fluid with which it is somewhat miscible. The solvent penetrates and plasticizes the part, which swells, softens, and gains weight and physical dimension, while strength and stiffness decrease. Even if the fluid is a good solvent for the plastic, solvation usually occurs very slowly because of the macromolecular nature of the plastic. Thus, the first signs of solvation are similar to plasticization, although surface softening may be pronounced.
- *Chemical reaction.* May occur between a plastic part and an environmental substance. A common example is the hydrolysis of polyamide when exposed to hot water, resulting in lowered molecular weight and degraded mechanical properties.
- *Environmental stress cracking.* A phenomenon in which a stressed part develops crazing and cracking when exposed to an aggressive substance. Often, the cracks propagate rapidly, resulting in sudden, brittle failure. A familiar example is the cracking developed by polystyrene when exposed to butter and other food fats.

In describing the compatibility of materials, not only is the obvious chemical environment important but also the other materials with which a part comes into proximity or contact. Chemical activity occurs from solid as well as gaseous or liquid environments. Some materials contain residual monomers or chemicals that are by-products of their polymerization, and others contain additives, such as flame retardants. These materials migrate or diffuse from the plastic over time and can provide a very chemically active environment. For example, it is known that two-stage phenol formaldehyde phenolic plastics give off gaseous ammonia or liquid phenols in very small amounts over extended periods of time. This volatilization is not apparent in applications that are commonly open to the atmosphere, because the chemicals can dissipate. However, in a confined space, the concentration can increase to a level that could affect adjoining materials. An example of this incompatibility occurs with phenolics and polycarbonate. At room temperature or in open applications, the two materials could be considered compatible. However, in a confined space at temperatures exceeding 60 °C (140 °F), the ammonia given off by the phenolic will cause the polycarbonate to soften and eventually revert to a liquid state. Although some volatilization information is available from material suppliers, designers should conduct their own compatibility testing whenever a new combination of materials is tried.

7.11 Categories of Engineering Plastics and Applications

The number of engineering plastics, which is already tremendous, continues to grow rapidly because of the desire to use materials that are lower in weight than metals and readily formable. These characteristics permit consolidation of parts and reduced assembly costs.

The commodity resins make up the greatest percent of tonnage in plastics production. They are typically low priced and competitively produced by a number of major vendors. Inherent in the progression from intermediate to engineering to high-performance plastics is an increasing ability to support higher loads, operate at higher temperatures, or resist the effects of the environment to a greater degree. These attributes do have an associated cost increase and involve more specificity in suppliers, who have greater product identification.

Commodity Thermoplastics

Commodity thermoplastics are usually inexpensive and available from a large number of suppliers. As previously discussed, their properties can often be significantly increased by fillers, such as short glass fibers. In some instances, a glass-filled commodity thermoplastic will then be competitive with a more expensive engineering thermoplastic.

Polyethylene (PE) plastics have the largest volume use of any plastics. There are three basic types of polyethylene: high-density polyethylene (HDPE), low-density polyethylene (LDPE), and linear low-density polyethylene (LLDPE). Polyethylene polymers are semicrystalline thermoplastics that exhibit toughness, near-zero moisture absorption, excellent chemical resistance, excellent electrical insulating properties, low coefficients of friction, and ease of processing. Their heat deflection temperatures are reasonable but not high. Branching in LLDPE and LDPE decreases the crystallinity. HDPE exhibits greater stiffness, rigidity, improved heat resistance, and increased resistance to permeability than LDPE and LLDPE. Specialty grades of PE include very low density, medium density, and ultrahigh-molecular-weight PE.

The major use of HDPE is in blow-molded bottles, shipping drums, carboys, automotive gasoline tanks; injection-molded materials handling pallets, crates, totes, trash and garbage containers; household and automotive parts; and extruded pipe products (corrugated, irrigation, sewer/industrial, and gas distribution pipes).

The LDPE/LLDPEs are used in film form for food packaging as a vapor barrier film that includes stretch and shrink wrap; for plastic bags such as grocery bags and laundry and dry cleaning bags; for extruded wire and cable insulation; and for bottles, closures, and toys.

Polypropylene (PP) is a highly semicrystalline thermoplastic that exhibits low density, rigidity, and good chemical resistance to hydrocarbons, alcohols, and oxidizing agents; negligible water absorption; excellent chemical properties; and has an excellent balance of impact and stiffness. Polypropylene is used in blow molding bottles and automotive parts and in injection-molded closures, appliances, housewares, automotive parts, and toys. Polypropylene can be extruded into fibers and filaments for use in carpets, rugs, and cordage.

Polystyrene (PS). Catalytic polymerization of styrene yields general-purpose polystyrene, often called crystal polystyrene. It is a clear, amorphous polymer that exhibits high stiffness, good dimensional stability, moderately high heat deflection temperature, and excellent electrical insulating properties. However, it is brittle under impact and exhibits very poor resistance to surfactants and solvents.

The ease of processing and its rigidity, clarity, and low cost combine to support applications in toys, displays, consumer goods, and housewares such as food packaging, audio/video consumer electronics, office equipment, and medical devices. Polystyrene foams can readily be prepared and are characterized by excellent low thermal conductivity, high strength-to-weight ratio, low water absorption, and excellent energy absorption. These attributes have made PS foam of special interest as insulation boards for construction, protective packaging materials, insulated drinking cups, and flotation devices.

Copolymerization of styrene with the rubber polybutadiene reduces the brittleness of PS but at the expense of rigidity and heat deflection temperature. Depending on the rubber level, impact PS or high-impact PS can be prepared. These materials are translucent to opaque and generally exhibit poor weathering characteristics.

Styrene-Acrylonitrile (SAN) Copolymer. Copolymerization of styrene with a moderate amount of acrylonitrile provides a clear, amorphous polymer with increased heat deflection temperature and chemical resistance compared to PS. However, impact resistance is poor. SAN is used in typical PS-type applications where a slight increase in heat deflection temperature and/or chemical resistance is needed, such as housewares and appliances.

Acrylonitrile-butadiene-styrene (ABS) is a ter-polymer prepared from the combination of acrylonitrile, butadiene, and styrene monomers. Compared to PS, ABS exhibits good impact strength, improved chemical resistance, and similar heat deflection temperature. ABS is also opaque. Properties are a function of the ratio of the three monomers. ABS is used for tough consumer products (refrigerator door liners); interior automotive trim; business machine housings; telephones and other consumer electronics; luggage; and pipe, fittings, and conduit.

Polyvinyl Chloride (PVC). The catalytic polymerization of vinyl chloride yields polyvinyl chloride. It is commonly referred to as PVC or vinyl and is second only to PE in volume use. Normally, PVC has a low degree of crystallinity and good transparency. The high chlorine content of the polymer produces advantages in flame resistance, fair heat deflection temperature, good electrical properties, and good chemical resistance. However, the chlorine also makes PVC difficult to process. The chlorine atoms have a tendency to separate out under the influence of heat during processing and heat and light during end use in finished products, producing discoloration and embrittlement. Therefore, special stabilizer systems are often used with PVC to retard degradation. There are two major subclassifications of PVC: rigid and flexible (plasticized). In addition, there are also foamed PVC and PVC copolymers.

PVC alone is a fairly good rigid polymer, but it is difficult to process and has low impact strength. Both of these properties are improved by the addition of elastomers or impact-modified graft copolymers, such as ABS and impact acrylic polymers. These improve the melt flow during processing and improve the impact strength without seriously lowering the rigidity or the heat deflection temperature. With this improved balance of properties, rigid PVCs are used in such applications as door and window frames; pipe, fittings, and conduit; building panels and siding; rainwater gutters and downspouts; credit cards; and flooring.

Flexible PVC is a plasticized material. The PVC is softened by the addition of compatible, nonvolatile, liquid plasticizers. The plasticizers lower the crystallinity in PVC and act as internal lubricants to give a clear, flexible plastic. Plasticized PVC is also available in liquid formulations known as plastisols or organosols. Plasticized PVC is used for wire and cable insulation, outdoor apparel, rainwear, flooring, interior wall covering, upholstery, automotive seat covers, garden hose, toys, clear tubing, shoes, tablecloths, and shower curtains. Plastisols are used in coating fabric, paper, and metal and are rotationally cast into balls and dolls.

Rigid PVC can also be foamed to a low-density cellular material that is used for decorative moldings and trim. Foamed plastisols add greatly to the softness and energy absorption already inherent in plasticized PVC, giving richness and warmth to leatherlike upholstery, clothing, shoe fabrics, handbags, luggage, and auto door panels, as well as energy absorption for quiet and comfort in flooring, carpet backing, and auto headliners.

Copolymerization of vinyl chloride with 10 to 15% vinyl acetate gives a vinyl polymer with improved flexibility and less crystallinity than PVC, making the copolymers easier to process without detracting seriously from the rigidity and heat deflection temperature. These copolymers find primary applications in flooring and solution coatings.

Poly(vinylidene chloride) (PVDC) is a semicrystalline polymer that exhibits high strength, abrasion resistance, a high melting point, better-than-ordinary heat resistance (100 °C, or 212 °F, maximum service temperature), and outstanding impermeability to oil, grease, water vapor, oxygen, and carbon dioxide. It is used for packaging films, coatings, and monofilaments. When the polymer is extruded into film, quenched, and oriented, the crystallinity is fine enough to produce high clarity and flexibility. These properties contribute to widespread use in packaging film, especially for food products that require impermeable barrier protection.

PVDC and/or copolymers with vinyl chloride, alkyl acrylate, or acrylonitrile are used in coating paper, paperboard, or other films to provide more economical, impermeable materials. A small amount of PVDC is extruded into monofilament and tape that is used in outdoor furniture upholstery. PVDC is used in food packaging where barrier properties are needed. Applications for injection-molded grades are fittings and parts in the chemical industry. PVDC pipe is used in the disposal of waste acids.

Poly(methyl methacrylate) (PMMA), also known as acrylic, is a strong, rigid, clear, amorphous polymer. It is an amorphous linear polymer. PMMA has excellent resistance to weathering, low water absorption, and good electrical resistivity. Its outstanding property is excellent transparency, which makes it competitive with glass in optical applications. Examples include automotive tail light lenses, optical instruments, and aircraft windows. Its limitation, when compared with glass, is a much lower scratch resistance. PMMA is also used for glazing, lighting diffusers, skylights, outdoor signs, and exterior lighting lenses in cars and trucks. Other uses of PMMA include floor waxes and emulsion latex paints. Another important use of acrylics is in fibers for textiles.

Engineering Thermoplastics

Engineering thermoplastics comprise a special high-performance segment of synthetic plastic materials that offer premium properties. When properly formulated, they may be shaped into mechanically functional, semiprecision parts or structural components. *Mechanically functional* implies that the parts may be subjected to mechanical stress, impact, flexure, vibration, sliding friction, temperature extremes, and hostile environments and continue to function. As substitutes for metal in the construction of mechanical apparatus, engineering plastics offer advantages such as transparency, light weight, self-lubrication, and economy in fabrication and decorating.

Acetals. Polymers of formaldehyde, correctly called polyoxymethylenes, acetals are among the strongest (tensile strength of 69 MPa, or 10 ksi) and stiffest (modulus in flexure of 2.8 GPa, or 410 ksi) thermoplastics. Acetals also have excellent fatigue life and dimensional stability. Other outstanding properties include low friction coefficients, exceptional solvent resistance, and high heat resistance for extended use up to 104 °C (220 °F).

There are two basic types of acetal: homopolymer and copolymer. The homopolymers are somewhat tougher and harder than the copolymers but suffer from instability in processing. However, stabilization improvements have reduced odor and mold deposit problems and widened the processing temperature range.

The fact that acetals are highly semicrystalline thermoplastics accounts for their excellent properties and long-time performance under load. In creep resistance, acetal is one of the best thermoplastics; however, the apparent modulus falls off consistently with long-term loading. Acetals also have an excellent fatigue endurance limit; at 100% relative humidity, it is 34 MPa (5 ksi) at 25 °C (77 °F), and it is still 21 MPa (3 ksi) at 66 °C (150 °F). In addition, lubricants and water have little effect on the fatigue life.

The impact strength of acetals does not fall off abruptly at subzero temperature like that of many other thermoplastics, and their hardness is only slightly reduced by moisture absorption or temperature below 102 °C (215 °F). Although not as good as nylons, the abrasion resistance of acetals is better than that of many other thermoplastics. Like nylons, acetals have a slippery feel.

Acetals are notable among thermoplastics because of their resistance to organic solvents. However, when in contact with strong acids, acetals will craze. In addition to good mechanical properties, acetals have good electrical properties. The dielectric constant and the dissipation factor are uniform over a wide frequency range and up to temperatures of 121 °C (250 °F). Aging has little effect on electrical properties.

Acetals are available as compounds for injection molding, blow molding, and extrusion. Filled, toughened, lubricated, and UV-stabilized versions currently are offered. Grades reinforced with glass fibers (higher stiffness, lower creep, improved dimensional stability), fluoropolymer, and aramid fibers (improved frictional and wear properties) are also available.

Many applications involve replacement of metals where the higher strength of metals is not required and costly finishing and assembly operations can be eliminated. Typical parts include gears, rollers, and bearings, conveyor chains, auto window lift mechanisms and cranks, door handles, plumbing components, and pump parts.

Fluoroplastics. Outstanding properties of fluorocarbon polymers or fluoroplastics include inertness to most chemicals and resistance to high temperatures. Fluoroplastics have a rather waxy feel, extremely low coefficients of friction, and excellent dielectric properties that are relatively insensitive to temperature and power frequency. Their mechanical properties are normally low but increase dramatically when they are reinforced with glass fibers or molybdenum disulfide fillers.

Numerous fluoroplastics are available, including polytetrafluoroethylene (PTFE), polychlorotrifluoroethylene (PCTFE), fluorinated ethylene propylene (FEP), ethylene chlortrifluoroethylene (ECTFE), ethylene tetrafluoroethylene (ETFE), polyvinylidene fluoride (PVDF), polyvinylfluoride (PVF), and perthoroalkoxy (PFA) resin.

PTFE is extremely heat resistant (up to 260 °C, or 500 °F) and has outstanding chemical resistance, being inert to most chemicals. Its coefficient of friction is lower than that of any other plastic, and it can be used unlubricated. TFE has a tensile strength on the order of 10 to 34 MPa (1.5 to 5 ksi). TFE also has outstanding low-temperature characteristics and remains usable even at cryogenic temperatures. PTFE is extremely difficult to process via melt extrusion and molding. The material usually is supplied in powder form for compression and sintering or in water-based dispersions for coating and impregnating.

CTFE has a heat resistance of up to 200 °C (390 °F) and is chemically resistant to all inorganic corrosive liquids, including oxidizing acids. It also is resistant to most organic solvents except certain halogenated materials and oxygen-containing compounds, which cause a slight swelling. In terms of processability, CTFE can be molded and extruded by conventional thermoplastic processing techniques. It can be made into transparent film and sheet with extremely low water vapor transmission.

FEP, a copolymer of TFE and hexafluoropropylene, shows some of the same properties as TFE (i.e., toughness, excellent dimensional stability, and outstanding electrical insulating characteristics over a wide range of temperature and humidity), but it exhibits a melt viscosity low enough to permit it to be molded by conventional thermoplastic processing techniques. FEP is used as a pipe lining for chemical process equipment, in wire and cable applications, and for glazing in solar collectors.

ECTFE, an alternating copolymer of ethylene with chlorotrifluoroethylene, is significantly stronger and more wear resistant than PTFE and FEP. Its ignition and flame resistance makes it suitable for wire and cable applications. Other uses include tank linings, tower packing, and pump and valve components for the chemical process industry and corrosion-resistant film and coatings.

ETFE is a related copolymer to ECTFE, consisting of ethylene and tetrafluoroethylene. Although its continuous-use temperature is somewhat higher (179 °C, or 355 °F) than that of ECTFE, it melts and decomposes when exposed to a flame.

PVDF is also a tough fluoroplastic with high chemical resistance and good weatherability. It is self-extinguishing and has a high dielectric loss factor. A melt-processible material, it is used in seals and gaskets for chemical process equipment and in electrical insulation.

PVF, available only in film form, is used as a protective lamination on plywood, hardboard, and other panel constructions, as well as for interior truck linings.

PFA is also melt processible, similar to FEP, but with higher temperature resistance.

When reinforced with such materials as molybdenum disulfide, graphite, or glass, the fluorocarbons exhibit increased stiffness, hardness, and compressive strength and reduced elongation and deformation under load. These materials are much better in compression than in tension.

For example, depending on crystallinity, TFE has a tensile yield strength at 23 °C (73 °F) of approximately 12.5 MPa (1.8 ksi) at 30% strain; however, in compression, this value can be as high as 25.5 MPa (3.7 ksi) at 25% strain.

Polyamides (Nylons). The nylon family members are identified by the number of carbon atoms in the monomers. Where two monomers are involved, the polymer will carry two numbers (e.g., nylon 6/6). Semi-crystalline nylons have high tensile strengths, flexural moduli, impact strengths, and abrasion resistance. Nylons resist nonpolar solvents, including aromatic hydrocarbons, esters, and essential oils. They are softened by and absorb polar materials such as alcohols, glycols, and water. Moisture pickup is a major limitation for nylons, because it results in dimensional changes and reduced mechanical properties. Although nylon 6 absorbs moisture more rapidly than type 6/6, both eventually reach equilibrium at approximately 2.7% moisture content in 50% relative humidity air, and approximately 9 to 10% in water. However, overly dry material can cause processing problems, and particular attention must be paid to moisture control.

Several different types of nylon are available, the two most widely used being nylon 6/6 (hexamethylene diamine adipic acid) and nylon 6 (poly-caprolactam). Other types of nylon include 6/9, 6/10, 6/12, 11, 12, 4/6, and 12/12.

Nylon 6/6 is the most common polyamide molding material. Its special grades include heat-stabilized grades for molding electrical parts, hydrolysis-stabilized grades for parts to be used in contact with water, light-stabilized grades for weather-resistant moldings, and higher-melt-viscosity grades for molding of heavy sections and for better extrudability.

Special grades of nylon 6 include grades with higher flexibility and impact strength; heat- and light-stabilized grades for resistance to outdoor weathering; grades that incorporate nucleating agents to promote consistent crystallinity throughout sections, thereby providing better load-bearing characteristics; and higher-viscosity grades for extrusion of rod, film, pipe, large shapes, and blow-molded products.

General-purpose nylon molding materials are available for extrusion, injection molding, blow molding, rotational molding, and (for the nylon 6 materials) casting or anionic polymerization.

Nylon sheet and film also are marketed.

The properties of nylon resins can be improved by filling and reinforcing. Mineral-filled and glass-fiber-reinforced compounds are widely available. Several manufacturers offer specially toughened grades where extra impact strength is required. For specific engineering applications, a number of specialty nylons have been developed, including molybdenum-disulfide-filled nylons to improve wear and abrasion resistance, frictional characteristics, flexural strength, stiffness, and heat resistance; glass-fiber-filled nylons to improve tensile strength, heat distortion temperatures, and, in some cases, impact strength; and sintered nylons. Sintered nylons are fabricated by processes similar to powder metallurgy (the same as those used for TFE fluorocarbons). The resulting materials have improved frictional and wear characteristics, as well as higher compressive strength.

Polyamide-Imide (PAI). Among the highest-temperature amorphous thermoplastics, polyamide-imide resins have a useful service temperature range from cryogenic to almost 260 °C (500 °F). Their heat resistance approaches that of polyimides, but their mechanical properties are distinctly better than those of the polyimides. PAI is inherently flame retardant. The material burns with very low smoke and passes Federal Aviation Administration (FAA) standards for aircraft interior use. The chemical resistance of PAI is excellent; it generally is unaffected by aliphatic and aromatic hydrocarbons, acids, bases, and halogenated solvents. At high temperatures, however, it is attacked by steam and strong acids and bases.

PAI can be injection molded, but some special modifications are needed because the material is reactive at processing temperatures. Special screws and accumulators are recommended. To develop the full physical properties of PAI, molded parts must be postconditioned by gradually raising their temperature. Applications for PAI include engine and generator components, hydraulic bushings and seals, and mechanical parts for electronics and business machines.

Polyarylates are amorphous, aromatic polyesters. They fall between polycarbonate and polysulfone in terms of temperature resistance, with a deflection temperature under load (DTUL) of 149 to 177 °C (300 to 350 °F) at 1.82 MPa (264 psi). Other features are toughness, dimensional stability, UV resistance, flame retardance, and good electrical properties. Polyarylate polymers are transparent but tend toward yellowness. One producer has an almost water-white material. Polyarylates are used for automotive parts such as headlamp and mirror housings, brackets, and door handles, and in electrical/ electronics components such as connectors, fuses, and covers. The temperature resistance of the material allows it to withstand the soldering temperatures encountered in making circuit boards.

Polycarbonates are among the strongest, toughest, and most rigid thermoplastics. In addition, they have a ductility normally associated with the softer, lower-modulus thermoplastics. These properties, together with excellent electrical insulating characteristics, are maintained over a wide range of temperatures (–51 to 132 °C, or –60 to 270 °F) and loading rates. Although there may be a loss of toughness with heat aging, the material still remains stronger than many thermoplastics. Polycarbonates are transparent materials and resistant to a variety of chemicals, but they are attacked by a number of organic solvents, including carbon tetrachloride solvents.

The creep resistance of these materials is among the best of the thermoplastics. With polycarbonates, as with other thermoplastics, creep at a given stress level increases with increasing temperature; yet even at temperatures as high as 121 °C (250 °F), their creep resistance is good. The characteristic ductility of polycarbonate provides it with very high impact strength. Their fatigue resistance is also very good. The moisture absorption for polycarbonates is low, with equilibrium reached rapidly. However, the materials are adversely affected by weathering; a slight color change and slight embrittlement can occur on exposure to UV rays.

Polycarbonate molding compounds are available for extrusion, injection molding, blow molding, and rotational molding. Film and sheeting with excellent optical and electrical properties also are available. Among the specialty grades, glass-reinforced polycarbonates exhibit improved ultimate tensile strength, flexural modulus, tensile modulus, and chemical resistance.

Typical applications include the following: safety shields, lenses, glazing, electrical relay covers, helmets, pump impellers, sight gages, cams and gears, interior aircraft components, automotive instrument panels, headlights, lenses, bezels, telephone switchgear, snowmobile components, boat propellers, water bottles, housings for handheld power tools and small appliances, and optical storage disks.

Polyesters, Thermoplastic. The two dominant materials in this family are polyethylene terephthalate (PET) and polybutylene terephthalate (PBT). The thermoplastic polyesters are similar in properties to nylon 6 and 6/6 but have lower water absorption and higher dimensional stability than the nylons. To develop the maximum properties of PET, the resin must be processed to raise its level of crystallinity and/or to orient the molecules. Orientation increases the tensile strength by 300 to 500% and reduces permeability.

PET is a water-white polymer and is made into fibers, films and sheets, and blow-molded and thermoformed containers for soft drinks and foods. Glass-reinforced PET compounds can be injection molded into parts for automotive, electrical/electronic, and other industrial and consumer products. On the other hand, PBT crystallizes readily, even in chilled molds. Most PBT is sold in the form of filled and reinforced compounds for engineering applications. Its uses include appliances, automotive, electrical/electronic, materials handling, and consumer products. Thermoplastic polyesters can be used as a component in alloys and blends. Widely used combinations include those with polycarbonate, polysulfone, and elastomers.

Poly-1-4-cyclohexylenedimethylene terephthalate has higher heat resistance than either PET or PBT and is used mainly as a component of blends. Polyethylene naphthalate (PEN) has a combination of higher heat resistance, mechanical strength, chemical resistance, and dimensional stability and can be processed into films, fibers, and containers. Compared to PET, PEN has up to five times better oxygen barrier resistance.

Also in the family of thermoplastic polyesters is the class of materials known as liquid crystal polymers (LCPs), aromatic copolyesters with a tightly ordered structure that is self-reinforcing. LCPs generally flow very well in processing, but they must be thoroughly dried and molded at high temperatures. Although molded structures tend to be quite anisotropic, they exhibit very high mechanical properties. The anisotropy can be reduced by proper gating and mold design and by incorporating mineral fillers and glass-fiber reinforcements.

LCPs are resistant to most organic solvents and acids. They are inherently flame resistant and meet federal standards for aircraft interior use. Some LCPs withstand temperatures over 538 °C (1000 °F) before decomposing. LCPs are used in aviation, electronics (connectors, sockets, chip carriers), automotive underhood parts, chemical processing, and household cookware for conventional and microwave ovens.

Polyetherimide (PEI). An amorphous, transparent amber polymer, PEI combines high temperature resistance, rigidity, impact strength, and creep resistance. PEI has a glass transition temperature of 216 °C (420 °F) and a DTUL of 199 °C (390 °F) at 1.82 MPa (264 psi). Thorough drying is required before processing, and typical melt temperatures for injection molding run from 343 to 427 °C (650 to 800 °F). PEI resins qualify for UL94 V-0 ratings at thicknesses as low as 0.25 mm (0.010 in.) and meet FAA standards for aircraft interiors. PEI is soluble in partially halogenated solvents but resistant to alcohols, acids, and hydrocarbon solvents. It performs well under humid conditions and withstands UV and gamma radiation.

Glass-fiber-reinforced PEI grades are available for general-purpose molding and extrusion. Carbon-fiber-reinforced and other specialty grades also are produced for high-strength applications, and PEI itself can be made into a high-performance thermoplastic fiber.

PEI is used in medical applications because of its heat and radiation resistance, hydrolytic stability, and transparency. In the electronics field, it is used to make burn-in sockets, bobbins, and printed circuit substrates. Automotive uses include lamp sockets and underhood temperature sensors. PEI sheeting is used in aircraft interiors, and extruded PEI has been used as a metal replacement for furnace vent pipe.

Polyimides, Thermoplastic. These linear, aromatic polymers are known for their high-temperature performance. While nominally thermoplastic, polyimides will degrade at temperatures below their softening point and thus must be processed in precursor form. Material is supplied in powder form for compression molding and cold forming, and some injection molding grades are available.

The outstanding physical properties of the polyimides make them valuable in applications involving severe environments such as high temperature and radiation, heavy load, and high rubbing velocities. In terms of heat resistance, the continuous service of polyimide in air is on the order of 260 °C (500 °F). At elevated temperatures, polyimide parts retain an unusually high degree of strength and stiffness, but prolonged exposure at high temperatures can cause a gradual reduction in tensile strength.

Polyimides are resistant to most dilute or weak acids and organic solvents. They are not attacked by aliphatic or aromatic solvents, ethers, esters, alcohols, and most hydraulic fluids. However, they are attacked by bases such as strong alkali and aqueous ammonia solutions, and the resin is not suitable for prolonged exposure to hydrazine, nitrogen dioxide, or primary or secondary amine components. Parts fabricated from unfilled polyimide resin have unusual resistance to ionizing radiation and good electrical properties.

Polyimides are used to mold high-performance bearings for jet aircraft, compressors, and appliances. In film form, the material is used for electric motor insulation and in flexible wiring. Polyimides also are used in making printed circuit boards.

Polyketones are semicrystalline polymers of aromatic ketones that exhibit exceptional high temperature performance. Polyetheretherketone (PEEK) is tough and rigid, with a continuous-use temperature rating of 246 °C (475 °F). The polymer is self-extinguishing (UL94 V-0 rated) and exhibits very low smoke emission on burning. Its crystallinity gives it resistance to most solvents and hydrolytic stability. PEEK also is resistant to radiation. Polyetherketone (PEK) polymers are similar to PEEK but have slightly higher (approximately 5.6 ° C or 10 °F) heat resistance.

Polyketones can be processed by conventional methods, with predrying a requirement. Injection molding melt temperatures range from 377 to 410 °C (710 to 770 °F), and mold temperatures must be over 149 °C (300 °F). The materials also can be extruded for wire coating and for slot cast films. Powdered resin is supplied for rotational molding and for coating.

Glass- and carbon-fiber-reinforced polyketone compounds are standard, and PEEK also is combined with continuous carbon fiber to produce a thermoplastic prepreg for compression molding.

Because of their high price, uses for polyketones generally are in the highest-performance applications, such as metal replacement parts for aircraft and aerospace structures, electrical power plants, printed circuits, oven parts, and industrial filters.

Polyphenylene Ether (PPE). Also referred to as polyphenylene oxide, these resins are combined with other polymers to make useful alloys. PPE is compatible with polystyrene (PS) and is blended with (usually high-impact) PS over a wide range of ratios, yielding products with DTUL ratings from 79 to 177 °C (175 to 350 °F). Because both PPE and PS are hydrophobic, the alloys have very low water absorption rates and high dimensional stability. They exhibit excellent dielectric properties over a wide range of frequencies and temperatures. The resin will soften or dissolve in many halogenated or aromatic hydrocarbons. If an application requires exposure to or immersion in a given environment, stressed samples should be tested under operating conditions.

PPE/PS alloys are supplied in flame-retardant, filled and reinforced, and structural foam molding grades. They are used to mold housings for appliances and business machines, automotive instrument panels and seat backs, and fluid-handling equipment. Blow molding sometimes is used to make large structural parts.

PPE can also be alloyed with nylon to provide increased resistance to organic chemicals and better high-temperature performance. PPE/nylon blends have been used to injection mold automobile fenders.

Polyphenylene sulfide (PPS) is characterized by excellent high-temperature performance, inherent flame retardance, and good chemical resistance. The crystalline structural of PPS is best developed at high mold temperatures (121 to 149 °C, or 250 to 300 °F).

Most PPS is sold in the form of filled and reinforced compounds for injection molding. Because of its heat and flame resistance, the material has been used to replace thermoset phenolics in electrical and electronic applications. PPS will withstand contact with metals at temperatures up to 260 °C (500 °F), and in the electrical/electronics market, its biggest use is in molding connectors and sockets. Because it can withstand vapor-phase soldering temperatures, it is used in molding surface-mounted components and circuit boards. Another major application for PPS is in industrial pump housing, valves, and downhole parts for the oil drilling industry. PPS is used to mold appliance parts that must withstand high temperatures; applications include heaters, dryers, microwave ovens, and irons. Its heat and chemical resistance have led to the use of PPS in automotive underhood components, electronics, and lighting systems, as well as medical uses.

Sulfone Polymers. Polysulfone (PSO) is a transparent, amorphous thermoplastic that is stable, heat resistant, self-extinguishing, and can be molded, extruded, and thermoformed into a wide variety of shapes. Drying is required before processing. Characteristics of special significance to the design engineer are its heat distortion temperature of 174 °C (345 °F) at 1.8 MPa (264 psi) and its UL-rated continuous service temperature of 160 °C (320 °F). Its glass transition temperature (T_g) is 190 °C (374 °F).

PSO is rigid, with a flex modulus of almost 2.8 GPa (400 ksi), and strong, with a tensile strength of 70 MPa (10.2 ksi) at yield. Its electrical properties are good; it has a high dielectric strength and a low dissipation factor. The electrical and mechanical properties are retained at temperatures up to 177 °C (350 °F) and after immersion in water or exposure to high humidity.

PSO is highly resistant to mineral acid, alkali, and salt solutions. Its resistance to detergents, oils, and alcohols is good even at elevated temperatures under moderate levels of stress. PSO is attacked by polar organic solvents such as ketones and chlorinated and aromatic hydrocarbons. Mineral-filled and glass-reinforced grades of polysulfone are offered where additional strength is required, and plating grades also are available.

PSO is used in medical instruments and in sterilizing equipment. Its high-temperature capabilities and transparency are useful in the food industry, where it is used for processing equipment, coffee carafes, beverage dispensers, and piping. PSO also is used to mold microwave oven cookware. In the electronics field, PSO goes into circuit boards, capacitor films, switches, and connectors.

Polyarylsulfone is an amorphous material consisting of phenyl and biphenyl groups linked by thermally stable ether and sulfone groups. It is distinguished from PSO polymers by the absence of aliphatic groups, which are liable to oxidative attack.

Polyarylsulfone is characterized by a high heat deflection temperature (204 °C, or 400 °F) at 1.8 MPa (264 psi) and maintains its mechanical properties almost up to that temperature. At normal ambient temperatures, polyarylsulfone is a strong, stiff, tough material that, in general, offers properties comparable to those of other engineering thermoplastics. Polyarylsulfone has good resistance to a wide variety of chemicals, including acids, bases, and common solvents. It is unaffected by practically all fuels, lubricants, hydraulic fluids, and cleaning agents used on or around electrical components.

Polyarylsulfone may be injection molded on conventional equipment with sufficient injection pressure and temperature capabilities. Polyarylsulfone also can be extruded on equipment of varying design. The material must be dried before processing.

Polyarylsulfone can be supplied in transparent or opaque colors and in filled and reinforced grades. Its high temperature-resistance and combustion characteristics have led to its use in many electrical/electronics applications, such as motor parts, lamp housings, connectors, and printed circuit boards.

Polyethersulfone (PES). This transparent, amorphous polymer is one of the highest-temperature engineering thermoplastics, carrying a UL rating for continuous service of 180 °C (356 °F). The polymer is self-extinguishing and, when burned, emits little smoke. PES is rigid, tough, and dimensionally stable over a wide temperature range. It is resistant to most solvents, including gasoline and aliphatic hydrocarbons; however, polar, aromatic hydrocarbons will attack the resin, as will methylene chloride, ketones, and esters.

PES can be processed on conventional injection molding, blow molding, extrusion, and thermoforming equipment. Mold shrinkage is low. The low flammability and low smoke generation of PES make it a good candidate to meet FAA regulations for aircraft interior parts. Transparent grades of PES are used in medical and food processing applications and in electrical components such as illuminated switches. Both neat and reinforced grades are used in appliances and power tools, pumps, sockets and connectors, and printed circuit boards.

Polyurethanes include a large family of polymers, all characterized by the urethane group (NHCOO) in their structure. The chemistry of the polyurethanes is complex, and there are many chemical varieties in the family. Through variations in chemistry, cross linking, and processing, polyurethanes can be thermoplastic, thermosetting, or elastomeric materials, the latter two being the most important commercially. The largest application of polyurethane is in foams. These can range between elastomeric and rigid, the latter being more highly cross linked. Rigid foams are used as a filler material in hollow construction panels and refrigerator walls. In these types of applications, the material provides excellent thermal insulation, adds rigidity to the structure, and does not absorb water in significant amounts. Many paints, varnishes, and similar coating materials are based on urethane systems.

Silicones are inorganic and semi-inorganic polymers, distinguished by the presence of the repeating siloxane link (–Si–O–) in their molecular structure. By variations in composition and processing, polysiloxanes can be produced in three forms: fluids, elastomers, and thermosetting resins. Fluids are low-molecular-weight polymers used for lubricants, polishes, waxes, and other liquids—not really polymers but important commercial products nevertheless. Silicone elastomers and thermosetting silicones are cross linked. When highly cross linked, polysiloxanes form rigid resin systems used for paints, varnishes, and other coatings, and laminates such as printed circuit boards. They are also used as molding materials for electrical parts. Curing is accomplished by heating or by allowing the solvents containing the polymers to evaporate. Silicones are noted for their good heat resistance and water repellence, but their mechanical strength is not as great as other cross-linked polymers.

Alloys and Blends. In addition to the variations on polymers created by copolymerization, there is also the possibility of combining finished thermoplastics to create new materials. When the combination of polymers has a single glass transition temperature and yields a synergistic effect (i.e., when the properties of the mix are significantly better than either of the individual components), the result is termed an alloy. When the resulting product has multiple glass transition temperatures and properties that are an average of the individual components, the product is referred to as a blend. To keep the phases of polymer blends from separating, especially where the components are chemically incompatible, compatibilizers are included in the mix.

Alloying and blending can be used to tailor-make materials with specific sets of properties for certain applications. The advantages of semicrystalline polymers (chemical resistance, easy flow) and amorphous polymers (low shrinkage, impact strength) can be combined in a single material. One of the earliest commercially successful blends was acrylonitrile-butadiene-styrene (ABS), which combines the chemical resistance, toughness, and rigidity of its components.

The range of materials that can be produced by alloying and blending is almost limitless, and this has become one of the fastest-growing segments of the engineering thermoplastics business. The following is a partial list of alloys and blends available commercially:

- ABS/nylon
- Nylon/olefin elastomer
- Polycarbonate/ABS
- Polycarbonate/polyester
- Polycarbonate/polypropylene
- Polycarbonate/thermoplastic polyurethane
- Polyphenylene ether/nylon
- Polyphenylene ether/impact polystyrene
- PVC/ABS
- Polyester/elastomer
- Polysulfone/ABS
- Polysulfone/polyester
- Shape memory alloy/polycarbonate

Thermosetting Resins

Among plastic materials, thermosetting materials generally provide one or more of the following advantages: high thermal stability, resistance to creep and deformation under load and high dimensional stability, and high rigidity and hardness. Typical processing methods include compression molding, transfer molding, and thermoset injection molding.

Thermosetting molding compounds consist of two major ingredients: a resin system, which generally contains such components as curing agents, hardeners, inhibitors, and plasticizers; and fillers and/or reinforcements, which may consist of mineral or organic particles, inorganic or organic fibers, and/or inorganic or organic chopped cloth or paper.

The resin system usually exerts the dominant effect, determining to a great extent the cost, dimensional stability, electrical qualities, heat resistance, chemical resistance, decorative possibilities, and flammability. Fillers and reinforcements affect all these properties to varying degrees, but their most dramatic effects are seen in strength and toughness and, sometimes, electrical qualities.

Alkyds. Primarily electrical materials, alkyds combine good insulating properties with low cost. They are available in granular and putty form, permitting incorporation of delicate complex inserts. Their moldability is excellent, cure time is short, and pressures are low. In addition, their electrical properties in the radio frequency and ultrahigh-frequency ranges are relatively heat stable up to a maximum use temperature of 121 to 149 °C (250 to 300 °F).

General-purpose grades normally are mineral filled; compounds filled with glass or synthetic fibers provide substantial improvements in mechanical strength, particularly impact strength. Short fibers and mineral fillers give lower cost and good moldability, while longer fibers give optimum strength. Typical uses for alkyds include circuit breaker insulation coil forms, capacitor and resistor encapsulation, cases, housings, and switch-gear components.

Allylics. Of the allylic family, diallyl phthalate (DAP) is the most commonly used molding material. A relatively high-priced, premium material, DAP has excellent dimensional stability and a high insulation resistance (5×10^6 MΩ), which is retained after exposure to moisture; the same value is retained after 30 days at 100% relative humidity and 27 °C (80 °F).

Both DAP and diallyl isophthalate (DAIP) compounds are available, the latter providing primarily improved heat resistance. The maximum continuous service temperature of DAIP is 177 to 232 °C (350 to 450 °F) compared to 149 to 177 °C (300 to 350 °F) for DAP. Compounds are available with a variety of reinforcements (e.g., glass, acrylic, and polyester fibers). The physical and mechanical properties are good, and the materials are resistant to acids, alkalis, and solvents.

DAP compounds are available as electrically conductive and magnetic molding grades, achieved by incorporating either carbon black or precious metal flake. Most DAP compounds are used in electrical and electronic applications because they can withstand vapor-phase soldering temperatures.

Bismaleimides (BMI) are high-temperature-resistant condensation polymers made from maleic anhydride and a diamine such as methylene diamine. BMIs are used in applications where the heat resistance of epoxies is not sufficient, and have the capability of performing at 204 to 232 °C (400 to 450 °F) continuous-use temperatures. They have similar processing characteristics to epoxy resins and can be handled on the same equipment used for epoxies. Composite structures made from BMI resins are used in military aircraft and aerospace applications. These resins also are used in the manufacture of printed circuit boards and as heat-resistant coatings.

Epoxies have an excellent combination of strength, adhesion, low shrinkage, and processing versatility. Commercial epoxy resins and adhesives can be as simple as one epoxy and one curing agent; however, most contain a major epoxy, one to three minor epoxies, and one or two curing agents. The minor epoxies are added to provide viscosity control, higher elevated-temperature properties, lower moisture absorption, or to improve toughness.

There are a wide variety of curing agents that can be used with epoxies, but the most common for adhesives and composite matrices include aliphatic amines, aromatic amines, and anhydrides. Aliphatic amines are very reactive, producing enough exothermic heat given off by the reaction to cure at room or slightly elevated temperature. However, because these are room-temperature-curing systems, their elevated-temperature properties are lower than the elevated-temperature-cured aromatic amine systems. These systems form the basis for many room-temperature-curing adhesive systems. Aromatic amines require elevated temperatures, usually 121 to177 °C (250 to 350 °F), to obtain full cure. These systems are widely used for curing matrix resins, filament winding resins, and high-temperature adhesives. Aromatic amines produce structures with greater strength, lower shrinkage, and better temperature capability but less toughness than aliphatic amines. Anhydride curing agents require high temperatures and long times to achieve full cure. They are characterized by their long pot lives and low exotherms. They yield good high-temperature properties, good chemical resistance, and have good electrical properties.

Epoxies are used by industry in several ways. One is in combination with high-strength carbon or glass fibers. Typical uses for glass/epoxy composite are aircraft components and filament-wound rocket motor casings for missiles, pipes, tanks, pressure vessels, and tooling jigs and fixtures. Epoxies also are used in the encapsulation or casting of various electrical and electronic components and in the coating of various metal substrates. Epoxy coatings are used on pipe, containers, appliances, and marine parts.

Another major application area for epoxies is adhesives. The two-part systems cure with minimal shrinkage and without emitting volatiles. Formulations can be created to cure at different rates at room or elevated temperatures. The automotive industry has started to use epoxy adhesives in place of welding and for assembling plastic body parts.

Melamine molding compounds are best known for their extreme hardness, excellent and permanent colorability, arc-resistant nontracking characteristics, and self-extinguishing flame resistance. Dishware and household goods and some electrical uses are the primary applications. General-purpose grades for dishes or kitchenware are usually cellulose filled. Mineral-filled grades provide improved electrical properties, while fabric and glass-fiber reinforcements provide higher shock resistance and strength. Melamines are also used as adhesives, coating resins, and laminating resins. Chemical resistance is relatively good, although melamine is attacked by strong acids and alkalis. Melamines are tasteless, odorless, and are not stained by pharmaceuticals and many foodstuffs.

Phenolics were the first plastic material developed and have been the workhorse of the thermoset molding compound family. In general, they provide low cost, good electrical properties, excellent heat resistance, fair mechanical properties, and excellent moldability. Unmodified phenolics are extremely hard and brittle. They are generally limited in color (usually black or dark brown) and color stability.

Many phenolic applications have been converted to high-temperature thermoplastics, which have advantages in processing efficiencies. General-purpose grades are usually wood flour, fabric, and/or fiber filled. They provide a good all-around combination of moderately good mechanical, electrical, and physical properties at low cost. They are generally suitable for use at temperatures up to 149 °C (300 °F). The materials are severely attacked by strong acids and alkalis, whereas the effects of dilute acids and alkalis and organic solvents vary with the reagents and with the resin formulation.

Impact grades vary with the type and the level of reinforcement. Phenolic molding compounds can be reinforced or filled at levels higher than 70%. In order of increasing impact strength, paper, chopped fabric or cord, and glass fibers are used. Glass-fiber grades also provide substantial improvement in strength and rigidity. Glass-containing grades can be combined with heat-resistant resin binders to provide a combination of impact and heat resistance. Their dimensional stability is also substantially improved by glass. Because phenolics are self-extinguishing and have good fire resistance and low smoke emission, they are used for interior parts in aircraft and buildings. Fire-resistant phenolic foams are used as insulation materials.

Phenolics are processed by thermoset injection molding, compression molding, transfer molding, pultrusion, and compression molding. Electrical grades are generally mineral- or flock-filled materials designed for improved retention of electrical properties at high temperature and humidity. Many specialty-grade phenolics are also available. These include chemical-resistant grades in which the resin is formulated particularly for improved stability to certain chemicals; self-lubricating grades incorporating dry lubricants such as graphite or molybdenum disulfide; grades with improved resistance to moisture and detergents; rubber-modified phenolics for improved toughness, resilience, and resistance to repeated impact; and ultrahigh-strength grades with glass contents over 60% by weight designed to provide mechanical strengths comparable to those of glass-cloth-reinforced laminates.

Polyesters are used extensively in commercial applications but are limited in use for high-performance composites. Although lower cost than epoxies, they generally have lower temperature capability, lower mechanical properties, inferior weathering resistance, and exhibit more shrinkage during cure. Polyesters cure by addition reactions in which unsaturated carbon-carbon double bonds (C=C) are the locations where cross linking occurs. A typical polyester resin consists of at least three ingredients: a polyester; a cross-linking agent such as styrene; and an initiator, usually a peroxide. Styrene acts as the cross-linking agent and also lowers the viscosity to improve processability. However, styrene has been designated as a potential carcinogen, and its emissions are being strictly regulated to lower and lower levels. Resin suppliers are helping processors to reduce emissions by incorporating additives that keep the styrene from evaporating. It also is possible to replace styrene with paramethyl styrene, which has a much lower vapor pressure than styrene.

Styrene is not the only curing agent (cross linker). Others include vinyl toluene, chlorostyrene (which imparts flame retardance), methyl methacrylate (improved weatherability), and diallyl phtalate, which has a low viscosity and is often used for prepregs. The properties of the resultant polyester are strongly dependent on the cross-linking or curing agent used. One of the main advantages of polyesters is that they can be formulated to cure at either room or elevated temperature, allowing great versatility in their processing.

The unsaturated polyesters are extremely versatile in terms of the forms in which they are used. Polyester resins can be formulated to be brittle and hard, tough and resilient, or soft and flexible. In combination with reinforcements such as glass fibers, they offer outstanding strength, high strength-to-weight ratios, chemical resistance, and other excellent mechanical properties.

Using lay-up, spray-up, filament winding, pultrusion, injection, compression, and resin-transfer molding techniques, fiberglass/polyester is used for such diverse products as boat hulls, automotive body parts, building panels, housings, bathroom components, tote boxes, pipes and pressure vessels, appliances, and electronic and electrical applications. Where polyesters are used in building applications, they normally are filled with gypsum or alumina trihydrate to make them flame retardant. In sheet molding compounds for automotive body panels, they are heavily filled with talc and reinforced with glass fibers. Thermoplastic additives are used to provide a smooth class A surface. Another major application for polyesters is in "cultured marble," highly filled materials that are made to resemble natural materials for sinks and vanities. Polymer concretes based on polyesters are used to patch highways and bridges. Their rapid cure minimizes the disruption of traffic. Polyester patching compounds also are used to repair automobile body damage.

Polyimides. Thermoset polyimide resins are among the highest-temperature-performance polymers in commercial use, with the ability to withstand 538 °C (1000 °F) for short periods. The materials also perform well under cryogenic conditions. They exhibit high tensile strengths and dimensional stability. Polyimides are inherently combustion resistant. Applications for thermoset polyimides include films for electric motors, various aircraft/aerospace uses (engine components, insulation for wire and cable, bearings), chip carriers for integrated circuits, printed wiring boards, and coatings.

Ureas. Excellent colorability, moderately good strength, and low cost are the primary attributes of urea molding materials. Their dimensional stability and impact strength are poor. The materials are used for such products as decorative housings, jewelry casings, lighting fixtures, closures, wiring devices, and buttons.

Vinyl ester resins are unsaturated esters of epoxy compounds. Vinyl esters are very similar to polyesters but only have reactive groups at the ends of the molecular chain. Because this results in lower cross-link densities, vinyl esters are normally tougher than the more highly cross-linked polyesters. In addition, because the ester group is susceptible to hydrolysis by water, and vinyl esters have fewer ester groups than polyesters, they are more resistant to degradation from water and moisture. The properties of vinyl ester resins are fatigue resistance, retention of properties at moderate temperatures, impact resistance, and creep resistance. Corrosion-resistant reinforced resins based on vinyl esters and other compounds have successfully replaced traditional materials such as glass, steel, aluminum, concrete, and brick.

Their excellent corrosion resistance is attributed to three basic factors:

- The corrosion-susceptible ester linkage is shielded by a methyl group.
- The vinyl groups are very reactive, and a complete cure of the backbone is easily accomplished.
- The epoxy backbone is very resistant to chemical attacks.

Vinyl esters can be formulated for both room-temperature and elevated-temperature cures. Based on the specific formulation, gel times from less than 1 min to over 3 h are possible. Some of the processes used to fabricate parts with vinyl ester resins at room temperature include open molding techniques such as hand lay-up, spray-up, filament winding, and centrifugal casting.

7.12 Plastic Fabrication Processes

A comparison of thermoplastic and thermoset fabrication processes is given in Table 7.8. Most thermoplastic processing operations involve heating, forming, and then cooling the polymer into the desired shape. Although there are a number of variants, the major thermoplastics processing operations are extrusion, injection molding, blow molding, calendering, thermoforming, and rotational molding. Thermoset processing requires a curing process in which the resin is reacted to form an infusible solid. Major thermoset processes are compression and transfer molding, injection molding, and casting. Thermoset composite production processes include hand lay-up, spray-up, reaction injection molding, structural reaction injection molding, resin-transfer molding, filament winding, and pultrusion.

Table 7.8   Thermoplastics and thermoset processing comparison

Process	Process pressure		Maximum equipment pressure		Maximum size		Pressure limited	Ribs	Bosses	Vertical walls	Spherical shape	Box sections	Slides/cores	Weldable	Good finish, both sides	Varying cross section
	MPa	ksi	MN	tonf	m²	ft³										
Thermoplastics																
Injection	15–45	2–7	30	3370	0.75	8.0	y	y	y	y	n	n	y	y	y	y
Injection compression	20	2.9	30	3370	1.5	16	y	y	y	n	n	n	y	y	y	y
Hollow injection	15	2.2	30	3370	2.0	20	y	y	y	y	n	y	y	y	y	y
Foam injection	5	0.7	15	1690	3.0	30	y	y	y	…	n	y	y	y	y	y
Sandwich molding	20	2.9	30	3370	1.5	16	y	y	y	y	n	n	y	y	y	y
Compression	20	2.9	30	3370	1.5	16	y	y	y	y	n	n	y	y	y	y
Stamping	20	2.9	30	3370	1.5	16	y	n	n	n	n	n	n	y	y	y
Extrusion	n/a	n/a	n/a	n/a	n/a	n/a	n/a	y	n	n/a	n	n	n	y	y	n
Blow molding	1	0.15	10	1120	2.0	20	n	n	n	y	y	y	y	y	n	yy
Twin-sheet forming	1	0.15	10	1120	6.0	65	n	n	n	y	y	y	n	y	n	n
Twin-sheet stamping	1	0.15	30	3370	6.0	65	n	n	n	y	n	y	n	y	n	n
Thermoforming	0.1	0.015	n/a	n/a	…	…	n	n	n	y	n	n	n	y	y	n
Filament winding	0	0	n/a	n/a	…	…	n/a	y	n	y	y	y	y	y	n	y
Rotational casting	0.1	0.015	n/a	n/a	…	…	n	n	n	y	y	n	n	y	n	n
Thermoset plastics																
Compression																
Powder	60	8.7	30	3370	0.5	5	y	y	y	y	n	n	y	n	y	y
Sheet molding compound	6–20	0.85–3	30	3370	4–5	45–55	y	y	y	y	n	n	y	n	y	y
Cold-press molding	1	0.15	30	3370	…	…	n	n	y	y	n	n	n	n	y	y
Hot-press molding	5	0.75	30	3370	6.0	65	y	n	y	y	n	n	n	n	y	y
High-strength sheet molding compound	4–10	0.60–1.5	30	3370	3.0	30	y	y	y	y	n	n	n	n	y	y
Prepreg	0.5–5	0.07–0.75	30	3370	6.0	65	y	n	n	y	n	n	n	n	y	y
Vacuum bag	0.1	0.015	n/a	n/a	…	…	n	n	y	y	n	y	n	n	n	y
Hand lay-up	0	0	n/a	n/a	…	…	n	n	y	y	n	y	n	n	n	y
Injection																
Powder	100	14.5	10	1120	0.1	1.1	y	y	y	y	n	n	y	n	y	y
Bulk molding compound	30	4.5	30	3370	1.0	11	y	y	y	y	n	n	y	n	y	y
Foam polyurethane	0.5	0.07	n/a	…	…	…	n	y	y	y	y	y	n	n	y	y
Reinforced foam	1	0.15	30	3370	3.0	30	y	y	y	y	n	n	y	n	y	y
Filament winding	n/a	n/a	n/a	n/a	…	…	n/a	n	n	y	y	y	n	n	(a)	y
Pultrusion	n/a	n/a	n/a	n/a	n/a	n/a	n/a	y	n	n/a	n	y	n	n	y	y

Note: y, yes; n, no; n/a, not applicable. One side of filament-wound article will exhibit a strong fiber pattern. Source: Ref 7.7

The composite processes are discussed in Chapter 8, "Polymer-Matrix Composites," in this book.

Extrusion is a continuous process used to manufacture plastics film, fiber, pipe, and profiles (Fig. 7.14). The single-screw extruder is most commonly used. In this extruder, a hopper funnels plastic pellets into the channel formed between the helical screw and the inner wall of the barrel

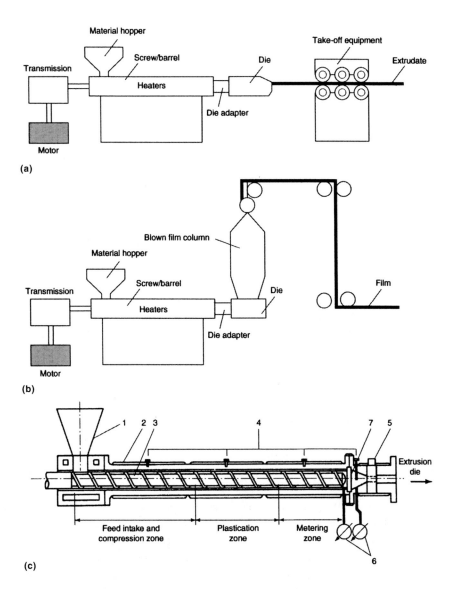

Fig. 7.14 Extrusion processes. (a) Profile/sheet extrusion. (b) Blown film extrusion. (c) Construction arrangement of the plastication barrel of an extruder. 1, feed hopper; 2, barrel heating; 3, screw; 4, thermocouples; 5, back pressure regulating valve; 6, pressure-measuring instruments; 7, breaker plate and screen pack. Source: Ref 7.7

that contains the screw. The extruder screw typically consists of three regions: a feed zone, a transition or compression zone, and a metering or conveying zone. The feed zone compacts the solid plastic pellets so that they move forward as the solid mass. As the screw channel depth is reduced in the transition zone, a combination of shear heating and conduction from the heated barrel begins to melt the pellets. The fraction of unmelted pellets is reduced until finally in the metering zone a homogeneous melt has been created. The continuous rotation of the screw pumps the plastic melt through a die to form the desired shape.

The die and ancillary equipment produce different extrusion processes. With blown film extrusion, air introduced through the center of an annular die produces a bubble of polymer film; this bubble is later collapsed and wound on a roll. In contrast, flat film is produced by forcing the polymer melt through a wide rectangular die and onto a series of smooth, cooled rollers. Pipes and profiles are extruded through dies of the proper shape and held in that form until the plastic is cooled. Fibers are formed when polymer melt is forced through the many fine, cylindrical openings of spinneret dies and then drawn (stretched) by ancillary equipment. In extrusion coating, low-viscosity polymer melt from a flat-film die flows onto a plastic, paper, or metallic substrate. However, in wire coating, wire is fed through the die and enters the center of the melt stream before or just after exiting the die. Finally, coextrusion involves two or more single-screw extruders that separately feed polymer streams into a single die assembly to form laminates of the polymers.

Injection molding is a batch operation used to rapidly produce complicated parts. Plastic pellets are fed through a hopper into the feed zone of a screw and melted in much the same way as occurs in a single-screw or ram extruder. However, rather than being forced through a die, in an injection molding machine, shown in Fig. 7.15, the melt is accumulated and subsequently forced under pressure into a mold by axial motion of the screw. The pressure is typically quite high and for rapid injection and/or thin-walled parts can exceed 100 MPa (14.5 ksi). Once the part has cooled sufficiently, the mold is opened, the part ejected, and the cycle recommences. The use of multiple-cavity molds allows for simultaneous production of a large number of parts, and often little finishing of the final part is required.

Blow molding operations generate hollow products, such as soda bottles and automobile fuel tanks. The three basic processes are continuous extrusion, intermittent extrusion, and injection blow molding. In continuous extrusion blow molding, a tube of polymer is continuously extruded. Pieces of this tube (called parisons) are cut off, inserted into the mold, and stretched into the cavity of the blow mold by air pressure. Although intermittent extrusion blow molding is similar, the tube of plastic is injected from the extruder rather than continuously extruded. In the injection blow molding process, a plastic preform is injection molded. Then this pre-form is brought to the forming temperature, either as part of the cooling from injection molding or after being reheated, and expanded into the blow mold. The process for blow molding a plastic bottle is illustrated in Fig. 7.16. Stretch blow molding is a variant of the blow molding process, in which the preform is stretched axially by mechanical action and then expanded in the transverse direction to contact the walls of the mold.

Calendering uses highly polished precision chromium rolls to trans-form molten plastic continuously into sheet >0.25 mm (>0.01 in.) or film (0.25 mm) for floor coverings. This process can also be used to coat a substrate, for example, cords coated with rubber for automotive tire use. Usually an extruder provides a reservoir of plastic melt, which is then passed between two to four calender rolls whose gap thickness and pressure

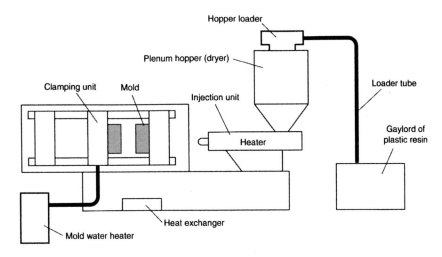

Fig. 7.15 Injection molding machine. Source: Ref 7.7

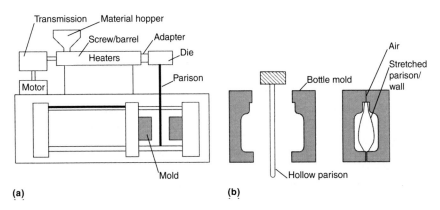

Fig. 7.16 Blow molding. (a) Equipment configuration. (b) Sequence at blow-ing station for a bottle mold. Source: Ref 7.7

profiles determine the final gage of the sheet being formed. Chill rolls are used to reduce the sheet temperature, and a windup station is generally required to collect the sheet product.

Thermoforming operations are used to produce refrigerator liners, computer housings, food containers, blister packaging, and other items that benefit from its low tooling costs and high output rates. In this process, infrared or convection ovens heat an extruded or calendered sheet to its rubbery state. Mechanical action and vacuum and/or air pressure force the heated sheet into complete contact with the cavity of the thermoforming mold, as shown in Fig. 7.17.

Rotational molding, or rotomolding, involves charging a polymeric powder or liquid into a hollow mold. The mold is heated, and then cooled, while being rotated on two axes in a machine similar to the one shown in Fig. 7.18. This causes the polymer to coat the inside of the mold. Because rotomolding produces hollow parts with low molded-in stresses, it is often used for chemical containers and related products where environmental stress crack resistance is required. It can also be used for hollow parts with complicated geometries that cannot be produced by blow molding.

Compression molding is a simple process that offers the manufacturer an excellent method for producing low-stress plastic parts. As shown in Fig. 7.19, the plastic material, usually in the form of a powder or preformed pill, is placed in the cavity or female section of the mold. Because most compression molding is associated with thermosetting plastics, the mold temperature will be relatively high (149 to 177 °C, or 300 to 350 °F). The male or core portion of the mold is located on the upper portion of the

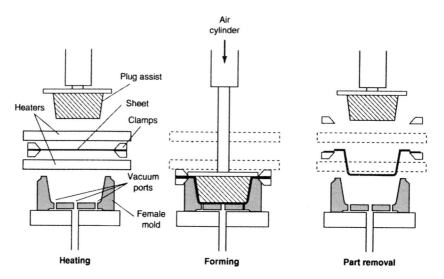

Fig. 7.17 Thermoforming (vacuum forming). Source: Ref 7.7

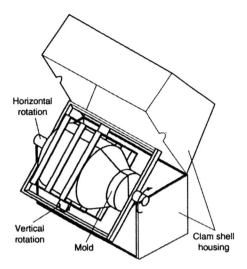

Fig. 7.18 Rotational molding equipment (clam shell type). Source: Ref 7.7

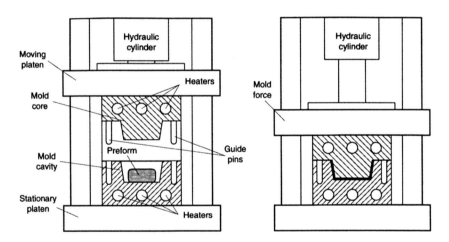

Fig. 7.19 Compression molding. Source: Ref 7.7

mold. After the plastic material is loaded, the male portion of the mold is lowered and compresses the heated plastic in the cavity. The heat and pressure cause the material to flow and fill the cavity details. In many compression molding processes, excess plastic forms a flash that will be removed in a secondary operation. The plastic material will then be allowed to cure or set, and the plastic part will be ejected from the mold. Because the thermosetting plastic flows only a short distance and the flow rate is relatively slow, little shear stress is developed in the process. This low stress level will result in low internal or molded-in stress in the part, which thus will have a high degree of dimensional stability.

Transfer molding is a variant of the compression molding process and is a precursor to the modern injection molding process. The main difference between compression and transfer molding is in the method by which the plastic material enters the mold. As stated previously, compression molding requires the material to be preloaded into the mold in the form of either a powder or a preform. Transfer molding uses a transfer ram or plunger where powder or preform is loaded into a chamber above the mold. The material in this chamber is forced through a sprue and runner/gate system to fill the mold.

The transfer process allows the plastic to enter the mold in a molten or fluid state; thus, the process is adaptable to insert or overmolding. The downside of the transfer process is that there is often a need for secondary operations to separate the product from the gates/runners. There is also a certain amount of waste with the sprue and runners. The ability to overmold or insert mold product allows many unique products to be transfer molded. These include electrical outlets, semiconductors such as integrated circuits, plastic handles for products such as knives, and plastic exteriors for metal parts.

Casting is the process of pouring liquid plastic into a mold. The recent development of a wide variety of casting resins and rapid tooling fabrication has allowed the casting process to be considered as a viable process for both prototyping and low-volume production. Typical casting resins include casting acrylic, casting polycarbonate, epoxy, polyurethane, and polyester. The casting process produces plastic parts with the lowest level of internal stress and a high degree of dimensional stability.

7.13 Joining Plastics

In plastic product design, a molded one-piece item is the ideal situation because it eliminates assembly operations. However, mechanical limitations and other considerations often make it necessary to join plastic parts, either to each other or to other plastic or metal parts. In such instances, the joining process can be an efficient production technique if a few precautions are taken and established procedures are followed. The effectiveness of the joining operation can have a large influence on the application of any polymer or composite material. A variety of plastic joining techniques are available, as outlined in Fig. 7.20.

Mechanical Fastening of Plastics. Mechanical fastening can be used to join both similar and dissimilar materials. For example, mechanical fastening is commonly used when joining a plastic to a metal, producing either permanent joints or allowing disassembly. The advantages of this approach are that no surface treatment is required, and disassembly of the components for inspection and repair is straightforward. The main

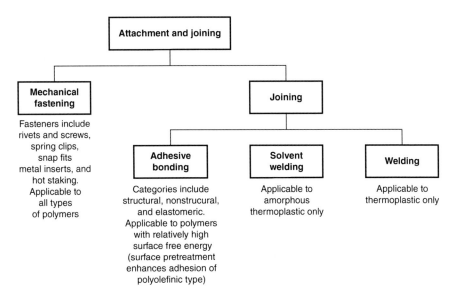

Fig. 7.20 Classification of different plastic joining processes. Source: Ref 7.8

limitations are increased weight, the presence of large stress concentrations around the fastener holes, and potential in-service corrosion problems. The typical applications of mechanical fastening are in the aerospace, automotive, and construction industries.

Thermoplastic resin molded parts are frequently secured together with machine screws or bolts, nuts, and washers (Fig. 7.21a).When the application involves infrequent assembly, molded-in threads can be used (Fig. 7.21b). Coarse threads can be molded into most materials more easily than fine threads. Self-threading screws are an economical means of securing separable joints in plastic. They can be either thread cutting or thread forming. A thread-forming screw displaces material as it is installed in the receiving hole. This type of screw induces high stress levels in the plastic part and is not recommended for parts made of some materials. Thread-cutting screws, such as type 23, type 25, or one that has a cutting edge on its point, actually remove material as they are installed, avoiding high stress buildup (Fig. 7.21c). However, these screws can cause problems if they are installed and removed repeatedly, because new threads can easily be cut each time they are reinstalled. Rivets are generally used to join thin sheet metal, such as electrical contacts, but they can also be used to join plastic parts. Care must be taken to minimize stresses induced during the fastening operation. To distribute the load, rivets with large heads (three times the shank diameter) should be used with washers under the flared end of the rivet. Standard rivet heads are shown in Fig. 7.21(d). In addition to self-locking steel fasteners that replace standard nuts and lock

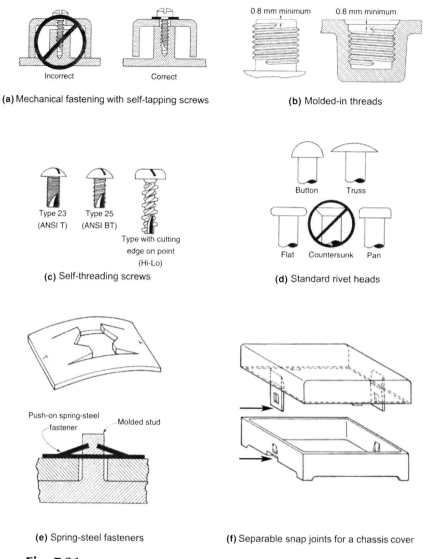

(a) Mechanical fastening with self-tapping screws

(b) Molded-in threads

(c) Self-threading screws

(d) Standard rivet heads

(e) Spring-steel fasteners

(f) Separable snap joints for a chassis cover

Fig. 7.21 Mechanical joints and fasteners used for plastics. Source: Ref 7.8

washers, push-on spring-steel fasteners can be used for holding light loads. As shown in Fig. 7.21(e), spring-steel fasteners are simply pushed on over a molded stud. Snap fits are a simple, economical, rapid way of joining components. Many different designs and configurations can be used with this technique. In all types of snap joints, a protruding part of one component, such as a hook, stud, or bead, is briefly deflected during the joining operation and is caught in a depression (undercut) in the mating component (Fig. 7.21f). After the joining operation, the joint should return to a stress-free condition. The joint may be separable or inseparable, depending on the shape of the undercut.

Solvent and Adhesive Bonding of Plastics. Solvent bonding and adhesive bonding are assembly processes in which two or more plastic parts are held together by the action of a separate agent. These techniques have found wide use by virtue of their low cost and adaptability to high-speed production. In addition, solvent and adhesive bonds provide a relatively uniform distribution of stresses over the assembled areas and a high strength-to-weight ratio. Solvent bonding is applicable only for joining of amorphous thermoplastics, whereas adhesive bonding can be used with almost all plastics.

Solvent bonding, also known as solvent cementing and solvent welding, is one of the simplest methods of assembling thermoplastic parts. When a solvent is applied to the surface of a plastic, the solvent dissolves some of the polymer. The polymer becomes "plasticized," or softened, allowing the polymer molecules to move. When two plastic surfaces are prepared with a solvent and then pressed together, the polymer molecules at the two surfaces are able to move across the interface of the surfaces and intertwine with the polymer molecules of the other surface. When the solvent evaporates, the molecules that moved across the interface are frozen or locked into place, thus creating a bond between the two materials.

Adhesive bonding can be used on all types of plastics and can be used to join plastics to other nonplastic materials. Bonding of plastic substrates depends primarily on chemical rather than mechanical adhesion because of the smooth surfaces of most plastic parts.

Surface Preparation of Thermosets. Many parts molded from thermosetting materials have a mold-release agent on the surface that must be removed before adhesive bonding can be accomplished. These agents can usually be removed by washing or wiping with detergent or solvent, followed by light sanding to break the surface glaze. Surface abrasion can be accomplished using fine sandpaper, carborundum or alumina abrasives, metal wools, or steel shot.

Surface Preparation of Thermoplastics. Thermoplastic surfaces, unlike those of thermosets, usually require physical and/or chemical modification to achieve acceptable bonding. This is especially true of semicrystalline thermoplastics such as polyolefins, linear polyesters, and fluoropolymers. Surface preparation of plastics before adhesive bonding is critical to bond reliability and integrity. After the surface is cleaned to remove obvious surface contamination, chemical or physical treatments can be used to ensure the complete removal of dirt, oils, and other contaminants. These include solvent cleaning and chemical treatment. Chemical treatment, as the name implies, involves a change in the chemical nature of the adherend surface to improve its adhesion characteristics. Etching promotes adhesion and wettability of polymer adherend surfaces.

Depending on the adherend surface and the surface effects desired, a variety of etchants can be used. Plasma treatment exposes the polymer surface to a gas plasma generated by glow discharge. This causes atoms to be expelled from the polymer surface, resulting in the formation of a strong, wettable surface with good adhesion characteristics. Adhesive bonds on plasma-treated surfaces can be two to four times stronger than those produced on untreated surfaces.

Selection of Adhesives for Joining Plastics. Selecting the proper adhesive involves consideration of many factors, including manufacturing conditions, substrates involved, joint configuration and size, end-use requirements, and cost. Thermosets are the most common adhesives used to join both thermoset and thermoplastic parts. Typical adhesives include epoxies, polyesters, and polyurethanes for moderate-temperature applications, and bismaleimides, cyanate esters, and polyimides for higher-temperature service.

Welding of Plastics. The weldability of plastics depends on whether they are thermosets or thermoplastics. In the case of thermoset resins, a chemical reaction occurs that causes curing, that is, an irreversible cross-linking reaction. Neither molded thermoset or vulcanized elastomer components can be reshaped by means of heating, because degradation occurs. It follows that thermoset and vulcanized rubber components can only be joined using adhesive bonding or mechanical fastening methods. Thermoplastic resins, on the other hand, can be softened as a result of the weakening of secondary van der Waals or hydrogen bonding forces between adjacent polymer chains. Therefore, thermoplastics can be remolded by the application of heat, and they can be fusion welded successfully.

During welding, the glass transition temperature (T_g) in amorphous polymers and the melting temperature (T_m) in semicrystalline polymers must be exceeded so that the polymer chains can acquire sufficient mobility to interdiffuse. A variety of methods exist for welding thermoplastics and thermoplastic composites, as outlined in Fig. 7.22. Thermal energy can be delivered externally through conduction, convection, and/or radiation methods, or internally through molecular friction caused by mechanical motion at the joint interface. In the case of external heating, the heat source is removed prior to the application of pressure, and longer welding times are balanced by the greater tolerance to variations in material characteristics. Internal heating methods depend markedly on the material properties. Heat and pressure are applied simultaneously, and shorter welding times are generally involved during the joining process. Welding is accomplished in stages. During the initial stage, the polymer chain molecules become mobile and surface rearrangement occurs. This is followed by wetting and the diffusion of polymer chains across the interface. The final stage involves cooling and solidification.

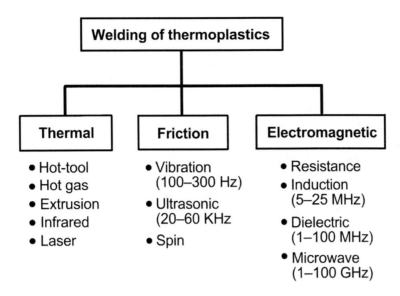

Fig. 7.22 Classification of different welding methods for thermoplastics. Source: Ref 7.8

ACKNOWLEDGMENTS

Portions of the text for this chapter came from "Introduction to Engineering Plastics" and "Mechanical Testing of Polymers," both in *Engineered Materials Handbook Desk Edition,* ASM International, 1995, and "Design for Plastics Processing" by E.A. Muccio in *Materials Selection and Design,* Vol 20, *ASM Handbook,* ASM International, 1997.

REFERENCES

7.1 R.B. Prime, Chapter 5 in *Thermal Characterization of Polymeric Materials,* E.A. Turi, Ed., Academic Press, 1981

7.2 Mechanical Testing of Polymers, *Engineered Materials Handbook Desk Edition,* ASM International, 1995

7.3 Manufacturing Handbook and Buyers' Guide, *Plast. Technol.,* 1987

7.4 Introduction to Engineering Plastics, *Engineered Materials Handbook Desk Edition,* ASM International, 1995

7.5 D.K. Felbeck and A.G. Atkins, *Strength and Fracture of Engineering Solids,* Prentice-Hall, 1984, p 520

7.6 N.M. Riddell, G.P. Koo, and J.L. O'Toole, *Polym. Sci. Eng.,* Vol 6, 1966, p 363

7.7 E.A. Muccio, Design for Plastics Processing, *Materials Selection and Design,* Vol 20, *ASM Handbook,* ASM International, 1997

7.8 Joining and Assembly of Plastics, *Engineered Materials Handbook Desk Edition,* ASM International, 1995

SELECTED REFERENCES

- Advanced Thermoplastics, *Engineered Materials Handbook Desk Edition,* ASM International, 1995
- A.-M.M. Baker and C.M.F. Barry, Effects of Composition, Processing, and Structure on Properties of Engineering Plastics, *Materials Selection and Design,* Vol 20, *ASM Handbook,* ASM International, 1997
- R.D. Beck, *Plastic Product Design,* 2nd ed., Van Nostrand Reinhold, 1980
- M.L. Berins, Ed., *Plastics Engineering Handbook of the Society of the Plastics Industry, Inc.,* 5th ed., Van Nostrand Reinhold, 1991
- D.G. Brady, Processing Thermoplastic Composites, *Proc. SME Conf.,* Society of Manufacturing Engineers, June 1985
- D.G. Brady, T.P. Murtha, J.H. Walker, A. South, and C.C. Ma, Long Fiber Reinforced PPS—A New Dimension in Toughness, *Proc. 41st Annual Tech. Conf. (ANTEC),* Society of Plastics Engineers, May 1984
- F.C. Campbell, Ed., *Joining: Understanding the Basics,* ASM International, 2011
- E.A. Campo, *Selection of Polymeric Materials,* William Andrew Publishing, 2008
- J.F. Carley, Ed., *Whittington's Dictionary of Plastics,* 3rd ed., Technomic, 1993
- R.E. Chambers, *Structural Plastics Design Manual,* No. 63, American Society of Civil Engineers, 1984
- M. Chanda and S.K. Roy, *Plastic Technology Handbook,* Marcel Dekker, 1987
- L.L. Clements, Polymer Science for Engineers, *Engineered Materials Handbook Desk Edition,* ASM International, 1995
- M.W. Darlington and S. Turner, in *Creep of Engineering Materials,* C.D. Pomeroy, Ed., Mechanical Engineering Publications, 1978
- E.N. Doyle, *The Development and Use of Polyester Products,* McGraw-Hill, 1969
- J.D. Ferry, *Viscoelastic Properties of Polymers,* John Wiley & Sons, 1980
- D.E. Floyd, Ed., *Polyamide Resins,* 2nd ed., Reinhold, 1966
- C.A. Harper, Ed., *Modern Plastics Handbook,* McGraw-Hill, 1999
- R. Juran, Ed., *Modern Plastics Encyclopedia,* McGraw-Hill, 1988
- G.F. Kinney, *Engineering Properties and Applications of Plastics,* John Wiley & Sons, 1957
- M. Kutz, *Handbook of Materials Selection,* John Wiley and Sons, 2002
- S. Levy and J.H. DuBois, *Plastic Product Design Engineering Handbook,* 2nd ed., Chapman and Hall, 1985
- J.M. Margolis, Ed., *Engineering Thermoplastics: Properties and Applications,* Marcel Dekker, 1985

- E. Miller, *Plastic Product Design Handbook, Part A,* Marcel Dekker, 1981
- E.A. Muccio, *Plastic Part Technology,* ASM International, 1991
- E.A. Muccio, *Plastics Processing Technology,* ASM International, 1994
- E. Passaglia and J.R. Knox, in *Engineering Design for Plastics,* E. Baer, Ed., Reinhold, 1964
- S. Schwartz and S. Goodman, *Plastics Materials and Processes,* Van Nostrand Reinhold Co., 1982
- R.B. Seymour, *Polymers for Engineering Applications,* ASM International, 1987
- Thermosets, *Engineered Materials Handbook Desk Edition,* ASM International, 1995
- G.G. Trantina, Design with Plastics, *Materials Selection and Design,* Vol 20, *ASM Handbook,* ASM International, 1997

CHAPTER **8**

Polymer-Matrix Composites

THE ADVANTAGES OF HIGH-PERFORMANCE COMPOSITES are many, including lighter weight, the ability to tailor lay-ups for optimum strength and stiffness, improved fatigue strength, corrosion resistance, and, with good design practice, reduced assembly costs due to fewer detail parts and fasteners. The specific strength (strength/density) and specific modulus (modulus/density) of high-strength fiber composites, especially carbon, are higher than other comparable aerospace metallic alloys. This translates into greater weight savings, resulting in improved performance, greater payloads, longer range, and fuel savings.

Composites do not corrode, and their fatigue resistance is outstanding. Corrosion of aluminum alloys is a major cost and a constant maintenance problem for both commercial and military aircraft. The corrosion resistance of composites can result in major savings in supportability costs. The superior fatigue resistance of composites, compared to high-strength metals, is shown in Fig. 8.1. As long as reasonable design strain levels are used, fatigue of carbon-fiber composites should not be a problem.

Assembly costs can account for up to 50% of the cost of an airframe. Composites offer the opportunity to significantly reduce the amount of assembly labor and fasteners. Detail parts can be combined into a single cured assembly, either during initial cure or by secondary adhesive bonding.

Disadvantages of composites include high raw material costs and high fabrication and assembly costs; composites are adversely affected by both temperature and moisture; composites are weak in the out-of-plane direction, where the matrix carries the primary load; composites are susceptible to impact damage and delaminations, or ply separations can occur; and composites are more difficult to repair than metallic structure.

The reinforcing phase provides the strength and stiffness. In most cases, the reinforcement is harder, stronger, and stiffer than the matrix. The

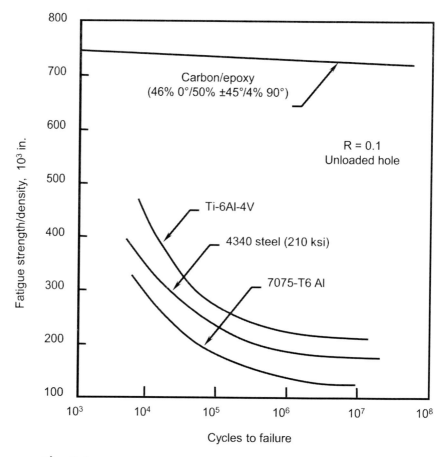

Fig. 8.1 Fatigue properties of aerospace materials. Source: Ref 8.1

reinforcement is usually a fiber or a particulate. Particulate composites have dimensions that are approximately equal in all directions. They may be spherical, platelets, or any other regular or irregular geometry. Particulate composites tend to be much weaker and less stiff than continuous fiber composites but are usually much less expensive. Particulate-reinforced composites usually contain less reinforcement (up to 40 to 50 vol%) due to processing difficulties and brittleness.

A fiber has a length that is much greater than its diameter. The length-to-diameter is known as the aspect ratio and can vary greatly. Continuous fibers have long aspect ratios, while discontinuous fibers have shorter aspect ratios. Continuous fiber composites normally have a preferred orientation, while discontinuous fibers generally have a random orientation. Examples of continuous reinforcements (Fig. 8.2) include unidirectional, woven cloth, and helical winding, while the discontinuous reinforcements shown are chopped fibers and random mat. Continuous fiber composites are often made into laminates by stacking single sheets of continuous fibers in different orientations to obtain the desired strength and stiffness properties, with fiber volumes as high as 60 to 70%. Fibers produce high-strength composites because of their small diameter; they contain much fewer defects, normally surface defects, than if the material were produced in bulk. As a general rule, the smaller the diameter of the fiber, the higher its strength, but often the cost increases as the diameter becomes smaller. In addition, smaller-diameter high-strength fibers have greater flexibility and are more amenable to fabrication processes such as weaving or forming over radii. Typical fibers include glass, aramid, and carbon, which may be continuous or discontinuous.

The continuous phase is the matrix, either a polymer, metal, or ceramic. Polymers have low strength and stiffness, metals have intermediate strength and stiffness but high ductility, while ceramics have high strength and stiffness but are brittle. The matrix, or continuous phase, performs several critical functions. It maintains the fibers in the proper orientation and spacing. It protects them from abrasion and the environment. In polymer- and metal-matrix composites that form a strong bond between the fiber and the matrix, the matrix transmits loads from the matrix to the fibers through shear loading at the interface. In ceramic-matrix composites, the objective

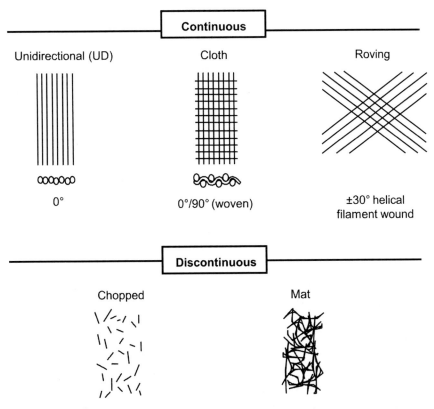

Fig. 8.2 Typical reinforcement options. Source: Ref 8.1

is often to increase the toughness rather than the strength and stiffness, and a low interfacial strength bond is therefore desirable.

The type and quantity of the reinforcement determines the final properties. As shown in Fig. 8.3, the highest strength and modulus are obtained with continuous fiber composites. There is a practical limit of approximately 70 vol% reinforcement that can be added to form a composite. At higher percentages, there is too little matrix to effectively support the fibers. While the theoretical strength of discontinuous fiber composites can approach those of continuous fiber composites if their aspect ratios are great enough and they are aligned, it is difficult in practice to maintain good alignment with discontinuous fibers. Discontinuous fiber composites are normally somewhat random in alignment, which dramatically reduces their strength and modulus. However, discontinuous fiber composites are generally much less costly than continuous fiber composites. Therefore, continuous fiber composites are used where higher strength and stiffness are required, but at a higher cost, and discontinuous composites are used where cost is the main driver and strength and stiffness are less important.

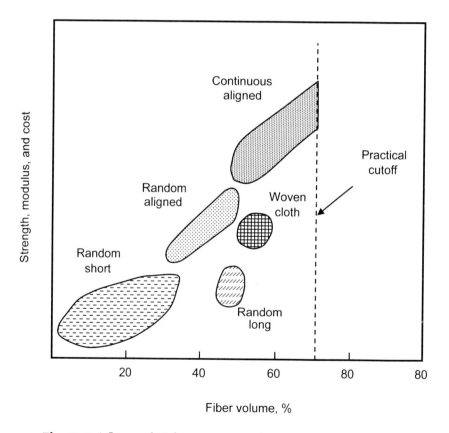

Fig. 8.3 Influence of reinforcement type and quantity on composite performance. Source: Ref 8.1

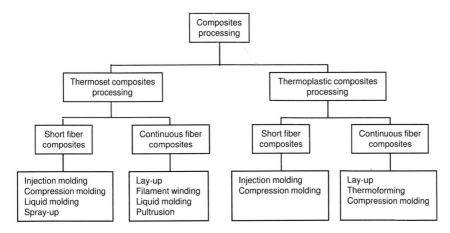

Fig. 8.4 Major polymer-matrix composite fabrication processes. Source: Ref 8.1

Both the type of reinforcement and the matrix affect processing. The major processing routes for polymer-matrix composites are shown in Fig. 8.4. Two types of polymer matrices are shown: thermosets and thermoplastics. A thermoset starts as a low-viscosity resin that reacts and cures during processing, forming an intractable solid, while a thermoplastic is a high-viscosity resin that is processed by heating it above its melting temperature. Because a thermoset resin sets up and cures during processing, it cannot be reprocessed by reheating. On the other hand, thermoplastics can be reheated above their melting temperature for additional processing. For both classes of resins, there are some processes that are more amenable to discontinuous fibers and some that are more amenable to continuous fibers. In general, because metal- and ceramic-matrix composites require very high temperatures and sometimes high pressures for processing, they are normally much more expensive than polymer-matrix composites. However, they have much better thermal stability, a requirement where the application is exposed to high temperatures.

8.1 Advantages and Disadvantages of Composite Materials

The advantages of composites are many, including lighter weight, the ability to tailor the lay-up for optimum strength and stiffness, improved fatigue life, corrosion resistance, and, with good design practice, reduced assembly costs due to fewer detail parts and fasteners. The specific strength (strength/density) and specific modulus (modulus/density) of high-strength fibers, especially carbon, are higher than other comparable aerospace metallic alloys.

Corrosion of aluminum alloys is a major cost and constant maintenance problem for both commercial and military aircraft. The corrosion resistance of composites can result in major savings in supportability costs. While carbon-fiber composites will cause galvanic corrosion of aluminum if the fibers are placed in direct contact with the metal surface, bonding a glass fabric electrical insulation layer on all interfaces that contact aluminum eliminates this problem.

The fatigue resistance of composites compared to high-strength metals is shown in Fig. 8.1. As long as reasonable strain levels are used during design, fatigue of carbon-fiber composites should not be a problem.

Disadvantages of composites include:

- Composites have high raw material costs and usually high fabrication and assembly costs.
- Composites are adversely affected by both temperature and moisture.
- Composites are weak in the out-of-plane direction where the matrix carries the primary load and should not be used where load paths are complex (e.g., lugs and fittings).
- Composites are susceptible to impact damage and delaminations, or ply separations can occur.
- Composites are more difficult to repair than metallic structure.

The major cost driver in fabrication for a conventional hand-layed-up composite part is the cost of laying-up or collating the plies. This cost generally consists of 40 to 60% of the fabrication cost, depending on part complexity. Assembly cost is another major cost driver, accounting for approximately 50% of the total part cost. Composites offer the opportunity to significantly reduce the amount of assembly labor and fasteners. Detail parts can be combined into a single cured assembly either during initial cure or by secondary adhesive bonding.

Temperature has an effect on composite mechanical properties. Typically, as the temperature increases, the matrix-dominated mechanical properties decrease. Fiber-dominated properties are somewhat affected by cold temperatures, but the amount of absorbed moisture depends on the matrix material and the relative humidity. Elevated temperatures speed the rate of moisture absorption. Absorbed moisture reduces the matrix-dominated mechanical properties. Absorbed moisture also causes the matrix to swell. This swelling relieves locked-in thermal strains from elevated-temperature curing. These strains can be large, and large panels fixed at their edges can buckle due to the swelling strains. During freeze-thaw cycles, the absorbed moisture expands during freezing and can crack the matrix. During thermal spikes, absorbed moisture can turn to steam. When the internal steam pressure exceeds the flatwise tensile (through-the-thickness) strength of the composite, the laminate will delaminate.

Composites are susceptible to delaminations (ply separations) during fabrication, assembly, and in service. During fabrication, foreign materials, such as prepreg backing paper, can be inadvertently left in the lay-up. During assembly, improper part handling or incorrectly installed fasteners can cause delaminations. When in service, low-velocity impact damage from dropped tools or fork lifts running into aircraft can cause damage. The damage may appear as only a small indentation on the surface but can propagate through the laminates, forming a complex network of delaminations and matrix cracks. Depending on the size of the delamination, it can reduce the static and fatigue strength and the compression buckling strength. If it is large enough, it can grow under fatigue loading.

Typically, damage tolerance is a resin-dominated property. The selection of a toughened resin can significantly improve the resistance to impact damage. In addition, S-2 glass and aramid fibers are extremely tough and damage tolerant. During the design phase, it is important to recognize the potential for delaminations and use conservative-enough design strains so that damaged structure can be repaired.

8.2 Applications

Applications (Fig. 8.5) include aerospace, transportation, construction, marine, sporting goods, and, more recently, infrastructure, with construction and transportation being the largest. In general, high-performance but more costly continuous carbon-fiber composites are used where high strength and stiffness along with light weight are required, and much lower-cost fiberglass composites are used for less-demanding applications where weight is not as paramount.

In military aircraft, weight is king for performance and payload reasons, and composites often approach 20 to 40% of the airframe weight. For decades, helicopters have used glass-fiber-reinforced rotor blades for improved fatigue resistance and in recent years have expanded into largely composite airframes. Military aircraft applications, the earliest users of high-performance continuous carbon-fiber composites, developed much of the technology now used by other industries. The U.S. Air Force B-2 bomber (Fig. 8.6) is a good example of the widespread use of carbon-fiber composites in military aircraft. Both small and large commercial aircraft rely on composites to decrease weight and increase fuel performance, the most striking example being the 50% composite airframe for the new Boeing 787. All future Airbus and Boeing aircraft will use large amounts of high-performance composites. Composites are also used extensively in both weight-critical reusable and expendable launch vehicles and satellite structures (Fig. 8.7). Weight savings due to the use of composite materials in aerospace applications generally ranges from 15 to 25%.

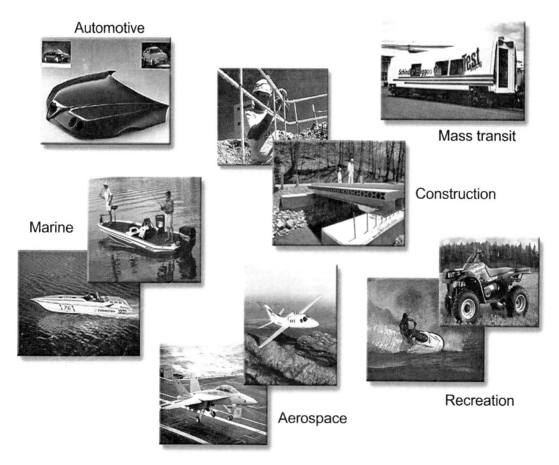

Fig. 8.5 Composite applications are varied and expanding. Source: Ref 8.1

The major automakers are increasingly turning toward composites to help them meet performance and weight requirements, thus improving fuel efficiency. Cost is a major driver for commercial transportation, and composites offer lower weight and lower maintenance costs, as for the applications shown in Fig. 8.8. Typical materials are fiberglass/polyurethane made by liquid or compression molding and fiberglass/polyester made by compression molding. Recreational vehicles have long used glass fibers, mostly for its durability and weight savings over metal. The product form is typically fiberglass sheet molding compound made by compression molding. For high-performance Formula 1 racing cars, where cost is not an impediment, most of the chassis, including the monocoque, suspension, wings, and engine cover, is made from carbon-fiber composites.

Corrosion is a major headache and expense for the marine industry. Composites help minimize these problems, primarily because they do not corrode like metals or rot like wood. Boat hulls ranging from small fishing boats to large racing yachts are routinely made from glass fibers and polyester or vinyl ester resins. Masts are frequently fabricated from carbon-fiber composites. Fiberglass filament-wound self-contained underwater

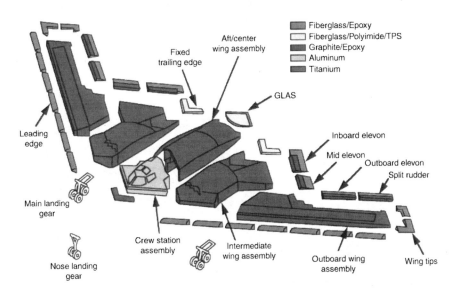

Fiberglass/Epoxy
Fiberglass/Polyimide/TPS
Graphite/Epoxy
Aluminum
Titanium

Aft/center wing assembly

Fixed trailing edge

GLAS

Leading edge

Inboard elevon

Mid elevon

Outboard elevon

Split rudder

Main landing gear

Crew station assembly

Intermediate wing assembly

Outboard wing assembly

Wing tips

Nose landing gear

Fig. 8.6 The U.S. Air Force B-2 advanced "stealth" bomber, which is constructed to a large extent of advanced composite materials. Source: Ref 8.2

Fig. 8.7 Composite launch and spacecraft structures. Source: Ref 8.1

Composites are used in both trucks and cars to reduce weight and Increase fuel efficiency.

Recreational vehicles have long used fiberglass, mostly for its durability and weight savings over metal.

Fig. 8.8 Composite transportation applications. Source: Ref 8.1

breathing apparatus tanks are another example of composites improving the marine industry. Lighter tanks can hold more air yet require less maintenance than their metallic counterparts. Jet skis and boat trailers often contain glass composites to help minimize weight and reduce corrosion. More recently, the topside structures of many naval ships are being fabricated from composites. Typical applications are shown in Fig. 8.9.

Using composites to improve the infrastructure of our roads and bridges is a relatively new but exciting application (Fig. 8.10). Many of the world's roads and bridges are badly corroded and in need of constant maintenance or replacement. In the United States alone, it is estimated that more than 250,000 structures, such as bridges and parking garages, need repair, retrofit, or replacement. Composites offer much longer life with less maintenance due to their corrosion resistance. Typical materials/processes

Rigid and flexible oil
gas tubulars

Maintenance and corrosion in either
fresh or salt water can be major
headaches and expenses.
Composites help minimize
those problems.

Racing sailboat hulls and equipment

More recently, composites are
being used for major components
In naval ships.

Fig. 8.9 Composite marine applications. Source: Ref 8.1

include wet lay-up repairs and corrosion-resistant fiberglass pultruded products. In construction, pultruded fiberglass rebar strengthens concrete, and glass fibers are used in some shingling materials. With the number of mature tall trees dwindling, the use of composites for electrical towers and light poles is greatly increasing. Typically, these are pultruded or filament-wound glass.

Wind power is the world's fastest growing energy source. The blades for large windmills are normally made from composites to improve electrical energy-generation efficiency. These blades (Fig. 8.11) can be as long as

Many of the world's roads and bridges are badly corroded and in need of constant maintenance or replacement.

Composites offer much longer life with less maintenance due to their corrosion resistance.

Repair, upgrading, and retrofit of bridges, buildings, and parking decks

Fig. 8.10 Composite infrastructure applications. Source: Ref 8.1

37 m (120 ft) and weigh up to 5200 kg (11,500 lb). In 2007, nearly 50,000 blades for 17,000 turbines were delivered, representing roughly 400 million pounds of composites. The predominant material is continuous glass fibers manufactured by either lay-up or resin infusion.

Tennis racquets have been made out of glass for years, and many golf club shafts are made of carbon. Processes can include compression molding for tennis racquets and tape wrapping or filament winding for golf shafts. Composites also make possible lighter, better, stronger skis and surfboards. Snowboards are another example of a composite application

Fig. 8.11 Composite energy-generation applications. Composites are being used for wind turbine blades to improve energy-generation efficiency and reduce corrosion problems. Source: Ref 8.1

that takes a beating yet keeps on performing. They are typically made using a sandwich construction (composite skins with a honeycomb core) for maximum specific stiffness.

Advanced composites are a diversified and growing industry due to their distinct advantages over competing metallics, including lighter weight, higher performance, and corrosion resistance. They are used in aerospace, automotive, marine, sporting goods, and, more recently, infrastructure applications. *The major disadvantage of composites is their high costs.* However, the proper selection of materials (fiber and matrix), product forms, and processes can have a major impact on the cost of the finished part.

8.3 Laminates

Continuous fiber composites are normally laminated materials (Fig. 8.12a) in which the individual layers or plies are oriented in directions that will enhance the strength in the primary load direction. Unidirectional (0°)

laminates are extremely strong and stiff in the 0° direction; however, they are also very weak in the 90° direction because the load must be carried by the much weaker polymeric matrix. While a high-strength fiber can have a tensile strength of 3450 MPa (500 ksi) or more, a typical polymeric matrix normally has a tensile strength of only 35 to 70 MPa (5 to 10 ksi) (Fig. 8.13). The longitudinal tension and compression loads are carried by the fibers, while the matrix distributes the loads between the fibers in tension and stabilizes and prevents the fibers from buckling in compression. The matrix is also the primary load carrier for interlaminar shear (i.e., shear between the layers) and transverse (90°) tension.

Because the fiber orientation directly impacts the mechanical properties, it would seem logical to orient as many of the layers as possible in the main load-carrying direction. While this approach may work for some structures, it is usually necessary to balance the load-carrying capability in a number of different directions, such as the 0°, +45°, −45°, and 90° directions. A micrograph of a cross-plied continuous carbon-fiber/epoxy laminate is shown in Fig. 8.14. A balanced laminate with equal

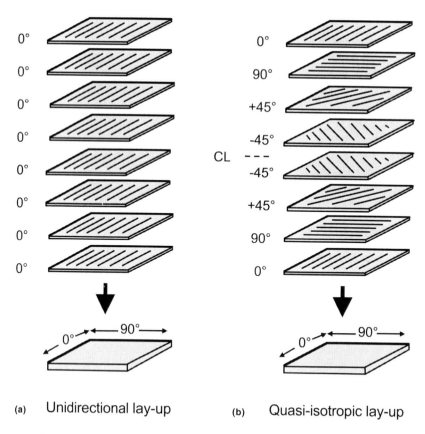

(a) Unidirectional lay-up   (b) Quasi-isotropic lay-up

Fig. 8.12 Unidirectional and quasi-isotropic laminate lay-ups. Source: Ref 8.1

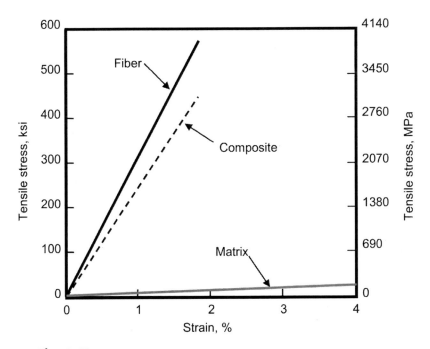

Fig. 8.13 Tensile properties of fiber, matrix, and composite. Source: Ref 8.1

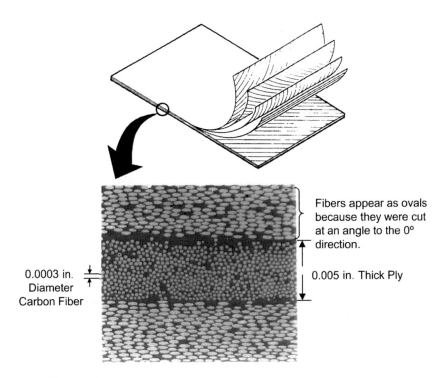

Fig. 8.14 Laminate construction. Source: Ref 8.1

numbers of plies in the 0°, +45°, −45°, and 90° directions is called a quasi-isotropic laminate (Fig. 8.12b) because it will carry equal loads in all four directions.

8.4 Fibers

Because fibers provide the strength and stiffness, it is appropriate to consider fiber selection first. Some comparative mechanical properties for unidirectional composites made from E-glass/epoxy, aramid/epoxy, high-strength carbon/epoxy, and intermediate-modulus carbon/epoxy are summarized in Table 8.1. With regard to this table, there are two words of caution: (1) the properties shown are for unidirectional composites, which are useful for comparing material systems but are not useful for determining overall structural performance because purely unidirectional composites are almost never used in real applications; they must be cross-plied to accept loads in several directions; and (2) mechanical properties are very dependent on the specific fiber and resin system and also the fabrication method used.

Glass fibers are the most widely used reinforcement due to their good balance of mechanical properties and low cost. E-glass is the most common glass fiber and is used extensively in commercial composite products. E-glass is a low-cost, high-density, low-modulus fiber that has good corrosion resistance and good handling characteristics. S-2 glass was developed in response to the need for a higher-strength fiber for filament-wound pressure vessels and solid rocket motor casings. While more expensive than E-glass, it has a density, performance, and cost between E-glass and carbon. Quartz fiber is used in many electrical applications due to its low dielectric constant; however, it is very expensive.

Table 8.1   Typical properties of unidirectional composite materials

Property	E-glass/epoxy	Aramid/epoxy	High-strength carbon/epoxy	Intermediate-modulus carbon/epoxy
Specific gravity	2.1	1.38	1.58	1.64
0° tensile strength, ksi	170	190	290	348
0° tensile modulus, 10^6 psi	7.60	12.0	18.9	24.7
90° tensile modulus, 10^6 psi	5.08	5.08	11.6	11.6
0° compression strength, ksi	1.16	1.16	1.31	1.31
0° compression modulus, 10^6 psi	131	36.3	189	232
In-plane shear strength, ksi	6.09	10.9	16.7	21.8
In-plane shear modulus, 10^6 psi	8.70	6.53	13.4	13.8
Interlaminar shear strength, ksi	0.580	0.304	0.638	0.638
Poisson's ratio	10.9	8.70	13.4	13.1
	0.28	0.34	0.25	0.27

Source: Ref 8.1

In addition to having a lower modulus and thus less stiffness than carbon-fiber composites, the fatigue properties of glass-fiber composites are not as good as carbon- or aramid-fiber composites (Fig. 8.15). In general, stiffer fibers, such as carbon and to some extent aramid, strain the matrix less during fatigue cycling and are thus more fatigue resistant. In addition, glass-fiber composites are subject to static fatigue or stress rupture. Static fatigue is the time-dependent fracture of a material under a constant load, as opposed to a conventional fatigue test where a cyclic load is employed. This phenomenon is illustrated in Fig. 8.16. When glass-fiber-reinforced composites are exposed to moist environments or other aggressive environments, they are also prone to degradation caused by weakening of the fiber-to-matrix interfacial bond. This generally occurs by chemical attack at the fiber surface. The degree of weakening depends on the matrix, the fiber coating, and the type of fiber. Weakening of the interface will result in significant loss in matrix-dominated mechanical properties such as

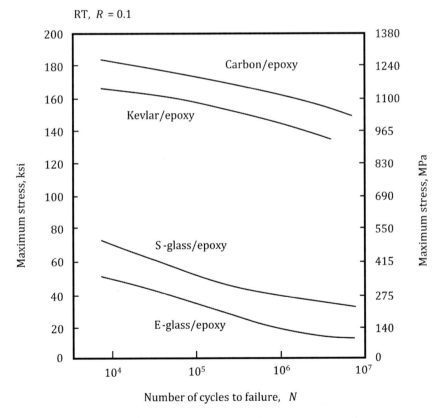

Fig. 8.15 Fatigue behavior of unidirectional composites. RT, room temperature. Source: Ref 8.1

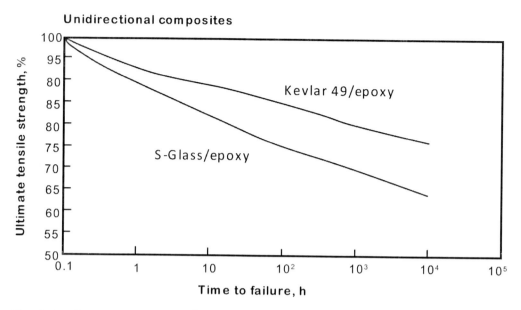

Fig. 8.16 Stress-rupture properties of Kevlar 49 and S-glass epoxies. Source: Ref 8.3

transverse tension, shear, and compression strength. Thus, environmental degradation is a significant concern for structural applications in which the ability to carry high loads is required, and particularly in applications under long sustained loading.

Aramid fiber (e.g., Kevlar, E.I. du Pont de Nemours and Company) is an organic fiber that has a low density and is extremely tough, exhibiting excellent damage tolerance. Aramid composites, introduced in the 1970s, originally had tensile properties somewhat similar to the carbon-fiber composites produced at that time. However, the properties of carbon fibers have improved dramatically over the past 35 years, to the point where aramid is not as competitive with carbon as it once was. Aramid-fiber composites offer good tensile properties at a lower density than glass-fiber composites but at a higher cost than glass-fiber composites. However, their compressive properties are extremely poor, which limits them to tension-dominated designs. The relative compression strengths of unidirectional composites, shown in Fig. 8.17, exhibit the poor compressive strength of aramid composites. In addition, due to the relatively poor fiber-to-matrix bond strength, the matrix-dominated properties, such as in-plane and inter-laminar shear and transverse tension, are somewhat lower than that of glass- and carbon-fiber composites. Even if a surface treatment is used to increase the fiber-to-matrix bond strength, the fibers themselves will then tend to fail through defibrillation, and no increase in properties is observed.

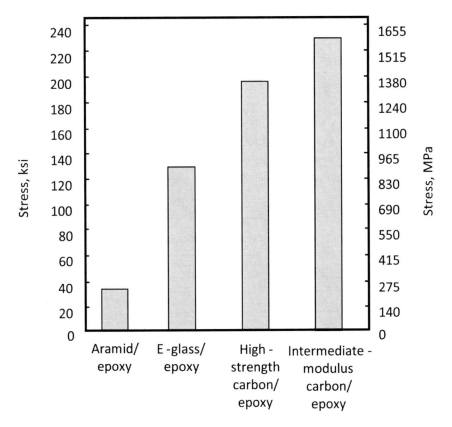

Fig. 8.17 Relative compression strengths of unidirectional composites. Source: Ref 8.1

Although unidirectional aramid composites respond elastically when loaded in tension, they exhibit a nonlinear, ductile behavior under compression. At a compression strain of only 0.3 to 0.5%, a yield is observed (Fig. 8.18). This corresponds to the formation of structural defects known as kink bands, which are related to compressive buckling of the aramid-fiber molecules. As a result of this compression behavior, the use of aramid composites in applications that are subject to high-strain compressive or flexural loads is limited. However, the compressive buckling characteristics have led to developments of crashworthy structures that rely on the fail-safe behavior of aramid composites under sustained high compressive loads.

Although not as good in fatigue as carbon-fiber composites (Fig. 8.15), aramid composites perform very well in fatigue. For aramids, tension-tension fatigue generally is not of significant concern in applications where an adequate static safety factor has been used. Aramid composites have been found to be superior to glass-fiber composites in both tensile-tensile

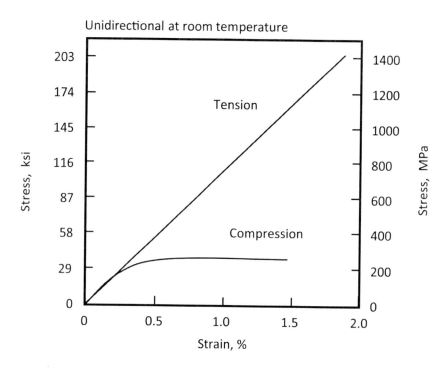

Fig. 8.18 Typical tension and compression behavior of aramid composite. Source: Ref 8.3

and flexural fatigue loading. For the same number of cycles to failure, Kevlar 49/epoxy composites can operate at a significantly larger percentage of their static strength than glass-fiber composites.

Fiber-dominated laminates of aramid composites exhibit very little creep. In general, creep strain increases with increasing temperature, increasing stress, and decreasing fiber modulus. However, under long-term loading, aramid composites, similar to glass-fiber composites, are subject to stress rupture, that is, failure of the fiber under sustained loading with little or no accompanying creep. Although aramids perform better than glass-fiber composites in stress rupture (Fig. 8.16), the susceptibility to stress rupture must be considered in any design where long-term loading is anticipated.

Aramid fibers are noted for their toughness and damage tolerance. The same microstructural characteristics that lead to the weakness of aramid fibers in buckling also make them very tough. During failure, the widespread bending, buckling, and other internal damage to the fibers absorbs a great deal of energy. The fibrillar structure and compressive behavior of aramid fibers contributes to composites that are less notch sensitive and that fail in a ductile, nonbrittle, or noncatastrophic manner, as opposed to carbon-fiber composites. In addition, the strength of aramid fibers is not very strain-rate sensitive; an increase in strain rate of more than 4 orders of magnitude decreases the tensile strength by only about 15%. The good energy-absorbing properties of aramid fibers make them useful for ballistics, tires, ropes, cables, asbestos replacement, and protective apparel.

Because aramid yarns and rovings are relatively flexible and nonbrittle, they can be processed in most conventional textile operations, such as twisting, weaving, knitting, carding, and felting. Yarns and rovings are used in the filament winding, prepreg tape, and pultrusion processes. Applications include missile cases, pressure vessels, sporting goods, cables, and tension members. Although continuous filament forms dominate composite applications, discontinuous or short fiber forms are also used, because the inherent toughness and fibrillar nature of aramid allows the creation of fiber forms not readily available with other fibers.

As a result of their high tensile strengths and superior damage tolerance, aramid composites are often used for lightweight pressure vessels, usually fabricated by filament winding. A comparison of Kevlar 49/epoxy and S-glass/epoxy pressure vessels is given in Table 8.2. The density is lower and the strength and modulus of the Kevlar pressure vessel are higher than those of the S-glass vessel. In pressure vessels, the relative performance is often measured by the parameter PV/W; that is, the burst pressure, P, times the volume, V, divided by the vessel weight, W. The PV/W index is higher for the Kevlar vessel. In addition, the fatigue performance is superior for the Kevlar vessel, an important parameter if the vessel is going to be subjected to multiple pressurization/depressurization cycles.

Being an organic fiber, aramid absorbs moisture. Aramid composites exhibit a linear decrease of both tensile strength and modulus when tested at elevated temperatures in air. The effects of temperature and moisture on a Kevlar 49/epoxy woven cloth are shown in Fig. 8.19. Aramid fiber has an equilibrium moisture content that is determined by the relative humidity. At 60% relative humidity, the equilibrium moisture of Kevlar 49 fiber is approximately 4%. The gain of moisture is completely reversible and, once removed, produces no permanent property changes. At cryogenic temperatures, the modulus increases slightly and strength is not degraded.

Table 8.2 Mechanical properties of filament-wound Kevlar 49/epoxy

Property	Kevlar 49/epoxy	S-glass/epoxy
Density, lb/in.$^{-3}$	0.044	0.069
Fiber volume, %	65	65
Tensile strength, ksi	223	197
Tensile modulus, 10^6 psi	13.2	8.8
PV/W, 10^6 psi	4.1	3.0
Relative fatigue performance at 90% of ultimate strength	10	1

Note: PV/W = burst pressure × volume ÷ vessel weight. Source: Ref 8.3

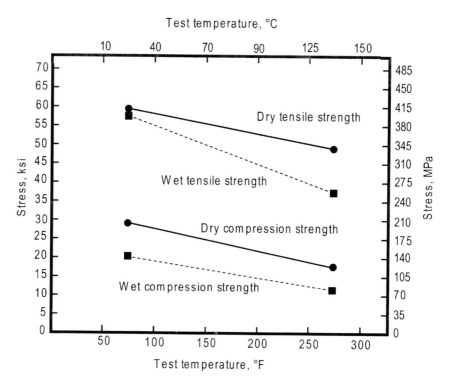

Fig. 8.19 Effect of temperature and moisture on Kevlar 49 style 120 cloth in 180 °C (350 °F) curing epoxy. Source: Ref 8.1

Ultraviolet radiation also can degrade bare aramid fibers. However, in composite form, they are protected by the matrix, which is normally painted before being placed in service. In polymeric composites, a strength loss of aramid composites has not been observed. Being an organic fiber that is susceptible to oxidation, aramid composites are generally not used for long times at temperatures above 149 °C (300 °F). In the transverse direction, aramid fibers are like most other materials in that they expand with increasing temperature. However, in the longitudinal direction, the fibers actually contract somewhat as temperature increases. The negative coefficient of thermal expansion (CTE) of aramid fibers can be used to advantage to design composites with tailored or zero CTE.

The poor off-axis and compressive properties of aramid fibers must be considered in any design. However, because of their high strength in axial tension and their toughness, aramid-fiber composites are often used in applications such as pressure vessels where the loading is almost totally in longitudinal tension. Although composites with carbon fibers have now supplanted aramid composites as having the highest specific strengths, aramids still offer combinations of properties not available with any other fiber. For example, aramids offer high specific strength, toughness, and creep resistance, combined with moderate cost. However, the applications of aramid composites continue to be limited by their poor compressive and off-axis properties and, in some applications, their tendency to absorb water. Nonetheless, aramids will continue to be a fiber of choice where outstanding impact resistance is critical.

Carbon fiber offers the best combination of properties but is also more expensive than either glass or aramid. Carbon fiber has a low density, a low CTE, and is conductive. It is structurally very efficient and exhibits excellent fatigue resistance. It is also brittle (strain-to-failure less than 2%) and exhibits low impact resistance. Being conductive, it will cause galvanic corrosion if placed in direct contact with more anodic metals such as aluminum. Carbon and graphite fiber is available in a wide range of strength and stiffness, with strengths ranging from 2 to 7 GPa (300 to 1000 ksi) and moduli ranging from 207 to 1000 GPa (30 to 145×10^6 psi). With this wide range of properties, carbon fiber is frequently classified as high strength, intermediate modulus, or high modulus. Both carbon and graphite fibers are produced as untwisted bundles called tows. Common tow sizes are 1, 3, 6, 12, and 24k where k = 1000 fibers. Immediately after fabrication, carbon and graphite fibers are normally surface treated to improve their adhesion to the polymeric matrix.

The mechanical properties of a typical second-generation toughened epoxy (Cycom 977-3) with an intermediate-modulus carbon fiber (IM-7) are summarized in Table 8.3. This material system cures at 177 °C (350 °F) and has an acceptable usage temperature up to 149 °C (300 °F), although it is often restricted in usage to 121 °C (250 °F). It is generally accepted that the critical environmental conditions for carbon-fiber composites are cold-dry tension and hot-wet compression. The cold-dry condition normally

Table 8.3 Properties of intermediate-modulus carbon/toughened epoxy

Property	−75 °F	RT	220 °F Dry	220 °F Wet	250 °F Dry	250 °F Wet	270 °F Dry	270 °F Wet	300 °F Dry	300 °F Wet
0° tensile strength, ksi	353	364	...	...	...	...	...	...	...	...
0° tensile modulus, 10^6 psi	22.9	23.5	...	...	...	...	...	...	...	...
90° tensile strength, ksi	...	9.3	...	...	...	...	...	...	...	...
90° tensile modulus, 10^6 psi	...	1.21	...	...	...	...	...	...	...	...
0° compression strength, ksi	...	244	...	221(a)	...	195(a)	...	180(a)	...	160(a)
0° compression modulus, 10^6 psi	...	22.3	21.4	21.2(a)	20.4	21.2(a)	20.2	22.6(a)	21.5	21.7(a)
0° flexural strength, ksi	...	256	246	173(a)	221	162(a)	218	140(a)	206	125(a)
0° flexural modulus, 10^6 psi	...	21.7	22.4	20.1(a)	20.8	21.2(a)	21.0	19.6(a)	21.0	18.9(a)
90° flexural strength, ksi	...	19.0	...	...	...	...	...	...	...	...
90° flexural modulus, 10^6 psi	...	1.19	...	...	...	...	...	...	...	...
In-plane shear modulus, 10^6 psi	...	0.72	...	0.61(b)	...	0.58(b)	...	0.50(b)	...	0.34(b)
Interlaminar shear strength, ksi	...	18.5	13.6	12.9(a)	13.3	11.4(a)	12.4	10.1(a)	11.4	9.0(a)
Open-hole comp. strength, ksi	...	46.7	...	37.0(c)	...	35.0(c)	...	...	...	...
Compression after impact, ksi	...	28.0	...	...	...	...	...	...	...	...

IM7/977-3 carbon/epoxy. Fiber-dominated properties normalized to 0.60 fiber volume. RT, room temperature. (a) Wet = 1 week immersion in 160 °F water. (b) Wet = 150 °F/85% relative humidity to equilibrium, approximately 1.1% weight gain. (c) Wet = 2 week immersion in 160 °F water. Source: Ref 8.4

produces less of a problem than the hot-wet condition. Note that the cold-dry condition of –59 °C (–75 °F) produces only a moderate reduction in 0° tensile strength, while the 0° wet compression strength progressively decreases on increasing temperatures up to 149 °C (300 °F) in the manner shown in Fig. 8.20. Part of this reduction in compression strength is due to a simple temperature increase and some is due to the absorbed moisture. The plot of interlaminar shear strength versus temperature for both dry and wet specimens in Fig. 8.21 illustrates the reduction due to both temperature and moisture. Other matrix-dependent properties affected by temperature and humidity include 90° tension and compression, the shear properties (in-plane and interlaminar), the flexural properties, and open-hole compression strength. While one may logically think that 0° compression and 0° flexural properties are fiber-dependent properties, the compression strength depends on the matrix to stabilize the fiber against microbuckling, and the 0° flexural testing of carbon/epoxy laminates with extremely high-strength carbon fibers results in failure modes that occur in compression, interlaminar shear, or combinations of compression and shear.

To date, most high-performance carbon/epoxy structures have been fabricated using prepreg that is autoclave cured at either 121 or 177 °C (250 or 350 °F). In the early 1990s, work began in developing lower-temperature-curing prepregs (<93 °C, or 200 °F) that could be cured under only vacuum bag pressure. The driver for this development was the need to be able to produce small numbers of parts without large expenditures for expensive autoclave-hardened tooling. Initially, the same prepregs that were being used to produce carbon/epoxy tools were used, but later developments have resulted in nonautoclave low-temperature/vacuum bag materials that have the same properties as autoclave-cured prepregs, at least for reasonably sized parts. The properties of one of these materials, Cycom 5215, are compared with a standard autoclave-cured carbon/epoxy in Table 8.4. The standard carbon/epoxy material was cured at 177 °C (350 °F) for 2 h under 585 kPa (85 psi) autoclave pressure. The low-temperature/vacuum bag 5215 material was initially cured for 14 h at 66 °C (150 °F) under vacuum bag pressure and then removed from the tool and given a freestanding postcure at 177 °C (350 °F) for 2 h. The properties of both materials in both unidirectional and woven cloth form are essentially equivalent. However, two points must be made concerning these materials. First, because there is no autoclave pressure to suppress void formation, and the removal of entrapped air is critical, it may be more difficult to produce consistent void-free parts, especially for larger and thicker parts where the air evacuation paths are longer. The second point is that currently these materials have compression strengths after impact (CAI) somewhat equivalent to the first-generation brittle epoxies. However, this toughness limitation is only temporary; a toughened version of this material (Cycom 5320) has recently been introduced.

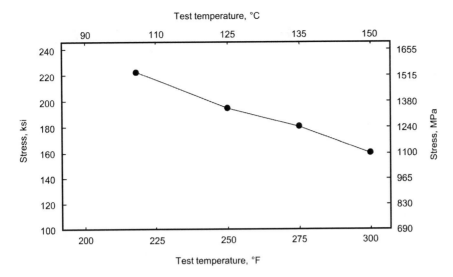

Fig. 8.20 Hot-wet compression strength of an intermediate-modulus carbon/toughened epoxy, wet = 1 week immersion in 70 °C (160 °F) water. Source: Ref 8.1

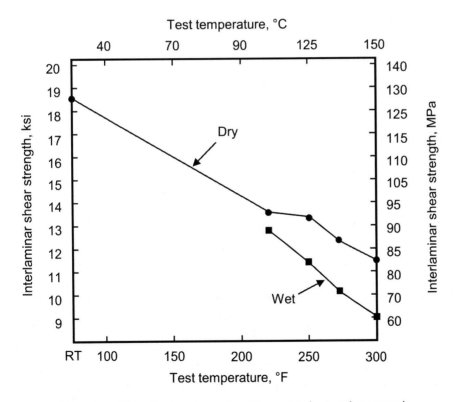

Fig. 8.21 Effect of temperature and moisture on interlaminar shear strength, wet = 1 week immersion in 70 °C (160 °F) water. Source: Ref 8.1

Table 8.4   Comparative properties of Cycom 5215 carbon/epoxy

Property	Unidirectional tape		6k five-harness satin woven cloth	
	G30-500 carbon/first-generation epoxy	G30-500 carbon/5215 epoxy	G30-500 carbon/first-generation epoxy	G30-500 carbon/5215 epoxy
RT dry 0° tensile strength, ksi	200	297	75.0	95.0
RT dry 0° compression strength, ksi	205	210	67.5	105
250 °F dry 0° compression strength, ksi	145	175	45.0	76.9
RT dry interlaminar shear strength, ksi	15.0	15.5	8.5	8.0
250 °F dry interlaminar shear strength, ksi	9.0	10.0	6.5	7.1
250 °F wet interlaminar shear strength(a), ksi	7.5	8.7	2.5	4.4
Compression after impact, ksi	17.5	16.0	25.0	23.7
Dry glass transition temperature (T_g), °F	350	378	350	366
Wet glass transition temperature (T_g)(b), °F	250	331	250	328

RT, room temperature. (a) Wet = 24 h water boil. (b) Wet = 48 h water boil. Source: Ref 8.5

Table 8.5 Comparative properties of high-strength and high-modulus composites

Property	AS-4/3501-6 carbon/epoxy	GY-70/934 carbon/epoxy
Specific gravity	1.58	1.59
Fiber volume, %	63	57
0° tensile strength, ksi	331	85.3
0° tensile modulus 10^6 psi	20.6	42.6
0° tension strain, %	0.015	0.002
90° tensile strength, ksi	8.27	4.26
90° tensile modulus, 10^6 psi	1.48	0.93
90° tensile strain, %	0.006	0.005
0° compression strength, ksi	209	71.2
In-plane shear strength, ksi	10.3	8.58
In-plane shear modulus, 10^6 psi	1.04	0.71
Poisson's ratio ($\mu12$)	0.27	0.23

Source: Ref 8.1

High-modulus pitch-based carbon/epoxy is used in spacecraft applications where extremely lightweight, stiff, and dimensionally stable structures are required. However, as shown in Table 8.5, all of the properties except for the longitudinal modulus are lower for high-modulus material compared to the standard high-strength material. High-modulus materials are also considerably more expensive than the standard-modulus materials. In addition, more expensive cyanate ester resins are often specified, because they absorb less moisture and are thus more dimensionally stable and less prone to outgassing in the near-vacuum atmosphere of space. Therefore, high-modulus carbon and graphite-fiber composites are somewhat restricted in usage to specialized applications.

One of the main attractions of thermoplastics is their improved damage tolerance compared to brittle epoxies. The majority of the development work has been conducted using polyetheretherketone (PEEK). More recently, the emphasis has shifted to polyetherketoneketone (PEKK) for two reasons: (1) the basic PEKK polymer is inherently less expensive to produce than the PEEK polymer, and (2) PEKK composites can be processed at lower temperatures (340 °C, or 645 °F) than PEEK composites (390 °C, or 735 °F). The properties of the two materials are compared in Table 8.6 and appear to be fairly equivalent when using both high-strength and intermediate-modulus carbon fibers, although the CAI strength for the PEKK composites, at least for this dataset, appears to be a little lower than for the PEEK composites. The tensile and flexural properties for several other thermoplastic composites are shown in Table 8.7. Polyetherimide is an amorphous thermoplastic with good high-temperature resistance to 121 °C (250 °F). Polyphenylene sulfide is a semicrystalline thermoplastic with a lower usage temperature (<93 °C, or 200 °F) than PEEK and PEKK (121 °C, or 250 °F). Finally, polypropylene is a semicrystalline thermoplastic that is limited to low usage temperatures (<66 °C, or 150 °F).

Epoxy-matrix composites are generally limited to service temperatures of approximately 121 °C (250 °F). For higher-usage temperatures, bismaleimides, cyanate esters, and polyimides are available. Bismaleimide and cyanate ester composites are useful at temperatures up to approximately 177 to 204 °C (350 to 400 °F). For even higher temperatures, up

Table 8.6   Comparative properties of unidirectional carbon/PEEK and carbon/PEKK

Property	AS-4/PEEK	AS-4/PEKK	IM-7/PEEK	IM-7/PEKK
0° tensile strength, ksi	341	340	421	400
0° tensile modulus, 10^6 psi	20.0	19.7	24.9	24.4
0° compression strength, ksi	197	234	190	177
0° compression modulus, 10^6 psi	18.0	17.8	22.0	21.8
In-plane shear strength, ksi	27.0	21.2	26.0	18.9
In-plane shear modulus, 10^6 psi	0.826	0.812	0.797	0.711
Interlaminar shear strength, ksi	15.2	14.2	...	...
Open-hole tensile strength, ksi	61.8	56.4	69.0	74.6
Open-Hole comp. strength, ksi	47.0	48.6	47.0	45.7
Compression after impact, ksi	51.1	36.4	53.1	44.5

PEEK, polyetheretherketone; PEKK, polyetherketoneketone. Fiber-dominated properties normalized to 0.60 fiber volume. Source: Ref 8.6

Table 8.7   Comparative properties of unidirectional thermoplastic composites

Property	AS-4/PEI	AS-4/PPS	E-glass/PP
Fiber volume fraction, %	59	59	60
0° tensile strength, ksi	278	297	108
0° tensile modulus, 10^6 psi	18.8	18.5	4.1
90° tensile strength, ksi	11.0	7.2	...
90° tensile modulus, 10^6 psi	1.3	1.3	...
0° flexural strength, ksi	269	243	85
0° flexural modulus, 10^6 psi	17.8	16.3	3.8

PEI, polyetherimide; PPS, polyphenylene sulfide; PP, polypropylene. Source: Ref 8.7

to 260 to 316 °C (500 to 600 °F), polyimides must be used. While cyanate esters and bismaleimides process in a similar manner to epoxies, polyimides are generally condensation-curing systems that require high processing temperatures and pressures. Because the condensation reaction gives off either water or alcohols, polyimides are very prone to developing voids and porosity. In addition, they are brittle systems that are subject to microcracking. The relative temperature performance of a number of these material systems is shown in the plot of interlaminar shear strength versus temperature in Fig. 8.22.

Several other fibers are occasionally used for polymeric composites. Boron fiber was the original high-performance fiber before carbon was developed. It is a large-diameter fiber made by pulling a fine tungsten wire through a long, slender reactor, where it is chemically vapor deposited with boron. Because it is made one fiber at a time rather than thousands of fibers at a time, it is very expensive. Due to its large diameter and high modulus, it exhibits outstanding compression properties. On the negative side, it does not conform well to complicated shapes and is very difficult to machine. Other high-temperature ceramic fibers, such as silicon carbide (Nicalon), aluminum oxide, and alumina boria silica (Nextel), are frequently used in ceramic-based composites but rarely in polymeric composites.

The following factors should be considered when choosing between glass, aramid, and carbon fibers:

- *Tensile strength.* If tensile strength is the primary design parameter, E-glass may be the best selection because of its low cost.
- *Tensile modulus.* When designing for tensile modulus, carbon has a distinct advantage over both glass and aramid.

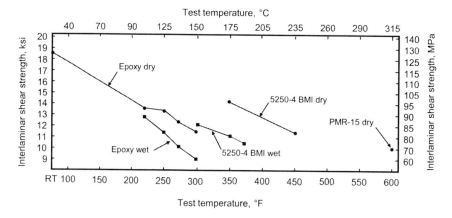

Fig. 8.22 Elevated-temperature interlaminar shear strength. BMI, bismaleimide. RT, room temperature. Source: Ref 8.1

- *Compression strength.* If compression strength is the primary requirement, carbon has a distinct advantage over glass and aramid. Due to its poor compression strength, aramid should be avoided.
- *Compression modulus.* Carbon fibers are the best choice, with E-glass having the lowest properties.
- *Density.* Aramid fibers have the lowest density, followed by carbon and then S-2 and E-glass.
- *Coefficient of thermal expansion.* Aramid and carbon fibers have a CTE that is slightly negative, while S-2 and E-glass are positive.
- *Impact strength.* Aramid and S-2 fibers have excellent impact resistance, while carbon is brittle and should be avoided. However, it should be noted that matrix selection also has a significant influence on impact strength.
- *Environmental resistance.* Matrix selection has the biggest impact on composite environmental resistance; however, (1) aramid fibers are degraded by ultraviolet light, and the long-term service temperature should be kept below 149 °C (300 °F); (2) carbon fibers are subject to oxidation at temperatures exceeding 370 °C (700 °F), although long-term 1000 h thermal oxidation stability tests in polyimides have shown strength decreases in the 260 to 316 °C (500 to 600 °F) range; and (3) glass sizings tend to be hydrophilic and absorb moisture.
- *Cost.* E-glass is the least expensive fiber, while carbon is the most expensive. In carbon, the smaller the tow size, the more expensive the fiber. Larger tow sizes help reduce labor costs because more material is deposited with each ply. However, large tow sizes in woven cloth can increase the chances of voids and matrix microcracking due to larger resin pockets.

8.5 Product Forms

There are a multitude of material product forms used in composite structures, some of which are illustrated in Fig. 8.23. The fibers can be continuous or discontinuous. They can be oriented or disoriented (random). They can be furnished as dry fibers or preimpregnated with resin (prepreg).

Not all fiber or matrix combinations are available in a particular material form, because the market drives availability. In general, the more operations required by the supplier, the higher the cost. For example, prepreg cloth is more expensive than dry woven cloth. While complex dry preforms may be expensive, they can translate into lower fabrication costs by reducing or eliminating hand lay-up costs. If structural efficiency and weight are important design parameters, then continuous reinforced product forms are normally used, because discontinuous fibers yield lower mechanical properties. Some common trade-offs when selecting composite materials are given in Table 8.8.

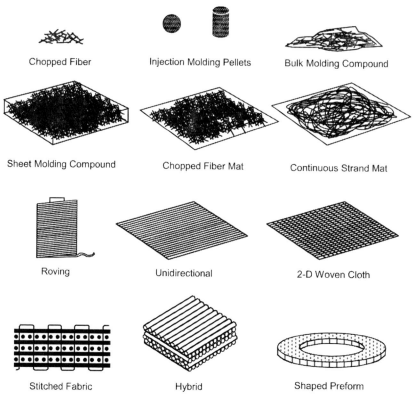

Fig. 8.23 Product forms used in composites. Source: Ref 8.1

Table 8.8 Common trade-offs when selecting composite materials

Design decision	Common trade-offs	Typically lowest cost	Typically highest performance
Fiber type	Cost, strength, stiffness, density (weight), impact strength, electrical conductivity, environmental stability, corrosion, thermal expansion	E-glass	Carbon
Tow size (if carbon is selected)	Cost, fiber volume, improved fiber wet-out, structural efficiency (minimize ply thickness), surface finish	12k tow	3k tow
Fiber modulus (if carbon is selected)	Cost, stiffness, weight, brittleness	Lowest-modulus carbon	Highest-modulus carbon
Fiber form (continuous vs. discontinuous)	Cost, strength, stiffness, weight, fiber volume, design complexity	Random/ discontinuous	Oriented and continuous
Matrix	Cost, service temperature, compressive strength, interlaminar shear, environmental performance (fluid resistance, ultraviolet stability, moisture absorption), damage tolerance, shelf life, processability, thermal expansion	Vinyl ester and polyester	High-temperature: polyimide(a) Low- to moderate-temperature: epoxy Toughness: toughened epoxy
Composite material forms	Cost (material and labor), process compatibility, fiber volume control, material handling, fiber wet-out, material scrap	Base form: neat resin/ rovings	Prepreg(b)

(a) Depends on how "highest performance" is defined: high temperature, toughness, and superior mechanical properties. (b) Material form is not driven by performance but typically defined by the manufacturing process. Source: Ref 8.1

Discontinuous Fiber Product Forms

Chopped fiber is made by mechanically chopping rovings, yarns, or tows into short lengths, typically 6 to 50 mm ($\frac{1}{4}$ to 2 in.) long. The minimum lengths of chopped fibers are very important to ensure maximum reinforcement efficiency. Milled fibers, typically 0.8 to 6 mm ($\frac{1}{32}$ to $\frac{1}{4}$ in.) long, have low aspect ratios (length/diameter) that provide minimal strength, and therefore, they should not be considered for structural applications. The strength of the composite will be greatly improved with increased fiber length. Stiffness properties are much less affected by fiber length. Chopped glass fibers (Fig. 8.24) are often embedded in thermoplastic or thermoset resins in the form of pellets for injection molding.

Mat. Chopped fibers or entangled continuous strands are combined with a binder to form mat (Fig. 8.25) or veil materials. A veil is a thin mat that is used to improve the surface finish of a molded composite. This material form is used extensively for automotive, industrial, recreational, and marine applications, where high-quality class A exterior surface finishes

Fig. 8.24 Typical chopped glass fibers. Source: Ref 8.1

Fig. 8.25 Glass fiber mat. Source: Ref 8.1

are required. These are very inexpensive product forms, with E-glass being the most common fiber. Other fibers may be available in mat form but are generally expensive due to the specialty nature of the product.

Presently, carbon mats are uncommon due to low demand. Some carbon veils for electromagnetic interference protection are available but are expensive. Historically, aerospace has not been interested in this product form due to the lack of structural efficiency and the associated weight penalty for its discontinuous reinforcement. On the other hand, the automotive industry, which widely uses mats, would not pay the cost penalty for the carbon-fiber material.

Sheet molding compound (SMC) consists of flat sheets of chopped, randomly oriented fibers, typically 25 to 51 mm (1 to 2 in.) in length, with a matrix staged to a paste-like consistency, usually containing E-glass fiber in either a polyester or vinyl ester resin. A typical sheet molding machine is shown in Fig. 8.26. The material is available either as rolls or precut sheets. Bulk molding compounds (BMCs) are also short, randomly oriented preimpregnated materials; however, the fibers are normally only 3.18 to 31.8 mm ($\frac{1}{8}$ to $1\frac{1}{4}$ in.) long, and the reinforcement percentage is lower. As a result, the mechanical properties for BMC composites are lower than for SMC composites. The BMCs are available in doughlike bulk form or extruded into logs for easier handling. Vinyl esters, polyesters, and phenolics are the most common matrices used for BMCs. Sometimes, for convenience in terminology, chopped prepreg, used for compression molding, is referred to as BMC.

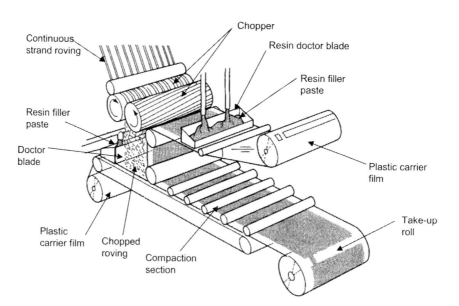

Fig. 8.26 Sheet molding compound machine. Source: Ref 8.8

Continuous Fiber Product Forms

Rovings, tows, and yarns are collections of continuous fiber. Continuous glass fiber roving is shown in Fig. 8.27. This is the basic material form that can be chopped, woven, stitched, or prepregged into other product forms. It is the least expensive product form and is available in all fiber types. Rovings and tows are supplied with no twist, while yarns have a slight twist to improve handling. While twist improves handling, it reduces the fiber strength. Some processes, such as wet filament winding and pultrusion, use rovings as their primary product form.

Prepreg. Continuous thermoset prepreg materials are available in many fiber and matrix combinations. A prepreg is a fiber form that has a predetermined amount of uncured resin impregnated on the fiber by the material supplier. Prepreg is made by melting the resin and then combining it with the fiber in a special machine, as shown in Fig. 8.28. Prepreg rovings and tapes are usually used in automated processes, such as filament winding and automated tape laying, while unidirectional tape and prepreg fabrics are used for hand lay-up. Unidirectional prepreg tapes offer improved structural properties over woven prepregs due to absence of fiber crimp and the ability to more easily tailor the designs. However, woven prepregs offer increased drapeability, making them attractive for complex substructure parts. With the exception of predominantly unidirectional designs, unidirectional tapes require placement of more individual plies during lay-up. For example, with cloth, for every 0° ply in the lay-up, 90° reinforcement is also included. With unidirectional tape, a 0° ply and a separate 90° ply must be placed onto the tool.

Prepregs are supplied with either a net resin (prepreg resin content ≈ final part resin content) or excess resin (prepreg resin content > final part resin content). The excess resin approach relies on the matrix flowing through the plies, removing entrapped air, while the extra resin is removed by impregnating bleeder plies on top of the lay-up. The amount of bleeder

Fig. 8.27 Glass-fiber roving. Source: Ref 8.1

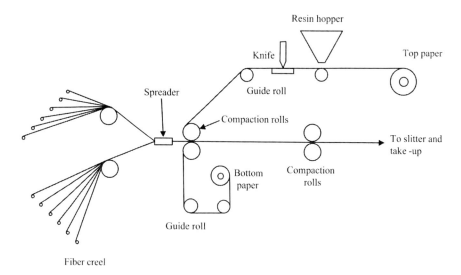

Fig. 8.28 Prepreg resin manufacturing process. Source: Ref 8.9

used in the lay-up will dictate the final fiber and resin content; however, as the laminate gets thicker, it becomes more difficult to remove the excess resin. Accurate calculations of the number and areal weight of bleeder plies for a specific prepreg are required to ensure proper final physical properties. Because the net resin approach contains the final resin content weight in the fabric, no resin removal is necessary. This is an advantage because the fiber and resin volumes can easily be controlled. Thermoset prepreg properties include volatile content, resin content, resin flow, gel time, tack, drape, shelf life, and out-time. Careful testing, evaluation, and control of these characteristics is necessary to ensure that the prepreg materials handling characteristics are optimal and that final part structural performance is obtained.

Woven fabric is the most common continuous dry material form. A woven fabric is made on modified textile equipment, as shown in Fig. 8.29, and consists of interlaced warp and fill yarns. The warp is the 0° direction as the fabric comes off the roll, and the fill, or weft, is the 90° fiber. Typically, woven fabrics are more drapeable than stitched materials; however, the specific weave pattern will affect their drapeability characteristics. The weave pattern will also affect the handleability and structural properties of the woven fabric. Many weave patterns are available. All weaves have their advantages and disadvantages, and consideration of the part configuration is necessary during fabric selection. Most fibers are available in woven fabric form. However, it can be very difficult to weave some high-modulus fibers due to their inherent brittleness. Advantages of woven fabric include drapeability, ability to achieve high fiber volumes, structural efficiency, and market availability. A disadvantage of woven fabric is the crimp that

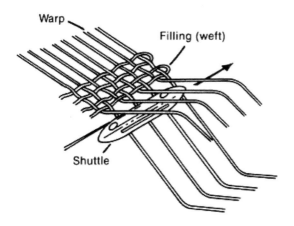

Fig. 8.29 Weaving glass cloth. Source: Ref 8.1

is introduced in the warp or fill fibers during weaving, which adversely affects the strength properties. Finishes or sizings are typically put on the fibers to aid in the weaving process and minimize fiber damage. It is important to ensure that the finish is compatible with the matrix selection when specifying a fabric.

Stitched fabric consists of unidirectional fibers oriented in specified directions that are then stitched together to form a fabric. A common stitched design includes 0°, +45°, 90°, and −45° plies in one multidirectional fabric, as shown in Fig. 8.30. Advantages include:

- The ability to incorporate off-axis orientations as the fabric is removed from the roll. Off-axis cutting is not needed for a multidirectional stitched fabric and can reduce scrap rates (up to 25%) when compared to conventional woven materials.
- Labor costs are also reduced when using multi-ply stitched materials because less plies are required to be cut or handled during fabrication of a part.
- Ply orientation remains intact during handling due to the z-axis stitch threads.

Disadvantages include:

- Availability of specific stitched ply set designs. Typically, a special order is required due to the tailoring requested by the customer, such as fiber selection, fiber volume, and stitching requirements.
- Not as many companies stitch as weave.
- Drapeability characteristics are reduced; however, this can be an advantage for parts with large, simple curvature. Careful selection of the stitching thread is necessary to ensure compatibility with the matrix and process temperatures.

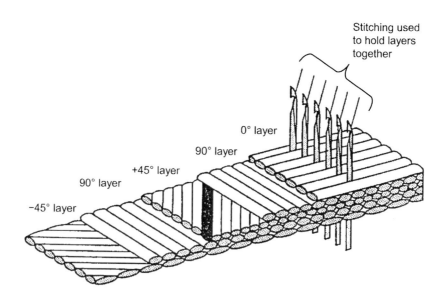

Fig. 8.30 Stitched dry preform

Hybrids are material forms that make use of two or more fiber types. Common hybrids (Fig. 8.31) include glass/carbon, glass/aramid, and aramid/carbon fibers. Hybrids are used to take advantage of properties or features of each reinforcement type. In a sense, a hybrid is a trade-off reinforcement that allows increased design flexibility. Hybrids can be interply (two alternating layers), intraply (present in one layer), or in selected areas. Hybridization with carbon in selected areas is usually done to locally strengthen or stiffen a part, while hybridization with glass in carbon laminates can be used to locally soften a laminate. Carbon/aramid hybrids have low thermal stresses compared to other hybrids due to their similar CTEs, increased modulus and compressive strength over an all-aramid design, and increased toughness over an all-carbon design. Carbon/E-glass hybrids have increased properties compared to an all-E-glass design but have lower cost than an all-carbon design. The CTE of each fiber type in a hybrid needs careful evaluation to ensure that high internal stresses are not introduced to the laminate during cure, especially at higher cure temperatures. The use of hybrids normally increases the amount of material that must be stocked and can increase the lay-up costs.

Preforms. A preform is a preshaped fibrous reinforcement that has been formed into shape, on a mandrel or in a tool, before placing into a mold. The shape of the preform closely resembles the final part configuration,

Fig. 8.31 Examples of hybrid weaves. Source: Ref 8.1

as shown in Fig. 8.32. A simple multi-ply stitched fabric is not a preform unless it is shaped to near its final configuration. The preform is the most expensive dry, continuous, oriented fiber form; however, using preforms can reduce fabrication labor. A preform can be made using rovings, chopped, woven, stitched, or unidirectional material forms. These reinforcements are formed and held in place by stitching, braiding,

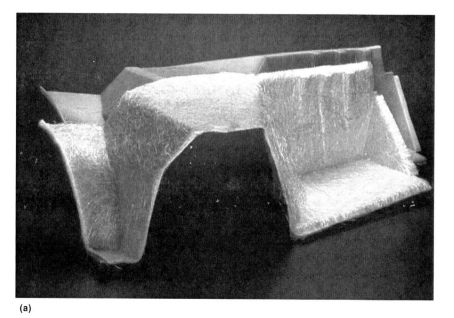

(a)

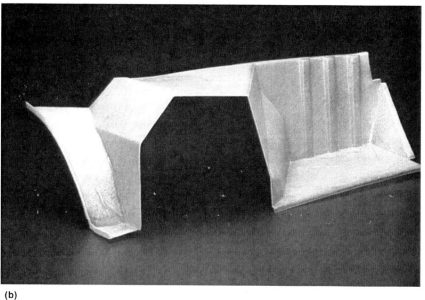

(b)

Fig. 8.32 Example of (a) fiberglass preform and (b) liquid molded part. Source: Ref 8.1

three-dimensional weaving, or with organic binders or tackifiers. Advantages include reduced labor costs, minimal material scrap, reduced fiber fraying of woven or stitched materials, improved damage tolerance for three-dimensional stitched or woven preforms, and the desired fiber orientations are locked in place. Disadvantages include high preform costs, fiber wettability concerns for thick complex shapes, tackifier or binder compatibility with the matrix, and limited flexibility if design changes are required. A common defect is an out-of-tolerance preform, which makes placement in the tool difficult or requires trimming. The use of preforms is not appropriate for all applications. Careful consideration of all issues must be evaluated before a preform is baselined. If the component is not a very complex configuration, the money saved in labor reduction may not offset the cost of the preform. Each application should be evaluated individually to determine if the preform approach offers cost and quality advantages.

8.6 Matrices

The matrix holds the fibers in their proper position, protects the fibers from abrasion, transfers loads between fibers, and provides the matrix-dependent mechanical properties. A properly chosen matrix will also provide resistance to heat, chemicals, and moisture; have a high strain-to-failure; cure at as low a temperature as possible and yet have a long pot or out-time life; and not be toxic. Matrices for polymeric composites can be either thermosets or thermoplastics. The most prevalent thermoset resins used for composite matrices are polyesters, vinyl esters, epoxies, bismaleimides, cyanate esters, polyimides, and phenolics. A relative comparison of the major thermoset resin systems is given in Table 8.9.

The first consideration in selecting a resin system is the service temperature required for the part. The glass transition temperature (T_g) is a good indicator of the temperature capability of the matrix. The T_g of a polymeric

Table 8.9 Comparison of thermoset matrix systems

Attribute	Polyester	Vinyl ester	Epoxy	Phenolic	Bismale-imide	Cyanate ester
Applications						
Typical applications	Marine, general	Marine, general	Aerospace, general	Fire, smoke, and toxicity applications	Aerospace, electrical	Aerospace
Performance						
Structural	Fair	Good	Good	Good, brittle	Good	Good
Corrosion and chemical resistance	Poor	Excellent	Good	Excellent	Good	Excellent
Moisture absorption	Poor	Excellent	Fair to good	Excellent	Fair to good	Excellent
Glass transition temperature (T_g)(a), °F	160	160–325 with postcure	200–350 with postcure	160–250+ with postcure	300–425 with postcure	350–450 with postcure
Fire, smoke, and toxicity	…	Requires additives	Requires additives	Excellent	Good	Good

(a) Actual T_g varies with each system and cure method. Source: Ref 8.10

material is the temperature at which it changes from a rigid glassy solid into a softer, semiflexible material. At this point, the polymer structure is still intact, but the cross links are no longer locked in position. A resin should never be used above its T_g unless the service life is very short (e.g., a missile body). A good rule of thumb is to select a resin in which the T_g is 28 °C (50 °F) higher than the maximum service temperature. Because most polymeric resins absorb moisture that lowers the T_g, it is not unusual to require that the T_g be as much as 56 °C (100 °F) higher than the service temperature. Therefore:

Maximum usage temperature = wet T_g − 28 °C (50 °F)

It should be noted that different resins absorb moisture at different rates, and the saturation levels can be different; therefore, the specific resin candidate must be evaluated for environmental performance. In general, thermoplastics absorb less moisture than thermosets. However, some thermoplastics, and in particular amorphous thermoplastics, have poor solvent resistance. Most thermoset resins are fairly resistant to solvents and chemicals.

In general, the higher the temperature performance required, the more brittle and less damage tolerant the matrix. Toughened thermoset resins are available but are more expensive, and their T_gs are typically lower. High-temperature resins are also more costly and more difficult to process. Temperature performance is difficult to quantify because it is dependent on time at temperature, but it is important to thoroughly understand the environment in which the matrix is expected to perform.

Higher-T_g resins also require higher cure temperatures and longer times. Epoxies generally have cure times of 2 to 6 h at elevated temperature. A postcure may not be required for some epoxies, polyesters, and vinyl esters; therefore, elimination of postcure requirements should be evaluated as a way to decrease processing costs, if the service temperature is low enough. Higher-T_g resins, such as bismaleimides and polyimides, require longer cure cycles and postcures. Postcuring further develops higher-temperature mechanical properties and improves the T_g of the matrix for some epoxies, bismaleimides, and polyimides. Very short cure times are desired for some processes, such as compression molding and pultrusion. Cure temperatures can range from 120 to 175 °C (250 to 350 °F) for epoxies. Bismaleimide cure temperatures typically range from 177 to 246 °C (350 to 475 °F) (including postcure). Polyimide cure and postcure temperatures range from 316 to 371 °C (600 to 700 °F).

Polyesters, vinyl esters, epoxies, cyanate esters, and bismaleimides are all addition-curing systems, while phenolics and polyimides cure by condensation reactions that evolve water or alcohols. Because the evolution of volatiles increases the propensity for voids and porosity, condensation-curing materials are more difficult to process. However, phenolics are often processed by compression molding processes where the high pressures can be applied to suppress void formation. In addition, phenolic-matrix composites are often used for interior furnishings for commercial aircraft because they tend to char instead of burn. Because these are only lightly loaded structures, the void content is not very important. Because polyimide-matrix composites are more difficult to process than the addition-curing systems, their usage should be restricted to high-temperature applications where they are really required.

Although the fiber selection usually dominates the mechanical properties of the composite, the matrix selection can also influence performance. Some resins wet-out and adhere to fibers better than others, forming a chemical and/or mechanical bond that affects the fiber-to-matrix load-transfer capability. The matrix can also microcrack during cure or in service. Resin-rich pockets and brittle resin systems are susceptible to microcracking, especially when the processing temperatures are high and the use temperatures are low (e.g., –54 °C, or –65 °F), because this condition creates a very large difference in thermal expansion between the fibers and the matrix. Toughened resins help to prevent microcracking but often at the expense of elevated-temperature performance.

Thermoplastic Composite Matrices

Polyetheretherketone (PEEK), polyetherketoneketone (PEKK), polyphenylene sulfide (PPS), and polypropylene (PP) are semicrystalline thermoplastics, while polyetherimide (PEI) is an amorphous thermoplastic. PEEK, PEKK, PPS, and PEI are normally used for continuous fiber-reinforced thermoplastic composites, while PP is a lower-temperature resin that is used quite extensively in the automotive industry as a discontinuous glass-fiber stampable sheet product form called glass mat reinforced thermoplastic. High-performance thermoplastics, such as PEEK, PEKK, PPS, and PEI, have high T_gs with good mechanical properties, much higher than conventional thermoplastics, but are also more costly. High-performance thermoplastics are usually aromatic, containing the benzene ring (actually the phenylene ring) that increases the T_g and provides thermal stability. Also, when the number of units in the molecular chain is large, there is a high degree of orientation in the liquid state that helps promote crystallinity during freezing. Highly aromatic thermoplastics exhibit good flame retardance because of their tendency to char and form a protective surface layer.

8.7 Fabrication Processes

Once the fiber, product form, and matrix are selected, the fabrication process selection is narrowed to those used to make discontinuous fiber composites and those used to make the higher-strength and -stiffness continuous fiber composites. Often, the material product form will be a major determining factor in the process selected, as shown in the examples in Table 8.10.

Discontinuous Fiber Processes

Injection molding is a high-volume process capable of making small- to medium-sized parts. The reinforcement is usually chopped glass fibers with either a thermoplastic or thermoset resin, although the majority of applications use thermoplastics because they process faster and have higher toughness. In the injection molding process (Fig. 8.33), pellets

Table 8.10   Typical material product forms versus process

Material form/process	Pultrusion	RTM	Compression molding	Filament winding	Hand lay-up	Auto tape laying
Discontinuous						
Sheet molding compound	...	...	•	...	...	...
Bulk molding compound	...	...	•	...	...	...
Random continuous						
Swirl mat/neat resin	•	•	•	•	•	...
Oriented continuous						
Unidirectional tape	...	...	•	...	•	•
Woven prepreg	...	...	•	...	•	...
Woven fabric/neat resin	•	•	...	•	•	...
Stitched material/neat resin	...	•	...	...	...	...
Prepreg roving	...	...	...	•	...	...
Roving/neat resin	•	...	...	•	•	...
Preform/neat resin	...	•	...	...	...	...

RTM, resin transfer molding. Source: Ref 8.1

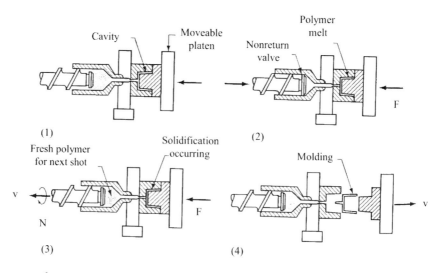

Fig. 8.33 Injection molding machine sequence. Source: Ref 8.11

containing embedded fibers or chopped fibers and resin are fed into a hopper. They are heated to their melting temperature and then injected under high pressure into a matched metal die. After the thermoplastic part cools or the thermoset part cures, they are ejected and the next cycle is started. Expensive matched-die molds are required, so the number of parts to be produced must be large enough to recover the tooling cost. Labor costs are minimal. Many injection molding facilities are highly automated, with only a minimal number of operators. While both thermosets and thermoplastic materials can be injected molded, by far the greatest amount are molded using thermoplastics. Because the fibers are short and randomly oriented, the amount of strength and stiffness improvement is only nominal, as shown for the properties of E-glass/nylon in Table 8.11. However, the improvement is substantial compared to the unreinforced polymer.

Spray-up is a more cost-effective process than wet lay-up, but the mechanical properties are much lower due to the use of randomly oriented chopped fibers. Normally, continuous glass roving is fed into a special gun (Fig. 8.34) that chops the fibers into short lengths and simultaneously

Table 8.11 Typical properties of injection molded E-glass/nylon

Property	Unreinforced	30% glass fibers
Specific gravity	1.14	1.39
Tensile strength, ksi	12.0	24.9
Tensile modulus, 10^6 psi	0.42	1.31
Elongation, %	60	4
Flexural strength, ksi	17.3	36.0
Flexural modulus, 10^6 psi	0.40	1.31
Notched Izod impact, J/m	53	107
Heat deflection temperature, °F	194	486

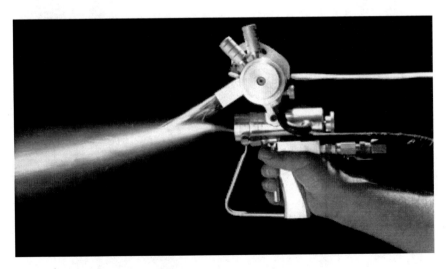

Fig. 8.34 Chopped fiber spray-up gun. Source: Ref 8.1

mixes them with either a polyester or vinyl ester resin that is then sprayed onto the tool. Manual compaction with rollers is used to compact the layup. Vacuum bag cures can improve part quality but are not normally used. Because the fibers are short and the orientation is random, this process is not used to make structural load-bearing parts.

Compression molding is another matched-die process that uses either discontinuous, randomly oriented SMC or BMC. A charge of predetermined weight is placed between the two dies, and then heat and pressure are applied. As shown in Fig. 8.35, the molding compound flows to fill the die and then rapidly cures in 1 to 5 min, depending on the type of polyester or vinyl ester used. Thermoplastic composites, usually consisting of glass fiber and polypropylene for the automotive industry, are compression molded using highly automated production lines.

Glass fibers, usually 25 to 51 mm (1 to 2 in.), are used extensively in SMCs that are compression molded in matched metal dies into structural parts. Although the reinforcements can be either random or continuous, random reinforcements are the most prevalent. Normally, polyester resins are used that contain fillers such as calcium carbonate. Some typical properties of several SMCs are shown in Table 8.12. The SMC-R material

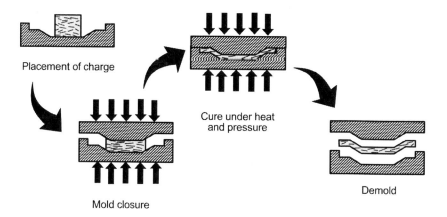

Fig. 8.35 Compression molding of thermosets. Source: Ref 8.1

Table 8.12 Typical sheet molding compound (SMC) properties

Material	Specific gravity	Glass, wt%	Tensile strength, ksi	Elongation, %	Flexural modulus, 10^6 psi
SMC-R25	1.83	25	12.0	1.34	1.70
SMC-R50	1.87	50	23.8	1.73	2.30
SMC-C20/R30	1.81	50	...	...	...
Longitudinal	...	...	41.9	1.73	3.73
Transverse	...	...	12.2	1.58	0.86
XMC-3 ±7.5° X-pattern	1.97	75	...	...	...
Longitudinal	...	...	81.4	1.66	4.95
Transverse	...	...	10.1	1.54	0.99

Source: Ref 8.12

contains randomly oriented short glass fibers and has somewhat isotropic properties in the two planes of the molded sheet. The designations SMC-R25 and SMC-R50 indicate that the fibers are random and the glass contents are 25 and 50%, respectively. The SMC-C20/R-30 material contains 20% continuous fibers and 30% random fibers to give a total glass content of 50%. Naturally, the strength is higher in the direction of the continuous fibers. The last material shown, XMC, contains continuous fibers in a ±7.5° X-pattern with a quite high glass fiber content of 75%. The materials with random fibers, SMC-R, are the most widely used because they flow better in the mold and are thus capable of making more complex part shapes than the materials containing continuous glass fibers.

Reaction injection molding (RIM) is a process for rapidly making unreinforced thermoset parts. A two-component highly reactive resin system is injected into a closed mold, where the resin quickly reacts and cures, as shown in Fig. 8.36. Polyurethanes are the most prevalent, but nylons, polyureas, acrylics, polyesters, and epoxies have also been used. The RIM resin systems have very low viscosities and can be injected at low pressures (<690 kPa, or 100 psi), which allows the use of inexpensive molds and low-force clamping systems. Typical tooling materials include steel, cast aluminum, electroformed nickel, and composites. Cycle times as short as 2 min for large automotive bumpers are typical. Reinforced reaction injection molding (RRIM) is similar to RIM except that short glass fibers are added to one of the resin components. The fibers must be extremely short (e.g., 0.76 mm, or 0.03 in.) or the resin viscosity will be too great. Short fibers, milled fibers, and flakes are commonly used. Again, polyurethanes are the predominant resin system. The addition of the fibers improves modulus, impact resistance, dimensional tolerance, and lowers the CTE. Structural reaction injection molding (SRIM) is similar to the previous two processes except that a continuous glass preform is placed in the die prior to injection. Again, this process is used almost exclusively with polyurethanes. Due to the highly reactive resins and short cycle times, SRIM cannot produce as large a part size as with resin transfer molding (RTM). Also, SRIM parts have lower fiber volumes and generally more porosity than RTM parts. Typical properties of milled and chopped glass-fiber RRIM and continuous glass SRIM composites are shown in Table 8.13.

Continuous Fiber Processes

Wet lay-up in open molds, shown schematically in Fig. 8.37, is probably the least expensive process for a relatively small number of units. This is a common process for large-production marine manufacturing, where thick, heavy, woven fiberglass mat and randomly oriented chopped fiber mat are hand or spray impregnated with polyester resin and cured at room temperature. After the ply is placed on the lay-up, hand rollers are used to remove excess resin and air and to compact the plies. After lay-up, cure

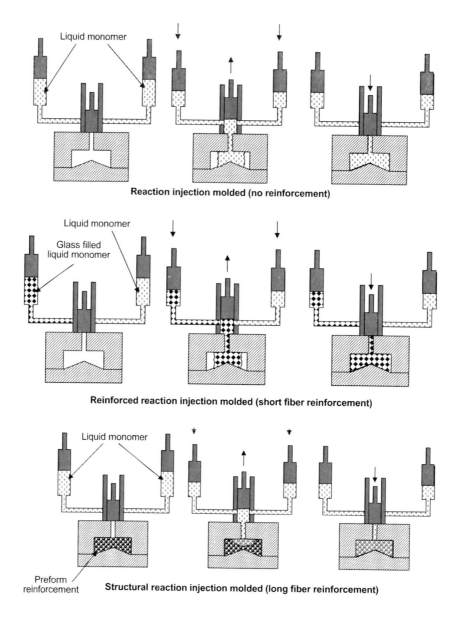

Liquid monomer

Reaction injection molded (no reinforcement)

Liquid monomer

Glass filled liquid monomer

Reinforced reaction injection molded (short fiber reinforcement)

Liquid monomer

Preform reinforcement

Structural reaction injection molded (long fiber reinforcement)

Fig. 8.36 Reaction injection molding, reinforced reaction injection molding, and structural reinforced injection molding processes. Source: Ref 8.1

can be done at room or elevated temperature. Frequently, cure is conducted without a vacuum bag, but vacuum pressure helps to improve laminate quality. Because cure is usually conducted at room or low temperatures, very inexpensive tooling (e.g., wood or fiberglass) can be used to minimize cost. Although conducive to large and inexpensive production, the optimization of fiber properties is not achieved, weight is often not minimized, and the level of precision is very operator dependent.

Table 8.13 Typical reinforced reaction injection molded (RRIM) and structural reinforced injection molded (SRIM) properties

Material	Specific gravity	Glass, wt%	Tensile strength, ksi	Elongation, %	Flexural modulus, 10^6 psi
RRIM-milled glass/polyurethane	1.08	15	...	...	...
Longitudinal	...	...	2.80	110	0.078
Transverse	...	...	2.80	140	0.048
RRIM-chopped glass/polyurethane	1.15	20	...	...	...
Longitudinal	...	...	3.50	25	0.194
Transverse	...	...	3.65	35	0.180
RRIM-chopped glass/polyurea	1.18	20	...	...	...
Longitudinal	...	...	4.83	31	0.244
Transverse	...	...	4.43	31	0.250
SRIM glass/polyurethane	1.5	37	24.9	4.2	1.80

Source: Ref 8.12

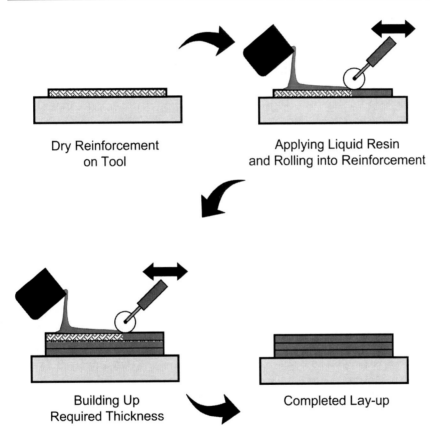

Dry Reinforcement on Tool

Applying Liquid Resin and Rolling into Reinforcement

Building Up Required Thickness

Completed Lay-up

Fig. 8.37 Typical wet lay-up operation. Source: Ref 8.1

The mechanical properties of a number of materials impregnated with polyester resins are shown in Table 8.14. Spray-up produces more or less randomly oriented fibers 25 to 51mm (1 to 2 in.) long. Because the fiber content is only moderate (30 to 35%), the properties are normally fairly low. Chopped strand mat (CSM) is also a randomly oriented product form

Table 8.14   Typical properties of glass chopped strand mat composites

Property	Spray roving	Chopped strand mat	Chopped strand mat	Chopped strand mat	Chopped strand mat/woven roving
Glass content, wt%	30–35	25–30	30–35	35–40	45–50
Glass content, vol%	16–20	14–16	16–20	20–24	28–32
Density, g/cm^3	1.45	1.40	1.45	1.50	1.68
Tensile strength, ksi	10	10	13	16	26
Tensile elongation, %	1.0	1.8	1.8	1.8	2.0
Tensile modulus, 10^6 psi	1.0	0.9	1.1	1.3	1.8
Flexural strength, ksi	20	20	22.5	25	35
Flexural modulus, 10^6 psi	0.9	0.8	0.95	1.2	1.5
Compressive strength, ksi	16	14.5	17.4	20	22
Compressive modulus, 10^6 psi	1.1	0.95	1.2	1.4	1.8

Source: Ref 8.13

in which glass fibers are either chopped and sprayed to produce a mat or are swirled into a mat product. Depending on the fiber content, again the properties are only low to moderate. Higher strengths and stiffness can be obtained by combining the CSM with continuous fiber woven roving, as shown in Table 8.15. In this table, CSM properties are compared with roving woven combi, which is a combination of CSM and woven cloth, a biaxial 0°/90° woven cloth, and a unidirectional woven cloth that has a high percentage of reinforcement in the 0° direction. Both the strength and stiffness are improved as the amount of continuous reinforcement is increased.

Autoclave curing of prepreg materials often offers the best mechanical properties. Prepreg provides a consistent product with a uniform distribution of reinforcement and resin, with high fiber volume fractions possible. Because laminated structures typically require a great amount of hand labor and autoclave processing, the manufacturing costs are relatively high. Prepreg lay-up is a process in which individual layers of prepreg are layed-up on a tool and then cured. The layers are layed-up in the required directions and to the correct thickness. A vacuum bag is then placed over the lay-up, and the air is evacuated to draw out air between the plies. The bagged part is placed in an oven or an autoclave and cured under the specified time, temperature, and pressure. If oven curing is used, the maximum pressure that can be obtained is atmospheric (14.7 psia or less). Autoclave processing (Fig. 8.38) offers the advantage that much higher pressures (e.g., 690 kPa, or 100 psi) can be used, resulting in better compaction, higher fiber volume percentages, and less voids and porosity. Presses can also be used for this process, but they have several disadvantages: the size of the part is limited by the press platen size, the platens may produce high- and low-pressure spots if the platens are not flat and parallel, and complex shapes are difficult to produce. Automated ply cutting, manual ply collation or lay-up, and autoclave curing is the most widely used process for high-performance composites in the aerospace industry. While manual ply collation is expensive, this process is capable of making high-quality complex parts.

To reduce the costs for expensive autoclave-hardened tooling, low-temperature/vacuum bag (LTVB) prepregs were developed. These materials are capable of producing autoclave-quality parts by initially curing at low temperatures on less-expensive tooling under vacuum bag pressure. The part is then removed from the tool and given a free-standing postcure in

Table 8.15 Typical properties of glass chopped strand mat/woven composites

Property	Chopped strand mat	Woven roving combi	Biaxial 0/90	Biaxial unidirectional
Fiber content, wt%	35	50	58	60
Fiber content, vol%	20	32	41	42
Density, g/cm^3	1.50	1.60	1.70	1.75
Tensile strength, ksi	18.1	29.7	50.7	87.0
Tensile elongation, %	1.9	1.9	2.5	2.4
Tensile modulus, 10^6 psi	1.1	2.3	2.9	4.1
Compressive strength, ksi	21.7	36.2	40.6	78.3
Compressive modulus, 10^6 psi	1.1	2.4	3.0	4.2

Source: Ref 8.13

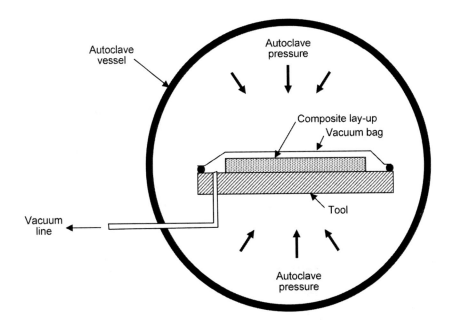

Fig. 8.38 Principle of autoclave curing. The autoclave vessel is pressurized with gas, usually nitrogen or carbon dioxide, at some pressure (e.g., 690 kPa, or 100 psi). Because the laminate inside the vacuum bag is either at atmospheric pressure or has an applied vacuum, there exists a pressure differential that provides compaction to the laminate plies. Source: Ref 8.1

an oven to increase its temperature resistance. However, even with these materials, the potential for voids and porosity exists, especially for larger parts where it may be more difficult to remove the entrapped air during the evacuation portion of the cure cycle.

The mechanical properties for autoclave-processed unidirectional carbon/epoxy composites are shown in Tables 8.3 and 8.5, and those for an LTVB material are given in Table 8.4.

Liquid Molding. The term *liquid molding* covers a fairly extensive set of processes. In the RTM process shown in Fig. 8.39, a dry preform or lay-up is placed in a matched metal die, and a low-viscosity resin is injected under pressure to fill the die. Because this is a matched-die process, it is capable of holding very tight dimensional tolerances. The die can contain internal heaters or be placed in a heated platen press for cure. Other variations of this process include vacuum-assisted resin transfer molding (VARTM), shown in Fig. 8.40, in which a single-sided tool is used along with a vacuum bag. Instead of injecting the resin under pressure, a vacuum pulls the resin through a flow medium that helps impregnate the preform. The biggest disadvantage of RTM is often the expense of large, complicated matched-die tools. Because VARTM uses single-sided tools and the pressures are lower, it is inherently less expensive than RTM.

Filament Winding. Higher-strength glass-fiber parts are also fabricated by processes such as filament winding and pultrusion. A comparison of

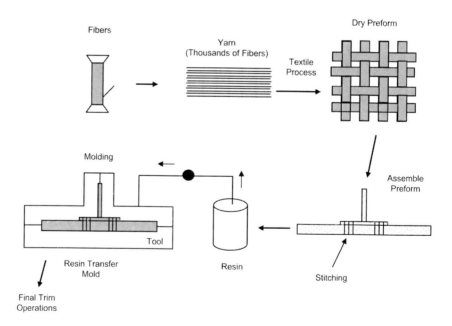

Fig. 8.39 Process flow for resin transfer molding. Source: Ref 8.14

filament-wound and pultruded part properties is given in Table 8.16. Filament winding (Fig. 8.41) is a process that has been used for many years to build highly efficient structures that are bodies of revolution or near bodies of revolution. Wet winding, in which dry fiber rovings are pulled through a resin bath prior to winding on the mandrel, is the most prevalent process, but prepregged roving or tows can also be filament wound. Wet filament winding is the primary manufacturing method for cylindrical or spherical pressure vessels, because continuous fibers can be easily placed in the hoop direction. Low-cost constant cross-section components can be made using a filament winder with two degrees of freedom: the rotation of the mandrel and the translation of the payout eye. Additional payout eyes can increase the speed of the process. Increasing the number of degrees of freedom increases its complexity as well as the complexity of parts that

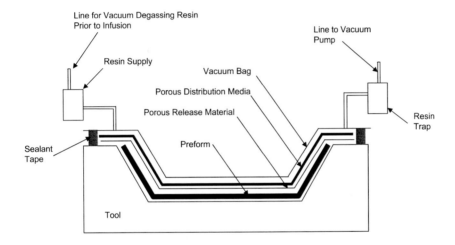

Fig. 8.40 Typical vacuum-assisted resin transfer molding process setup. Source: Ref 8.1

Table 8.16 Typical properties of filament-wound and pultruded glass-fiber composites

		Property			
Processing method	Material	Tensile strength, ksi	Tensile modulus, 10^6 psi	Flexural strength, ksi	Compression strength, ksi
Filament winding	30–80 wt% glass roving-epoxy resin, variable angle	40–80	3.0–6.0	40–80	45–70
Pultrusion rod and bar	60–80 wt% glass roving only	60–100	4.5–6.0	50–80	40–60
Pultrusion profile	40–55 wt% glass roving/ continuous strand mat	12–30	1.0–2.5	15–35	15–30
Pultrusion profile	50–65 wt% glass roving/ continuous strand mat/fabric	30–45	3.9–4.5	20–50	14–55

Source: Ref 8.15

can be created. Hoop (circumferential) winding involves laying fibers at nearly 90° with respect to the longitudinal axis of the mandrel. This results in a cylinder with high hoop strength. However, helical (off-axis) or polar (axial, near-longitudinal) windings provide reinforcement along the length of the structure. As these fibers are wound, they have a tendency to slip. One exception is a geodesic path, which is a straight line on the flattened equivalent surface. As the part is formed, its thickness will increase, thus changing the geometry of the process. Curing is usually conducted in an oven with or without a vacuum bag.

Pultrusion is a rather specialized composite fabrication process that is capable of making long, constant-thickness parts. Dry E-glass rovings are normally pulled through a wet resin bath and are then preformed to the desired shape before entering a heated die (Fig. 8.42). Mats and veils are

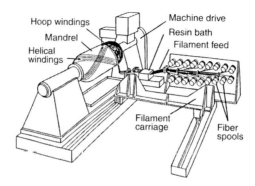

Fig. 8.41 Typical filament winding machine. Source: Ref 8.16

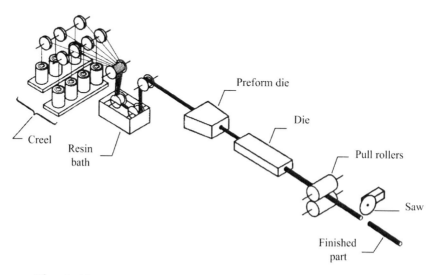

Fig. 8.42 Pultrusion process. Source: Ref 8.11

frequently incorporated into the part. Cure occurs inside the die, and the cured part is pulled to the desired length and cut off. Quick-curing polyesters and vinyl esters are the predominant resin systems. Design limitations include the need for a large proportion of reinforcement in the longitudinal (axial) direction.

8.8 Thermoplastic Consolidation

Consolidation of melt-fusible thermoplastics consists of heating, consolidation, and cooling, as depicted schematically in Fig. 8.43. Heating can be accomplished with infrared heaters, convection ovens, heated platen presses, or autoclaves. Because time for chemical reactions is not required, the time required to reach consolidation temperature is a function of the heating method and the mass of the tooling. The consolidation temperature depends on the specific thermoplastic resin but should be well above the T_g for amorphous resins, or above the melt temperature (T_m) for semicrystalline materials.

Press Consolidation. There are several methods to consolidate thermoplastic composites. Flat sheet stock can be preconsolidated for subsequent forming in a platen press. Two press processes are shown in Fig. 8.44. In the platen press method, precollated ply packs are preheated in an oven and then rapidly shuttled into the pressure-application zone for consolidation. If the material requires time for resin flow for full consolidation or crystallinity control, the press may require heating. If a well-consolidated

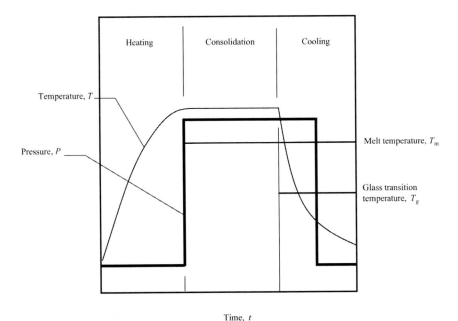

Fig. 8.43 Typical thermoplastic composite process cycle. Source: Ref 8.17

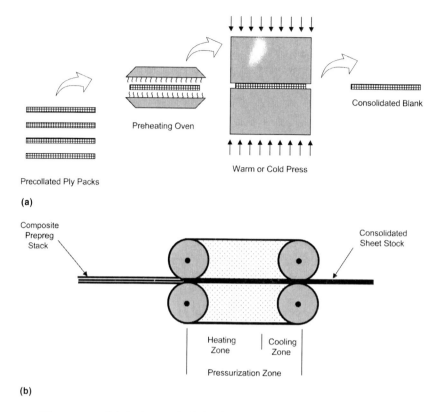

Precollated Ply Packs

Preheating Oven

Consolidated Blank

Warm or Cold Press

(a)

Composite Prepreg Stack

Consolidated Sheet Stock

Heating Zone

Cooling Zone

Pressurization Zone

(b)

Fig. 8.44 Fabrication methods for thermoplastic composite sheet stock. (a) Platen press. (b) Double belt press. Source: Ref 8.1

prepreg is used, then rapid cooling in a cold platen press may suffice. It should be pointed out that this process still requires collation of the ply packs or layers, usually a hand lay-up operation. Because the material contains no tack, soldering irons heated to 427 to 649 °C (800 to 1200 °F) are frequently used to tack the edges to prevent the material from slipping. Handheld ultrasonic guns have also been used for ply tacking. A continuous consolidation process is the double-belt press that contains both pressurized heating and cooling zones. This process is widely used in making glass mat thermoplastic prepreg for the automotive industry, with polypropylene as the resin and glass fiber as the reinforcement.

Autoclave Consolidation. If the part configuration is complex, an autoclave is certainly an option for part consolidation. However, there are several disadvantages to autoclave consolidation. First, it may prove difficult to even find an autoclave that is capable of attaining the 343 to 399 °C (650 to 750 °F) temperatures and 689 to 1378 kPa (100 to 200 psi) pressures required for some advanced thermoplastics. Second, at these temperatures, the tooling is going to be expensive and may be massive, dictating slow heat-up and cool-down rates. Third, because high processing temperatures are required, it is very important that the CTE of the tool match that of the part. Monolithic graphite, cast ceramic, steel, and Invar 42 are normally used for carbon-fiber thermoplastics. Fourth, the bagging materials must be capable of withstanding the high temperatures and pressures.

Autoconsolidation, or in situ placement of melt-fusible thermoplastics, is a series of processes that include hot tape laying, filament winding, and fiber placement. In the autoconsolidation process, only the area that is being immediately consolidated is heated above the melt temperature; the remainder of the part is held at temperatures well below the melt temperature. Two processes are shown in Fig. 8.45: a hot tape-laying process that

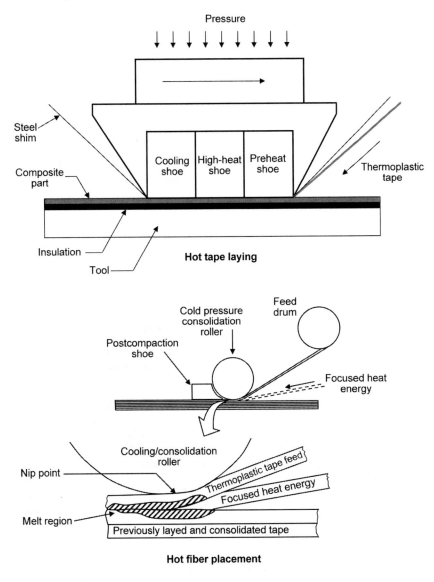

Fig. 8.45 Principle of autoconsolidation. Source: Ref 8.1

relies on conduction heating and cooling from hot shoes, and a fiber placement process that uses a focused laser beam at the nip point for heating. Other forms of heating include hot gas torches, quartz lamps, and infrared heaters. The mere fact that autoconsolidation is possible illustrates that the contact times for many thermoplastic polymers at normal processing temperatures can be quite short. Provided full contact pressure is made at the ply interfaces, autoconsolidation can occur in less than 0.5 of a second.

8.9 Thermoforming

One of the main advantages of thermoplastic composites is their ability to be rapidly processed into structural shapes by thermoforming. The term *thermoforming* encompasses quite a broad range of manufacturing methods. However, thermoforming is essentially a process that uses heat and pressure to form a flat sheet or ply stack into a structural shape. A typical thermoforming setup is shown in Fig. 8.46. The steps for thermoforming would be to first collate the ply pack. The collated ply pack could then be preconsolidated into a flat laminate, or the thermoforming operation could be conducted using the unconsolidated ply pack. There is some disagreement as to whether it is better to use preconsolidated blanks or loose unconsolidated ply packs for thermoforming. Preconsolidated blanks

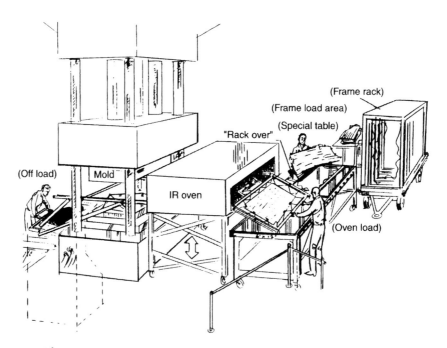

Fig. 8.46 Typical thermoforming setup. IR, infrared. Source: Ref 8.18

offer the advantage of being well consolidated with no voids or porosity, but they do not slip as well during the forming operation as loose unconsolidated ply packs. The laminate or unconsolidated ply pack would then be heated for forming, such as in the infrared oven shown in the figure. After reaching the forming temperature, it would be rapidly transferred to the thermoforming press and formed. The part must be held under pressure until it cools below its T_g to avoid inducing residual stresses and part warpage.

8.10 Sandwich and Cocured Structure

Sandwich and cocured structures provide the opportunity to reduce weight and assembly costs, because much of the weight and cost of mechanical fasteners is reduced or eliminated. The term *sandwich structure* normally applies to a structure that is adhesively bonded together with skins on the outside, with some type of lightweight core material on the inside. Sandwich structure can be fabricated by first curing the separate composite details and then adhesively bonding them to form a completed assembly, or it may be cocured in which the skins are cured at the same time they are bonded to the interior sandwich. Another option is cocured unitized structure in which all of the details are cured together at the same time without an interior sandwich to produce a one-piece structure.

Sandwich construction is used extensively in both the aerospace and commercial industries, because it is an extremely lightweight structural approach that exhibits high stiffness and strength-to-weight ratios. The basic concept of a sandwich panel is that the facings carry the bending loads (tension and compression), while the interior sandwich or core carries the shear loads. Sandwich construction, especially honeycomb core construction, is extremely structurally efficient, particularly in stiffness-critical applications. Doubling the thickness of the core increases the stiffness over 7 times with only a 3% weight gain, while quadrupling the core thickness increases stiffness over 37 times with only a 6% weight gain. Little wonder that structural designers like to use sandwich construction whenever possible. Sandwich panels are typically used for their structural, electrical, insulation, and/or energy-absorption characteristics.

Facesheet materials that are normally used include aluminum, glass, carbon, or aramid. Typical sandwich structure has relatively thin facing sheets (0.3 to 3 mm, or 0.010 to 0.125 in.), with core densities in the range of 16 to 480 kg/m^3 (1 to 30 lb/ft^3). Core materials include metallic and nonmetallic honeycomb core, balsa wood, open- and closed-cell foams, and syntactics. In general, the honeycomb cores are more expensive than the foam cores but offer superior performance. This explains why many commercial applications use foam cores while aerospace applications use the higher-performance but more expensive honeycombs. It should also be noted that the foam materials are normally much easier to work with than the honeycombs.

The details of a typical honeycomb core panel are shown in Fig. 8.47. Typical facesheets include aluminum, glass, aramid, and carbon. Structural film adhesives are normally used to bond the facesheets to the core. It is important that the adhesive provide a good fillet at the core-to-skin interface. The honeycomb itself can be manufactured from aluminum, glass fabric, aramid paper, aramid fabric, or carbon fabric. Honeycomb manufactured for use with organic-matrix composites is bonded together with an adhesive, called the node bond adhesive.

Cocured Honeycomb Assemblies. Honeycomb assemblies can also be made by cocuring the composite plies onto the core. In this process, the composite skin plies are consolidated and cured at the same time they are bonded to the core. Although a film adhesive is normally used at the skin-to-core interface, self-adhesive prepreg systems are available that do not require a film adhesive. To prevent core crushing and migration, this process is normally conducted at approximately 276 to 345 kPa (40 to 50 psi), as opposed to the normal 689 kPa (100 psi) used for regular laminate processing.

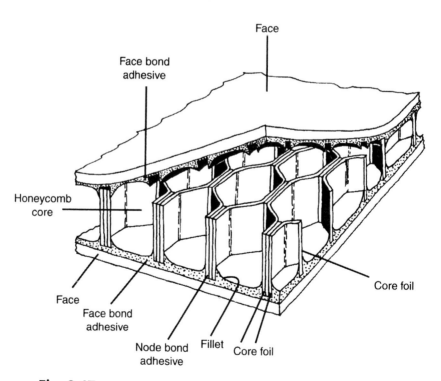

Fig. 8.47 Honeycomb panel construction. Source: Ref 8.19

There are three basic approaches to producing honeycomb cocured structure. In the first method, both skins are cocured at the same time they are bonded to the core. Because this process is limited to approximately 345 kPa (50 psi) autoclave pressure to prevent core crush, the skins will contain more porosity than skins cured at full autoclave pressure. This process has the shortest cycle time but is the riskiest. The second method consists of precuring one of the skins at full autoclave pressure and then bonding it to the core at the same time the other skin is cocured. This method takes longer, but at least one of the skins should have minimal porosity. In the third method, one skin is precured in an autoclave and then adhesively bonded to the core. Then, in a separate operation, the other skin is cocured and bonded to the core. The main advantage of this third method is that the core is stabilized during the initial bond cycle in which the first precured skin is bonded to the core. This method has been used when core edge migration and crushing are problems. Although this method takes longer, it is the lowest-risk approach.

Foam Cores. A second type of core material frequently used in an adhesively bonded structure is foam core. While the properties of foam cores are not as good as honeycomb core, they are used extensively in commercial applications such as boat building and light aircraft construction. The term *polymer foam* or *cellular polymer* refers to a class of materials that are two-phase gas-solid systems in which the polymer is continuous and the gaseous cells are dispersed through the solid. Polymeric foams can be produced by several methods, including extrusion, compression molding, injection molding, reaction injection molding, and other solid-state methods. Foam cores are made by using a blowing or foaming agent that expands during manufacture to give a porous, cellular structure. The cells may be open and interconnected or closed and discrete. Usually, the higher the density, the greater the percentage of closed cells. Almost all foams used for structural applications are classified as closed cell, meaning almost all of their cells are discrete. Open-cell foams, while good for sound absorption, are weaker than the higher-density closed-cell foams, and they also absorb more water, although water absorption in both open- and closed-cell foams can be problematic. Both uncross-linked thermoplastic and cross-linked thermoset polymers can be foamed, with the thermoplastic foams exhibiting better formability and the thermoset foams better mechanical properties and higher temperature resistance. Almost any polymer can be made into a foam material by adding an appropriate blowing or foaming agent.

The blowing agents used to manufacture foams are usually classified as either physical or chemical blowing agents. Physical blowing agents are usually gases, mixed into the resin, that expand as the temperature is increased, while chemical blowing agents are often powders that decompose on heating to give off gases, usually nitrogen or carbon dioxide.

Although there are foams that can be purchased as two-part liquids that expand after mixing for foam-in-place applications, the majority of structural foams are purchased as pre-expanded blocks that can be bonded together to form larger sections. Sections may be bonded together using either paste or adhesive films. Sections can also be heat formed to contour using procedures similar to those for nonmetallic honeycomb core. Although the uncross-linked thermoplastic foams are easier to thermoform, many of the thermoset foams are only lightly cross linked and exhibit some formability. Core densities normally range from approximately 32 to 641 kg/m³ (2 to 40 lb/ft³). The most widely used structural foams are summarized in Table 8.17. It is important to thoroughly understand the chemical, physical, and mechanical properties of any foam considered for a structural application, particularly with respect to solvent and moisture resistance and long-term durability. Depending on their chemistry, foam core materials can be used in the temperature range of 66 to 204 °C (150 to 400 °F).

Syntactic core consists of a matrix (e.g., epoxy) that is filled with hollow spheres (e.g., glass or ceramic microballoons), as shown in Fig. 8.48. Syntactics can be supplied as pastes for filling honeycomb core or as B-staged formable sheets for core applications. Syntactic cores are generally much higher density than honeycomb, with densities in the range of 480 to 1280 kg/m³ (30 to 80 lb/ft³). The higher the percentage of the microballoon filler, the lighter but weaker the core becomes. Syntactic core

Table 8.17 Characteristics of select foam sandwich materials

Name and type of core	Density, lb/ft³	Maximum temperature, °F	Characteristics
Polystyrene (Styrofoam)	1.6–3.5	165	Low-density, low-cost, closed-cell foam capable of being thermoformed. Used for wet or low-temperature lay-ups. Susceptible to attack by solvents
Polyurethane foam	3–29	250–350	Low- to high-density closed-cell foam capable of thermoforming at 425–450 °F. Both thermoplastic and thermoset foams are available. Used for cocured and secondarily bonded sandwich panels with both flat and complex curved geometries
Polyvinyl chloride foam (Klegecell and Dinvinycell)	1.8–26	150–275	Low- to high-density foam. Low density can contain some open cells. High density is closed cell. Can be either thermoplastic (better formability) or thermoset (better properties and heat resistance). Used for secondarily bonded or cocured sandwich panels with both flat and complex curved geometries
Polymethacrylimide foam (Rohacell)	2–18.7	250–400	Expensive high-performance closed-cell foam that can be thermoformed. High-temperature grades (WF) can be autoclaved at 350 °F/100 psi. Used for secondarily bonded or cocured high-performance aerospace structures

Source: Ref 8.1

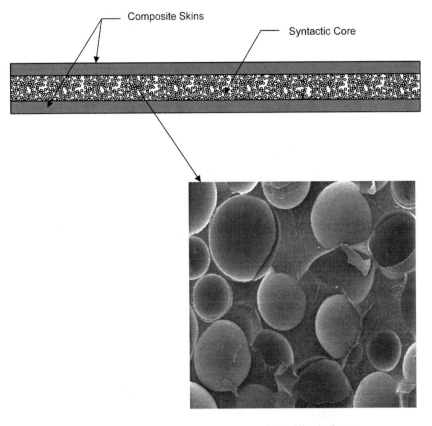

Glass Microballoons

Fig. 8.48 Syntactic core construction. Source: Ref 8.1

sandwiches are used primarily for thin, secondary composite structures where it would be impractical or too costly to machine honeycomb to thin gages. When cured against precured composite details, syntactics do not require an adhesive. However, if the syntactic core is already cured and requires adhesive bonding, it should be scuff sanded and then cured with a layer of adhesive.

8.11 Integrally Cocured Unitized Structure

Cocuring is a process in which uncured composite plies are cured and bonded simultaneously during the same cure cycle to either core materials or to other composite parts. Integrally cocured or unitized structure is another manufacturing approach that can greatly reduce the part count and final assembly costs for composite structures. An example of a cocured control surface is shown in Fig. 8.49. In this structure, the ribs were cocured to the lower skin. The upper skin was cured at the same time, but a release film was used so it could be removed after cure to allow the installation of a metallic hinge fitting.

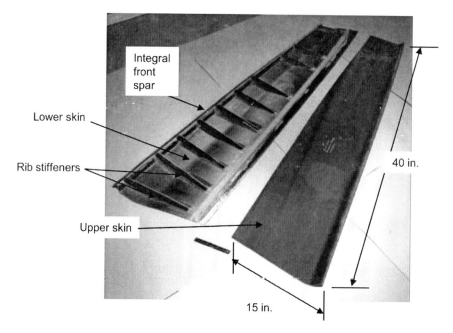

Fig. 8.49 Cocured unitized control surface. Courtesy of The Boeing Company. Source: Ref 8.1

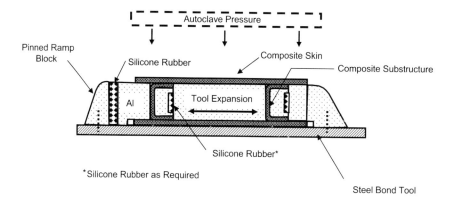

Fig. 8.50 Pressure application for unitized cocured structure. Source: Ref 8.1

During cure, pressure is provided both by the autoclave and the expansion of aluminum substructure blocks, as shown in Fig. 8.50. The autoclave applies pressure to the skins and rib caps, while the expansion of the aluminum substructure blocks applies pressure to the rib webs. If required, the expansion of the aluminum substructure blocks can be supplemented by the presence of silicone rubber intensifiers.

The advantages of this type of structure are obvious: fewer detail parts, fewer fasteners, and fewer problems with part fit-up on final assembly. The main disadvantages are the cost and accuracy of the tooling required, and the complexity of the lay-up that requires a highly skilled workforce.

8.12 Machining

Composites are more prone to damage during trimming and machining than conventional metals. Composites contain strong and very abrasive fibers held together by a relatively weak and brittle matrix. During machining, they are prone to delaminations, cracking, fiber pullout, fiber fuzzing (aramid fibers), matrix chipping, and heat damage. It is important to minimize forces and heat generation during machining. During metallic machining, the chips help to remove much of the heat generated during the cutting operation. Due to the much lower thermal conductivity of the fibers (especially glass and aramid), heat buildup can occur rapidly and degrade the matrix, resulting in matrix cracking and even delaminations. When machining composites, generally high speeds, low feed rates, and small depths of cuts are used to minimize damage.

Conventional machining methods such as milling are not normally performed on composite parts because they cut through the continuous fibers and reduce the strength. However, most composite parts require peripheral edge trimming after cure. Edge trimming is usually done either manually with high-speed cut-off saws or automatically with numerically controlled abrasive waterjet machines. Lasers have often been proposed for trimming of cured composites, but the surfaces become charred due to the intense heat and are unacceptable for most structural applications.

8.13 Mechanical Fastening

Hole drilling of composites is more difficult than in metals, again due to their relatively low sensitivity to heat damage and their weakness in the through-thickness direction. Composites are very susceptible to surface splintering, particularly if unidirectional material is present on the surface. Note that splintering can occur at both the drill entrance and exit side of the hole. As shown in Fig. 8.51, when the drill enters the top surface, it

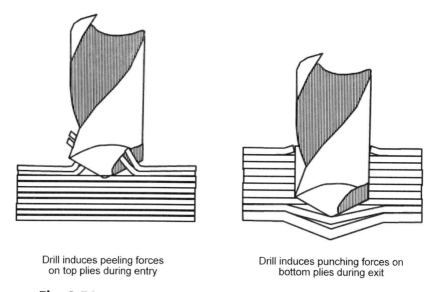

Drill induces peeling forces
on top plies during entry

Drill induces punching forces on
bottom plies during exit

Fig. 8.51 Drilling forces on composite laminate. Source: Ref 8.20

creates peeling forces on the matrix as it grabs the top plies. When it exits the hole, it induces punching forces that again create peel forces on the bottom surface plies. If top surface splintering is encountered, it is usually a sign that the feed rate is too fast, while exit surface splintering indicates that the feed force is too high. It is common practice to cure a layer of fabric on both surfaces of composite parts, which will largely eliminate the hole-splintering problem; that is, woven cloth is much less susceptible to splintering than unidirectional material.

There are many types of fasteners used in aerospace structural assembly, the most prevalent being solid rivets, pins with collars, bolts with nuts, and blind fasteners, with examples shown in Fig. 8.52. Rivets are rarely used in composites for two reasons: aluminum rivets will galvanically corrode when in contact with carbon fibers, and the vibration and expansion of the rivet during the driving process can cause delaminations. If rivets are used, they are usually a bimetallic rivet consisting of a Ti-6Al-4V pin with a softer titanium-niobium tail that are installed by squeezing rather than vibration driving.

Mechanical fastener material selection for composites is important in preventing potential corrosion problems. Aluminum- and cadmium-coated steel fasteners will galvanically corrode when in contact with carbon fibers. Titanium (Ti-6Al-4V) is usually the best fastener material for carbon-fiber composites, based on its high strength-to-weight ratio and corrosion resistance. When higher strength is required, cold-worked A286 iron-nickel or the iron-nickel-base alloy Inconel 718 can be used. If extremely high

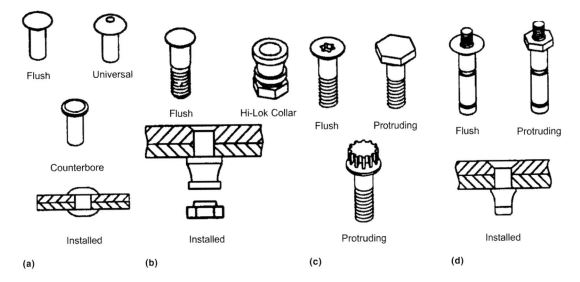

Fig. 8.52 Typical aerospace structural fasteners, (a) rivets, (b) pin and collar, (c) bolts and nuts, (d) blind fasteners. Source: Ref 8.1

strengths are required for very highly loaded joints, the Ni-Co-Cr multiphase alloys MP35N and MP159 are available. It should be noted that glass and aramid fibers, being nonconductive, do not cause galvanic corrosion with metallic fasteners.

8.14 Thermoplastic Joining

Another unique advantage of thermoplastic composites is the rather extensive joining options available. While thermosets are restricted to either cocuring, adhesive bonding, or mechanical fastening, thermoplastic composites can be joined by melt fusion, dual-resin bonding, resistance welding, ultrasonic welding, or induction welding, as well as by conventional adhesive bonding and mechanical fastening.

Adhesive Bonding. In general, structural bonds using thermoset (e.g., epoxy) adhesives produce lower bond strengths with thermoplastic composites than with thermoset composites. This is believed to be due primarily to the differences in surface chemistry between thermosets and thermoplastics. Thermoplastics contain rather inert, nonpolar surfaces that impede the ability of the adhesive to wet the surface. A number of different surface preparations have been evaluated, including sodium hydroxide etching, grit blasting, acid etching, plasma treatments, silane coupling agents, corona discharge, and Kevlar (aramid) peel plies. While a number of these surface preparations, or combinations of them, give acceptable bond strengths, the long-term service durability of thermoplastic adhesively bonded joints has not yet been established.

Mechanical Fastening. Thermoplastic composites can be mechanically fastened in the same manner as thermoset composites. Initially, there was concern that thermoplastics would creep excessively, resulting in a loss of fastener preload and thus lower joint strengths. Extensive testing has shown that this was an unfounded fear, and mechanically fastened thermoplastic composite joints behave very similar to thermoset composite joints.

Melt Fusion. Because thermoplastics can be processed multiple times by heating above their T_g for amorphous or T_m for semicrystalline resins with minimal degradation, melt fusion essentially produces joints as strong as the parent resin. An extra layer of neat (unreinforced) resin film can be placed in the bondline for gap-filling purposes and to ensure that there is adequate resin to facilitate a good bond. However, if the joint is produced in a local area, adequate pressure must be provided over the heat-affected zone to prevent the elasticity of the fiber bed from producing delaminations at the ply interfaces.

Dual-Resin Bonding. In this method, a lower-melting-temperature thermoplastic film is placed at the interfaces of the joint to be bonded. As shown in Fig. 8.53, in a process called amorphous bonding or the Thermabond process, a layer of amorphous polyetherimide (PEI) is used to bond two polyetheretherketone (PEEK) composite laminates together. To provide the best bond strengths, a layer of PEI is fused to both PEEK laminate surfaces prior to bonding to enhance resin mixing. In addition, an extra layer of film may be used at the interfaces for gap-filling purposes. Because the processing temperature for PEI is below the melt temperature of the PEEK laminates, the danger of ply delamination within the PEEK substrates is avoided. Like the melt fusion process, dual-resin bonding would normally be used to join large sections together, such as bonding stringers to skins.

Resistance Welding. In resistance welding, a metallic heating element is embedded in a thermoplastic film and placed in the bondline (Fig. 8.54). This process is used to weld the ribs to the skin for the Airbus A380 wing leading edges. For all fusion welding operations with thermoplastic composites, it is necessary to maintain adequate pressure at all locations that are heated above the melt temperature. If pressure is not maintained at all locations that exceed the melt temperature, deconsolidation due to fiber bed relaxation will likely occur. The pressure should be maintained until the part is cooled below its T_g. Typical processing times for resistance welding are 30 s to 5 min at 689 to 1379 kPa (100 to 200 psi) pressure.

Ultrasonic welding is used extensively in commercial processes to join lower-temperature unreinforced thermoplastics and can also be

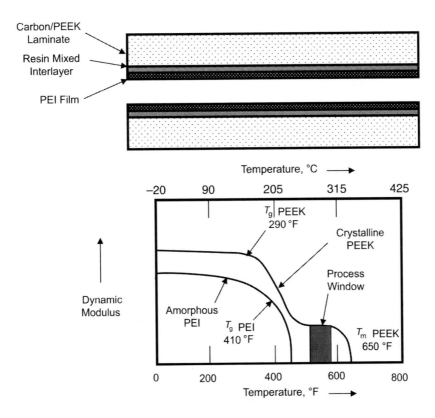

Fig. 8.53 Amorphous or dual-resin bonding. PEEK, polyetheretherketone; PEI, polyetherimide; T_g, glass transition temperature; T_m, melt temperature. Source: Ref 8.1

used for advanced thermoplastic composites. An ultrasonic horn, also known as a sonotrode, is used to produce ultrasonic energy at the composite interfaces to convert electrical energy into mechanical energy. The sonotrode is placed in contact with one of the pieces to be joined. The second piece is held stationary while the vibrating piece creates frictional heating at the interface. Ultrasonic frequencies of 20 to 40 kHz are normally used. The process works best if one of the surfaces has small asperities (protrusions) that act as energy directors or intensifiers. The asperities have a high energy per unit volume and melt before the surrounding material. The quality of the bond is increased with increasing time, pressure, and amplitude of the signal. Again, it is common practice to incorporate a thin layer of neat resin film to provide gap filling. Typical weld parameters are less than 10 s at 483 to 1379

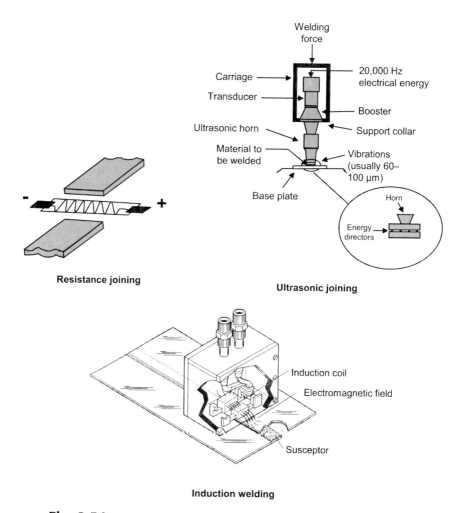

Fig. 8.54 Thermoplastic welding methods. Source: Ref 8.18

kPa (70 to 200 psi) pressure. This process is somewhat similar to spot welding of metals.

Induction welding techniques have been developed in which a metallic susceptor may, or may not, be placed in the bondline. However, it is generally accepted that the use of a metallic susceptor produces superior joint strengths. A typical induction setup uses an induction coil to generate an electromagnetic field that results in heating due to eddy currents in the conductive susceptor and/or hysteresis losses in the susceptor. Susceptor materials evaluated include iron, nickel, carbon fibers, and copper meshes. Typical welding parameters are 5 to 30 min at 345 to 1379 kPa (50 to 200 psi) pressure.

ACKNOWLEDGMENT

Portions of the text for this chapter came from *Structural Composite Materials* by F.C. Campbell, ASM International, 2010.

REFERENCES

8.1 F.C. Campbell, *Structural Composite Materials,* ASM International, 2010

8.2 Introduction to Composites, *Composites,* Vol 21, *ASM Handbook,* ASM International, 2001

8.3 M.W. Wardle, Aramid Fiber Reinforced Plastics—Properties, *Comprehensive Composite Materials,* Vol 2, *Polymer Matrix Composites,* Elsevier Science Ltd., 2000

8.4 "Cycom 977-3 Toughened Epoxy Resin," Cytec Engineered Materials Data Sheet, 1995

8.5 "Cycom 5215 Modified Epoxy Resin," Cytec Engineered Materials Data Sheet

8.6 J.-M. Bai and D. Leach, "High Performance Thermoplastic Polymers and Composites," paper presented at a SAMPE conference

8.7 "CETEX Thermo-Lite Thermoplastic Composites for Automated Fiber Placement & Rapid Lamination Processes," TenCate Advanced Composites Data Sheet, 2008

8.8 Molding Compounds, *Composites,* Vol 21, *ASM Handbook,* ASM International, 2001, p 141–149

8.9 C. Smith and M. Gray, "ICI Fiberite Impregnated Materials and Processes—An Overview," unpublished white paper

8.10 J. Bottle, F. Burzesi, and L. Fiorini, Design Guidelines, *Composites,* Vol 21, *ASM Handbook,* ASM International, 2001

8.11 M.P. Groover, *Fundamentals of Modern Manufacturing—Materials, Processes, and Systems,* Prentice-Hall, Inc., 1996

8.12 J.C. Reindl, Commercial and Automotive Applications, *Composites,* Vol 1, *Engineered Materials Handbook,* ASM International, 1987

8.13 F.R. Andressen, Open-Molding: Hand Lay-Up and Spray-Up, *Composites,* Vol 21, *ASM Handbook,* ASM International, 2001

8.14 B.N. Cox and G. Flanagan, "Handbook of Analytical Methods for Textile Composites," NASA Contractor Report 4750, March 1997

8.15 J.E. Sumerak and J.D. Martin, Pultrusion, *Composites,* Vol 21, *ASM Handbook,* ASM International, 2001

8.16 S.T. Peters and J.L. McLarty, Filament Winding, *Composites,* Vol 21, *ASM Handbook*, ASM International, 2001.

8.17 J. Muzzy, L. Norpoth, and B. Varughese, Characterization of Thermoplastic Composites for Processing, *SAMPE J.,* Vol 25 (No. 1), Jan/Feb 1989, p 23–29

8.18 D.A. McCarville and H.A. Schaefer, Processing and Joining of Thermoplastic Composites, *Composites,* Vol 21, *ASM Handbook,* ASM International, 2001, p 633–645

8.19 F. C. Campbell, Secondary Adhesive Bonding of Polymer-Matrix Composites, *Composites,* Vol 21, *ASM Handbook,* ASM International, 2001

8.20 B.T. Astrom, *Manufacturing of Polymer Composites,* Chapman & Hall, 1997

SELECTED REFERENCES

- S.W. Beckwith and C.R. Hyland, Resin Transfer Molding: A Decade of Technology Advances, *SAMPE J.,* Vol 34 (No. 6), Nov/Dec 1998, p 7–19
- J. Bottle, F. Burzesi, and L. Fiorini, Design Guidelines, *Composites,* Vol 21, *ASM Handbook,* ASM International, 2001
- F.C. Campbell, A.R. Mallow, and C.E. Browning, Porosity in Carbon Fiber Composites: An Overview of Causes, *J. Adv. Mater.,* Vol 26 (No. 4), July 1995, p 18–33
- D.O. Evans, Fiber Placement, *Composites,* Vol 21, *ASM Handbook,* ASM International, 2001, p 477–479
- R.E. Fields, Overview of Testing and Certification, *Composites,* Vol 21, *ASM Handbook,* ASM International, 2001
- J.M. Griffith, F.C. Campbell, and A.R. Mallow, "Effect of Tool Design on Autoclave Heat-Up Rates," Composites in Manufacturing 7 Conference and Exposition, Society of Manufacturing Engineers, 1987
- M.N. Grimshaw, Automated Tape Laying, *Composites,* Vol 21, *ASM Handbook,* ASM International, 2001, p 480–485
- R.C. Harper, Thermoforming of Thermoplastic Matrix Composites— Part I, *SAMPE J.,* Vol 28 (No. 2), March/April 1992, p 9–17
- R.C. Harper, Thermoforming of Thermoplastic Matrix Composites— Part II, *SAMPE J.,* Vol 28 (No. 3), May/June 1992, p 9–17
- S.C. Mantel and D. Cohen, Filament Winding, *Processing of Composites,* Hanser, 2000
- J.D. Muzzy and J.S. Colton, The Processing Science of Thermoplastic Composites, *Advanced Composites Manufacturing,* John Wiley & Sons, Inc., 1997
- M.J. Paleen and J.J. Kilwin, Hole Drilling in Polymer-Matrix Composites, *Composites,* Vol 21, *ASM Handbook,* ASM International, 2001
- R.T. Parker, Mechanical Fastener Selection, *Composites,* Vol 21, *ASM Handbook,* ASM International, 2001
- S.T. Peters, W.D. Humphrey, and R.F. Foral, *Filament Winding Composite Structure Fabrication,* 2nd ed., SAMPE, 1995

- C.C. Poe, H.B. Dexter, and I.S. Raju, Review of NASA Textile Composites Research, *J. Aircr.,* Vol 36 (No. 5), 1999, p 876–884
- *Polymer Matrix Composites,* Vol 3, *Materials Usage, Design, and Analysis,* MIL-HNBK-17-3F, U.S. Department of Defense, 2001
- L. Repecka and J. Boyd, "Vacuum-Bag-Only-Curable Prepregs That Produce Void-Free Parts," SAMPE 2002, May 12–16, 2002
- B. Thorfinnson and T.F. Bierrinann, "Production of Void Free Composite Parts without Debulking," 31st International SAMPE Symposium and Exposition, April 1986
- B. Thorfinnson and T.F. Bierrnann, "Measurement and Control of Prepreg Impregnation for Elimination of Porosity in Composite Parts," Society of Manufacturing Engineers, Fabricating Composites '88, Sept 1988

CHAPTER **9**

Metal-Matrix Composites

METAL-MATRIX COMPOSITES (MMCS) are capable of providing higher-temperature operating limits than their base metal counterparts, and they can be tailored to give improved strength, stiffness, thermal conductivity, abrasion resistance, creep resistance, or dimensional stability. Unlike polymer-matrix composites, they are nonflammable, do not outgas in a vacuum, and suffer minimal attack by organic fluids such as fuels and solvents. Unfortunately, MMCs are expensive, especially continuous fiber MMCs, and their applications are therefore rather limited.

In an MMC, the matrix phase is a monolithic alloy, usually a low-density nonferrous alloy, and the reinforcement consists of high-performance carbon, metallic, or ceramic additions. Reinforcements, either continuous or discontinuous, may constitute from 10 to 70 vol% of the composite. Continuous fiber or filament (f) reinforcements include graphite, silicon carbide (SiC), boron, aluminum oxide (Al_2O_3), and refractory metals. Discontinuous reinforcements consist mainly of SiC in whisker (w) form, particulate (p) types of SiC, Al_2O_3, and titanium diboride (TiB_2), and short or chopped fibers (c) of Al_2O_3 or graphite.

A relative comparison of composite performance with materials and process technologies for a number of MMCs is shown in Fig. 9.1. Continuous fiber MMCs offer the highest performance but also the highest cost, while discontinuous MMCs, and in particular those made by casting processes, have lower properties but also much lower costs.

9.1 Applications of Metal-Matrix Composites

Metal-matrix composites are primarily used in the ground transportation and aerospace industries. The current applications are primarily discontinuously reinforced aluminum (DRA)-matrix composites.

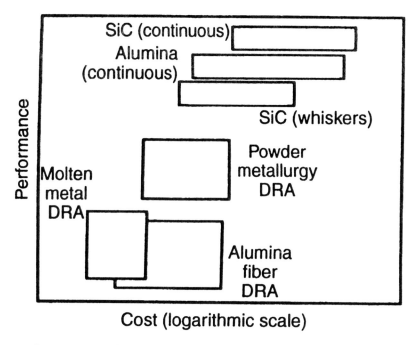

Fig. 9.1 Material cost versus performance of various aluminum-matrix composites. DRA, discontinuously reinforced aluminum. Source: Ref 9.1

Ground Transportation/Automotive. The automotive and rail industries account for the largest application of MMCs by volume. By far the single most prevalent composite used in this sector is DRA. A selectively reinforced piston head was introduced by Toyota Motor Manufacturing in 1983. Produced by squeeze casting, this was a reasonably low-cost and high-rate-production process. The reinforcements provide improved wear resistance and lower thermal conductivity, so that more of the heat generated by the combustion gases is available for producing work. Further, the MMC provides a lower coefficient of thermal expansion (CTE) than unreinforced aluminum, so that tighter tolerances and hence higher pressure and better performance are obtained. Squeeze-cast piston liners have been used in the Honda Prelude since 1990, displacing cast iron inserts (Fig. 9.2). In the manufacturing process, the engine block casting and piston liner preform infiltration are performed simultaneously, eliminating the cost of assembly associated with the cast iron inserts. More importantly, the MMC liners provide improved wear resistance, reducing overall liner thickness. This yields an increase in engine displacement, so that more horsepower is obtained from the same overall power plant weight and volume. Finally, the thermal conductivity of the MMC is much higher than the cast iron liner, decreasing operating temperature and resulting in extended engine life.

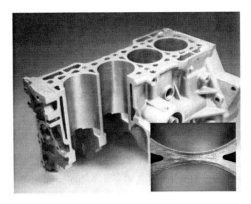

Fig. 9.2 Cutaway section of the Honda Prelude 2000 cm^3 cast aluminum engine block with integral metal-matrix composite (MMC) piston liners. A cross section of the MMC liners is shown in the inset. These piston liners have been in production since 1990. Source: Ref 9.2

Automotive driveshafts represent a component application motivated primarily by structural properties. Driveshaft design is limited by rotational instability, which is controlled by the specific stiffness of the driveshaft material. The higher specific stiffness of DRA relative to steel or aluminum allows a longer driveshaft of a given diameter. This is important in trucks and large passenger cars, where two-piece metal driveshafts are often used. Replacement with DRA allows a single-piece driveshaft. Not only is significant weight saved as a result of material substitution, but elimination of the central support for the two-piece unit provides additional weight savings. Finally, MMC driveshafts require less counterweight mass compared to steel. In all, as much as 9 kg (20 lb) have been saved by this application. Metal-matrix composite driveshafts were first introduced in the Chevrolet S-10 and GMC Sonoma in 1996 and have been used in the Chevrolet Corvette since 1997. In 1999, Ford Motor Company introduced MMC driveshafts in the "Police Interceptor" version of the Crown Victoria. Other significant MMC components include automotive and rail brake components, DRA snow tire studs, and discontinuously reinforced titanium automotive intake and exhaust valves in the Toyota Altezza.

Thermal Management and Electronic Packaging. Materials for thermal management and electronic packaging require high thermal conductivity to dissipate large, localized heat loads, and a controlled CTE to minimize thermal stresses with semiconductor and ceramic baseplate materials. Aluminum MMCs, such as DRA, provide better performance (higher thermal conductivity) relative to previous preferred materials. Unlike the automotive market, MMCs for the electronic packaging sector are high-value-added. Although this is the second-largest MMC market in terms of volume, it is by far the largest in terms of value. Current applications of MMCs include radio frequency packaging for microwave transmitters in commercial low-Earth orbit communications satellites (Fig. 9.3) and for power semiconductors in geosynchronous Earth orbit. Metal-matrix composites are also used as power semiconductor baseplates for electric motor controllers and for power conversion in cell phone ground station transmitting towers. Finally, MMCs are being more widely used as thermal management materials for commercial flip-chip packaging of computer chips.

Aerospace. Several DRA applications emerged in the early 1990s as a result of defense investment in the United States. The ventral fin on the F-16 aircraft was experiencing a high incidence of failure as a result of unanticipated turbulence. Of the several materials and design options considered, DRA sheet was chosen as the best overall alternative. The higher specific strength and stiffness, good supportability, and affordability were considerations in the final selection of DRA. As a result of the successful experience with DRA in this application, the F-16 project office selected DRA to solve a cracking problem of the fuselage at the corners of fuel access doors. Again, the mechanical properties of DRA and retention of the same form, fit, and function led to the selection of DRA as the final solution. Discontinuously reinforced aluminum was also qualified and entered service in a commercial gas turbine application (fan exit guide vanes) as a result of this program. The DRA double-hollow extrusions replaced solid carbon/epoxy to resolve an issue with poor erosion and ballistic impact response. Discontinuously reinforced aluminum also resulted in a cost-savings to the manufacturer of well over $100 million.

Continuously reinforced titanium-matrix composites (TMCs) are included in the bill of material for nozzle actuator piston rods for the Pratt and Whitney F119 engine in the F-22 aircraft. Specific strength and specific stiffness, along with good fatigue response at a maximum operating temperature of 450 °C (850 °F), are the requirements. The hollow TMC rod replaced a solid rod of precipitation-hardened stainless

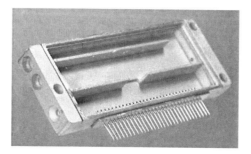

Fig. 9.3 A SiC/Al radio frequency microwave package used in commercial low-Earth orbit communications satellites. Courtesy of General Electric Company. Source: Ref 9.3

steel and has resulted in a direct weight savings of 3.4 kg (7.5 lb) per aircraft. This was the first aerospace application of TMC materials. Following the successful specification of TMCs for this part, TMCs are now specified for nozzle actuator links in the General Electric F110 engine for the F-16 aircraft.

Industrial, Recreational, and Infrastructure. Metal-matrix composites are used for a range of applications in these sectors, including DRA for bicycle frames and iron-base MMCs reinforced with TiC for wear-resistant tool and die coatings and industrial rollers. A continuously reinforced aluminum MMC produced by 3M is currently used in high-performance automotive applications and has completed certification testing for overhead power transmission conductors.

Steel cable is typically used as the core for high-tension power conductors. The steel bears the weight of the aluminum conductor but carries little of the current, due to a low electrical conductivity (only one-eighth that of aluminum). Depending on the specifics of the transmission installation, the conductor can heat to temperatures in excess of 200 °C (400 °F) during peak use. Sagging due to thermal expansion is an issue. The important properties for conductor cores are specific strength, electrical conductivity, CTE, high-temperature capabilities, and cost. Increased demand for electricity and the effect of deregulation requires utility companies to consider means for increasing the ampacity (i.e., the maximum current flow in the line). Higher ampacity produces higher conductor temperatures, resulting in line sagging that requires significant tower modifications to maintain needed line clearance. Tower construction is the major cost associated with new or increased power transmission and includes considerations such as purchasing right of way, satisfying environmental impacts, and design and construction costs. The ability to avoid the costs associated with tower construction provides the opportunity for new conductor materials. The aluminum MMC conductor provides the strength of steel cable at less than half the density. More importantly, the MMC conductor carries four times more current than steel and has a CTE that is one-half that of steel. Although the aluminum MMC conductor is more expensive than conventional conductors on a unit-length basis, the ampacity gains are significant, with projected increases of 200 to 300% with no tower modifications required. Thus, at a system level, the use of MMC conductors may provide an attractive cost-savings. A cross section of a conductor with an aluminum MMC core is shown in Fig. 9.4.

9.2 Discontinuously Reinforced Metal-Matrix Composites

Discontinuously reinforced MMCs offer a range of attractive material properties, both mechanical and physical, that cannot be achieved using conventional engineering alloys. These enhanced material properties are

Fig. 9.4 Cross section of an electrical conductor for power transmission. The core consists of 19 individual wires made from a continuously reinforced aluminum metal-matrix composite (MMC) produced by 3M. The MMC core supports the load for the 54 aluminum wires and also carries a significant current, unlike competing steel cores. Courtesy of 3M. Source: Ref 9.3

the direct result of the interaction between the metallic matrix and the reinforcement.

In a discontinuous MMC materials system, the reinforcement strengthens the metal matrix by load transfer to the ceramic reinforcement, and by increasing the dislocation density of the metal matrix. Discontinuously reinforced metal-matrix composite properties can be tailored to meet specific engineering requirements by selecting a particular reinforcement and by varying the amount added to the metal matrix. Thus, the physical and mechanical properties of the composite materials system can be controlled with some independence. Increasing the reinforcement volume in a discontinuous MMC system increases mechanical properties, such as elastic modulus, ultimate strength, and yield strength, while reducing the thermal expansion and, in some cases, the density of the composite system. Unfortunately, ductility and fracture toughness typically decrease with increasing reinforcement volume.

The increase in both the elastic modulus and strength (ultimate and yield) is believed to be due to the difference in thermal expansion between the ceramic reinforcement particles and the metallic matrix during processing. During production, both the reinforcement and matrix are heated to the processing temperature, brought to thermomechanical equilibrium, and then allowed to cool. The thermal contraction of the metallic matrix during cooldown is typically much greater than that of the reinforcement, which leads to a geometric mismatch. At the ceramic-metal interface, this geometrical disparity creates mismatch strains that are relieved by the generation of dislocations in the matrix originating from sharp features on the ceramic reinforcement.

Discontinuous MMC systems are commonly used in applications that require high specific materials properties, enhanced fatigue resistance, improved wear resistance, controlled expansion, or the ability to absorb neutron radiation (boron carbide). Additionally, discontinuous MMCs can be designed to yield a materials system that offers multiple roles. Some examples of multiple roles are systems that offer high strength and fatigue resistance for aerospace and mechanical applications, thermal management coupled with expansion control for spaceborne applications, moderate strength and neutron-absorption capabilities for nuclear applications, high strength and wear resistance for heavy equipment applications, and impact/energy dissipation for armor applications. The correct selection of reinforcement is very important in yielding the desired resultant materials properties. An improper reinforcement selection may lead to less-than-desirable composite materials properties, difficulty in fabrication, and high cost.

The most common discontinuous MMCs used for current aerospace structural applications are silicon carbide (SiC) and boron carbide (B$_4$C) particulate reinforcement in an aluminum alloy matrix. Aluminum oxide particles are a lower-cost alternative most commonly used for casting applications. Titanium carbide is being investigated for higher-temperature applications. The mechanical and physical properties of various ceramic reinforcements commonly used in discontinuous MMCs are listed in Table 9.1.

Silicon Carbide. Discontinuously reinforced MMC materials systems based on SiC are the most commonly used. The benefits of using SiC as reinforcement are improved stiffness, strength, thermal conductivity, wear resistance, fatigue resistance, and reduced thermal expansion. Additionally, SiC reinforcements are typically low cost and are relatively low

Table 9.1 Mechanical and physical properties of various ceramic particulate reinforcements commonly used in the manufacture of discontinuously reinforced metal-matrix composites

Ceramic	Density, g/cm^3	Elastic modulus GPa	Elastic modulus 10^6 psi	Knoop hardness	Compressive strength MPa	Compressive strength ksi	Thermal conductivity W/m K	Thermal conductivity Btu ft/h ft^2 °F	Coefficient of thermal expansion 10^{-6}/K	Coefficient of thermal expansion 10^{-6}/°F	Specific thermal conductivity, W m^2/kg K
SiC	3.21	430	62.4	2480	2800	406.1	132	76.6	3.4	6.1	41.1
B$_4$C	2.52	450	65.3	2800	3000	435.1	29	16.8	5.0	9.0	11.5
Al$_2$O$_3$	3.92	350	50.8	2000	2500	362.6	32.6	18.9	6.8	12.2	8.3
TiC	4.93	345	50.0	2150	2500	362.6	20.5	11.9	7.4	13.3	4.2

Source: Ref 9.4

density. Particle size and shape are important factors in determining materials properties. Fatigue strength is greatly improved with the use of fine particles, and the uniform distribution of reinforcement is improved by matching the size of the reinforcement to the size of the matrix particles. The effect of variation in particle size on several composite properties is shown in Fig. 9.5.

The shape of a particle is characterized by its aspect ratio, the ratio of its longest to shortest linear dimension. Most ceramic reinforcement particles have a low aspect ratio, being blocky with sharp edges. They are easy to produce by simple milling and therefore comparatively inexpensive, yielding composites with approximately isotropic properties. Whiskers and platelets are particles with higher aspect ratios. A high-aspect-ratio SiC particle is shown in Fig. 9.6. Although they are typically more expensive and harder to work with than the blocky particles, high-aspect-ratio particles are used when anisotropic properties are desired, concentrating the benefits of the reinforcement into a limited direction.

Boron carbide is a commonly used reinforcement when low density is important, when low reactivity with the matrix is needed, and when neutron absorption is required. The morphology for B_4C particulates is shown in Fig. 9.7. When anisotropic composite properties are necessary, B_4C whiskers are used (Fig. 9.8); however, this form is expensive and rarely used. Some naturally occurring boron atoms (B^{10}) have a high neutron-absorption cross-sectional area. Consequently, B_4C, having four boron atoms per structural unit, is an important reinforcement for use in nuclear-containment applications. Boron carbide is more inert in the presence of aluminum at high temperatures than SiC, making it more suitable for applications involving welding or casting.

Aluminum oxide particulate is another ceramic powder commonly used in reinforcement of discontinuous MMC materials systems. The resultant benefits are not as great as SiC and B_4C. Aluminum oxide (Al_2O_3) reinforcement powders possess very low reactivity in molten metal baths

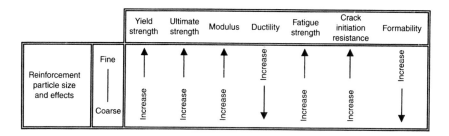

Fig. 9.5 Materials properties and formability as a function of reinforcement particle size. Source: Ref 9.4

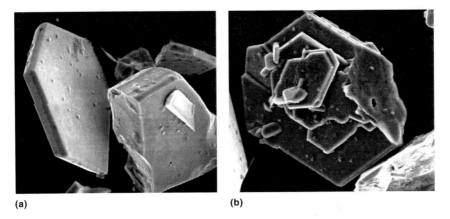

(a)　　　(b)

Fig. 9.6 Silicon carbide platelet reinforcement showing the basal plane morphology in the β phase. Original magnification: 1000×. Source: Ref 9.4

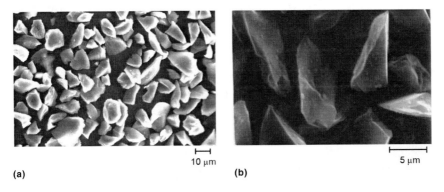

(a)　　　(b)

Fig. 9.7 Morphology of B_4C particulate. (a) Original magnification: 1000×. (b) Original magnification: 5000×. Source: Ref 9.4

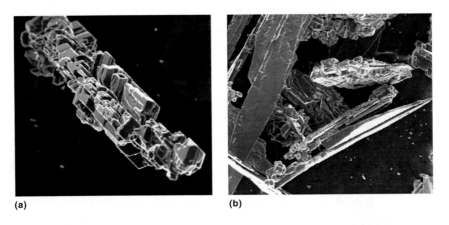

(a)　　　(b)

Fig. 9.8 Boron carbide whisker reinforcement showing polycrystalline microstructure for different whisker morphologies. (a) Original magnification: 200×. (b) Original magnification: 100×. Source: Ref 9.4

(casting) and are relativity low cost. The resultant composite properties, such as stiffness, strength, and fracture toughness, are not as high as those of either the SiC or B_4C reinforcement. However, the low reactivity and low cost make this reinforcement very attractive for the production of cast MMCs that require moderate strengths and stiffness improvements while retaining good wear resistance.

Titanium carbide (TiC) is not a widely used ceramic reinforcement powder. However, its inherent high-temperature stability is attractive for use in elevated-temperature applications, where high strength, stiffness, and creep resistance are required. This reinforcement is used primarily in titanium- and nickel-base alloy MMC materials systems that require stability at very high temperatures (up to 1100 °C, or 2000 °F). In these applications, all previously discussed reinforcements would rapidly react into brittle intermetallics, leading to less-than-desirable MMC properties. The use of this reinforcement comes at a price: TiC is very dense when compared to SiC, B_4C, and Al_2O_3 and, as a result, tends to be used only when very demanding high-temperature composites are necessary. The tensile strength of 20 vol% TiC in a matrix of nickel-base 718 alloy has been shown to be greatly improved over that of the monolithic 718 alloy between temperatures of 650 and 1100 °C (1200 and 2000 °F). Iron-base materials systems reinforced with TiC are candidates for extrusion dies, where extremely high strength and wear resistance are necessary at elevated temperatures, such as extrusion of discontinuous MMCs.

9.3 Discontinuously Reinforced Aluminum

Because more aluminum MMCs are produced than MMCs of all other matrix alloys combined, the Aluminum Association (AA) developed a standard designation system for MMCs that has since been adopted by the American National Standards Institute (ANSI). ANSI 35.5-1992 provides that aluminum MMCs be identified as follows:

Matrix/reinforcement/vol% form

For example, 2124/SiC/25w describes the AA-registered alloy 2124 reinforced with 25 vol% of silicon carbide whiskers; $7075/Al_2O_3/10p$ is the AA-registered alloy 7075 reinforced with 10 vol% of alumina particles; 6061/SiC/47f is the AA-registered alloy 6061 reinforced with 47 vol% of continuous SiC fibers; and A356/C/05c is an AA-registered casting alloy with 5 vol% of chopped graphite fibers.

In general, however, the influence of hard particle reinforcement (e.g., SiC) on the relevant mechanical and physical properties of DRAs can be summarized as follows:

- Both the ultimate tensile strength and yield strength increase with an increase in reinforcement fraction. (It should be noted that these properties are decreased with an increase in volume fraction of soft particles, e.g., graphite.)
- The fracture toughness and ductility (percent elongation and strain to failure) decrease with an increase in reinforcement volume fraction.
- The elastic modulus increases with an increase in reinforcement volume fraction.
- The thermal and electrical conductivities, as well as the CTE, decrease with increasing reinforcement volume fraction.

The effects of SiC volume fraction on the properties of DRAs are shown in Fig. 9.9, and the strength retention at elevated temperatures is illustrated in Fig. 9.10.

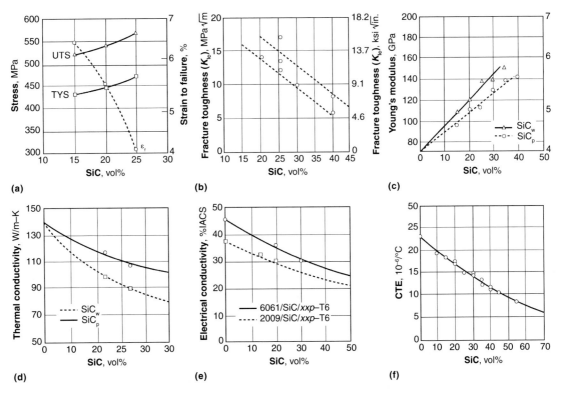

Fig. 9.9 Effect of reinforcement volume fraction on the properties of aluminum metal-matrix composites. (a) The ultimate tensile strength (UTS), tensile yield strength (TYS), and strain-to-failure (εf) for 6013/SiC/xxp–T6. (b) Fracture toughness as a function of SiC volume fraction. (c) Young's modulus as a function of SiCw and SiCp volume fraction. (d) Thermal conductivity for 2009/SiC/xx-T6. (e) Electrical conductivity for 2009/SiC/xxp-T6 and 6061/SiC/xxp-T6. (f) Coefficient of thermal expansion (CTE) for 6061/SiC/xxp-T6. Source: Ref 9.5

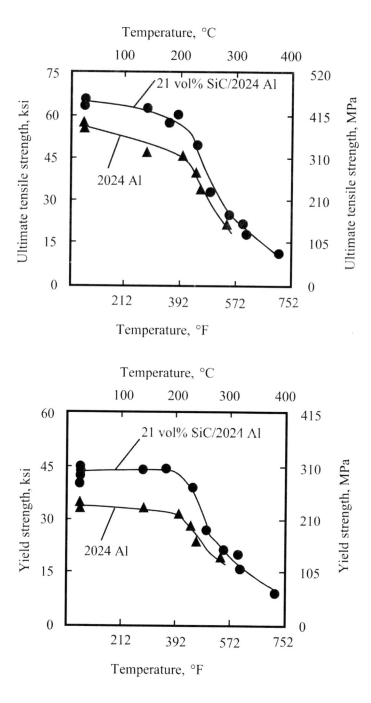

Fig. 9.10 Elevated-temperature properties of SiC$_p$ discontinuously reinforced aluminum composites. Source: Ref 9.6

9.4 Processing of Discontinuously Reinforced Aluminum

The combination of light weight, environmental resistance, and useful mechanical properties has made aluminum alloys very popular. These properties also make aluminum well suited for use as a matrix metal. The melting point of aluminum is high enough to satisfy many application requirements yet low enough to render composite processing reasonably convenient. Also, aluminum can accommodate a variety of reinforcing agents. Although much of the early work on aluminum MMCs concentrated on continuous fiber types, most of the present work is focused on discontinuously reinforced (particle or whisker) aluminum MMCs because of their greater ease of manufacture, lower production costs, and relatively isotropic properties. However, as shown in Fig. 9.1, higher-performance composites are produced by more expensive, continuous fiber reinforcements. At the opposite end of the cost-performance spectrum are the particle-reinforced molten (or cast) metal composites. In between lie medium-priced composites, including those produced by preform infiltration and powder metallurgy (PM) techniques.

Processing methods for discontinuous aluminum MMCs include various casting processes, liquid metal infiltration, spray deposition, and PM. Each of these processes is briefly reviewed in the following sections.

Casting. Virtually every casting method that can be used with unreinforced aluminum has been used with aluminum MMCs. These include the sand, gravity die (permanent mold), investment, squeeze, and lost foam casting processes as well as high-pressure die casting. However, experience has shown that several modifications must be made to the normal melting and casting practice in order to produce high-quality castings from a composite. Some of the differences in composite foundry practice are:

- Melting under an inert cover gas is at the discretion of the caster. Conventional degassing techniques, such as plunging tablets or gas injection, can cause nucleation of gas bubbles on the SiC particles and subsequent dewetting of the ceramic. Salt fluxing removes the SiC. On the other hand, a rotary injection system is available that can successfully flux and degas the melt. It uses an argon-SF_6 gas mixture. Additionally, simply bubbling argon through the melt using a diffuser tube can be used to remove hydrogen that has been absorbed by the melt.
- Close control of melt temperature is needed to avoid overheating and subsequent formation of aluminum carbide.
- The melt must be gently stirred during casting to maintain a uniform dispersion of SiC particles. The ceramic particles do not melt and dissolve in the matrix alloy, and because they are denser than the host alloy, there is a tendency for the particulate to sink to the bottom of the furnace or crucible.
- Turbulence during casting must be minimized to avoid entrapping gas.

By heeding these general guidelines, MMC ingots can be successfully remelted and cast using the current casting methods mentioned previously.

Melting. Aluminum MMCs are melted in a manner very similar to that used for unreinforced aluminum alloys. Conventional induction, electric resistance, and gas- and oil-fired crucible furnaces are suitable. If a protective gas is used, the crucible should be charged with the gas prior to adding dry, preheated ingots. The ingots are dried at a temperature above 200 °C (390 °F) to drive off unwanted moisture, which could contaminate the melt. All furnace tools, such as skimmers, ladles, and thermocouples, also must be coated and thoroughly dried and preheated before use.

Melt temperature control is standard, and, in general, pouring temperatures are similar to those used for unreinforced alloys. However, care must be taken to avoid overheating, which can cause the formation of aluminum carbide by the reaction $4Al + 3SiC \rightarrow Al_4C_3 + 3Si$. This reaction occurs very slowly at temperatures to approximately 750 °C (1380 °F) but accelerates with increasing temperatures from 780 to 800 °C (1435 to 1470 °F) for matrices containing a nominal 9% Si. The Al_4C_3 precipitates as crystals that adversely affect melt fluidity, weaken the cast material, and decrease the corrosion resistance of the casting.

Stirring. Because the SiC particles are completely wetted by liquid aluminum, they will not coalesce into a hard mass but will instead concentrate on the bottom of the furnace if the melt is not stirred. The density of most aluminum alloys is approximately 2.7 g/cm^3 (0.10 $lb/in.^3$), while the density of SiC is ~3.2 g/cm^3 (~0.12 $lb/in.^3$). Therefore, the use of a stirrer that disperses the particles throughout the melt is recommended. The stirring action must be slow to prevent the formation of a vortex at the surface of the melt, and care must be taken not to break the surface, which could contaminate the bath with dross. A scooping action, whereby the lower portion of the melt is gently but firmly made to rise, is the best method of hand stirring. Use of a slowly rotating, propeller-like mechanical stirrer (Fig. 9.11) is preferred by some foundries. In fact, studies indicate that the mechanical properties of the casting are maximized by continuous stirring versus intermittent (hand) stirring. When induction melting, the natural eddy current stirring action of the furnace usually is sufficient to disperse the particles, although supplementary hand stirring (with the power off) is recommended to ensure that no particles have congregated in potential "dead" zones.

After stirring, the settling rate of the SiC particles is quite slow, due partially to thermal currents in the melt and to the "hindered settling" phenomenon. After 10 to 15 min, however, approximately the top 30 mm (1.2 in.) of the unstirred bath becomes devoid of SiC particles, although the distribution remains uniform throughout the balance of the melt. Consequently, it is important to stir the metal immediately before pouring, regardless of whether it was stirred during the melting and holding stages. Aluminum MMCs produced by stir casting are also commonly extruded. Room-temperature properties of extruded aluminum MMCs are given in Table 9.2.

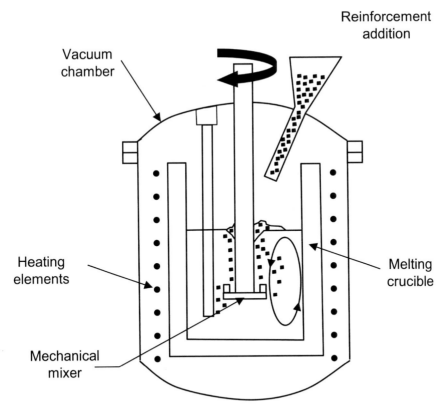

Fig. 9.11 Stir casting. Source: Ref 9.7

Table 9.2 Room-temperature properties of extruded stir-cast aluminum metal-matrix composites in the T6 condition

Material	Al₂O₃ content, vol%	Ultimate strength				Yield strength				Tensile elongation, %	Elastic modulus		Fracture toughness	
		Typical		Minimum(a)		Typical		Minimum(a)						
		MPa	ksi	MPa	ksi	MPa	ksi	MPa	ksi		GPa	10⁶ psi	MPa m¹ᐟ²	ksi in.¹ᐟ²
6061/Al₂O₃/xp	0	310	45	260	38	275	40	240	35	20	69.0	10.0	29.6	27.0
	10	350	51	325	47	295	43	260	38	10	81.4	11.8	24.0	21.9
	15	365	53	340	49	325	47	290	42	6	89.0	12.9	22.0	20.0
	20	370	54	345	50	350	51	315	46	4	97.2	14.1	21.5	19.6
2014/Al₂O₃/xp	0	525	76	470	68	475	69	415	60	13	73.1	10.6	25.3	23.0
	10	530	77	495	72	495	72	455	66	3	84.1	12.2	18.0	16.4
	15	530	77	495	72	505	73	460	67	2	93.8	13.6	18.8	17.1
	20	515	75	485	70	505	73	460	67	1	101	14.7	...	...

Data given for extruded bar or rod (extrusion ratio of ~20 to 1). (a) Values represent 99% confidence interval. Source: Ref 9.5

Fluxing and Degassing. A method of fluxing and degassing composite melts has been developed that uses a rotating impeller-like device to both stir the bath and inject a blend of argon and SF_6 gases. It also can be used to keep the SiC particles in suspension. The system uses a six-blade graphite impeller, which is connected to a threaded graphite drive shaft.

A 610 mm (24 in.) diameter crucible requires the use of a 205 mm (8 in.) diameter impeller. Ten minutes of operation at a speed of 200 rpm usually is sufficient to shear the argon-SF_6 bubbles into an effective size. The thick, foamy dross that results is removed after the cycle is completed. The melt can also be degassed by simply bubbling argon through the melt by a diffuser tube designed to provide a distribution of small bubbles. The melt should be stirred while bubbling the argon through the melt. A degassing time of 20 min at an argon flow rate of approximately 0.30 m³/h (10 ft³/h) can effectively reduce the hydrogen content of the melt to less than 0.10 mL/100 g for a 225 kg (500 lb) batch of molten composite. Some degree of melt cleansing or oxide removal is observed due to the attachment of oxides to the rising argon bubbles. The thick, frothy dross generated by bubbling argon through the melt should be continuously skimmed from the melt surface during the degassing period. Subsequent to degassing, the melt should be allowed to "rest" for 30 to 45 min without stirring, to allow any remaining bubbles to float to the melt surface.

Pouring. It is neither practical nor necessary to maintain either an inert gas cover or stirring action while the liquid metal is transferred from furnace to pouring station. The recommended sequence of operations is to stir the bath thoroughly, skim off dross in the furnace, and then transfer the liquid to the pouring ladle (or remove the crucible). If the metal transfer involves pouring from, for example, a tilting furnace into a ladle, it is important to minimize turbulence in the metal stream to avoid entrapping gas. However, tilting furnaces are not generally recommended for use with composite melts. Pouring practice is the same as for unreinforced aluminum.

Gating Systems. The basic rules of running and feeding also apply to castable MMCs, including the use of filters that pass the SiC particles but trap oxides. However, the viscous melt behaves as though partially solidified, and the ceramic particles impede the free flow of gases. The composite is far less forgiving of turbulence than conventional aluminum. Thus, a poorly designed gating system can cause the formation and entrapment of gas bubbles in the liquid that remain in the solidified casting. Optimal running and feeding systems for castable MMCs are being developed, although many sound castings have been produced using existing schemes.

High-Pressure Die Casting. One casting process that DRAs have proved to be especially adaptable to is high-pressure die casting. The extremely rapid solidification rates achieved in die casting produce a very fine dendritic structure that is nearly pore free and yields excellent mechanical properties. Aluminum MMCs have been high-pressure die cast without any changes to the die cast machines or dies. The dimensional shrinkage factor for these composite materials is in the range of 0.6%, which is similar to that of unreinforced aluminum alloys. In some cases, there is no need for alterations to the gating and venting of the dies that are currently used for conventional aluminum alloys. These composite materials can also be run with or without vacuum. The microstructure of a typical die-cast DRA is shown in Fig. 9.12.

The DRAs may be up to 50 times more viscous than their matrix alloy because they are semisolid materials (molten metal/solid particle mixtures). The thixotropic behavior of these aluminum semisolid composites can best be used in high-pressure die casting. This is because shear rates that are applied to the semisolid melt to inject it into the die cavity improve its fluidity. This, combined with the fact that semisolid materials are more viscous than their aluminum-matrix alloy, means that they enter the die cavity with much less turbulence and less entrapped air.

To best use the thixotropic behavior of aluminum MMCs, a minimum gate velocity of 30 m/s (100 ft/s) is recommended. With this velocity, the composite material enters the die in an almost laminar flow, which has produced castings that are of a higher quality (less porous) than castings from unreinforced aluminum alloys poured under the same conditions. High-pressure die casters have observed that productivity can be increased due to shorter cycle times. The shorter cycle times may be attributable to the unique properties of these materials. There are several factors that contribute to these shorter cycle times.

Because up to 20% of the melt is solid ceramic, less heat of fusion must be removed during solidification, and the thermal conductivity is greater due to the addition of the SiC particles. Therefore, with less heat being removed more quickly, the casting will solidify faster, resulting in shorter cycle times. Aluminum composites can be poured at low temperatures (as low as 650 to 675 °C, or 1200 to 1250 °F), which is much lower than the normal pouring temperature of 705 to 730 °C (1300 to 1350 °F) for the

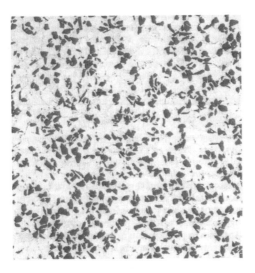

Fig. 9.12 Typical microstructure of an aluminum-matrix composite containing 20 vol% SiC. Original magnification: 125×. Source: Ref 9.5

high-silicon (16 to 18 wt% Si) A390 alloy. These lower temperatures result in less thermal shock to the die and decreased die casting cycle times. A cycle time reduction of up to 20% has been observed. These factors can also result in increased die life due to less thermal stress on the die.

The chemical compositions of typical die casting and sand casting DRA alloys are listed in Table 9.3, and typical mechanical properties of high-pressure die-cast DRAs are given in Table 9.4. The automotive industry is behind the development of cast aluminum MMCs. Current or potential applications include brake rotors and drums, brake calipers, brake pad backing plates, and cylinder liners.

Compocasting. When a liquid metal is vigorously stirred during solidification, it forms a slurry of fine, spheroidal solids floating in the liquid. Stirring at high speeds creates a high shear rate, which tends to reduce the viscosity of the slurry even at solid fractions as high as 50 to 60% volume. The process of casting such a slurry is called rheocasting. The slurry can also be mixed with particulates, whiskers, or short fibers before casting.

Table 9.3 Matrix alloys for cast discontinuously reinforced aluminum metal-matrix composites

| Alloy | Composition, wt% | | | | | | | | | |
	Si	Fe	Cu	Mn	Mg	Ni	Ti	Zn	Other	Al
High-pressure die castings										
F3D(a)	9.50–10.5	0.80–1.20	3.00–3.50	0.50–0.80	0.30–0.50	1.00–1.50	0.20 max	0.03 max	0.03 max; 0.10 total	bal
F3N(b)	9.50–10.5	0.80–1.20	0.20 max	0.50–0.80	0.50–0.70	...	0.20 max	0.03 max	0.03 max; 0.10 total	bal
Sand and permanent mold castings										
F3S(c)	8.50–9.50	0.20 max	0.20 max	...	0.45–0.65	...	0.20 max	...	0.03 max; 0.10 total	bal
F3K9(d)	9.50–10.5	0.30 max	2.80–3.20	...	0.80–1.20	1.00–1.50	0.20 max	...	0.03 max; 0.10 total	bal

(a) Duralcan F3DxxS composites (xx = volume percent SiC particulate) are general-purpose, die casting composites. They are similar to 380/SiC/xxp (Aluminum Association metal-matrix composite, or MMC, nomenclature). (b) Duralcan F3NxxS composites, containing virtually no copper or nickel, are designed for use in corrosion-sensitive applications. They are similar to 360/SiC/xxp. (c) Duralcan F3SxxS composites (xx = volume percent SiC particulate) are general-purpose composites for room-temperature applications. They are similar to 359/SiC/xxp (Aluminum Association MMC nomenclature). (d) Duralcan F3KxxS composites, containing significant amounts of copper and nickel, are designed for use at elevated temperatures. They are similar to 339/SiC/xxp. Source: Duralcan USA and Ref 9.5

Table 9.4 Typical mechanical properties for high-pressure die-cast F3D/SiC composites

| Material | Ultimate strength | | Yield strength | | Elongation(a), % | Elastic modulus | | Hardness, HRB | Impact energy | |
	MPa	ksi	MPa	ksi		GPa	10⁶ psi		J	ft · lbf
A380-F(b)	317	46.0	159	23.1	3.5	71.0	10.3	40	3.4	2.5
A390-F(b)	283	41.0	241	34.9	1.0	81.4	11.8	76	1.4	1.0
F3D.10S-F(c)	345	50.0	241	34.9	1.2	93.8	13.6	77	1.4	1.0
F3D.10S-O(d)	276	40.0	152	22.0	1.7	93.8	13.6	55	2.7	2.0
F3D.10S-T5(d)	372	53.9	331	48.0	0.7	93.8	13.6	84	1.4	1.0
F3D.20S-F(c)	352	51.0	303	43.9	0.4	113.8	16.5	82	0.7	0.5
F3D.20S-O(d)	303	43.9	186	27.0	0.8	113.8	16.5	62	1.4	1.0
F3D.20S-T5(e)	400	58.0	(f)	(f)	(f)	113.8	16.5	87	0.7	0.5

See Table 9.3 for composition of matrix alloy. (a) Measured by direct reading from stress-strain plot. (b) Handbook values. (c) Cast-to-size tensile bars. (d) Cast-to-size tensile bars, annealed at 343 °C (649 °F) for 4 h. (e) Cast-to-size tensile bars, aged at 177 ° C (351 °F) for 5 h. (f) Test bars fractured before yielding. Source: Duralcan USA and Ref 9.5

This modified form of rheocasting to produce near-net shape MMC parts is called compocasting.

The melt reinforcement slurry can be cast by gravity casting, die casting, centrifugal casting, or squeeze casting. The reinforcements have a tendency to either float to the top or segregate near the bottom of the melt, because their densities differ from that of the melt. Therefore, a careful choice of casting technique as well as of mold configuration is of great importance in obtaining uniform distribution of reinforcements in a compocast MMC.

Compocasting allows a uniform distribution of reinforcement in the matrix as well as a good wet-out between the reinforcement and the matrix. Continuous stirring of the slurry creates intimate contact between them. Good bonding is achieved by reducing the slurry viscosity as well as by increasing the mixing time. The slurry viscosity is reduced by increasing the shear rate as well as by increasing the slurry temperature. Increasing the mixing time provides a longer interaction between the reinforcement and the matrix.

Compocasting is one of the most economical methods of fabricating a composite with discontinuous fibers. It can be performed at temperatures lower than those conventionally employed in foundry practice during pouring, resulting in reduced thermochemical degradation of the reinforced surface.

Pressure Infiltration Casting. Pressure infiltration casting (PIC) is an MMC fabrication technique that involves infiltrating an evacuated particulate or fiber preform with molten metal subjected to an isostatically applied gas pressure. As in more traditional casting processes, PIC produces components to near-net shape. This is an important attribute for producing MMC components, because it reduces the amount of machining needed for these often difficult-to-machine materials.

One of the most common composite materials fabricated using the PIC process is DRA. Although there are several variations to the PIC process for fabrication of near-net shape DRA components, all involve the infiltration of molten aluminum into a free-standing, evacuated preform by an external isostatically applied inert gas. An alumina fiber preform is shown in Fig. 9.13. The preform acts as a nucleation site for solidification and inhibits grain growth during solidification and cooldown, resulting in a very fine cast microstructure. In addition, because the preform is evacuated and the mold is directionally cooled, properly designed and processed components can be produced without porosity. Preforms have been fabricated and pressure infiltration cast in a range of reinforcement levels varying from 30% to greater than 70%. Current technology does not allow for lower reinforcement volume fractions, because the preform must have sufficient reinforcement content to produce a stable geometry. A number of particulate reinforcements have been used to make preforms for the PIC process. The most common reinforcements used are SiC and Al_2O_3

particulates. With a density only slightly lower than that of aluminum, boron carbide (B_4C) has also been investigated as a reinforcement—the close match in density means that settling of the reinforcement is avoided. However, at this time, boron carbide is not widely used in PIC, primarily due to its cost.

The PIC process begins by inserting a reinforcement preform into a mold, which is placed inside a metal canister, as shown in Fig. 9.14. The

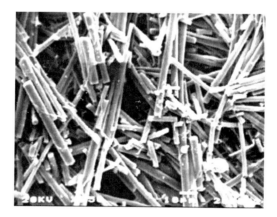

Fig. 9.13 Saffil alumina fiber preform

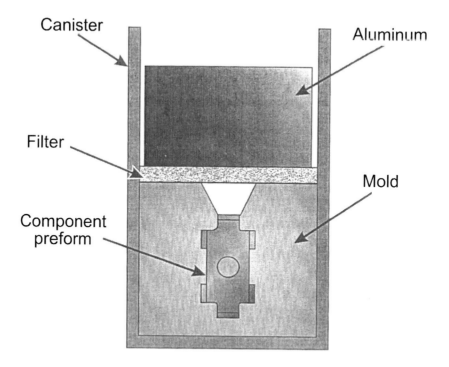

Fig. 9.14 Schematic of assembly for top-fill pressure infiltration casting. Source: Ref 9.8

metal canister is used to maintain vacuum inside of the mold and preform during the pressurization stage of the process. For the top-fill version of the process, the canister typically extends above the mold to contain the molten aluminum charge. The entry port to the mold is sealed with molten aluminum through the use of a filter with sufficient surface tension to maintain the molten aluminum head.

An early experimental variation of the process is known as bottom fill. In this variation, the mold assembly is sealed within the metal canister, and a tube extends down to a separate crucible containing the aluminum charge. The entry to the mold is sealed when the aluminum melts, and the end of the tube is submerged in the aluminum.

In both variations, the mold and the preform within it are evacuated. The most common method of evacuating the mold and preform is to place the entire mold assembly in a vacuum-pressure vessel and evacuate both the mold and vacuum-pressure chamber. The aluminum charge is then melted under vacuum, thus sealing the entry port to the evacuated mold.

In a variation of the top-fill approach, the aluminum is melted in a separate crucible and poured into the metal canister and onto the filter outside of the vacuum-pressure vessel. A tube leading through the filter and into the mold is then used to evacuate the mold and preform. After sufficient vacuum is attained, the tube is crimped and sealed. The entire assembly of both the mold and molten aluminum is placed inside a pressure vessel.

Regardless of the setup used, the mold and preform are heated to a predetermined temperature, so that the molten aluminum does not solidify upon contact and choke off infiltration. After the mold is preheated and the aluminum melt has reached the desired superheat temperature, inert gas pressure in the range of 2 to 10 MPa (300 to 1500 psig) is applied. The pressure difference between the gas outside of the mold and the vacuum within the mold overcomes the surface tension forces of the reinforcement and drives the aluminum into the preform. As liquid aluminum infiltrates into the preform, the pressure acting on the mold quickly approaches the isostatic state. Hence, the mold only supports the pressure difference for a very short period of time so that large, expensive, and cumbersome molds are not needed. To aid filling shrinkage porosity, the mold is cooled directionally, and pressure is maintained until the entire casting is solidified.

The PIC process can produce DRA parts with a wide range of reinforcement types and volume fractions using both conventional casting and wrought aluminum alloys as the matrix. Most traditional casting alloys increase their fluidity through the addition of elements such as silicon. Because PIC is a pressure casting process, it does not need to rely on the inherent fluidity of the molten aluminum to fully fill the mold and preform. This allows the use of off-the-shelf commercial alloys, which helps lower the cost of DRA parts due to the widespread availability of the alloys. Additionally, using commercial alloys enables the performance gain of the DRA to be easily ascertained. Discontinuously reinforced aluminum components have been successfully fabricated using familiar aluminum alloys, such as 2024, 6061, and A356, as the matrix.

The greatest number of components fabricated by PIC are DRA electronic packages used to house and mount integrated circuits and multichip modules. The benefits of DRA to this application include a controlled CTE, which closely matches that of directly mounted integrated circuits, and a high thermal conductivity, which aids removal of the heat generated by these components. Components have also been prototyped for a number of applications in the automotive (e.g., brake rotors, connecting rods), aerospace (e.g., hydraulic manifolds, control links), gas turbine engines (e.g., stator vanes), and space propulsion (e.g., pump housings, flanges) industries.

Pressureless Liquid Metal Infiltration. The pressureless metal infiltration process is based on materials and process controls that allow a metal to infiltrate substantially nonreactive reinforcements without the application of pressure or vacuum. Reinforcement level can be controlled by the starting density of the material being infiltrated. As long as interconnected porosity and appropriate infiltration conditions exist, the liquid metal will spontaneously infiltrate into the preform.

Key process ingredients for the manufacture of reinforced aluminum composites include the aluminum alloy, a nitrogen atmosphere, and magnesium present in the system. During heating to infiltration temperature (~750 °C, or 1380 °F), the magnesium reacts with the nitrogen atmosphere to form magnesium nitride (Mg_3N_2). The Mg_3N_2 is the infiltration enhancer that allows the aluminum alloy to infiltrate the reinforcing phase without the necessity of applied pressure or vacuum. During infiltration, the Mg_3N_2 is reduced by the aluminum to form a small amount of aluminum nitride (AlN). The AlN is found as small precipitates and as a thin film on the surface of the reinforcing phase. Magnesium is released into the alloy by this reaction. The pressureless infiltration process can produce a wide array of engineered composites by tailoring of alloy chemistry and particle type, shape, size, and loading. Particulate loading in cast composites can be as high as 75 vol%, given the right combination of particle shape and size. A typical microstructure is shown in Fig. 9.15.

The most widely used cast composite produced by liquid metal infiltration is an Al-10Si-1Mg alloy reinforced with 30 vol% SiC. The 1% Mg present in this alloy is obtained during infiltration by the reduction of the Mg_3N_2. This composite system is being used for all casting processes except die casting. The composite most used for die casting is based on this system, with the addition of 1% Fe. Alloy modifications can be made to the alloy prior to infiltration or in the crucible prior to casting. The only universal alloy restriction for this composite system is the presence of magnesium to allow the formation of the Mg_3N_2. For the SiC-containing systems, silicon must also be present in sufficient quantity to suppress the formation of Al_4C_3. Composites consisting of Al_2O_3-reinforced aluminum that exhibit low wear rates are also produced.

An important application area for pressureless molten metal infiltration is Al/SiC_p packages, substrates, and support structures for electronic components. Typical requirements include a low CTE to reduce mechanical stresses imposed on the electronic device during attachment and operation, high thermal conductivity for heat dissipation, high stiffness to minimize distortion, and low density for minimum weight. Compared with conventional aluminum alloys, composites having high loadings of SiC particles feature greatly reduced CTEs and significantly higher elastic moduli, with little or no penalty in thermal conductivity or density (Table 9.5).

Spray deposition involves atomizing a melt and, rather than allowing the droplets to solidify totally as for metal powder manufacture, collecting the semisolid droplets on a substrate. The process is a hybrid rapid solidification process, because the metal experiences a rapid transition through the liquidus to the solidus, followed by slow cooling from the solidus to

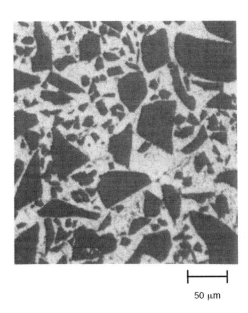

├────────┤
50 μm

Fig. 9.15 Discontinuous SiC/Al metal-matrix composite (60 vol% SiC) produced by the liquid metal infiltration process. Source: Ref 9.8

Table 9.5 Physical properties of an Al/SiC/xxp metal-matrix composite for electronic applications

Property	Composite, SiC loading		Typical aluminum alloys
	55 vol%	70 vol%	
Coefficient of thermal expansion, $10^{-6}/°C$ ($10^{-6}/°F$)	8.5 (4.7)	6.2 (3.4)	22–24 (12–13)
Thermal conductivity, W/m · K (Btu/h · ft · °F)	160 (93)	170 (99)	150–180 (87–104)
Density, g/cm^3 (lb/in.^{3})	2.95 (0.106)	3.0 (0.0108)	2.7 (0.097)
Elastic modulus, GPa (10^6 psi)	200 (29)	270 (39)	70 (10)

Source: Ref 9.5

room temperature. This results in a refined grain and precipitation structure with no significant increase in solute solubility.

The production of MMC ingot by spray deposition (Fig. 9.16) can be accomplished by introducing particulate into the standard spray deposition metal spray, leading to co-deposition with the atomized metal onto the substrate. Careful control of the atomizing and particulate feeding conditions is required to ensure that a uniform distribution of particulate is produced within a typically 95 to 98% dense aluminum matrix.

Spray deposition was developed commercially in the late 1970s and throughout the 1980s by Osprey, Ltd. (United Kingdom) as a method of building up bulk material by atomizing a molten stream of metal with jets of cold gas. Most such processes are covered by their patents or licenses and are now generally referred to as Osprey processes. The potential for adapting the procedure to particulate MMC production by injection of ceramic powder into the spray was recognized at an early stage and has been developed by a number of primary metal producers.

A feature of much MMC material produced by the Osprey route is a tendency toward inhomogeneous distribution of the ceramic particles. It is common to observe ceramic-rich layers approximately normal to the

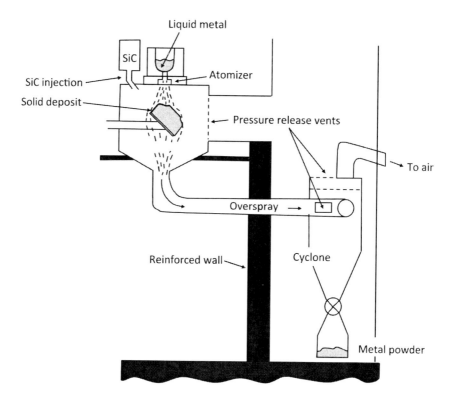

Fig. 9.16 Osprey spray deposition process. Source: Ref 9.9

overall growth direction. Among the other notable microstructural features of Osprey MMC material are a strong interfacial bond, little or no interfacial reaction layer, and a very low oxide content. Porosity in the as-sprayed state is typically about 5%, but this is normally eliminated by secondary processing. A number of commercial alloys have been explored for use in Osprey-route MMCs.

Spray forming in the Osprey mode involves the sequential stages of atomization and droplet consolidation to produce near-net shape preforms in a single processing step. The as-sprayed material has a density greater than 98% of the theoretical density and exhibits a uniform distribution of fine, equiaxed grains and no prior particle boundaries or discernible macroscopic segregation. Mechanical properties are normally isotropic and meet or exceed those of counterpart ingot-processed alloys. A major attraction of the process is its high rate of metal deposition, typically in the range of 0.2 to 2.0 kg/s (0.4 to 4.4 lb/s). Commercial viability mandates close tolerances in shape and dimensions, as well as consistency in microstructure and product yield. This requires an understanding of and control over the effects of several independent process parameters, namely, melt superheat, metal flow rate, gas pressure, spray motion (spray scanning frequency and angle), spray height (distance between the gas nozzles and the substrate), and substrate motion (substrate rotation speed, withdrawal rate, and tilt angle).

A number of aluminum alloys containing SiC particulate have been produced by spray deposition. These include aluminum-silicon casting alloys and the 2*xxx*, 6*xxx*, 7*xxx*, and 8*xxx* (aluminum-lithium) series wrought alloys. Significant increases in specific modulus have been realized with SiC-reinforced 8090 alloy (Table 9.6). Products that have been produced by spray deposition include solid and hollow extrusions, forgings, sheet, and remelted pressure die castings.

Powder metallurgy processing of aluminum MMCs involves both SiC particulates and whiskers, although Al_2O_3 particles and Si_3N_4 whiskers have also been employed. Processing involves (1) blending of the gas-atomized matrix alloy and reinforcement in powder form; (2) compacting (cold pressing) the homogeneous blend to roughly 80% density;

Table 9.6 Properties of conventionally processed aluminum alloys (ingot metallurgy) and spray-deposited aluminum metal-matrix composites

Material	Elastic modulus		Density		Specific modulus	Improvement, %
	GPa	10⁶ psi	g/cm³	lb/in.³		
2014	72	10	2.8	0.101	25.7	0
8090	80	12	2.55	0.092	31.4	22
2014/SiC/15p(a)	95	14	2.84	0.103	33.5	30
8090/SiC/15p(a)	100	15	2.62	0.095	38.2	49

(a) Spray co-deposited, extruded, and peak aged. Source: Ref 9.5

(3) degassing the preform (which has an open, interconnected pore structure) to remove volatile contaminants (lubricants and mixing and blending additives), water vapor, and gases; and (4) consolidation by vacuum hot pressing or hot isostatic pressing. The hot-pressed cylindrical billets can be subsequently extruded, rolled, or forged.

Whisker-reinforced aluminum MMCs may experience some whisker alignment during extrusion or rolling (Fig. 9.17). Control of whisker alignment enables production of aluminum MMC product forms with directional properties needed for some high-performance applications. Cross rolling of sheet establishes a more planar whisker alignment, producing a two-dimensional isotropy. The mechanical properties of whisker-reinforced aluminum MMCs are superior to particle-reinforced composites at any common volume fraction (Fig. 9.18).

Powder metallurgy methods involving cold pressing and sintering, or hot pressing, to produce MMCs are shown in Fig. 9.19 and 9.20. The matrix and the reinforcement powders are blended to produce a homogeneous distribution. The blending stage is followed by cold pressing to produce a green body, which is approximately 80% dense and can be easily handled. The cold-pressed green body is canned in a sealed container and degassed to remove any absorbed moisture from the particle surfaces. The final step is hot pressing, uniaxial or isostatic, to produce a fully dense composite. The hot pressing temperature can be either below or above that of the matrix alloy solidus.

The PM hot pressing technique generally produces properties superior to those obtained by casting and by liquid metal infiltration (squeeze casting) techniques. Because no melting and casting are involved, the powder

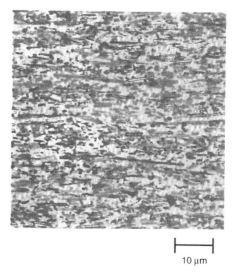

10 μm

Fig. 9.17 SiC whisker-reinforced (20 vol% SiC) aluminum alloy sheet with the whiskers aligned in the direction of rolling. Source: Ref 9.8

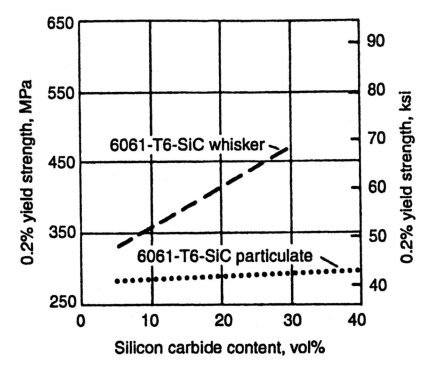

Fig. 9.18 Yield strength comparison between whisker- and particulate-reinforced aluminum metal-matrix composite. Source: Ref 9.8

process is more economical than many other fabrication techniques and offers several advantages, including the following:

- A lower temperature can be used during preparation of a PM-based composite compared to preparation of a fusion metallurgy-based composite. This results in less interaction between the matrix and the reinforcement, consequently minimizing undesirable interfacial reactions, which leads to improved mechanical properties.
- The preparation of particulate- or whisker-reinforced composites is, generally speaking, easier using the PM blending technique than it is using the casting technique.

This method is popular because it is reliable compared with alternative methods, but it also has some disadvantages. The blending step is a time-consuming, expensive, potentially dangerous operation. In addition, it is difficult to achieve an even distribution of particulate throughout the product, and the use of powders requires a high level of cleanliness; otherwise, inclusions will be incorporated into the product, with a deleterious effect on fracture toughness and fatigue life.

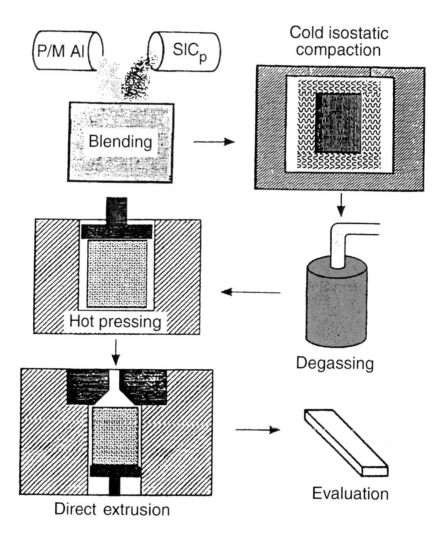

Fig. 9.19 Schematic interpretation of the processing route for powder metallurgy (PM) Al-SiC$_p$ composites. Source: Ref 9.10

9.5 Processing of Discontinuously Reinforced Titanium

Particle-reinforced TMCs are processed by PM methods. Although a variety of materials have been studied, the most common combination is Ti-6Al-4V reinforced with 10 to 20 wt% TiC. These composites offer increased hardness and wear resistance over conventional titanium alloys. Properties of unreinforced and reinforced Ti-6Al-4V are compared in Table 9.7.

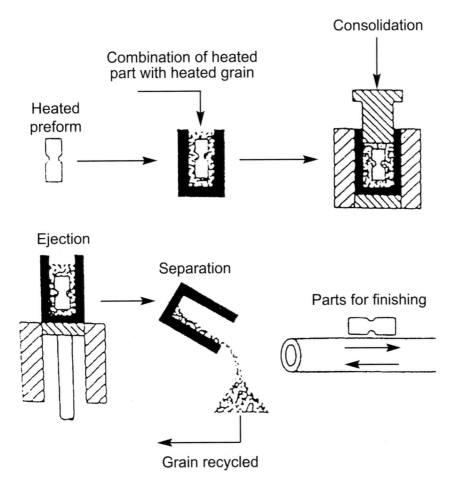

Fig. 9.20 Schematic illustration of the Ceracon technique for fabricating powder metallurgy metal-matrix composites. Source: Ref 9.11, 9.12

Table 9.7 Mechanical properties of discontinuous TiC$_p$/Ti-matrix composites

Property	Ti-6Al-4V	10 wt% TiC/Ti-6Al-4V	20 wt% TiC/Ti-6Al-4V
Density, lb/in.^{3}	0.160	0.16	0.16
Tensile strength, ksi, at:			
Room temperature	130	145	153
1000 °F	65	80	90
Modulus, 106 psi, at:			
Room temperature	16.5	19.3	21
1000 °F	13	15.3	16
Fatigue limit (10^6 cycles), ksi	75	40	...
Fracture toughness, ksi · in.$^{1/2}$	50	40	29
Hardness, HRC	34	40	44

Source: Ref 9.5

As opposed to adding reinforcements as distinct and separate constituents, there are some systems in which the reinforcement can be formed within the metallic matrix during solidification, referred to as in situ composites. Directional solidification of certain eutectic systems can produce fibers or rodlike structures, such as shown for the niobium/niobium carbide system in Fig. 9.21. While these structures can be produced, the growth rates are slow (~1 to 5 cm/h, or 0.4 to 2.0 in./h) due to the need to maintain a stable growth front, which requires a large temperature gradient. In addition, all of the reinforcement is aligned in the growth or solidification direction. There are also limitations in the nature and volume fraction of the reinforcement.

Another in situ method is to use exothermic reactions, such as the exothermic dispersion (XD) process. In the XD process, an exothermic reaction between two components is used to produce a third component. Generally, a master alloy is produced that contains a high volume fraction of a ceramic reinforcing phase. The master alloy is then mixed and remelted with a base alloy to produce the desired amount of particle reinforcement. In one process, aluminum, titanium, and boron are heated together to produce an exothermic reaction between the aluminum and titanium, which produces a mixture of TiB_2 ceramic particles distributed in a titanium aluminide matrix. Strength levels of greater than 690 MPa (100 ksi) were measured at room temperature and at 800 °C (1470 °F) for a gamma titanium aluminide (Ti-45 at.% Al) alloy reinforced with TiB_2 ceramic particulates.

Compared to the conventional MMCs, in situ MMCs exhibit several possible advantages: the in-situ-formed reinforcements are thermodynamically stable, leading to less degradation at elevated temperature; the

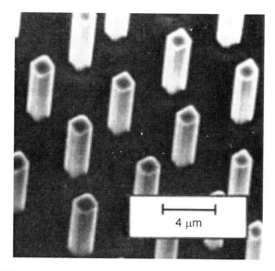

Fig. 9.21 Rod-reinforced NbC/Ni eutectic. Source: Ref 9.13

reinforcement-to-matrix interfaces are clean, resulting in a strong interfacial bonding; and the in-situ-formed reinforcements are finer in size and their distribution in the matrix is more uniform.

9.6 Secondary Processing of Discontinuous Metal-Matrix Composites

After full consolidation, many discontinuously reinforced MMCs are subjected to additional deformation processes to improve their mechanical properties and produce part shapes. Secondary deformation processing of discontinuously reinforced MMCs helps to break up reinforcement agglomerates, reduce or eliminate porosity, and improve the reinforcement-to-matrix bond.

Extrusion can generate fiber alignment parallel to the extrusion axis, but often at the expense of progressive fiber fragmentation. Due to the presence of 15 to 25% hard nondeformable particulates or whiskers, fracture during extrusion can occur with conventional extrusion dies. Therefore, streamlined dies that gradually neck down are used to help minimize fracture. The degree of fiber fracture decreases with increasing temperature and decreasing local strain rate. Particulate segregation can also be a problem, in which particulate clusters occur next to regions containing little or no reinforcement. For whisker-reinforced composites, whisker rotation and alignment with the metal flow direction can occur. In addition, the aspect ratio of whiskers is generally reduced due to the shearing action. To prevent extrusion-related defects, it is important that the material remain in compression during the extrusion process. Another microstructural feature of extruded MMCs is the formation of ceramic-enriched bands parallel to the extrusion axis. The mechanism of band formation is still unclear, but it appears to involve the concentration of shear strain in regions where ceramic particles or fibers accumulate. However, extrusion of consolidated MMCs, such as castings, can reduce the level of clustering and inhomogeneties in the material.

Hot rolling is used after extrusion to produce sheet and plate products. Because the compressive stresses are lower in rolling than extrusion, edge cracking can occur during rolling. To minimize edge cracking, discontinuous composites are rolled at temperatures approximately $0.5T_m$, where T_m is the absolute melting temperature, using relatively slow speeds. Isothermal rolling using light reductions and large-diameter rolls can also reduce cracking. Isothermal rolling is often conducted as a pack rolling operation in which the MMC is encapsulated by a stronger metal. During rolling, the outside of the pack cools during rolling, while the interior containing the MMC remains at a fairly constant hot temperature. When the rolling temperature is low, light reductions and intermediate anneals may be required. As for extrusion, rolling further breaks up the particulate agglomerates. In heavily rolled sheet that has undergone approximately a 90% reduction in thickness, the particulate clusters are completely broken up and the matrix has flowed between the individual particles.

Processes such as rolling and forging, which involve high strains applied quickly, can cause damage such as cavitation, particle fracture, and macroscopic cracking, particularly at low temperatures. In addition, very high temperatures increase the chance of matrix liquation, resulting in hot tearing or hot shortness. In contrast, hot isostatic pressing (HIP) generates uniform stresses and so is unlikely to give rise to either microstructural or macroscopic defects. It is an attractive method for removing residual porosity. However, it can be very difficult to remove residual porosity in regions of very high ceramic content, such as within particle clusters, and the absence of any macroscopic shear stresses means that such clusters are not readily dispersed during HIP.

Machining discontinuous MMCs can be accomplished using circular saws and router bits. For straight cuts, diamond-grit-impregnated saws with flood coolant produce good edges. For contour cuts, solid carbide router bits with a diamond-shaped chisel-cut geometry or diamond-coated router bits can be used with good results. Due to the hard reinforcement particles, speeds and feeds for all machining operations must be adjusted. In general, the speed is reduced to minimize tool wear, while the feed is increased to obtain productivity before the tools wear. The higher the reinforcement content, the greater the tool wear, with wear generally being a bigger problem than excessive heat generation. Polycrystalline diamond cutting tools exhibit longer lives than either solid or diamond-coated carbide tools but are also more expensive. They also require a rigid setup to prevent edge chipping. Other methods, such as abrasive waterjet cutting and wire electrical discharge machining, also produce acceptable cuts.

9.7 Continuous Fiber-Reinforced Metal-Matrix Composites

As previously shown in Fig. 9.1, MMCs reinforced with continuous fibers provide the highest strength and stiffness. Some mechanical properties of several aluminum-matrix continuous fiber MMCs are given in Table 9.8. In this table, only unidirectional properties are shown, which is a

Table 9.8 Properties of continuous fiber aluminum-matrix composites

Property	B/6061 Al	SCS-2/6061 Al(a)	P100 Gr/6061 Al	Nextel 610/ Al(b)
Fiber content, vol%	48	47	43.5	60
Longitudinal modulus, 10^6 psi	31	29.6	43.6	35
Transverse modulus, 10^6 psi	...	17.1	7.0	23.2
Longitudinal strength, ksi	220	212	79	232
Transverse strength, ksi	...	12.5	2	17.4

(a) SCS-2 is a silicon carbide fiber. (b) Nextel 610 is an alumina fiber. Source: Ref 9.5

little misleading because cross-plied constructions are normally required for real applications. Of the reinforcements shown, boron and silicon carbide (SCS-2) are monofilaments, and graphite and Nextel 610 are multifilaments. The smaller and more numerous multifilament tows are difficult to impregnate using solid-state processing techniques, such as diffusion bonding, because of their small size and the tightness of their tow construction. One of the main advantages for both continuous fiber aluminum- and titanium-matrix composites is their ability to be used at higher temperatures than their base metals (Fig. 9.22). However, because of their high cost, most applications have been limited, even in the aerospace industry.

Aluminum Oxide Fibers. Today the most common approach for producing polycrystalline tow-based ceramic fibers is by spinning and heat treating chemically derived precursors, which, in the case of oxide fibers, are sol gel precursors. The use of chemical precursor technology allows the commercial preparation of ceramic fibers with properties not accessible by traditional fiber-forming technology, such as spinning of molten glasses. Traditional slurry processing methods can also be used to spin fibers (Almax, FP). A key characteristic of ceramic fibers is their ultrafine microstructure, with grain sizes sometimes in the nanometer range. Fine grains are required for good tensile strength (>2 GPa, or 290 ksi). These processes also allow fiber production in the form of continuous-length multifilament tows or rovings, which are typically coated with a thin polymer-based sizing and then supplied to customers on spools. These sized tows are flexible and easily handled so that they can be woven or braided into fabrics, tapes, sleeves, and other complex shapes. Most small-diameter

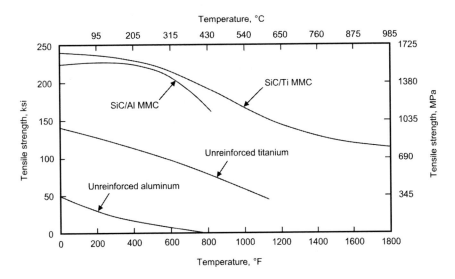

Fig. 9.22 Strength retention at elevated temperature for continuous fiber SiC/Al and SiC/Ti metal-matrix composites (MMCs). Source: Ref 9.14

fiber manufacturers also supply fibers in fabric form. Monofilament alumina is also available as a 125 μm (5 mils) diameter single-crystal fiber, which is grown directly from the molten oxide. This fiber has been studied for reinforcing some of the intermetallic-matrix composites such as titanium aluminides and nickel aluminides.

Key properties of commercially available alumina-base fibers are listed in Table 9.9. Tow-based alumina fibers are used extensively for aluminum alloy reinforcement, and while generally nonreactive with molten aluminum, there is the danger of a degrading reaction with alloying elements in some aluminum alloys (e.g., those containing magnesium). These fibers are used principally where low density, high tensile and compression strengths, and high stiffness are desired in the composite. Additionally, good off-axis properties result from the strong bonding between aluminum and alumina, and the chemical stability of the fiber aids in producing good elevated-temperature properties.

Silicon Carbide Fibers. For fibers based on silicon carbide (SiC), the most common approach for producing polycrystalline tow-based ceramic fibers is by spinning and heat treating chemically derived precursors based on polymer precursors. As with the alumina-base tow materials, fine grain sizes are produced that promote high strength, and the ability to make multifilament rovings permits woven, braided, and fabric fiber forms. The multifilament form of SiC has been used to reinforce aluminum alloys.

There are also commercially available forms of monofilament SiC. These are made using a continuous chemical deposition process in which a mixture of hydrogen and chlorinated alkyl silanes are reacted at the surface of a heated substrate wire. The substrate (fiber core) can be either tungsten or a spun carbon filament. The bulk fiber then consists of fine-grained β-SiC. Key properties are listed in Table 9.10. These fibers are used principally to reinforce titanium alloys where the low density, high

Table 9.9   Commercial alumina-base fibers for reinforcement of metal-matrix composites

Tradename (manufacturer)	Composition	Density, g/cm³	Average diameter, μm	Tow count	Average tensile strength at room temperature GPa	ksi	Tensile modulus at room temperature GPa	10⁶ psi	Thermal expansion, ppm/°C (to 1000 °C)
Alumina-silica-base fibers									
Altex (Sumitomo)	85% Al_2O_3 + 15% SiO_2	3.3	15	500, 1000	2.0	290	210	30	7.9
Nextel 312 (3M)	62% Al_2O_3 + 24% SiO_2 + 14% B_2O_3	2.7	10–12	420, 780	1.7	247	150	22	3
Nextel 440 (3M)	70% Al_2O_3 + 28% SiO_2 + 2% B_2O_3	3.05	10–12	420, 780	2.0	290	190	28	5.3
α-alumina-base fibers									
Almax (Mitsui Mining)	99.5% Al_2O_3	3.6	10	1000	1.8	260	330	48	8.8
Nextel 610 (3M)	>99% Al_2O_3	3.9	12	420, 780, 2600	3.1	450	373	54	7.9
Saphikon (Saphikon)	100% Al_2O_3	3.98	125	Monofilament	3.5	507	460	67	9

Source: Ref 9.15

tensile strength, high stiffness, and elevated-temperature capability offer the potential of great weight savings in high-speed rotating jet engine structures. To prevent excessive fiber embrittling reactions between the SiC monofilaments and titanium matrices during high-temperature processing and elevated-temperature service, the protective coatings shown in Table 9.11 are applied to the surfaces.

Boron fibers are also made using chemical vapor deposition. A fine tungsten wire provides a substrate, and boron trichloride gas is the boron source. Continuous production provides for boron filaments of 100 and 140

Table 9.10 **Commercial boron- and silicon carbide-base fibers for reinforcement of metal-matrix composites**

Trade name (manufacturer)	Composition	Density, g/cm³	Average diameter, μm	Tow count	Average tensile strength at room temperature		Tensile modulus at room temperature		Thermal expansion, ppm/°C (to 1000 °C)
					GPa	ksi	GPa	10⁶ psi	
Boron-base fibers									
Boron (Textron Specialty Materials)	Boron	2.57/2.50	100/140	Monofilament	3.6	522	400	58	4.5
SiC-base fibers									
Nicalon, NL200 (Nippon Carbon)	Si-C-O, 10 O	2.55	14	500	3.0	435	220	32	3.2
Hi-Nicalon (Nippon Carbon)	SiC, 30 C, 0.5 O	2.74	14	500	2.8	406	270	39	3.5
Hi-Nicalon-S (Nippon Carbon)	SiC, 0.2 O	3.05	13	500	2.5	363	400–420	58–61	...
Tyranno Lox M (Ube Industries)	Si-Ti-C-O, 10 O, 2 Ti	2.48	11	400, 800	3.3	479	187	27	3.1
Tyranno SA 1-3 (Ube Industries)	SiC, 0.3 O, <2 Al	3.02	10–7.5	800, 1600	2.8	406	375	54	...
Sylramic (Dow Corning)	SiC, 3 TiB₂, 2 B	3.05	10	800	3.2	464	400	58	5.4
SCS-6, SCS-9 (Textron Specialty Materials)	SiC, trace Si/trace C	3.02.8	140	Monofilament	3.5	507	380	55	4.6
Ultra SCS (Textron Specialty Materials)	SiC, trace Si/trace C	3.0	140	Monofilament	6.2	900	410	59	4.6
Trimarc (Atlantic Research Corp.)	SiC, trace Si/trace C	3.3	125	Monofilament	3.5	507	425	62	...
Sigma 1140, 1240 (DERA)	SiC, trace Si/trace C	3.4	100	Monofilament	3.5	507	400	58	...

Source: Ref 9.15

Table 9.11 **Protective coatings for SiC monofilament fibers**

Fiber	Diameter, μm	Fiber core	Fiber coating
SCS-6	140	33 μm carbon core	3 μm carbon layer + graded C-Si layer
Ultra SCS	140	33 μm carbon core	3 μm carbon layer + graded C-Si layer
Trimarc	125	12.5 μm W + C layer	3 μm carbon layer + (H-S-H)² layer
Sigma 1140+	100	14 μm tungsten core	3–5 μm carbon layer
Sigma 1240	100	14 μm tungsten core	1 μm carbon layer + TiB$_x$ layer

Source: Ref 9.15

μm (4 and 5.6 mils) diameter. Properties are listed in Table 9.10. Boron fibers have been used principally in reinforcing aluminum alloys where low density, high tensile and compression strengths, and high stiffness result. The bare fiber may be used with solid-state composite forming, but with liquid processing routes, the fiber must be coated with either B_4C or SiC to eliminate reaction with the fiber and the accompanying loss of fiber properties.

Carbon Fibers. Two classes of carbon fiber, polyacrylonitrile (PAN) and pitch-based fiber, derive quite different structures and properties, with the PAN being higher strength, lower modulus with little graphitization of the fiber surface, and the pitch being moderate strength, high modulus with greater levels of surface graphitization. Properties of some carbon fibers are shown in Table 9.12. The combination of high strength, modulus, low CTE, low density, and low cost seemingly make them a desirable choice to reinforce metals. However, structural carbon fibers are still not widely used in MMCs, for example, aluminum and magnesium alloys, despite much research and development activity. Several technical issues remain to be solved, including the reactivity during composite processing, the reactivity at elevated temperatures, the oxidation resistance of the fiber at elevated temperatures, the corrosion resistance due to the inherent galvanic coupling between carbon and the metal matrix (aluminum or magnesium), and the corrosion resistance related to the leaching of any fiber-matrix reaction products (e.g., aluminum carbide).

Carbon fibers have also been used to reinforce copper for thermal management applications, but the difficulties encountered relate to both the poor bond strength and high wetting angle between carbon and copper, necessitating the use of coatings or reactive elements to increase the bond strength. Generally, the low bond strength between carbon fibers and any of the common metals (aluminum, magnesium, copper) is a difficulty, because poor off-axis properties result. In aluminum, the pitch fibers appear to be less reactive, but their handling characteristics (high modulus, brittle) make fabrication a more difficult task. However, the high-modulus

Table 9.12 Commercial carbon and graphite fibers for reinforcement of metal-matrix composites

Trade name (manufacturer)	Composition	Density, g/cm³	Average diameter, µm	Tow count	Average tensile strength at room temperature		Tensile modulus at room temperature		Thermal expansion, ppm/°C (to 1000 °C)
					GPa	ksi	GPa	10⁶ psi	
T300 (Amoco)	92 C, 8 N	1.76	7	1,000–12,000	3.65	530	231	34	−0.6
IM7 (Hercules/ Hexcel)	C, trace N	1.77	5	6,000, 12,000	5.30	770	275	40	−0.2
UHM (Hercules/ Hexcel)	C, trace N	1.87	4.5	3,000, 12,000	3.45	500	440	64	−0.5
P120 (Amoco)	99+% C	2.17	10	2,000	2.41	350	830	120	−1.45

Source: Ref 9.15

fibers also have a high thermal conductivity, which makes them attractive for use in combined structural/thermal management applications.

9.8 Processing of Continuous Fiber-Reinforced Aluminum Composites

Boron/aluminum (B/Al) was one of the first systems evaluated. A typical cross section of a unidirectional B/Al composite is shown in Fig. 9.23. The boron carbide (B_4C) coating on the fibers is used to minimize the reaction between the boron fiber and the aluminum matrix during hot pressing. Applications include tubular truss members in the midfuselage structure of the Space Shuttle orbiter and cold plates in electronic microchip carrier multilayer boards. Fabrication processes for B/Al composites are based on hot press diffusion bonding of alternating layers of aluminum foil and boron fiber mats (foil-fiber-foil processing) or plasma spraying methods.

Continuous SiC fibers have largely replaced boron fibers, because they have similar properties but are not degraded by the aluminum matrix during processing, and they are less expensive. The most prevalent SiC fiber is the SCS series, which can be manufactured with any of several surface chemistries to enhance bonding with a particular matrix, such as aluminum or titanium.

Hot molding is a low-pressure, hot pressing process designed to fabricate SiC/Al parts at significantly lower cost than is possible with solid-state diffusion bonding. Because the SCS-2 fibers can withstand molten aluminum for long periods, the molding temperature can be raised into the liquid plus

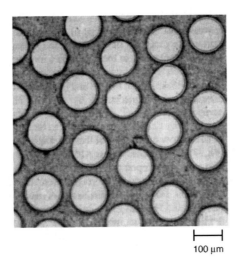

100 μm

Fig. 9.23 Cross section of a continuous fiber-reinforced boron/aluminum composite. Shown here are 142 μm diameter boron filaments coated with B_4C in a 6061 aluminum alloy matrix. Source: Ref 9.8

solid region of the alloy to ensure aluminum flow and consolidation at low pressure, thereby eliminating the need for high-pressure die molding equipment. The hot molding process is analogous to the autoclave molding of graphite/epoxy, in which components are molded in an open-faced tool. The mold in this case is a self-heating, slip-cast ceramic tool that contains the profile of the finished part. A plasma-sprayed aluminum preform is laid into the mold, heated to near-molten aluminum temperature, and pressure consolidated in an autoclave by a metallic vacuum bag. SiC/Al MMCs exhibit increased strength and stiffness as compared with unreinforced aluminum, with no weight penalty. In contrast to the base metal, the composite retains its room-temperature tensile strength at temperatures up to 260 °C (500 °F).

Graphite/aluminum (Gr/Al) MMC development was initially prompted by the commercial appearance of strong and stiff graphite fibers in the 1960s. Graphite fibers offer a range of properties, including an elastic modulus up to 966 GPa (140×10^6 psi) and a negative CTE down to $-1.62 \times 10^{-6}/°C$ ($-0.9 \times 10^{-6}/°F$). However, graphite and aluminum in combination are difficult materials to process into a composite. A deleterious reaction between graphite and aluminum, poor wetting of graphite by molten aluminum, and oxidation of the graphite are significant technical barriers to the production of these composites. Two processes are currently used for making commercial aluminum MMCs: liquid metal infiltration of the matrix on spread tows, and hot press bonding of spread tows sandwiched between sheets of aluminum. With both precursor wires and metal-coated fibers, secondary processing, such as diffusion bonding or pultrusion, is needed to make structural elements. Squeeze casting also is feasible for the fabrication of this composite. Precision aerospace structures with strict tolerances on dimensional stability need stiff, lightweight materials that exhibit low thermal distortion. Gr/Al MMCs have the potential to meet these requirements. Unidirectional P100 Gr/6061 aluminum pultruded tube exhibits an elastic modulus in the fiber direction significantly greater than that of steel, and it has a density approximately one-third that of steel.

The Al_2O_3/Al MMCs can be fabricated by a number of methods, but liquid or semisolid-state processing techniques are commonly used. A fiber-reinforced aluminum MMC is now used in pushrods for high-performance racing engines. The 3M Company produces the material by infiltrating Nextel 610 alumina fibers with an aluminum matrix at 60 vol%. Hollow pushrods of several diameters are made, where the fibers are axially aligned along the pushrod length. Hardened steel end caps are bonded to the ends of the MMC tubes.

Fabrication of complex-shaped components is commonly achieved by diffusion bonding of monolayer composite tapes. The tapes may be prepared by a number of methods, but the most commonly used is filament winding, where the matrix is incorporated in a sandwich construction by laying up thin metal sheets between filament rows or by an arc spraying technique. In the arc spray process, molten matrix alloy droplets are sprayed onto a cylindrical drum wrapped with fibers. The operation is carried out in a controlled atmosphere chamber to avoid oxidation and corrosion. The drum is rotated and passed in front of an arc spray head to produce a controlled porosity monotape. When lamination of the fibers between matrix alloy sheets is used to produce the monotape, a binder is used to hold the fibers in place prior to consolidation. This technique can also be combined with consolidation by HIP. Filament winding and monotape lay-up into multiple layers permit close control over the position and orientation of the fibers in the final composite.

A combination of the arc spray technique and diffusion bonding using SCS-6 SiC monofilament fibers was used to produce composites for various prototype applications in the aerospace and defense sectors of the market. In one example, a rocket motor shell was produced by filament winding of SCS-6 fibers and plasma spraying with an aluminum alloy. The rocket motor was reported to have been successfully fired and to have withstood the extreme temperatures, pressures, and vibrations associated with missile firing. In addition, the assembled motor, using the MMC shell, weighed approximately 10% less than a conventional steel motor.

9.9 Processing of Continuous Fiber-Reinforced Titanium Composites

Continuous fiber-reinforced titanium-matrix composites (TMCs) offer the potential for strong, stiff, lightweight materials for usage temperatures as high as 800 to 1000 °C (1500 to 1800 °F). The principal applications for these materials would be for hot structure, such as hypersonic airframe structures, and for replacing superalloys in some portions of aerospace engines. The use of continuous fiber TMCs has been somewhat restricted by both the high cost of the materials and the fabrication and assembly procedures. Typical properties of a unidirectional SiC/Ti composite are shown in Table 9.13.

Foil-Fiber-Foil Process. One method that has been used to fabricate continuous fiber TMCs is the foil-fiber-foil method, depicted in Fig. 9.24. In this method, a silicon carbide fiber mat (Fig. 9.25) is held together

Table 9.13 Properties of unidirectional SiC/Ti metal-matrix composite

Property	SCS-6/Ti-6Al-4V
Fiber content, vol%	37
Longitudinal modulus, 10^6 psi	32
Transverse modulus, 10^6 psi	24
Longitudinal strength, ksi	210
Transverse strength, ksi	60
Source: Ref 9.5	

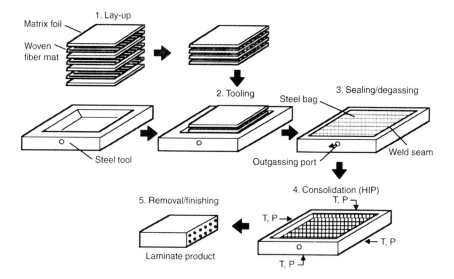

Fig. 9.24 Foil-fiber-foil method for titanium-matrix composite fabrication. HIP, hot isostatic pressing; P, pressure; T, temperature. Source: Ref 9.8

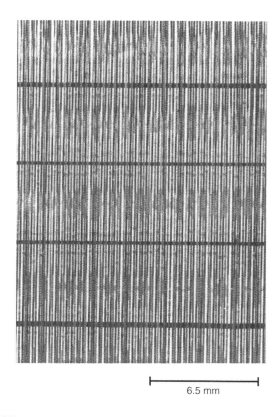

6.5 mm

Fig. 9.25 SiC uniweave fabric showing cross-weave. Source: Ref 9.8

with a cross-weave of molybdenum, titanium, or titanium-niobium wire or ribbon. The fabric is a uniweave system in which the relatively large-diameter SiC monofilaments are straight and parallel and held together by a cross-weave of metallic ribbon. Ti-15V-3Cr-3Sn-3Al foil is normally cold rolled to a thickness of 0.11 mm (0.0045 in.). The plies are cut, laid up on a consolidation tool, and consolidated by either vacuum hot pressing or HIP. High-temperature, short-time roll bonding was used some years ago, but only to a very limited extent. Typical fiber contents for SiC/Ti laminates range from 35 to 40 vol%. A micrograph of a consolidated SCS-6 TMC laminate is shown in Fig. 9.26.

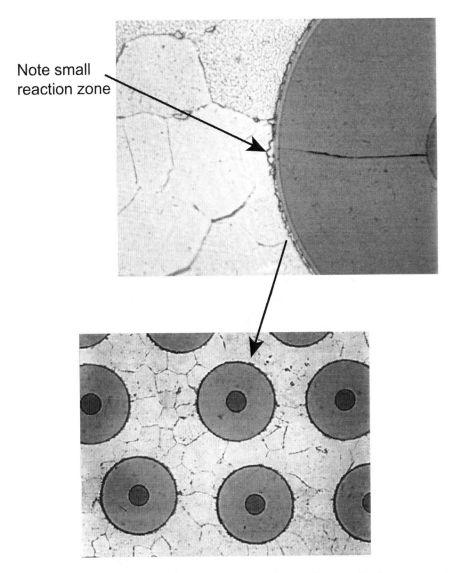

Fig. 9.26 Silicon carbide monofilament/titanium-matrix composite. Source: Ref 9.16

In the vacuum hot pressing technique, the lay-up is sealed in a stainless steel envelope and placed in a vacuum hot press. After evacuation, a small positive pressure is applied via the press platens. This pressure acts to hold the filaments in place during the initial 450 to 550 °C (800 to 1000 °F) soak used to decompose the binder and remove it under the action of a dynamic vacuum. The temperature is then gradually increased to a level where the titanium flows around the fibers under an increased pressure, and the foil interfaces are diffusion bonded together. Each fabricator uses a specific set of consolidation parameters, although a typical range is 900 to 950 °C (1650 to 1750 °F) at 41 to 69 MPa (6 to 10 ksi) pressure for 60 to 90 min.

Hot isostatic pressing has largely replaced vacuum hot pressing as the consolidation technique of choice. The primary advantages of HIP consolidation are that the gas pressure is applied isostatically, alleviating the concern about uneven platen pressure, and the HIP process is much more amenable to making complex structural shapes. Typically, the part to be hot isostatically pressed is canned (or a steel bag is welded to a tool), evacuated, and then placed in the HIP chamber. For titanium, typical HIP parameters are 850 to 950 °C (1600 to 1700 °F) at 103 MPa (15 ksi) gas pressure for 2 to 4 h. Because HIP is a fairly expensive batch processing procedure, it is normal practice to load a number of parts into the HIP chamber for a single run.

Green Tape Process. Another method for making TMCs involves placing a layer of titanium foil on a mandrel and filament winding the silicon carbide fiber over the foil in a collimated manner to produce a unidirectional single ply. An organic fugitive binder, such as an acrylic adhesive, is used to maintain the fiber spacing and alignment once the preform is cut from the mandrel. In this method (Fig. 9.27), often called the green tape method, the fibers are normally wound onto a foil-covered rotating drum, oversprayed with resin, and the layer cut from the drum to provide a flat sheet of monotape. The organic binder is burned off prior to the HIP cycle or during the early portions of the vacuum hot pressing cycle.

Plasma spraying replaces the resin binder with a plasma-sprayed matrix. Plasma spraying removes the potential of an organic residue causing contamination problems during the consolidation cycle and speeds the process by not having to outgas an organic binder. One potential disadvantage of plasma spraying is that titanium, being an extremely reactive metal, can pick up oxygen from the atmosphere, potentially leading to embrittlement problems. This method has been primarily evaluated for titanium aluminide-matrix composites, due to the difficulty of rolling these materials into thin foil.

Metal Wire Process. A continuous fiber TMC manufactured by a novel metal wire process has been developed to replace stainless steel in a piston actuator rod in a Pratt and Whitney F119 engine for the F-22 aircraft. A

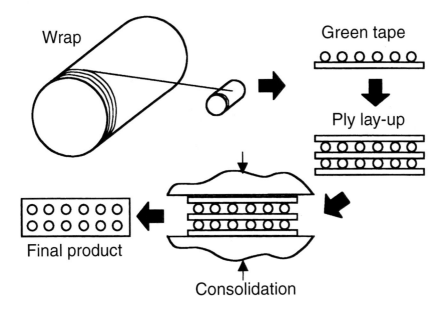

Fig. 9.27 Green tape method for titanium-matrix composite fabrication. Source: Ref 9.8

Ti-6Al-2Sn-4Zr-2Mo (Ti-6242) alloy is hot drawn in a conventional wire drawing process to a diameter of 178 μm (0.007 in.). Trimarc-1 SiC monofilament reinforcement is used. It has a diameter of 129 μm (5.1 mils) and is produced by a chemical vapor deposition process on a tungsten wire core. An outer carbon-base coating protects the fiber from chemical interaction with the matrix during consolidation and service. The Ti-6242 metal wire is combined with 34 vol% of the SiC monofilament by wrapping on a rotating drum. The wires are held together with an organic binder and are then cut and removed from the drum to make a preform "cloth." This cloth is wrapped around a solid titanium mandrel, so that the SiC and Ti-6242 wires are parallel to the long axis of the mandrel. A short Ti-6242 cylinder, with an outer diameter of 10 cm (4 in.) and 5 cm (2 in.) high, is placed around one end of the cloth-wrapped mandrel, from which the piston head will be formed. After HIP consolidation, the piston head is machined from the stocky cylinder at one end of the piston, and a threaded connection is machined from the titanium mandrel at the other end. The remainder of the mandrel is removed by gun drilling. The final component, shown in Fig. 9.28(a), is 30.5 cm (12 in.) long, and the shaft is 3.79 cm (1.49 in.) in diameter. The use of a TMC in the actuator piston is a landmark application and represents the first production aerospace application of TMCs.

Based on the experience gained from the success of the TMC actuator piston rod, TMCs have been certified as nozzle links on the General

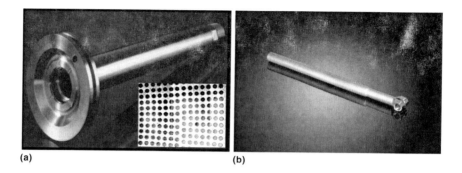

Fig. 9.28 Jet engine applications of titanium-matrix composites. (a) A nozzle actuator piston rod used on the Pratt & Whitney F119 engine for F-22 aircraft. The part is made of a Ti-6Al-2Sn-4Zr-2Mo alloy reinforced with SiC monofilaments that are 129 µm (5.1 mils) in diameter. The inset shows the typical distribution of the SiC reinforcements (inside diameter of the rod is to the left in this inset micrograph). (b) A nozzle actuator link from the General Electric F110 engine, made of a similar metal-matrix composite, used for F-16 aircraft. Component dimensions are provided in the text. Courtesy of Atlantic Research Corporation. Source: Ref 9.17

Electric F110 engine, which is used for F-16 aircraft. The original link was produced from a square tube of Inconel 718, which was formed from sheet and welded along its length. The manufacturing process for the TMC nozzle links begins by winding Trimarc-2 SiC monofilament, which is similar to Trimarc-1 but is produced on a carbon core, on a drum as with the piston rod. Rather than using metal wire, Ti-6242 powder is sprayed with an organic binder over the wound SiC fibers to produce a preform cloth. This cloth is then cut and removed from the drum, as before, and wrapped around a mandrel. Ti-6242 fittings are added at each end for a clevis attachment and a threaded end, and the entire assembly is consolidated via HIP. After machining the clevis and the threaded end, the mandrel is removed. A finished part is shown as Fig. 9.28(b).

9.10 Secondary Fabrication of Titanium-Matrix Composites

Successful joining of TMC components by diffusion bonding can be accomplished at pressures and temperatures lower than normal HIP runs. To process preconsolidated C-channels into spars, as shown in Fig. 9.29, they can be joined together in a back-to-back fashion. For secondary diffusion bonding in a HIP chamber, the parts are assembled and encapsulated in a leak-free steel envelope or bag. Because the bag is subjected to the same high pressure and temperature as the parts, the ability of the steel bag to withstand the HIP forces is critical for success. If the bag develops a leak, the isostatic pressure is lost, and so is the bonding pressure. To minimize the risk associated with the extreme pressures and temperatures required for initial consolidation, lower temperatures and pressures can be used for secondary diffusion bonding. Lower temperatures and pressures reduce the risk of bag failures during the secondary HIP bonding cycle. However, because the TMC is inherently stiffer than conventional sheet metal components, it generally requires higher pressures to achieve secondary diffusion bonding. A machined filler is placed in the corner to fill the void created by the C-channels. By not pinning the TMC components together, and allowing them to slide a little in relation to one another, complete bonding is achieved. Incomplete bonding or damage can occur if movement is not permitted. Successful bonding can be achieved with HIP pressures as low as 35 MPa (5 ksi) at 870 °C (1600 °F) for 2 h.

Superplastic forming and diffusion bonding (SPF/DB) can be used to take advantage of elevated-temperature characteristics inherent in certain titanium alloys. Structural shapes, which combine superplastic forming and bonding of the parent metal, can be fabricated from thin titanium alloy sheets to achieve high levels of structural efficiency. Components fabricated from TMCs do not lend themselves to superplastic forming. However, because they retain the diffusion bonding capability of the matrix alloy, TMC components can be readily bonded with superplastically formed sheet metal substructures. A TMC-reinforced SPF/DB structural panel retains the structural efficiency of the TMC, while possessing the fabrication simplicity of a superplastically formed part. Two potential methods of fabricating TMC-stiffened SPF/DB panels are shown in Fig. 9.30. The core pack, which forms the substructure, can be fabricated from a higher-temperature titanium alloy, such as Ti-6Al-2Sn-4Zr-2Mo, rather than Ti-6Al-4V, which is normally used in SPF/DB parts. Ti-6Al-2Sn-4Zr-2Mo has good superplastic forming characteristics at a moderate processing temperature (900 °C, or 1650 °F). The final shape and size of the substructure is determined by the resistance seam welding pattern of

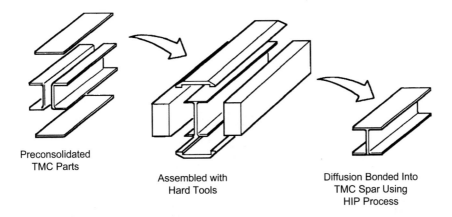

Preconsolidated
TMC Parts

Assembled with
Hard Tools

Diffusion Bonded Into
TMC Spar Using
HIP Process

Fig. 9.29 Secondary diffusion bonding of titanium-matrix composite (TMC) spars. HIP, hot isostatic pressing. Source: Ref 9.16

the core pack prior to the SPF/DB cycle, which is inflated during the SPF/DB cycle with argon gas to form the substructure.

Conventional nondestructive testing techniques, including both through-transmission and pulse echo ultrasonics, can be used to detect manufacturing defects. Conventional ultrasonic and x-ray have the ability to find many TMC processing defects, including lack of consolidation, delaminations, fiber swimming, and fiber breakage. Ultrasonics can find defects as small as 1.2 mm ($\frac{3}{64}$ in.) in diameter, with through transmission giving the best resolution. Normal radiographic x-ray inspection can also be used to examine TMC components.

A TMC is extremely difficult to machine. The material is highly abrasive, and tool costs can be high. Improper cutting not only damages tools but can also damage the part as well. Abrasive waterjet cutting has been found to work well on TMCs. Typical machining parameters are 310 MPa (45 ksi) water pressure, dynamically mixed with #80 garnet grit, at a feed rate of 13 to 30 mm/min (0.5 to 1.2 in./min). The waterjet cutter has multiaxis capability and can make uninterrupted straight and curved cuts. A diamond cutoff wheel also produces excellent cuts. However, this method is limited to straight cuts and is rather slow. The diamond cutoff wheel is mounted on a horizontal mill, and cutting is done with a flood coolant using controlled speeds and feeds. Because this method is basically a grinding operation, cutting rates are typically slow; however, the quality of the cut edge is excellent.

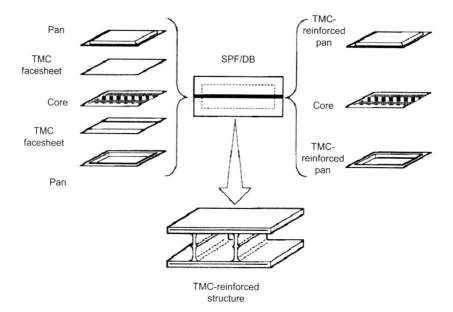

Fig. 9.30 Methods for making superplastic formed and diffusion bonded (SPF/DB) titanium-matrix composite (TMC)-reinforced parts. Source: Ref 9.16

Wire electrical discharge machining (EDM) is also a flexible method of cutting TMCs. The EDM is a noncontact cutting method that removes material through melting or vaporization by high-frequency electric sparks. Brass wire that is 0.25 mm (0.010 in.) in diameter can be used at a feed rate of 0.5 to 1.3 mm/min (0.020 to 0.050 in./min). The EDM unit is self-contained, with its own coolant and power system. EDM has the advantage of being able to make small-diameter cuts, such as small scallops and tight radii. Like waterjet cutting, the EDM method is programmable and can make uninterrupted straight or curved cuts.

Several methods can be used to generate holes in TMCs. For thin components (e.g., three plies), with only a few holes, the cobalt grades of high-speed steel (e.g., M42) twist drills can be used with power feed equipment; however, tool wear is so rapid that several drills may be required to drill a single hole. Punching has also been used successfully on thin TMC laminates. Using conventional dies, punching is fast and clean, with no coolant required. Punching results in appreciable fiber damage and metal smearing. However, most of the disturbed metal can be removed and the hole cleaned up by reaming. However, some holes may require several passes, and several reamers, to sufficiently clean up the hole. In some instances, the diameter of the reamer is actually reduced due to wear of the reamer cutting edges, rather than increasing the diameter of the hole. Punching is not used extensively because of the large number of fastener holes required in the internal portion of structures. In addition, load-bearing sections are normally too thick for punching.

Neither conventional twist drills, nor punching, will consistently produce high-quality holes in TMCs, especially in thicker material. The use of diamond core drills has greatly improved hole-drilling quality in thick TMC components. High-quality holes can consistently be produced with diamond core drills. The core drills are tubular, with a diamond matrix built up on one end. This construction is similar to some grinding wheels. A typical core drill and coolant chuck is shown in Fig. 9.31. The important parameters for successful diamond core drilling are drill design, coolant-delivery system, drill plate design, type of power feed drilling equipment used, and the speeds and feeds used during drilling. During drilling, the core drill abrasively grinds a cylindrical core plug from the material. Some fabricators even mix an abrasive grit with the water coolant to improve the material-removal rate. A properly drilled hole can hold quite good tolerances, depending on the thickness of the TMC: ±0.05 mm (0.0021 in.) in 4 plies, ±0.08 mm (0.0030 in.) in 15 plies, and ±0.1 mm (0.0055 in.) in 32 plies.

Another method for joining thin TMC components into structures is resistance spot welding. Conventional 50 kW resistance welding equipment, with water-cooled copper electrodes, has been successfully used to spot weld thin TMCs. Fabricators often use conventional titanium (e.g., Ti-6Al-4V) to set the initial welding parameters. As with any spot welding

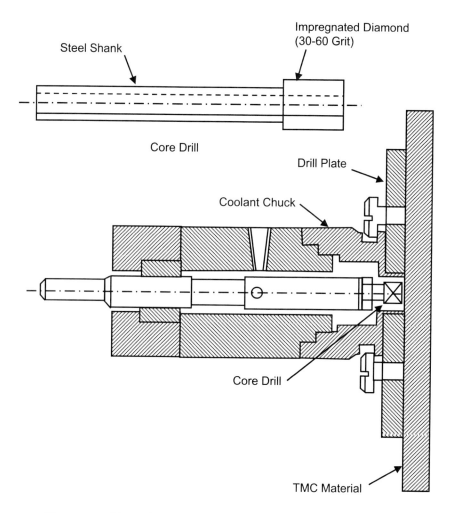

Fig. 9.31 Diamond-impregnated core drill and coolant chuck. TMC, titanium-matrix composite. Source: Ref 9.16

operation, it is important to thoroughly clean the surfaces before welding. Initial welding parameters should be verified by metallography and lap shear testing.

9.11 Processing of Other Metal-Matrix Composites

In addition to aluminum and titanium, several other metals and alloys have been investigated as matrix alloys for MMCs. This section describes work done to develop magnesium, copper, and superalloy MMCs. These MMCs have found only limited commercial application to date.

Magnesium-matrix composites are being developed to exploit essentially the same properties as those provided by aluminum MMCs: high stiffness, light weight, and low CTE. In practice, the choice between aluminum and magnesium as a matrix is usually made on the basis of weight versus corrosion resistance. Magnesium is approximately two-thirds as dense as aluminum, but it is more active in a corrosive environment. Magnesium has a lower thermal conductivity, which is sometimes a factor in its selection. Magnesium MMCs include continuous fiber graphite/magnesium for space structures, short-staple-fiber Al_2O_3/Mg for automotive engine components, and discontinuous SiC or B_4C/Mg for engine components and low-expansion electronic packaging materials. Matrix alloys include AZ31, AZ91, ZE41, QE22, and EZ33. Processing methods parallel those used for the aluminum MMC counterparts.

Copper-matrix composites have been produced with continuous tungsten, silicon carbide, and graphite-fiber reinforcements. Of the three composites, continuous graphite/copper MMCs have been studied the most. Interest in continuous graphite/copper MMCs gained impetus from the development of advanced graphite fibers. Copper has good thermal conductivity, but it is heavy and has poor elevated-temperature mechanical properties. Pitch-based graphite fibers have been developed that have room-temperature axial thermal conductivity properties better than those of copper. The addition of these fibers to copper reduces density, increases stiffness, raises the service temperature, and provides a mechanism for tailoring the CTE. One approach to the fabrication of graphite/copper MMCs uses a plating process to envelop each graphite fiber with a pure copper coating, yielding MMC fibers flexible enough to be woven into fabric. The copper-coated fibers must be hot pressed to produce a consolidated component. Graphite/copper MMCs have the potential to be used for thermal management of electronic components, satellite radiator panels, and advanced airplane structures.

Superalloy-Matrix Composites. In spite of their poor oxidation resistance and high density, refractory metal (tungsten, molybdenum, and niobium) wires have received a great deal of attention as fiber-reinforcement materials for use in high-temperature superalloy MMCs. Although the theoretical specific strength potential of refractory alloy fiber-reinforced composites is less than that of ceramic fiber-reinforced composites, the more ductile metal fiber systems are more tolerant of fiber-matrix reactions and thermal expansion mismatches. When refractory metal fibers are used to reinforce a ductile and oxidation-resistant matrix, they are protected from oxidation, and the specific strength of the composite is much higher than that of superalloys at elevated temperatures.

Fabrication of superalloy MMCs is accomplished by solid phase, liquid phase, or deposition processing. The methods include investment casting, the use of matrix metals in thin sheet form, the use of matrix metals in powder sheet form made by rolling powders with an organic binder, PM techniques, slip casting of metal alloy powders, and arc spraying.

9.12   Fiber-Metal Laminates

Fiber-metal laminates are laminated materials consisting of thin layers of metal sheet and unidirectional fiber layers embedded in an adhesive system. Glass laminate aluminum reinforced (GLARE) is a type of aluminum fiber-metal laminate in which unidirectional S-2 glass fibers are embedded in FM-34 epoxy structural film adhesive. A typical construction is shown in Fig. 9.32. The S-2 glass is bonded to the aluminum sheets with the film adhesive. The aluminum metal layers are chemically cleaned (chromic acid anodized or phosphoric acid anodized) and primed with BR-127 corrosion-inhibiting primer. The adhesion between the FM-34 adhesive and treated metal surface, and between FM-34 and S-2 glass fibers, is so high that these bondlines often remain intact until cohesive adhesive failure occurs.

GLARE is normally available in six different standard grades, as outlined in Table 9.14. They are all based on unidirectional S-2 glass fibers embedded with FM-34 adhesive, resulting in a 0.13 mm (5 mils) thick prepreg with a nominal fiber volume fraction of 59%. The prepreg is layed-up in different orientations between the aluminum alloy sheets. From 1990 to 1995, GLARE laminates were produced only as flat sheets. It was believed that the aircraft manufacturer would use these flat sheets to manufacture fuselage panels by applying the curvature, thickness steps, and joints, using conventional methods developed for metal structures (forming, bonding, riveting, etc.). Several studies showed the benefits in performance and weight of these GLARE shells, but they also indicated the high cost of these parts in comparison with conventional aluminum structures.

To reduce manufacturing costs, a self-forming technique was developed in which thin aluminum sheets are drape-formed onto a tool containing the moldline contour. Autoclave heat and pressure is then used to consolidate the aluminum sheet, the S-2 glass reinforcement, and the FM-34 film adhesive sandwich to the tool contour. Additional adhesive is also added at certain locations to adhere interrupted metal sheets to each other, adhere thin aluminum internal or external doublers to the aluminum layers of the laminate, and to fill gaps in the laminate that would otherwise remain unfilled. To join the aluminum sheets to each other, overlap splices are

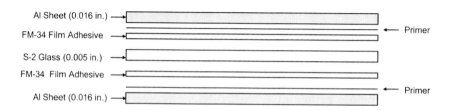

Fig. 9.32   Typical glass laminate aluminum-reinforced ply construction. Source: Ref 9.16

Table 9.14   Commercial glass laminate aluminum-reinforced (GLARE) grades

GLARE grade	Subgrade	Material sheet thickness, in., and alloy	Prepreg orientation in each layer	Principal benefits
GLARE 1	...	0.020–0.016, 7475-T761	0/0	Fatigue, strength, yeild strength
GLARE 2	GLARE 2A	0.0008–0.020, 2024-T3	0/0	Fatigue, strength
	GLARE 2B	0.0008–0.020, 2024-T3	90/90	Fatigue, strength
GLARE 3	...	0.0008–0.020, 2024-T3	0/90	Fatigue, impact
GLARE 4	GLARE 4A	0.0008–0.020, 2024-T3	0/90/0	Fatigue, strength in 0° direction
	GLARE 4B	0.0008–0.020, 2024-T3	90/0/90	Fatigue, strength in 90° direction
GLARE 5	...	0.0008–0.020, 2024-T3	0/90/90/0	Impact
GLARE 6	GLARE 6A	0.0008–0.020, 2024-T3	+45/–45	Shear, off-axis properties
	GLARE 6B	0.0008–0.020, 2024-T3	–45/+45	Shear, off-axis properties

Source: Ref 9.17

used. Testing has shown that the splices are not critical during static or fatigue loading. In other words, the splice is not the weakest link in the panel strength.

The primary advantages of GLARE are better fatigue crack propagation resistance than aluminum, superior damage tolerance compared to aluminum, higher bearing strengths than carbon/epoxy, 10% lighter weight than aluminum, and lower cost than carbon/fiber composite but higher cost than aluminum. In addition to GLARE, the introduction of titanium foils in carbon/epoxy laminates (TiGr) has been evaluated, primarily for improved bearing strengths.

ACKNOWLEDGMENTS

Portions of the text for this chapter came from "Metal-Matrix Composites," *Metals Handbook Desk Edition,* 2nd ed., ASM International, 1998; and "Discontinuous Reinforcements for Metal-Matrix Composites" by C.A. Smith and C. McCullough, "Continuous Fiber Reinforcements for Metal-Matrix Composites" by C. McCullough, and "Processing of Metal-Matrix Composites" in *Composites,* Vol 21, *ASM Handbook,* ASM International, 2001.

REFERENCES

9.1   Processing of Metal-Matrix Composites, *Composites,* Vol 21, *ASM Handbook,* ASM International, 2001

9.2   W.H. Hunt and D.B. Miracle, Automotive Applications of Metal-Matrix Composites, *Composites,* Vol 21, *ASM Handbook,* ASM International, 2001

9.3   Introduction to Composites, *Composites,* Vol 21, *ASM Handbook,* ASM International, 2001

9.4   C.A. Smith, Discontinuous Reinforcements for Metal-Matrix Composites, *Composites,* Vol 21, *ASM Handbook,* ASM International, 2001

9.5   Metal-Matrix Composites, *Metals Handbook Desk Edition*, 2nd ed., ASM International, 1998

9.6   S.V. Nair, J.K. Tien, and R.C. Bates, SiC Reinforced Aluminium Metal Matrix Composites, *Int. Met. Rev.*, Vol 30 (No. 275), 1985

9.7   D.R. Herling, G.J. Grant, and W. Hunt, Low-Cost Aluminum Metal Matrix Composites, *Adv. Mater. Process.*, 2001, p 37–40

9.8   Processing of Metal-Matrix Composites, *Composites*, Vol 21, *ASM Handbook*, ASM International, 2001

9.9   T.W. Clyne and P.J. Withers, *An Introduction to Metal Matrix Composites*, Cambridge University Press, 1993

9.10   C.W. Brown, Particulate Metal Matrix Composite Properties, *Proc. P/M Aerosp. Def. Technol. Conf. Exhibit.*, F.H. Froes, Ed., 1990, p 203–205

9.11   B.L. Ferguson and O.D. Smith, Ceracon Process, *Powder Metallurgy*, Vol 7, *Metals Handbook*, 9th ed., ASM International, 1984, p 537

9.12   B.L. Ferguson, A. Kuhn, O.D. Smith, et al., *Int. J. Powder Metall. Powder Technol.*, Vol 24, 1984, p 31

9.13   D.F. Stefanescu and R. Ruxanda, Fundamentals of Solidification, *Metallography and Microstructures*, Vol 9, *ASM Handbook*, ASM International, 2004

9.14   J.V. Foltz and C.M. Blackmon, Metal-Matrix Composites, *Properties and Selection: Nonferrous Alloys and Special-Purpose Materials*, Vol 2, *ASM Handbook*, ASM International, 1990

9.15   C. McCullough, Continuous Fiber Reinforcements for Metal-Matrix Composites, *Composites*, Vol 21, *ASM Handbook*, ASM International, 2001

9.16   F.C. Campbell, *Structural Composite Materials*, ASM International, 2010

9.17   D.B. Miracle, Aeronautical Applications of Metal-Matrix Composites, *Composites*, Vol 21, *ASM Handbook*, ASM International, 2001

SELECTED REFERENCES

- G.B. Bilow and F.C. Campbell, "Low Pressure Fabrication of Borsic/Aluminum Composites," Sixth Symposium of Composite Materials in Engineering Design, May 1972

- J.U. Ejiofor and R.G. Reddy, Developments in the Processing and Properties of Particulate Al-Si Composites, *J. Met.*, Vol 49 (No. 11), 1997, p 31–37

- M.R. Ghomashchi and A. Vikhrov, Squeeze Casting: An Overview, *J. Mater. Process. Technol.*, Vol 101, 2000, p 1–9

- Z.X. Guo and B. Derby, Solid-State Fabrication and Interfaces of Fibre Reinforced Metal Matrix Composites, *Prog. Mater. Sci.*, Vol 39, 1995, p 411–495

- J. Hashim, L. Looney, and M.S.J. Hashmi, Metal Matrix Composites: Production by the Stir Casting Method, *J. Mater. Process. Technol.*, Vol 92–93, 1999, p 1–7
- J. Hashim, L. Looney, and M.S.J. Hashmi, The Wettability of SiC Particles by Molten Aluminum Alloy, *J. Mater. Process. Technol.*, Vol 119, 2001, p 324–328
- J.W. Kaczmar, K. Pietrzak, and W. Wlosinski, The Production and Application of Metal Matrix Composites, *J. Mater. Process. Technol.*, Vol 106, 2000, p 58–67
- V.K. Lindroos and M.J. Talvitie, Recent Advances in Metal Matrix Composites, *J. Mater. Process. Technol.*, Vol 53, 1995, p 273–284
- J.A. McElman, Continuous Silicon Carbide Fiber MMCs, *Composites,* Vol 1, *Engineered Materials Handbook,* ASM International, 1987
- M.A. Mittnick, Continuous SiC Fiber Reinforced Materials, *21st International SAMPE Technical Conference,* Sept 1989, p 647–658
- S.V. Nair, J.K. Tien, and R.C. Bates, SiC Reinforced Aluminium Metal Matrix Composites, *Int. Met. Rev.,* Vol 30 (No. 275), 1985
- R.A. Shatwell, Fibre-Matrix Interfaces in Titanium Matrix Composites Made with Sigma Monofilament, *Mater. Sci. Eng. A,* Vol 259, 1999, p 162–170
- T.S. Srivatsan, T.S. Sudarshan, and E.J. Lavernia, Processing of Discontinuously-Reinforced Metal Matrix Composites by Rapid Solidification, *Prog. Mater. Sci.,* Vol 39, 1995, p 317–409
- D.M. Stefanescu and R. Ruxanda, Fundamentals of Solidification, *Metallography and Microstructures,* Vol 9, *ASM Handbook,* ASM International, 2004
- S. Sullivan, "Machining, Trimming and Drilling Metal Matrix Composites for Structural Applications," ASM International Materials Week 92, Nov 2–5, 1992 (Chicago, IL)
- A. Vassel, Continuous Fibre Reinforced Titanium and Aluminum Composites: A Comparison, *Mater. Sci. Eng. A,* Vol 263, 1999, p 305–313

CHAPTER **10**

Structural Ceramics

CERAMICS ARE INORGANIC NONMETALLIC MATERIALS that consist of metallic and nonmetallic elements bonded together with either ionic and/or covalent bonds. Although ceramics can be crystalline or noncrystalline, the important engineering ceramics are all crystalline. Due to the absence of conduction electrons, ceramics are usually good electrical and thermal insulators. In addition, due to the stability of their strong bonds, they normally have high melting temperatures and excellent chemical stability in many hostile environments. However, ceramics are inherently hard and brittle materials that, when loaded in tension, have almost no tolerance for flaws. As a material class, few ceramics have tensile strengths above 172 MPa (25 ksi), while the compressive strengths may be 5 to 10 times higher than the tensile strengths.

Under an applied tensile load at room temperature, ceramics almost always fracture before any plastic deformation can occur. Stress concentrations leading to brittle failure can be minute surface or interior cracks (microcracks) or internal pores, which are virtually impossible to eliminate or control. Plane-strain fracture toughness (K_{Ic}) values for ceramic materials are much lower than for metals, typically below 10 MPa (9 ksi), while for metals they can exceed 110 MPa $\sqrt{m}$ (100 ksi $\sqrt{in.}$). There is also considerable scatter in the fracture strength for ceramics, which can be explained by the dependence of fracture strength on the probability of the existence of a flaw that is capable of initiating a crack. Therefore, size or volume also influences fracture strength; the larger the size, the greater the probability for a flaw and the lower the fracture strength.

In metals, plastic flow takes place mainly by slip. Due to the nondirectional nature of the metallic bond, dislocations move under relatively low stresses. However, ceramics form either ionic or covalent bonds, both of which restrict dislocation motion and slip. One reason for the hardness and brittleness of ceramics is the difficulty of slip or dislocation motion. While ceramics are inherently strong, they cannot slip or plastically deform to accommodate even small cracks or imperfections; therefore, their strength is never realized in practice. They facture in a premature brittle manner long before their inherent strength is approached.

The nature of the ionic and covalent bonds is shown in Fig. 10.1. In the ionic bond, the electrons are shared by an electropositive ion (cation) and an electronegative ion (anion). The electropositive ion gives up its valence electrons, and the electronegative ion captures them to produce ions having full electron orbitals or suborbitals. As a consequence, there are no free electrons available to conduct electricity. In ionically bonded ceramics, there are very few slip systems along which dislocations may move. This is a consequence of the electrically charged nature of the ions. For slip in some directions, ions of like charge must be brought into close proximity to each other, and because of electrostatic repulsion, this mode of slip is very restricted. This is not a problem in metals, because all atoms are electrically neutral. In covalently bonded ceramics, the bonding between atoms is specific and directional, involving the exchange of electron charge between pairs of atoms. Thus, when covalent crystals are stressed to a sufficient extent, they exhibit brittle fracture due to a separation of electron pair bonds, without subsequent reformation. It should be noted that ceramics are rarely either all ionically or covalently bonded; they usually consist of a mix of the two types of bonds. For example, silicon nitride (Si_3N_4) consists of approximately 70% covalent bonds and 30% ionic bonds.

Advanced ceramics are ceramic materials that exhibit superior mechanical properties, corrosion/oxidation resistance, or electrical, optical, and/or magnetic properties. This group includes many monolithic ceramics as well as particulate-, whisker-, and fiber-reinforced glass-, glass ceramic-, and ceramic-matrix composites. The classification of advanced ceramics according to function is shown in Table 10.1. The general term *structural ceramics* refers to a large family of ceramic materials used in an extensive range of applications, including both monolithic ceramics and

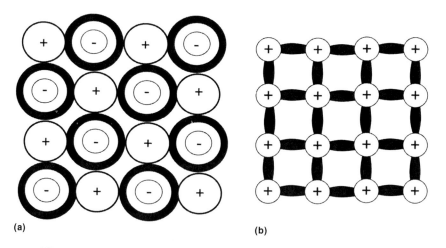

(a) (b)

Fig. 10.1 (a) Ionic and (b) covalent bonding. Source: Ref 10.1

Table 10.1   Classification of advanced ceramics

Main function	Properties required	Applications (examples)
Thermal	High-temperature and thermal shock resistance, thermal conductivity (high or low, respectively)	High-temperature components, burner nozzles, heat exchanger heating elements, noniron metallurgy insulating parts, thermal barrier coatings
Mechanical	Long-term, high-temperature resistance; fatigue, thermal shock, wear resistance	Wear parts; sealings; bearings; cutting tools; engine, motor, and gas turbine parts; thermal barrier coatings
Chemical, biological	Corrosion resistance, biocompatibility	Corrosion protection, catalyst carriers, environmental protection, sensors, implants (joints, teeth, etc.)
Electrical, magnetic	Electrical conductivity (high or low, respectively), semiconducting, piezo-thermoelectricity, dielectric properties	Heating elements, insulators, magnets, sensors, integrated circuit packages, substrates, solid electrolytes, piezoelectrics, dielectrics, superconductors
Optical	Low absorption coefficient	Lamps, windows, fiber optics, infrared optics
Nuclear	Irradiation resistance, high absorption coefficient, high-temperature resistance, corrosion resistance	Fuel and breeding elements, absorbers, shields, waste conditioning

Source: Ref 10.2

ceramic-matrix composites. Chemically, structural ceramics include oxides, nitrides, borides, and carbides.

10.1   Structural Ceramics

One of the growing uses for advanced ceramics is in the area of structural or load-bearing applications. These applications require materials that have high strengths at room temperature and/or retain high strength at elevated temperatures, resist deformation (slow crack growth or creep), are damage tolerant, and resist corrosion and oxidation. Ceramics appropriate to these applications offer a significant weight savings over currently used metals. Applications include heat exchangers, automotive engine components such as turbocharger rotors and roller cam followers, power generation components, cutting tools, biomedical implants, and processing equipment used for fabricating a variety of polymer, metal, and ceramic parts. Major obstacles to their more widespread use are their reliability and cost.

10.2   Applications for Structural Ceramics

The selection process for ceramic materials is generally more complex than that for metals because of the ceramic characteristics of low-strain tolerance, low fracture toughness, and considerable scatter in strength properties—all manifestations of a brittle material. These characteristics dictate the need for consistent materials, flaw-detection methods, and the appropriate design methodology to ensure component reliability. In considering the use of structural ceramics for engineering applications, the engineer needs to know the key physical and mechanical properties, principally by comparison with equivalent metal properties.

Industrial uses, required properties, and examples of specific applications for structural ceramics are summarized in Table 10.2. These applications take advantage of the temperature resistance, corrosion resistance, hardness, chemical inertness, thermal and electrical insulating properties, wear resistance, and mechanical properties of the structural ceramic materials. Ceramics offer advantages for structural applications because their density is approximately one-half the density of steel, and they provide very high stiffness-to-weight ratios over a broad temperature range. The high hardness of structural ceramics can be used in applications where mechanical abrasion or erosion is encountered. The ability to maintain mechanical strength and dimensional tolerances at high temperature makes them suitable for high-temperature use. For electrical applications, ceramics have high resistivity, low dielectric constants, and low loss factors that, when combined with their mechanical strength and high-temperature stability, make them suitable for extreme electrical insulating applications.

Engine Components. The materials used in engine applications must exhibit some combination of good strength, damage tolerance, creep resistance, and oxidation and corrosion resistance at temperatures up to approximately 1375 °C (2500 °F). Typical automotive engine components

Table 10.2   Industry, use, properties, and applications for structural ceramics

Industry	Use	Property	Application
Fluid handling	Transport and control of aggressive fluids	resistance to corrosion, mechanical erosion, and abrasion	Mechanical seal faces, meter bearings, faucet valve plates, spray nozzles, micro-filtration membranes
Mineral processing power generation	Handling ores, slurries, pulverized coal, cement clinker, and flue gas neutralizing compounds	Hardness, corrosion resistance, and electrical insulation	Pipe linings, cyclone linings, grinding media, pump components, electrostatic precipitator insulators
Wire manufacturing	Wear applications and surface finish	Hardness, toughness	Capstans and draw blocks, pulleys and sheaves, guides, rolls, dies
Pulp and paper	High-speed paper manufacturing	Abrasion and corrosion resistance	Slitting and sizing knives, stock-preparation equipment
Machine tool and process tooling	Machine components and process tooling	Hardness, high stiffness-to-weight ratio, low inertial mass, and low thermal expansion	Bearings and bushings, close tolerance fittings, extrusion and forming dies, spindles, metal-forming rolls and tools, coordinate-measuring machine structures
Thermal processing	Heat recovery, hot-gas cleanup, general thermal processing	Thermal stress resistance, corrosion resistance, and dimensional stability at extreme temperatures	Compact heat exchanges, heat exchanger tubes, radiant tubes, furnace components, insulators, thermocouple protection tubes, kiln furniture
Internal combustion engine components	Engine components	High-temperature resistance, wear resistance, and corrosion resistance	Exhaust port liners, valve guides, head faceplates, wear surface inserts, piston caps, bearings, bushing, intake manifold liners
Medical and scientific products	Medical devices	Inertness in aggressive environments	Blood centrifuge, pacemaker components, surgical instruments, implant components, lab ware

Source: Ref 10.3

include turbocharger rotors, valves, turbine rotors, nozzles, vanes, and piston rings. A silicon nitride radial rotor for a gas turbine engine is shown in Fig. 10.2.

Power generation systems include stationary engines, heat-recovery systems, burners and combustors, waste incineration systems, separation and filtration systems, and insulating refractoriness. The property requirements for power generation systems are similar to those needed for the automotive components. The individual components for these systems require a variety of shapes and sizes, including tubes, plates, and mixed thin- and thick-section parts. Ceramic honeycomb structures are used in heat exchangers (Fig. 10.3) to maximize efficiency. Maximum application temperatures can range from approximately 500 to 1400 °C (900 to 2500 °F).

Fig. 10.2 Silicon nitride radial rotor for gas turbine engine. Source: Ref 10.4

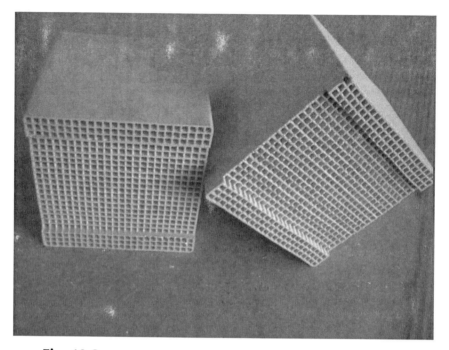

Fig. 10.3 Ceramic honeycomb used in heat exchanger

Materials for tribological applications require resistance to wear in the form of erosion, abrasion, and adhesion. They include bearings, seals, nozzles, brakes, clutches, wear pads, liners, and other rotating or reciprocating parts. Typical ceramic pump parts for handling abrasive slurries are shown in Fig. 10.4. These parts can operate at low temperatures, for example, cam followers, or at elevated temperatures, such as high-temperature seals. Processing equipment includes chemical reactors and process equipment used for manufacturing a number of products. Ceramic-based components that go into this type of equipment include reactor vessels, pumps, valves, piping, tanks, heat exchangers, manufacturing or fabrication dies, and storage vessels for toxic and/or corrosive liquids. Obviously, chemical and temperature resistance are key requirements.

Biomedical applications encompass orthopaedic and dental implants and prostheses and knee and hip joint replacements. The key to using advanced ceramics in these applications is the ability to mimic natural bones and teeth. They must have good specific strength and stiffness, be biologically compatible with body fluids, provide a visual appearance that matches the natural system (for teeth), and resist fatigue. The candidate materials for these applications include high-purity Al_2O_3, synthetic hydroxyapatites, borosilicate glasses, and glass-ceramics.

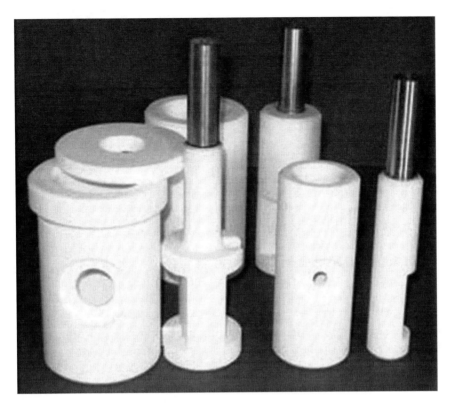

Fig. 10.4 Ceramic corrosion- and wear-resistant pump parts. Courtesy of MIKERON Kft.

Primary properties of interest for structural applications are strength (room and elevated temperature), modulus (elastic, shear, and bulk), fracture toughness, Poisson's ratio, creep and creep-rupture behavior, hardness, tribological properties such as abrasion resistance and friction, chemical resistance, and thermal shock resistance. Optical absorption and index of refraction may also be important where appearance is a consideration, and electrical properties may be important in a high-temperature sensor application. The favorable thermal, chemical, and tribological properties of some of the structural ceramics can also be achieved by the use of ceramic coatings on other materials, such as metals.

Ceramic Armor. The major advantage of ceramic armor lies in its significantly lower areal weight, which allows weight savings of more than 50% over conventional metallic solutions. The most important ceramic materials today (2012) for ballistic protection are alumina (Al_2O_3), silicon carbide (SiC), and boron carbide (BC). As a result of its price-efficiency ratio, alumina is the preeminent ceramic armor material for vehicular applications. Silicon and boron carbide materials are used when extremely low weight is required, such as for personal protection or for helicopters.

In general, the construction of lightweight composite system is based on four main components:

* Spall foil
* Ceramic
* Adhesive
* Composite substrate

In a composite armor system such as the one shown in Fig. 10.5, the ceramic is normally placed on the strike face, preferably perpendicular to the expected threat. Polymer fibers composed of polyamide, polyethylene, or polypropylene form the composite backing. The stiffening and structural enhancement of the individual polymer layers is achieved by impregnation and subsequent curing of the adhesive. Proper selection of adhesives, such as rubber, polyurethane, or epoxies, results in the desired hardness.

This chemical bond between ceramic and composite substrate and/or between the individual polymer layers is significant for the performance of the entire system. In addition, spall protection is applied on the front side of the ceramic; glass fiber laminates are used for the spall protection layer.

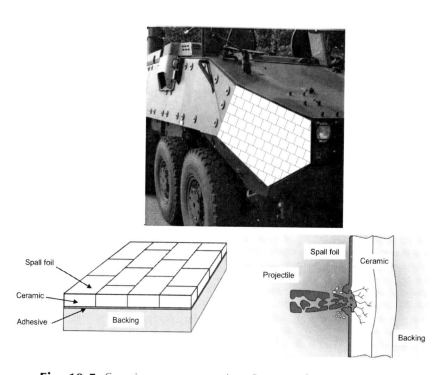

Fig. 10.5 Ceramic armor construction. Courtesy of ©CeramTec-ETEC. Source: Ref 10.5

Each component within the composite system has a specific fun
The hard ceramic layer reduces the speed of the projectile and frag
the projectile. The resulting low mass and the significantly reduced
of these residual fragments are completely absorbed by the elastic/
deformation in the composite substrate.

When the projectile impacts the surface of the ceramic, its
energy is greatly reduced without penetrating the ceramic. This is
by the dwell effect. In that phase, the projectile experiences a highly ductile deformation. After ~15 to 20 µs, the projectile actually penetrates the ceramic body, and in the process the kinetic energy of the projectile is reduced further by erosion. The shattered fragments of the projectile completely penetrate the ceramic after ~30 µs. The residual energy of these fragments amounts to only about 15%, which can be fully absorbed by the backing.

10.3 Properties of Structural Ceramics

The properties of some structural ceramics are shown in Table 10.3. In addition, the properties of several metallic systems are listed for comparison.

Alumina. Aluminum oxide, Al_2O_3 (often referred to as alumina), is perhaps the material most commonly used in the production of technical ceramics. The reasons for its wide acceptance are many; alumina has a high hardness, excellent wear and corrosion resistance, and low electrical conductivity. It is also fairly economical to manufacture, involving low-cost alumina powders.

Alumina ceramics actually include a family of materials, typically having alumina contents from 85 to $\geq 99\%$ Al_2O_3, the remainder being a grain-boundary phase. The different varieties of alumina stem from diverse application requirements. For example, 85% alumina ceramics such as milling media are used in applications requiring high hardness, yet they are economical. Aluminas having purities in the 90 to 97% range are often found in electronic applications as substrate materials, due to their low electrical conductivity. The grain-boundary phase in these materials also allows for a strong bond between the ceramic and the metal conduction paths for integrated circuits. High-purity alumina (>99%) is often used in the production of translucent envelopes for sodium-vapor lamps.

The microstructure and resulting properties of alumina ceramics greatly depend on the percentage of alumina present. For example, high-purity aluminas typically have a fairly simple microstructure of equiaxed alumina grains, whereas 96% alumina ceramic will have a more complicated microstructure consisting of alumina grains (often elongated in shape) surrounded by a grain-boundary phase. Depending on processing, this grain-boundary phase may be amorphous, crystalline, or both. The properties of this family of materials vary widely, as shown in Table 10.4.

perties of structural ceramic materials

	Young's modulus		Poisson's ratio	Thermal conductivity, W/m · K		Thermal expansion, $\times 10^{-6}$/K	Specific heat, J/g · °C	Density, g/cm³	Strength, MPa (ksi)		Weibull modulus (m) at RT	Maximum use temperature	
al	GPa	10⁶ psi		RT	600 °C (1110 °F)				RT	600 °C (1110 °F)		°C	°F
icon nitride													
Hot-pressed (HPSN)	290	42.1	0.3	29	22	2.7	0.75	3.3	830 (120)	805 (117)	7	1400	2550
Sintered (SSN)	290	42.1	0.28	33	18	3.1	1.1	3.3	800 (116)	725 (105)	13	1400	2550
Reaction-bonded (RBSN)	200	29.0	0.22	10	10	3.1	0.87	2.7	295 (43)	295 (43)	10	1400	2550
Silicon carbide													
Hot-pressed (HPSN)	430	62.4	0.17	80	51	4.6	0.67	3.3	550 (80)	520 (75)	10	1500	2730
Sintered (SSN)	390	56.6	0.16	71	48	4.2	0.59	3.2	490 (71)	490 (71)	9	1500	2730
Reaction-bonded (RBSN)	413	59.9	0.24	225	70	4.3	1.0	3.1	390 (57)	390 (57)	10	1300	2370
Partially stabilized zirconia (PSZ)	205	29.7	0.30	2.9	2.9	10.2	0.5	5.9	1020 (148)	580 (84)	14	950	1740
Lithium-aluminum-silicate	68	9.9	0.27	1.4	1.9	0.5(a)	0.78	2.3	96 (14)	96 (14)	10	1200	2190
Common metals(b)													
Cast iron	170	24.7	0.28	49	40	12	0.45	7.1	620 (90)	100 (14.5)	...	500	930
Steel	200	29.0	0.28	38		14	0.45	7.8	1500 (218)	140 (20)	...	600	1110
Aluminum	70	10.2	0.33	160		22.4	0.96	2.7	370 (54)	0	...	350	660

These properties are typical of these classes of materials, but in many cases, large variations may exist between various formulations. Strengths are for four-point bending spans of 19.05/9.525 mm (0.745/0.375 in.) and bar cross sections of 6.35 × 3.175 mm (0.25 × 0.125 in.). RT, room temperature. (a) Maximum excursion from 350 to 800 °C (660 to 1470 °F); initial expansion is negative. (b) Data provided for comparison. Source: Ref 10.6

Table 10.4 Properties of various alumina ceramics

Alumina content, %	Bulk density, g/cm³	Flexure strength		Fracture toughness		Hardness		Elastic modulus		Thermal conductivity		Linear coefficient of thermal expansion	
		MPa	ksi	MPa	ksi	GPa	10⁶ psi	GPa	10⁶ psi	W/m · K	Btu/ft · h · °F	ppm/°C	ppm/°F
85	3.41	317	46	3–4	2.8–3.7	9	1.3	221	32	16.0	9.24	7.2	4
90	3.60	338	49	3–4	2.8–3.7	10	1.5	276	40	16.7	9.65	8.1	4.5
94	3.70	352	51	3–4	2.8–3.7	12	1.7	296	43	22.4	12.9	8.2	12.9
96	3.72	358	52	3–4	2.8–3.7	11	1.6	303	44	24.7	14.3	8.2	4.6
99.5	3.89	379	55	3–4	2.8–3.7	14	2.0	372	54	35.6	20.6	8.0	4.4
99.9	3.96	552	80	3–4	2.8–3.7	15	2.2	386	56	38.9	22.5	8.0	4.4

Source: Ref 10.3

Aluminum titanate, Al_2TiO_5, is a ceramic material that has recently received much attention because of its good thermal shock resistance. Aluminum titanate has an orthorhombic crystal structure, which results in a very anisotropic thermal expansion. The coefficient of thermal expansion

(CTE) normal to the c-axis of the orthorhombic crystal is $-2.6 \times 10^{-6}/°C$ ($-1.4 \times 10^{-6}/°F$), whereas the CTE parallel to the c-axis is approximately $11 \times 10^{-6}/°C$ ($6.1 \times 10^{-6}/°F$). The resulting CTE for a polycrystalline material is very low ($0.7 \times 10^{-6}/°C$, or $0.4 \times 10^{-6}/°F$), as shown in Table 10.5.

The excellent thermal shock resistance of aluminum titanate derives from this considerable thermal expansion anisotropy. During cooling from the densification temperature, the aluminum titanate grains shrink more in one direction than the other, which results in small microcracks in the microstructure as the grains actually pull away from each other. Subsequent thermal stresses (either by fast cooling or heating) are thereby dissipated by the opening and closing of these microcracks. Unfortunately, a consequence of the microcracks is that aluminum titanate does not have particularly high strength (25 MPa, or 3 ksi). However, the microcracks do impart very low thermal conductivity, making it an excellent candidate for thermal insulation devices.

The excellent thermal shock resistance of aluminum titanate offers the potential for many applications. For example, aluminum titanate is used as funnels and ladles in the foundry industry, because aluminum, magnesium, zinc, and iron do not wet aluminum titanate. The automotive industry is also investigating aluminum titanate for exhaust port liners and exhaust manifolds.

Silicon carbide, SiC, is a ceramic material that has been in existence for decades but has recently found many applications as an advanced ceramic. There are actually two families of silicon carbide, one known as direct-sintered SiC, and the other known as reaction-bonded SiC (also referred to as siliconized SiC). In direct-sintered SiC, submicrometer SiC powder is compacted and sintered at temperatures in excess of 2000 °C (3600 °F), resulting in a high-purity product. Reaction-bonded SiC, on the other hand, is processed by forming a porous shape comprised of SiC and carbon-powder particles. The shape is then infiltrated with silicon metal; the silicon metal acts to bond the SiC particles together.

Table 10.5 Physical properties of various ceramics

Material	Bulk density, g/cm³	Flexure strength		Fracture toughness		Hardness		Elastic modulus		Thermal conductivity		Linear coefficient of thermal expansion	
		MPa	ksi	MPa √m	ksi √in.	GPa	10⁶ psi	GPa	10⁶ psi	W/m · K	Btu/ft · h · °F	ppm/°C	ppm/°F
Aluminum titanate	3.10	25	3.6	...	...	...	...	5	0.7	1.0	0.6	0.7	0.4
Sintered SiC	3.10	550	80	4	3.6	29	4.2	400	58	110.0	63.6	4.4	2.4
Reaction-bond SiC	3.10	462	67	3–4	2.7–3.6	25	3.6	393	57	125.0	72.2	4.3	2.4
Silicon nitride	3.31	906	131	6	5.5	15	2.2	311	45	15.0	8.7	3.0	1.7
Boron carbide	2.50	350	51	3–4	2.7–3.6	29	4.2	350	51	...	...	...	...

Source: Ref 10.3

The properties of the two families of SiC are similar in some ways and quite different in others. Both materials have very high hardness and high thermal conductivities. The fracture toughness of both materials is generally low, of the order of 3 to 4 MPa $\sqrt{m}$ (2.7 to 3.6 ksi $\sqrt{in.}$). However, the major differences are in strength, wear, and corrosion resistance. Due to the gaseous products formed during reaction bonding, the strength of reaction-bonded SiC is lower than direct-sintered SiC (Table 10.3). In addition, direct-sintered SiC has a greater ability to withstand severely corrosive and erosive environments, because a limiting factor for reaction-bonded SiC is the unreacted or free silicon metal.

Applications for SiC ceramics are typically in the areas where wear and corrosion are problems. For example, SiC is often found as pump seal rings and automotive water pump seals. The high thermal conductivity of silicon carbide also allows it to be used as radiant heating tubes in metallurgical heat treatment furnaces.

Silicon Nitride. An intense interest in silicon nitride (Si_3N_4) ceramics has emerged over the past few decades. The motivation for such interest lies in the automotive industry, where use of ceramic components in engines would greatly improve operating efficiency. Silicon nitride offers great potential in these applications because of its excellent high-temperature strength of 900 MPa (130 ksi) at 1000 °C (1830 °F), high fracture toughness of 6 to 10 MPa $\sqrt{m}$ (5.5 to 9 ksi $\sqrt{in.}$), and good thermal shock resistance. It also has very good oxidation resistance, a property of particular importance in automotive applications. The automotive components of interest are turbocharger rotors, pistons, piston liners, and valves. However, the greatest application of Si_3N_4 is as a cutting tool material in metal machining applications, where machining rates can be dramatically increased due to the high-temperature strength of Si_3N_4.

Boron Carbide. The chief advantages of B_4C are its exceptionally high hardness and low density (2.50 g/cm^3, or 0.09 lb/in.^{3}). However manufacturing B_4C is difficult because of the high temperatures necessary for densification (>>2000 °C, or 3600 °F). Thus, in most cases, B_4C is densified with pressure, as in hot pressing. This limits the complexity of shapes possible without excessive grinding and machining. Another disadvantage of B_4C is the high cost of the powders and subsequent processing. As such, B_4C has found use only in applications that demand the unique properties of B_4C, namely, military armor.

SiAlON is an acronym for silicon-aluminum-oxynitride. SiAlON is fabricated in several ways but is typically made by reacting Si_3N_4 with Al_2O_3 and AlN at high temperatures. SiAlON is a generic term for the family of compositions that can be obtained by varying the quantities of the original constituents. The advantages of SiAlONs are their low CTE (2 to 3 × 10^{-6}/°C, or 1 to 1.7 × 10^{-6}/°F) and good oxidation resistance. The array of potential applications is similar to that of Si_3N_4, namely, automotive components and machine tool bits. However, the chemistry of SiAlON is complex, and reproducibility is a major problem to its becoming more commercially successful.

Zirconia. Pure zirconia cannot be fabricated into a fully dense ceramic body using existing conventional processing techniques. The 3 to 5% volume increase associated with the tetragonal-to-monoclinic phase transformation causes any pure ZrO_2 body to completely destruct upon cooling from the sintering temperature. Additives such as calcia (CaO), magnesia (MgO), yttria (Y_2O_3), or ceria (CeO_2) must be mixed with ZrO_2 to stabilize the material in either the tetragonal or cubic phase. Applications for cubic-stabilized ZrO_2 include various oxygen sensor devices (cubic ZrO_2 has excellent ionic conductivity), induction heating elements for the production of optical fibers, resistance heating elements in new high-temperature oxidizing kilns, and inexpensive diamondlike gemstones. Partially stabilized or tetragonal-stabilized ZrO_2 systems are discussed subsequently.

10.4 Toughened Ceramics

Decades ago, ceramics were characterized as hard, high-strength materials with excellent corrosion and electrical resistance in addition to high-temperature capability. However, low fracture toughness limited its use in structural applications. The development of toughened ceramics coincided with industrial applications requiring high-temperature capability, high strength, and an improvement in fracture resistance over existing ceramic materials (Table 10.6). The primary driving force toward developing toughened ceramics was the promise of an all-ceramic engine. Several of the materials discussed in this section were or are being considered as ceramic engine component materials.

Table 10.6 Typical mechanical properties of various ceramics

Material	Bulk density, g/cm³	Flexure strength		Fracture toughness		Hardness		Elastic modulus	
		MPa	ksi	MPa $\sqrt{m}$	ksi $\sqrt{in.}$	GPa	10⁶ psi	GPa	10⁶ psi
Zirconia-toughened alumina (ZTA)	4.1–4.3	600–700	87–101	5–8	4.6–7.3	15–16	2–2.3	330–360	48–52
Magnesia-partially stabilized zirconia (Mg-PSZ)	5.7–5.8	600–700	87–101	11–14	10–13	12	1.7	210	30
Yttria-tetragonal zirconia polycrystalline (Y-TZP)	6.1	900–1200	130–174	8–9	7.3–8.2	12	1.7	210	30
Alumina-SiC	3.7–3.9	600–700	87–101	5–8	4.6–7.3	15–16	2–2.3	380–430	55–62
Silicon nitride-SiC	32. –3.3	800–1000	116–145	6–8	5.5–7.3	15–16	2–2.3	300–380	43–55

Source: Ref 10.3

One remarkable method of microstructural control to yield improved properties is transformation (martensitic) toughening, first observed in zirconia (ZrO_2) ceramics. Zirconia exhibits a phase transformation from cubic to triclinic to monoclinic as its temperature is lowered. If the zirconia is stabilized with materials such as yttria, calcia, or magnesia, the triclinic phase can be preserved in a metastable state at ambient (or other temperature), depending on the amount of additive, heat treatment, cooling rate, grain size, and grain size distribution. As shown in Fig. 10.6, the tetragonal zirconia can transform to the monoclinic phase under the influence of a crack tip stress. The transformation may cause the crack tip to deflect due to the shear stress (approximately 1 to 7%) or to be impeded by the local compressive stress (>3%). A very significant toughening effect results (Fig. 10.7), which can as much as triple K_{Ic}. A fracture toughness similar to high-strength aluminum alloys results, which makes the ceramic very resistant to impact failure. Flexural strengths as great as 2000 MPa (300 ksi) have been observed. It was found that these structural ceramics exhibit R-curve behavior in which the K_{Ic} increases as a crack grows in slow crack growth.

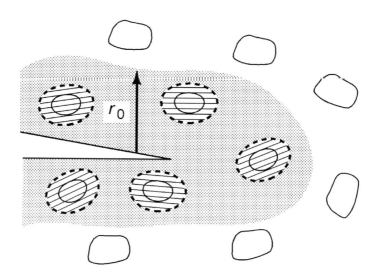

○ Original metastable zirconia particle (tetragonal)

⊝ Martensitically transformed zirconia particle (monoclinic)

░ Stress field around crack tip

Fig. 10.6 Stress-induced transformation of metastable ZrO_2 particles in the elastic stress field of a crack. Source: Ref 10.7

Zirconia-toughened ceramics also show remarkable resistance to abrasion and wear because surface layers transform under stress (Fig. 10.8), creating a highly compressive surface layer. This layer resists abrasion, and the state of compression will renew if it is worn away. Zirconia cutting tools, thread guides, and wear surfaces have been found to last ten or more times longer than their tool steel counterparts in industrial applications. Some consumer applications such as household knives and scissors as well as breakproof buttons have been commercialized. Metastable tetragonal zirconia particles can also be added to various engineering ceramics and provide a substantial toughening effect. This is shown for alumina, silicon nitride, and silicon carbide in Fig. 10.9.

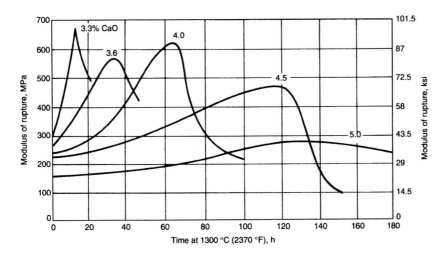

Fig. 10.7 Strength/aging curves obtained by heat treatment at 1300 °C (2370 °F) for various compositions of CaO-partially stabilized zirconia materials. Source: Ref 10.7

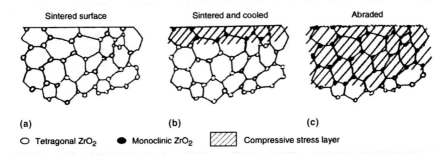

Fig. 10.8 Diagram of a section through a free surface at (a) the sintering temperature. On cooling, particles of ZrO_2 near the surface (b) transform due to reduced constraint, developing a compressive stress in the matrix. The thickness of this compressively stressed layer can be increased (c) by abrasion or machining. Source: Ref 10.7

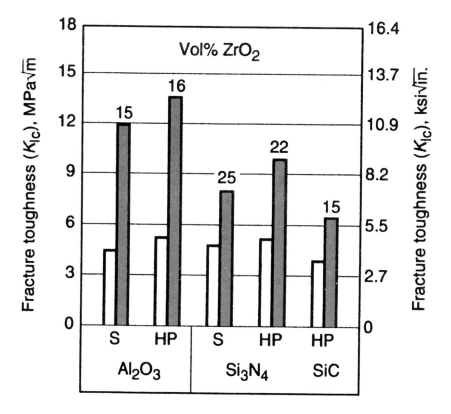

Fig. 10.9 Increase in fracture toughness (K_{Ic}) observed upon inclusion of zirconia particles in the ceramic matrix. The volume of zirconia added is shown in the shaded histogram. The toughness of the matrix material is shown in the adjacent white histogram bars. S, sintered. HP, hot pressed. Source: Ref 10.7

Zirconia-toughened alumina (ZTA) is the generic term applied to alumina-zirconia systems where alumina is considered the primary or continuous (70 to 95%) phase. Zirconia particulate additions (either as pure ZrO_2 or as stabilized ZrO_2 from 5 to 30%) represent the second phase. The solubility of ZrO_2 in Al_2O_3 and Al_2O_3 in ZrO_2 is negligible. The ZrO_2 is present either in the tetragonal or monoclinic symmetry. Zirconia-toughened alumina is a material of interest primarily because it has a significantly higher strength and fracture toughness than alumina.

The microstructure and subsequent mechanical properties can be tailored to specific applications. Higher ZrO_2 contents lead to increased fracture toughness and strength values, with little reduction in hardness and elastic modulus, provided most of the ZrO_2 can be retained in the tetragonal phase. Strengths up to 1050 MPa (152 ksi) and fracture toughness values as high as 7.5 MPa $\sqrt{m}$ (6.8 ksi $\sqrt{in.}$) have been measured. Wear properties in some applications may also improve due to mechanical property enhancement compared to alumina. These types of ZTA compositions have been used in some cutting tool applications.

Zirconia-toughened alumina has also seen some use in thermal shock applications. Extensive use of monoclinic ZrO_2 can result in a severely microcracked ceramic body. This microstructure allows thermal stresses to be distributed throughout a network of microcracks where energy is expended opening and/or extending microcracks, leaving the bulk ceramic body intact. However, commercial success has been limited, partly due to the failure of industry to produce a low-cost ZTA with improved properties and its failure to identify markets allowing immediate penetration. One exception has been the use of ZTA in some cutting tool applications.

Transformation-toughened zirconia is a generic term applied to stabilized zirconia systems in which the tetragonal symmetry is retained as the primary zirconia phase. The four most prominent tetragonal-phase stabilizers are CeO_2, Y_2O_3, CaO, and MgO. The use of these four additives results in two distinct microstructures. MgO- and CaO-stabilized ZrO_2 consist of 0.1 to 0.25 μm tetragonal precipitates within 50 to 100 μm cubic grains. Firing usually occurs within the single cubic-phase field, and phase assemblage is controlled during cooling.

Interest in CaO-stabilized ZrO_2 has waned in recent years. MgO-stabilized ZrO_2 (Mg-PSZ), on the other hand, has been a commercial success. Its combination of moderate-high strength of 600 to 700 MPa (87 to 100 ksi), high fracture toughness of 11 to 14 MPa $\sqrt{m}$ (10 to 13 ksi $\sqrt{in.}$), and flaw tolerance enables the use of Mg-PSZ in the most demanding structural ceramic applications. The elastic modulus is approximately 210 GPa (30×10^6 psi), and the hardness is approximately 12 to 13 GPa (1.7 to 1.9 $\times 10^6$ psi). Among the applications for this material are extrusion nozzles in steel production, wire-drawing cap stands, foils for the paper-making industry, and compacting dies. Among the toughened or high-technology ceramic materials, Mg-PSZ exhibits the best combination of mechanical properties and cost for room- and moderate-temperature structural applications.

Yttria-stabilized ZrO_2 (Y-TZP) is a fine-grained, high-strength, and moderate-high-fracture-toughness material. High-strength Y-TZPs are manufactured by sintering at relatively low sintering temperatures (1400 °C, or 2550 °F). Nearly 100% of the zirconia is tetragonal, and the average grain size is approximately 0.6 to 0.8 μm. The tetragonal phase is very stable. Higher firing temperatures (1550 °C, or 2800 °F) result in a high-strength (1000 MPa, or 145 ksi), high-fracture-toughness (8.5 MPa $\sqrt{m}$, or 7.7 ksi $\sqrt{in.}$), fine-grained material with excellent wear resistance. The microstructure consists of a mixture of 1 to 2 μm tetragonal grains (90 to 95%) and 4 to 8 μm cubic grains (5 to 10%). The tetragonal phase is more readily transformable to the larger tetragonal grain size and a lower yttria content in the tetragonal phase, resulting in a tougher material.

Among the applications for Y-TZP are ferrules for fiber optic assemblies. Material requirements include a very fine-grained microstructure, grain size control, dimensional control, excellent wear properties, and high strength. The fine-grained microstructure and good mechanical properties lend the Y-TZP as a candidate material for knife edge applications, including scissors, slitter blades, knife blades, and scalpels. However, compared to Mg-PSZ, Y-TZP is more expensive, has a lower fracture toughness, and is not nearly as flaw tolerant.

There are some temperature limitations in these materials. Mechanical strength of both Mg-PSZ and Y-TZP may start to deteriorate at temperatures as low as 500 °C (930 °F). Also, the Y-TZP ceramic is susceptible to severe degradation at temperatures between 200 to 300 °C (400 to 570 °F).

10.5 Design and Selection Considerations

The materials selection process for structural ceramic components involves consideration of a variety of material characteristics. Most obvious are strength and maximum operating temperature capability. Other characteristics that are important to varying degrees are thermal conductivity, specific heat, elastic modulus, CTE, thermal shock resistance, method of fabrication, and cost.

Data regarding the reliability of advanced ceramics are closely associated with the complex relations among structures and properties. Certain material properties depend on composition, structure, and measurement details. Thus, details of materials specifications and measurement methods are essential to the interpretation of materials property values for advanced ceramics. Materials specification and measurement method details must be considered integral parts of materials property databases and of the use of the databases for engineering designs.

Fast Fracture Reliability. Ceramic materials have two basic properties that tend to make their successful application to structural components difficult. First, the material has a very low strain tolerance. Second, there is frequently a large scatter in the strength data. The first property manifests itself in that the material has a linear stress-strain relationship all the way from zero stress to failure. This linear relationship by itself is attractive to analysts because it makes the modeling process easier. However, the absence of plastic flow means that the mechanism by which the local stress concentrations are relieved around the ever-present flaws is eliminated. In turn, this means that the analysis must be refined enough to locate and eliminate local high-stress areas. This frequently makes the analysis more complex than would be required for a similar part made of metal. The fact that the material strength shows considerable variability means that the design must be based on probabilistic methods.

For years, ceramics have been used in simple structures by designing them to stay in compression. This is because ceramics have been shown to be much stronger in compression than in tension. However, today (2012) there are many uses for which ceramics could offer performance benefits that are not conducive to this design-for-compression philosophy. Therefore, to confidently design components subjected to tensile loading, the design methodology had to be reevaluated. Weibull developed a probabilistic system of design based only on the tensile stresses in the component. Compressive failure is not considered in this technique, because components made from brittle materials usually fail from tensile stresses. His methodology has been used extensively in the design of ceramic components for engines and has been shown to work quite well for a number of applications.

Weibull Theory. Determining ceramic component reliability based on Weibull's weakest-link theory is based on an assumption that a part is like a chain of many links. If any link (small element of the part) fails, then the whole chain (part) has failed. Similarly, if any small volume in a ceramic part is stressed sufficiently to cause a crack to propagate, the part will generally fail. Thus, the key is to determine whether any of the elements in a part are likely to fail. Because ceramic material properties are variable due to the random distribution of flaws in ceramics, there is a variation of strengths for various elements of a part. Thus, the strength of the various elements can be considered to have a statistical distribution with values above and below some characteristic strength. If an element is subjected to some value of stress, there is a certain probability that the local strength of the material will be exceeded. As the number of elements in a chain is increased, the probability that a weak link will occur and cause the chain to fail also increases. Similarly, the probability of failure of a ceramic component is a function of the volume (or area) of material subjected to the various stress levels. By considering the probability of failure of all the elements, the probability of failure of the total part is determined.

The reliability of a ceramic component is a function of the material properties and the stress conditions within the part. The material properties are a function of the process controls established during the fabrication of the component, starting with powder manufacturing. In general, the more detailed and more rigorous the process controls, the greater the consistency of the material in the component and thus the higher the Weibull modulus (i.e., reduced variability in strength). A basic understanding of the material coupled with close process controls can lead to components that have good microstructure and high strength. Obviously, high-strength materials with a high Weibull modulus coupled with good component design have the potential for the highest reliabilities. Producing the highest-quality components can also mean producing the lowest-cost components by elimination of waste. Thus, paradoxically, producing the highest-quality parts can also mean producing the most cost-effective parts. However, because of design constraints and material limitations, even the highest-quality parts may not demonstrate extremely high reliabilities in all applications. This can occur when the designer is trying to push a material design combination to the very limits of its capabilities. In these cases, the question arises as to what may be acceptable reliability.

Choosing an acceptable reliability in these cases depends on the consequence of failure. For example, if failure could result in a life-threatening situation, then the reliability must be very high. Initially, design iterations must be incorporated to ensure the highest possible component reliability. Then, if possible, each component should be proof tested to the full design stress condition to force the reliability to 1 (naturally, this assumes that tests have been performed to verify that the proof test does not damage the component). For cases in which failure would result in factory downtime, warranty costs, loss of customers, or damage to the company image, high reliabilities will also be required. For these parts, a probability of failure of one in a million may be reasonable. For low-volume parts that could be replaced easily and with minimal inconvenience to the user, a probability of failure of one in a thousand may be acceptable. For prototype parts for experimental evaluation to determine the benefits of using the component, a failure rate of one in ten could be acceptable.

Lifetime Reliability. In testing and evaluation of ceramics in actual operating environments, it has been discovered that in some cases, even though the component survives initial tests, it fails in subsequent tests. This occurs even though the severity of the test has not increased. This means that the fast fracture reliability determined for the new part is not an accurate prediction for the part during service. There are several reasons why reliability may decrease during operation. These may include one or more of the following: physical change of the material properties, surface degradation from corrosion or erosion, slow crack growth, or creep. Fracture mechanism maps, such as that shown in Fig. 10.10, illustrate the relative contribution of various failure modes as a function of temperature and stress.

Material Surface Degradation. A physical change in the properties of a material can occur when the material goes through a compositional change or a phase change. When the material properties are changing with time, it is extremely difficult to get a good estimate of the component reliability versus time. One approach would be to submit material to an actual service environment for the desired life and then measure the material properties. These properties could then be used in the fast fracture reliability calculation. This approach is probably an improvement over using the "new" material properties, but it can still lead to difficulties. The material in the actual component may not change uniformly, so there may

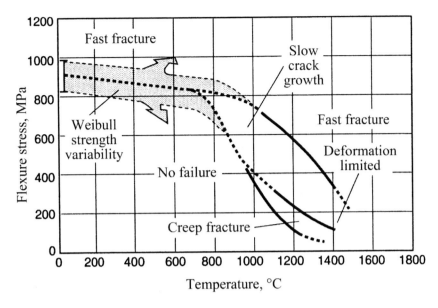

Fig.10.10 Fracture mechanism map for hot-pressed silicon nitride flexure bars. Source: Ref 10.8

be internal stresses that cannot be accounted for in the normal thermal and stress analyses. Also, it has been shown that oxidation of ceramics can be enhanced by the addition of stress. Thus, simply soaking a material in a furnace will not lead to the same degradation of properties that may occur in the part that will be in a stressed state.

When the surface of the component is subjected to corrosion or erosion, the size and nature of surface flaws may change, which can significantly reduce the strength of the surface material. Once again, the strength of test samples subjected to an actual service environment could be determined and used for reliability calculations. This approach will again be better than using virgin material properties but may suffer from the same problem as mentioned in the previous paragraph; that is, damage to a part under stress may be more severe than that occurring in the unstressed condition. Generation of strength from test bars subjected simultaneously to stress and erosion or corrosion may give reasonable values for reliability calculations, but this is only conjecture and has not been proven by actual tests.

Slow crack growth, also called subcritical crack growth, is another factor that can reduce reliability over time. Eventually, even a slow-growing crack can reach a critical length at which sudden and catastrophic crack propagation ensues. In ceramic components, it is difficult to detect which flaw will cause component failure, because there can be a large number of flaws, the critical flaw size is small, and rupture times vary widely, even under the same temperature and load conditions. However, using probabilistic design techniques, predictions can be made about whether a ceramic component will fail within a given amount of service time.

Joints, Attachments, and Interfaces. Frequently, the most difficult problem facing the user of structural ceramics is the design of the ceramic-metal interface. The reason is that usually the ceramic and the metal have very different thermal expansions. Thus, as the system changes temperature, the ceramic and metal components try to move relative to each other. This causes stresses in the ceramic and can lead to failure or debonding between metal and ceramic. Ceramic failure can occur even if the components are allowed to slide relative to each other. These differential thermal-expansion-induced stresses can occur whenever components go through changes in temperature, but they are obviously a much more serious problem when large temperature fluctuations are involved.

Thermal Shock Considerations. In designing with ceramics, thermally induced stresses are frequently critical. Of these, rapidly induced transient thermal stresses, known as thermal shock, are usually the worst, and thermal downshock (rapid cooling from a hot condition) is the worst of all. Thermal downshock is so critical because it induces a high tensile stress in a very thin surface layer. This can initiate a crack that can propagate through the part, causing catastrophic failure.

The best way to successfully design for thermal shock is by a complete transient thermal and stress analysis coupled with a reliability analysis. This type of analysis is very expensive because of the many solution iterations that must be examined. Also, because it is frequently difficult to obtain accurate boundary conditions, this type of analysis must be calibrated with actual component data. However, this is still the best approach, especially if several components must be developed that will have different but similar designs.

There are some initial design guidelines on how to minimize thermal stresses. First, try to minimize any cross-sectional thickness changes in the component. This will tend to let the component respond to any thermal inputs more uniformly. Second, try to separate into two or more components any regions of the structure that will be subjected to vastly different temperatures. This will tend to keep each part at a more uniform temperature. Third, avoid any sharp edges or very thin sections in regions subjected to high heat flux. This will prevent a rapid temperature change in a small region of the part.

When developing new components, it will sometimes happen that actual hardware changes and evaluations on simple rigs can be completed more rapidly than a revised thermal, stress, and reliability analysis. An example of this was the development of a turbine stator vane design. This component was subjected to the hot gases exiting from a combustor and was thus subjected to severe thermal transients. During startup-shutdown tests with the original stator design, cracks frequently developed in the stator vanes. When segments of the stator were mounted on a thermal shock rig, similar cracks developed in the vanes. A design change was made to eliminate the thin trailing edge from the vane. When these stator segments were tested on the thermal shock rig, no vane cracks developed. Subsequent light-off qualification and engine testing verified this design change. Thus, rig tests can be useful in sorting out design changes.

Proof Testing. As mentioned earlier, the design of ceramics is based on probabilistic methods. A statistical distribution of the material strengths is generated from test bar data. These data are then used to predict the probability of failure of a component. In most cases, it is desired that the probability of failure of the component be very much less than the failure probabilities used with the test bar strength data to generate the material characteristic strength and Weibull modulus. For example, if 30 data points were used, the lowest probability of failure used in the materials property generation is 0.022. Thus, the failure distribution curve from the test data is extrapolated well beyond the limits of the data to the extreme tails of the distribution. This obviously limits the confidence in the projected reliabilities. One way of improving the confidence in the predicted reliabilities is to proof test the component. This is a form of testing that will weed out the weakest components and thus truncate the tail of the distribution.

Another reason why one may want to proof test the component is to raise the allowable design stress. For very high desired reliabilities, the allowable design stress can become a small percentage of the test bar strength. By proof testing, the allowable design strength can be increased.

Probabilistic Design (Weibull Analysis). The most rapidly growing approach to ceramic design is the probabilistic method. It deals with the wide variability of strengths dictated by the distribution of flaws in a brittle ceramic. The method accounts for the wide distribution of strengths in conducting the design protocol. It is often integrated with computer finite-element analysis (FEA) models to allow the effect(s) of materials modification, change in flaw population, and part design to be iterated to a successful design in terms of the stresses calculated by FEA. The finite-element net is combined with the known statistical failure behavior (flaw distribution) of the ceramic. This protocol is most frequently based on Weibull statistics, which consider the cumulative probability of failure versus failure stress for each element (usually each sample or part).

A typical Weibull plot constructed on a log-probability scale is shown in Fig. 10.11(a). This plot considers 20 reaction-bonded silicon nitride samples. While 20 elements appear adequate for this plot, it would usually be considered a small database for design. A common rule of convenience is to test more than 50 samples. The plot is constructed by ordering the n strength values and assigning each element a probability equal to its rank (from 1 to n) divided by $[n + 1]$. This approximation to true probability usually leads to less than 5% relative error and allows for simple treatment of a moderate number of failure samples. The slope

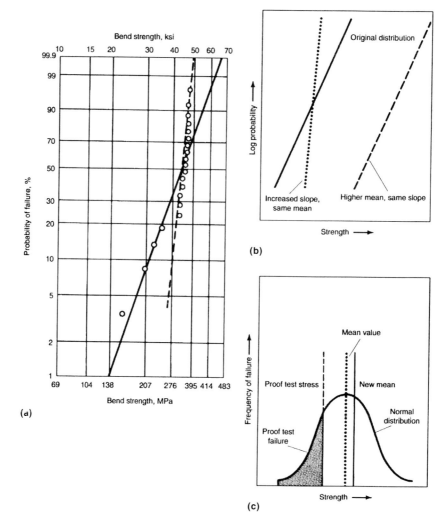

Fig. 10.11 Statistical aspects of material design. (a) Weibull plot for reaction-bonded silicon nitride. (b) Improved Weibull distributions. (c) Modification of normal distribution by proof testing. Source: Ref 10.9

of the line in the plot (two lines in Fig. 10.11a) is termed the Weibull modulus, m. (This modulus should not be confused with other mechanical moduli, such as the modulus of elasticity.) A higher slope indicates a narrower distribution of failure stresses, that is, a narrower distribution of flaws or a more fracture-tough (damage-tolerant) material. A mean stress for failure can also be determined from the plot. The desired failure probability, for example, one part per million, is used as a design criterion, and the Weibull probability with strength is used to design to this performance requirement. Changes in selected material (fracture toughness), fabrication (intrinsic flaw distribution), finishing (extrinsic flaw distribution), and part configuration or applied forces (stress distribution) can be treated iteratively to yield a final design in terms of material, production requirements, and part configuration for a particular set of applied forces and environment.

The Weibull plot can also be used in several other approaches to improved material or part performance. In Fig. 10.11(a), two line segments are shown (solid and dashed). The high-strength failures (dashed line) have a higher Weibull modulus (slope) and mean strength than those elements on the solid line, which exhibit greater variability in strength and a lower mean strength. Usually, the lower slope and mean indicates a different, more severe flaw for corresponding samples. The overall performance can be improved by examining these severe failure-inducing flaws and/or the specifics of production for the low-m samples and correcting matters.

There are two ways to improve performance by material improvement and/or flaw reduction, as shown in Fig. 10.11(b). The mean failure value can be increased to yield a higher overall strength, or the Weibull modulus can be increased, improving the reliability by decreasing the distribution of failure values. The improvement in modulus decreases the low-strength portion of the population and allows a higher-use stress at the same failure probability. Some caution should be taken because a high modulus is not necessarily good—severely abraded material may have very low strength but a high modulus because similar large flaws have been introduced throughout all parts. Naturally, a change in material or production method that increases both the slope and mean would be most desirable.

It should be noted that the Weibull modulus approach provides useful tools for improved design of materials and processes. Nondestructive evaluation methods can be used to eliminate samples with large flaws and consequent low strength. If the flaws belong to a separable group of low strength values such that the solid line in Fig. 10.11(a) can be drawn, the Weibull model can be used unmodified. Another application of this concept is to mechanical proof testing. As diagrammed in Fig. 10.11(c), all produced parts are subjected to stress at a predetermined proof-test level. Parts with large, effective flaws fail (shaded portion) and are eliminated from the population. A new, higher mean value of strength results.

There are two cautions:

- If the initial distribution is normal, as shown (one flaw type), normal statistics can no longer be applied.
- If flaws, particularly cracks, propagate during proof testing, the strength and/or Weibull modulus will deteriorate relative to application.

Fortunately, there is significant evidence that little or no deterioration need occur during proof tests if nonimpact types of proof testing in dry environments (prevents chemical attack of flaws) are employed.

CARES/LIFE. To optimize design, it is necessary to determine the reliability of a loaded component. Methods of quantifying this reliability and the corresponding failure probability have been incorporated into a public domain computer program known as CARES/LIFE (Ceramics Analysis and Reliability Evaluation of Structures Life Prediction). A flowchart for the program is shown in Fig. 10.12. Two earlier versions of this program were called SCARE (Structural Ceramics Analysis and Reliability Evaluation) and simply CARES.

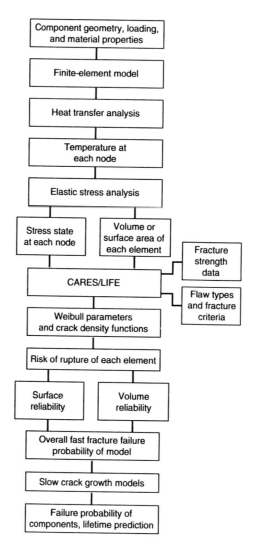

Fig. 10.12 Block diagram for the analysis and reliability evaluation of ceramic components. Source: Ref 10.6

The basic purpose of CARES/LIFE is to predict the lifetime reliability of a ceramic component. It does this by modeling slow crack growth and calculating the probability of failure as a function of service time. In particular, it can:

- Predict the strength degradation of certain materials (Fig. 10.13, 10.14)
- Predict the reliability of a component that has undergone proof testing
- Predict the effect of multiaxial stresses on reliability
- Estimate fatigue parameters from naturally flawed specimens ruptured under static, cyclic, or dynamic loading
- Generate a data file that can be used to draw the critical regions of a structure

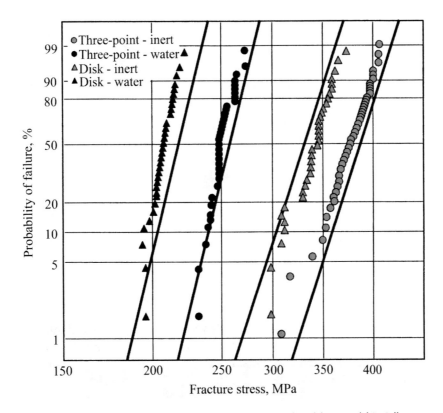

Fig. 10.13 Weibull plot of alumina three-point bend bars and biaxially loaded disks fractured in inert and water environments. The solid lines show CARES/LIFE predictions made using Weibull and power law fatigue parameters obtained from four-point bend bar fast fracture and dynamic fatigue (constant stress-rate loading) data. Strength degradation in water is predicted for a dynamic load of 1 MPa/s. A mixed-mode fracture criterion was chosen to account for the change in surface flaw reliability for multiaxial stress states. Source: Ref 10.10

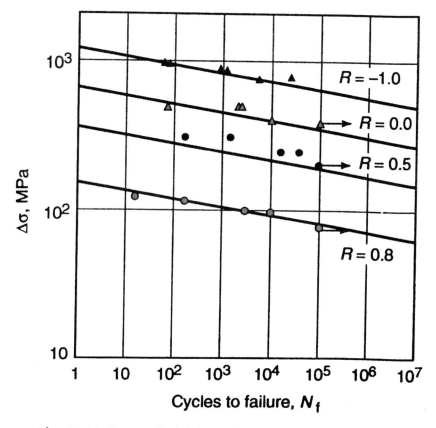

Fig. 10.14 Stress amplitude-failure cycles (*S-N*) plot of 3 mol%-yttria-stabilized zirconia tensile specimens for various *R*-ratios. Solid lines show CARES/LIFE predictions at 50% reliability using the Walker slow crack growth law to predict strength degradation due to cyclic fatigue. Source: Ref 10.10

- Couple to a number of finite-element programs, including ABAQUS, ANSYS, COSMOS/M, MARC, and MSC/NASTRAN

CARES/LIFE retains the capabilities of the earlier CARES program, which was developed principally to calculate the fast fracture reliability of macroscopically isotropic ceramic components. A more limited program, CARES/PC, is available for statistical analysis of data obtained from the fracture of simple, uniaxial tensile or flexural ceramic specimens. The Weibull and Baldorf material parameters are calculated from these data.

A wide range of applications makes use of ceramic parts designed with CARES and CARES/LIFE, including parts for turbine and internal combustion engines, bearings, laser windows on test rigs, radomes, radiant heating tubes, spacecraft activation valves and platforms, cathode ray tubes, rocket launcher tubes, and packaging for microprocessors. Both programs are also used to reduce data from specimen tests and provide statistical parameters for material characterization. This allows engineers and material scientists to compare materials without regard to specimen geometry and loading.

10.6 Processing of Structural Ceramics

The processing steps for producing structural ceramics are shown in the flow chart given in Fig. 10.15. The steps can be grouped into six general categories:

1. Raw material preparation
2. Forming and fabrication
3. Drying
4. Green-body machining

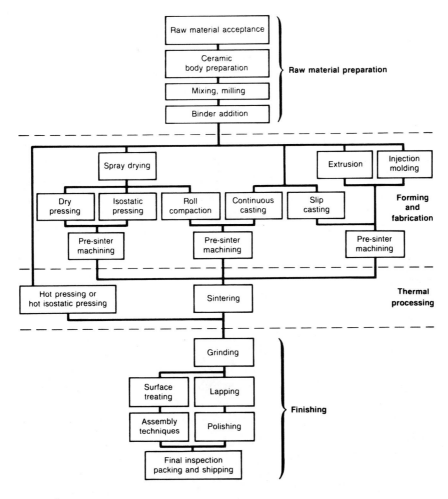

Fig. 10.15 Flow chart for ceramic processing. Source: Ref 10.3

5. Sintering
6. Final finishing

Raw Material Preparation

Raw material preparation includes materials selection, ceramic-body preparation, mixing and milling, and the addition of processing additives such as binders. Materials selection is important because structural ceramics require high-quality starting materials. For example, the Al_2O_3 powder for alumina ceramics is usually obtained as calcined alumina from the Bayer process, which uses bauxite as the starting material. Zirconia is obtained by processing zircon ($ZrSiO_4$) to produce 99% pure ZrO_2. Silicon carbide (SiC) is produced by the Acheson process in which silica, SiO_2, and coke are placed in an arc furnace and reacted at 2200 to 2500 °C (4000 to 4500 °F). Silicon nitride (Si_3N_4) is produced by reacting silicon with nitrogen. Chemical techniques are used to produce powders where extremely high purity and very fine particle sizes are required. The assurance of final product quality starts with well-defined and strict material acceptance criteria. For oxide systems, the starting materials are generally mixed in aqueous systems and milled to obtain the specified particle size distribution for the body. If necessary, organic binders are added after the milling and mixing. This results in a slurry or slip, which is the starting material for part forming and fabrication.

Advanced ceramics tend to use processed, high-purity powders such as alumina, zirconia, silicon nitride, aluminum nitride, and silicon carbide. As the performance requirements become more critical, chemical purity, controlled particle size (distribution), reactivity, and freedom from agglomeration become more important. The optimal particle size distribution is still a matter of disagreement. An extremely fine, uniform, submicrometer particle size can yield ordered packing, provide freedom from pore defects, provide high sintering or reaction rates, lower sintering time, and prevent exaggerated grain growth defects. However, achieving full density may be difficult because uniform spheres pack with approximately one-third void space. In addition, forming times are long, it is difficult to obtain good forming rheology, and prevention of agglomeration is critical. A broad distribution of particle sizes with many orders of magnitude of particle size down to extremely fine (nanometer) scale will provide denser green packing, and desirable forming rheology is far easier to achieve. Forming time is shortened, although defects, such as large pores and exaggerated grain growth, require careful control to prevent.

The various raw materials are typically ground together to provide intimate mixing and particle size reduction. The configuration, speed, and time duration of milling can significantly alter the intimacy of mixing and ultimate particle size distribution. As milling time increases, the relative improvement in mixing or particle size reduction decreases in a quasi-logarithmic fashion. Milling is usually done wet with water or another other liquid that becomes part of the final formulation. In some cases, dry milling is done, although many materials are dried after milling.

Ceramic materials are crushed and ground to a fine powder. The mechanical methods consist of the same ball mill grinding operations used for traditional ceramics. The trouble with these methods is that the ceramic particles can become contaminated from the materials used in the balls and walls of the mill. This compromises the purity of the ceramic powders and results in microscopic flaws that reduce the strength of the final product. The material then is screened or undergoes flotation, filtering, or magnetic separation to remove out-of-size or undesirable components. The resulting powder is mixed with various other materials, including various metallic oxide fluxes, organic binders, plasticizers, and, often, water. Ball milling or hammer milling equipment is used to remove any agglomerated material and ensure thorough mixing. Powders for structural ceramics may be made with the same processes used for standard ceramics but with stricter standards for material purity, particle size, and mixing.

Structural ceramics are also prepared by several sophisticated processes that provide a still higher purity, better particle size distribution, and more uniformity of characteristics from lot to lot. Raw materials may undergo chemical reactions that create intermediate compounds that are more easily purified and then chemically converted to the oxide or other final compound. Processes that are used include vapor deposition, precipitation, and calcination.

Vapor-phase deposition uses reactant gases that are mixed and heated in a suitable chamber. Particles nucleate and grow from the gas phase.

Precipitation is carried out with one of several process variations. In addition to conventional precipitation, precipitation under pressure (hydrothermal precipitation), and co-precipitation may be used. The latter approach provides better mixing if several oxides or salts are included in the powder mix. Precipitation is a key procedure in the production of high-purity alumina (Al_2O_3) from bauxite. The aluminum hydroxide in bauxite is first dissolved in caustic soda. The liquid is then separated from nonsoluble material by filtering, and the resulting aluminum trihydrate is precipitated by the addition of seed crystals and by changing the pH. Very fine alumina powders result.

Calcination is common in the production of structural ceramic powders. With this process, the raw material is heated to a high temperature to decompose salts or hydrates or to remove volatile ingredients, chiefly water. After the heating operation (calcination), the powders are ball milled to break the bonds between the crystallites that are formed by diffusion during the calcination. A very fine powder can be produced. Magnesium oxide, MgO, is often made by extracting magnesium hydroxide from seawater and calcining it to convert it to the oxide. Magnesium oxide is used to make high-temperature electrical insulators and refractory bricks.

Dry pressed powders must usually be granulated or spray dried to provide free-flowing powders that will readily fill a mold. Spray drying usually provides spherical aggregates of powder particles ideal for flow and die filling. Spray drying also permits the uniform incorporation of multiphase materials, binders, lubricants, and other additives.

Forming and Fabrication

Structural ceramics are formed from powders, stiff pastes, or slurries. The slurry from the preparation procedures is converted to an agglomerated flowable powder by spray drying or to a stiff paste by filter pressing. Structural components are formed by pressing of powders, extrusion of stiff pastes, or by slip casting of slurries. In some cases, presinter machining (green-body machining) is required. Evaluation factors for forming of ceramic powders are shown in Table 10.7.

Many forming processes used in ceramics, such as dry pressing, extrusion, plastic forming, and injection molding, are similar in principle to processes used in the metals, plastics, and food industries. Experience in these areas can be used insofar as the rheological behavior is similar. The abrasiveness of ceramics requires that production equipment be hard-faced to avoid wear and contamination. Slip casting and tape casting are forming methods developed specifically for ceramic processing. The great variability of practices and methods for those described and the number of specialty processes used means that there is significant variation from

Table 10.7 Evaluation factors for different ceramic forming methods

Forming methods	Large component sizes	Complex component shape	Independence of plasticity	Tolerance	Ceramic problems (defects)	Production volume	Production speed	Required plant space	Equipment cost
Wet forming									
Slip casting	1	1	3	5	5	5	5	5	1
Pressure casting	2	2	3	4	4	5	4	5	2
Tape casting	3	5(a)	4	1	2	1	1	2	4
Plastic forming									
Extrusion	2	3(b)	5	3	3	2	2	4	3
Plastic pressing	3	3	4	2	3	2	3	3	2
Jiggering	2	4(c)	5	4	4	3	4	3	2
Injection molding	5	5(d)	3	1	2	2	2	3	5
Dry forming									
Dry pressing	4	5(e)	1	1	1	1	1	3	3
Cold isostatic pressing	3(f)	4	1	1	2	4	3	4	4

Note: Under usual circumstances, 1 is best and 5 is worst. Variability in processing methods may deviate from common factors indicated. (a) Thin sheets only. (b) Uniform cross section. (c) Axially symmetric shapes. (d) Size limited by binder burnout. (e) Small to medium size, short aspect ratio. (f) Some very large shapes made. Source: Ref 10.9

Table 10.7, and the descriptions in both Table 10.7 and this section should be used only as a preliminary guide.

Wet processing methods include slip casting, pressure casting, and tape casting.

Slip casting provides great flexibility in both part shape and size. The major incremental cost for a new shape is development of a master mold. Multiple-use molds made of plaster and/or porous plastics are produced from the master with controlled mold porosity—fine pores provide capillarity, while coarse pores provide a larger liquid reservoir. The liquid ceramic mixture contains fine particles with deflocculants and dispersants to ensure uniform suspension of the ceramic particles. Ceramic powders are usually dispersed in water or nonaqueous media, resulting in a slip, a fluid slurry with thixotropic rheology. It is desirable to have a material that flows when stirred yet maintains its shape after discharge into a mold. A clear understanding of colloid chemistry and rheological control is vital to proper slip preparation. Factors that affect slip (or plastic working) rheology include solids fraction, liquid fraction, dispersant(s) type and quantity, particle size distribution, particle surface chemistry, pH, and order (timing) of additions. Structural ceramics require greater additive and time control than traditional clay-based ceramics because they have no natural plasticity.

A layer of the slurry is deposited on the mold surface as water is drawn off by the porous nature of the mold wall. Capillary suction continues to draw off water and deposit material from the slurry, building up the thickness of the cast structure. The part gradually assumes a leathery consistency as the water is absorbed by the mold. After a suitable time, the part can be removed from the mold, handled, and finished, prior to drying and firing. Finishing usually involves careful wiping with wet sponges to remove mold flash and to smooth any areas where needed.

Fine fractions may migrate through the drying cake so that the surface microstructure may differ from the bulk. This difference may be desirable for surface finish or fired densification. However, inappropriate chemical segregation or drying/firing stresses can result if segregation is not controlled. Molds must be conditioned with slip before use and dried between casts to maintain casting behavior. After a number of casts, either mold definition is lost or capillaries clog, preventing indefinite mold use. Vacuum suction can be applied to the outside of the mold to speed the capillary effect somewhat. Usually, many similar molds are used in sequence to maintain productivity. As a result, space usage is high, considerable labor is required, and cycle times are long, although the molds are low-cost forming equipment with great shape and composition flexibility.

Mold time depends on the liquid content of the slurry and the thickness of the cross section cast. The ideal slip maximizes solids content so that there is as little water to draw off as possible. Thick cross sections can take a very long time to produce because a parabolic decrease in casting rate occurs with time. For solid casting, a liquid reservoir must be provided in the mold design to provide additional slurry as water is drawn off and shrinkage occurs. Drain casting (Fig. 10.16) is used to provide a hollow internal cavity for shapes. After the cake has built up to a controlled thickness, the residual slurry is poured off and reused. For all shapes, edges and mold seams are trimmed green (unfired) with a knife and smoothed with a wet sponge. Shapes can be attached by applying a small amount of slurry to cast pieces and pressing pieces together until attached.

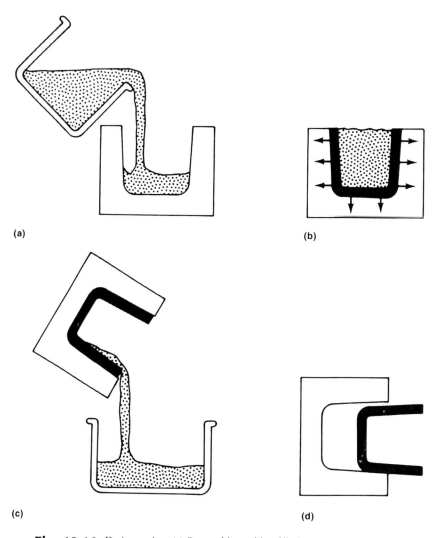

(a)

(b)

(c)

(d)

Fig. 10.16 Drain casting. (a) Permeable mold is filled with slip. (b) Liquid is extracted from the mold, while forming compacts along mold walls. (c) Excess slip is drained. (d) Casting is removed after partial drying. Source: Ref 10.11

Pressure casting applies pressure to the slip in the slip casting process to partly or completely replace the capillary action of the porous mold. Pressure casting can be used to accelerate the casting speed of thick-walled or solid pieces. The pressure also allows a slightly lower fluid content in the slip. More importantly, it results in a better part. Dewatering of the cast material is greater, there is less postcasting shrinkage, and the dimensions of the finished piece are more accurate. The part has an improved surface finish and greater strength.

Gas pressure is used to inject the slurry into the mold and help to force the liquid through the mold capillaries. Pressures up to approximately 4 MPa (580 psi) are used. The pressure also causes an intimate conformance with the mold wall to provide superior definition and surface quality. If even greater precision and definition is desired, fine particulate can be sprayed on a solid mold made of wax or plastic. This is done in layers with alternate furnace bakeouts and eventual buildup with coarser particulate slurry. Molds are made from porous plastic polyelectrolytes, which can withstand the pressures involved. Finally, the mold is dissolved or burned out.

Tape casting produces thin sheets approximately 0.1 to 2 mm (4 to 100 mils) thick of ceramic, particularly for electronic substrates and multilayer capacitors. A volatile organic liquid is usually used with appropriate binders and plasticizers that provide rheological control, green strength, and flexibility to the tape produced. As shown in Fig. 10.17, the material is

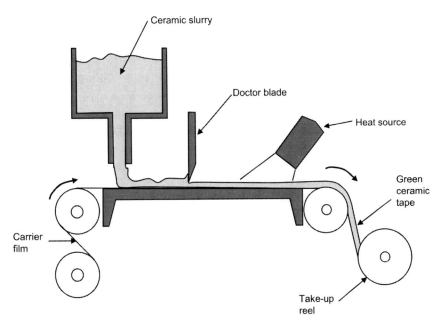

Fig. 10.17 Ceramic tape casting

drawn across a thin sheet of flexible, nonporous plastic using a doctor blade to control thickness. Excess material is held back by the blade while the belt carries the other material in a sheet of uniform thickness. The continuous ribbon of slurry on plastic passes through a drying oven to evaporate the vehicle. The sheet is then taken up on a reel for later processing. Further drying, blanking, punching to the shape desired, and firing then take place. The resulting green ceramic can be cut into pieces and fired after removal from the tape. It can also be metallized, stacked, and punched to produce multilayer capacitors and electronic substrates. Very fine particle sizes, usually of materials such as high alumina, titanates, and aluminum nitride, are the typical materials produced in this way. Fine particle size increases strength and permits a thinner tape to be produced.

Extrusion. Stiff pastes are used in extrusion and injection molding to produce ceramic preforms. Ceramic powders used in plastic forming employ similar additives and liquids to those used in slip casting, but the fluid content is reduced and additive formulation adjusted. A stiff plastic body must be prepared suitable to the particular plastic process, applied stress, and forming rate. Preparation is usually performed in pug mills or sigma blade mixers that provide high shear. The body must show a suitable yield point (stress at which flow begins) and maintain its shape after forming. Bingham viscoelastic behavior (linear relation between shear stress and shear rate with a yield stress) is usually desired. For structural ceramics that contain no clay fraction organic binders, plasticizers and lubricants are added to obtain appropriate plastic rheology.

Extrusion rapidly forms a uniform cross section, such as for circular or square pipe or thermocouple tubes, by forcing the material out of a die aperture of appropriate shape using a ram or screw auger. The feed angles of the die must be appropriate to provide sound and efficient flow of the plastic mass, but this depends on the rheology of the ceramic at operating temperature and the extrusion rate. If the die angle is improper, either inefficient production or defects such as central burst and surface cracks can occur. Aside from the use of abrasion-resistant linings, equipment is virtually identical to that used in plastic and food extrusion. Prior to extrusion, the material is prepared by mixing in enough water to produce a stiff paste. Trapped air or gases may be removed during material preparation, or the ceramics extruder may have a vacuum chamber for this purpose. If the extruded shape is to have a hollow interior, a mandrel of the desired shape is incorporated in the die. After extrusion, the green part is cut to length. Because extruded sections are of a uniform cross section, they are sometimes machined or shaped by other methods prior to firing to provide the desired external shape. Thermocouple insulation tubes and automotive emission catalyst supports are among products made by extrusion.

Injection Molding. Complex objects can be made to precise, near-net shapes by using injection molding. Injection molding machines (Fig. 10.18), similar to those used for plastics injection molding, are often used

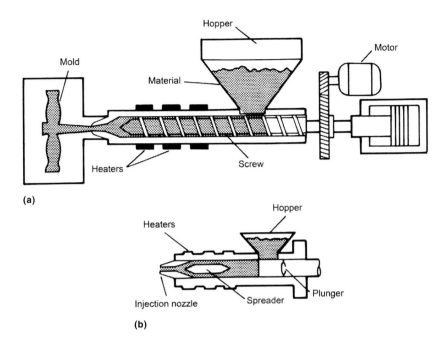

Fig. 10.18 Cross-sectional side views of injection molding machines. (a) Screw type. (b) Plunger type. Source: Ref 10.11

to mold green parts from ceramic paste. The process is useful for a large-quantity production of parts with greater complexity than those produced by pressing. Relatively wet material mixtures with organic binders are used, but the binders must be removed before final firing. Tolerances are broader than those for dry-pressed parts. A ram or screw feed forces the plastic mass into a mold. Injection molding has been used particularly for making precise, complex, structural ceramic shapes, such as turbocharger rotors, by incorporating materials such as silicon nitride into an organic plastic and solvent. Subsequent to forming, the binder material must be burnt out before sintering. This limits the object size to relatively thin cross sections.

Applications of ceramic injection-molded parts are currently inhibited by difficulties in debinding and sintering. Burning off the polymer is relatively slow, and its removal significantly weakens the green strength of the molded part. Warping and cracking often occur during sintering. Further, ceramic products made by powder injection molding are especially vulnerable to microstructural flaws that limit their strength. Work has been done to minimize the amount of binder and to design binders that will burn out rapidly. New water-based systems are being developed that lower production costs and decrease binder content, burnout cost, and environmental impact of nonaqueous solvents.

Powder pressing is the most common ceramics forming operation and is the preferred method, if it is feasible. Most pressing is similar to pressing of glass or compression molding of plastics in that the material is placed in a metal mold and then compressed with strong force from a descending punch. Mechanical or hydraulic presses are used. Lubricants and binders may be included in the material. Binders may be organic or inorganic. If the ceramic mixture is relatively dry (0 to 5% water), the method is known as dry pressing, and the operation is very similar to the compaction of powder metal parts. High pressures and precision dies are used. If the ceramic mixture is relatively wet (wet pressing), the material contains more water and is capable of greater flow when pressed. The operation becomes similar to the compression molding of plastics where the plastic material flows into remote areas of the mold. Wet pressing is used for somewhat more complex parts than dry pressing, but the greater water content of the material leads to greater dimensional variations and larger tolerances than those for dry-pressed parts. Generally, pressing, especially dry pressing, is used when the part shape is relatively simple.

In dry pressing, a nearly dry, free-flowing powder (usually spray dried) fills a metal mold and is compacted uniaxially. This method is most suitable for high-volume production of small, simple, low-aspect-ratio shapes with fairly uniform cross section. Die and interparticle friction preclude forming large, long, and complex objects. Internal (in the ceramic) and external (on the die) lubricants are used to decrease friction, thereby facilitating forming and reducing density variations. Multisegment dies, multidirectional pressing, floating die cavities, and high time/pressure cycles can reduce pressure gradients and increase the flexibility of size and shape, with high capital costs and part cycle times.

To minimize density variations and produce more complex (or larger) shapes, isostatic pressing is used. This is often termed cold isostatic pressing to distinguish it from hot isostatic pressing, a firing process. The advantage of isostatic pressing is that pressing occurs in all directions, providing more uniform compaction. Isostatic pressing is used for larger components and where extensive presinter machining is required. The nominally dry powder with internal lubricants is filled into a rubbery mold with the desired shape features. The mold is sealed and placed into a hydraulic fluid, which is then pressurized. A part is produced with very uniform density because of the multidirectional pressure and because significantly higher pressure can be achieved compared to that obtained by uniaxial pressing. Pressure in the vessel, transferred by a medium of water, oil, or glycerin, compacts the powder in the mold. This approach is used with spark plug insulators, which have a central metal mandrel to form the center hole and flexible elements to rough form the outer surfaces. They are then finish machined in the green state before the part is fired. However, in many other applications of isostatic pressing, the part can be finished to final shape without machining or further forming.

Drying

Drying involves the removal of free, nonstructural water or other solvents from a formed ceramic. It is a two-step process controlled both by internal flow of liquid to the surface and surface evaporation of the liquid. The internal flow can be increased by increasing permeability with a coarser structure, decreasing the viscosity of the liquid, or by increasing the concentration gradient to the surface (more rapid removal of surface liquid). Evaporation can be accelerated by air motion (removal of saturated vapor), increased temperature, lower vapor pressure liquid, increased drying surface area, decreased air humidity, or reduced pressure.

As shown in Fig. 10.19, drying occurs in three stages. At high moisture content, liquid is the continuous phase both in bulk and on the surface and evaporates almost as if it were free, continuous liquid. This is the constant rate period, region A of Fig. 10.19. In this region, evaporation dominates the drying rate. Most drying shrinkage occurs during this stage. An excessive drying rate can cause differential shrinkage in thick cross sections. Warping or cracking may result, depending on whether the body can accommodate the drying stresses or not. When the surface film of liquid becomes discontinuous, the first falling-rate period with a linear drying rate begins, region B. Liquid is still continuous in the ceramic because large pores are being drained. There is a small amount of shrinkage in this

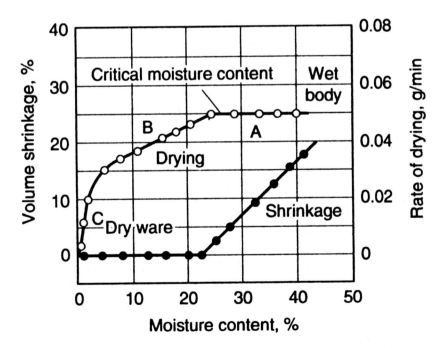

Fig. 10.19 Change in bulk volume on drying a ceramic body. Source: Ref 10.9

region, so that the transition from the constant rate to the first falling-rate period is often considered the critical liquid content for shrinkage damage. However, some care must also be taken not to dry too rapidly in the first falling-rate period. In the second falling-rate period, region C, the drying rate is nonlinear. Air is the continuous phase, and liquid pockets are removed by evaporation. If liquid is distributed uniformly in the body, there is no shrinkage in the second falling-rate period. Additional causes of damage due to differential shrinkage are mechanical restraint of the body, preferred particle orientation, uneven drying, and varying cross sections of the body.

Several methods are available to remove moisture and other additives from ceramic workpieces before they are sintered. Room-temperature drying is the simplest. After forming, some free water may be removed from workpieces that have been made with wet forming methods, by allowing them to stand for a period at room temperature. For further drying, workpieces may be allowed to dry for several days in a drying room or chamber. These facilities are kept at a high humidity or the workpieces are wrapped with plastic film so that the drying is gradual and even.

Under more elaborate production conditions, the workpieces are subjected to a drying operation under conditions of controlled heat and humidity. The ideal cycle in a batch drying chamber starts with high-humidity cool air being blown onto the workpieces and ends with high-temperature dry air after a gradual change over a period of time. High velocity of airflow is needed throughout the cycle. Various methods may be used to provide heat to facilitate the operation while still controlling the moisture-removal rate. If the drying is a continuous operation, a drying tunnel is used and the workpieces proceed through it on a conveyor or on a series of material handling buggies. The starting high-temperature air enters the tunnel at the exit end. As the air flows toward the entrance end, it gradually becomes more moist and cooler, so the first air blast hitting the entering workpieces is cool and moist and the air blast at the end is warm and dry.

Drying must be carried out in a controlled manner to avoid cracking, distortion, or excess internal stresses. The control is needed because ceramics shrink when moisture is removed and, at a certain point, are brittle enough for differential shrinkage to cause cracking. Drying is most important for workpieces formed with one of the wet forming methods, such as casting, extruding, injection molding, and wet pressing. Drying helps remove organic binders and plasticizers as well as moisture from the workpiece. The degree of drying depends on the accuracy required in the final dimensions and what subsequent forming or shaping operations, if any, are required. Controlled drying is sometimes the first stage of the firing process.

In some cases, drying of wet-formed ceramic parts is performed in two stages. The first is a preliminary partial drying that puts the workpieces in the green state so that they can be handled and trimmed, or otherwise processed, before the final drying. Workpieces requiring trimming are often partially dried before the trimming operation. Organic materials as well as moisture must be removed prior to sintering. Otherwise, gases formed in the workpiece from the decomposition of organic materials or from steam may cause voids or other problems with the workpieces. Final drying, whether a separate operation or the first stage of sintering, is intended to remove these materials completely.

Microwave drying (also called dielectric or radio-frequency drying) is used in production shops to speed the drying process. The process is similar to heating and cooking food in a microwave oven in that radio-frequency energy provides the heat that dries the workpieces. The process is much quicker than other drying methods because heat is generated within the workpiece and does not have to penetrate from the surface as it would if radiant, convection, or conduction heat sources were used. Greater heat is generated in those portions of the workpiece that are wettest. Hence, uniformity of drying is promoted. Because the operation is faster than other methods, it minimizes the quantity of work in process in the drying cycle. Both batch and through-feed ovens are available.

Green-Body Machining

Whenever possible, it is desirable to machine a ceramic while it is still green (unfired). Conventional machining methods can be used and the piece may be machined either before or after drying, depending on the strength, shape, and integrity of the part. In some cases, parts can be bisque (partially) fired to provide enough strength for the machining operation while avoiding firing a fully matured piece. Green and bisque firing may not provide critical tolerances, although careful control of both the machining and firing shrinkage may prove adequate. Machining after firing is quite costly (as much as 80% of total cost) and can introduce critical strength-limiting flaws in the material.

Machining methods include turning, milling drilling, boring, threading, tapping, and other operations. These operations are normally carried out before the part is sintered, when it is in the "leather-hard" condition, after drying. Carbide cutting tools are used because of the abrasive nature of the ceramic material. Such machining is made difficult by the weak and fragile nature of unsintered ceramics. There are also dimensional changes from firing that may make it difficult to achieve close dimensions of machined surfaces after sintering.

Sintering

The functions of sintering are to bond individual grains into a solid mass, increase density, and reduce or eliminate porosity. Temperatures approximately 80 to 90% of the melting temperature of the material are commonly used in sintering ceramics. In sintering, the dried ceramic workpiece is heated to a high temperature for a specified period. Typical temperatures are approximately 1400 to 1650 °C (2500 to 3000 °F). Some ceramic materials may require a temperature over 2000 °C (3600 °F). Alumina without glassy material additions is fused at 1930 °C (3500 °F).

There are two prime sintering mechanisms: solid-state diffusion, where particles fuse together without melting (as with the pure alumina), and liquid-phase sintering. When glassy materials are included in the ceramic, liquid-phase sintering takes place. The glasses melt and fuse the particles together at a lower temperature. Some surface melting of other particles in the material may also occur, providing a bond between the particles. Also, chemical reactions at the elevated temperature may create liquids that provide liquid-phase sintering. When glasses are not included, diffusion is the usual means of sintering. When sintering is completed with either mechanism, the particles fuse into a hard, strong, homogeneous, and dense state. Both heating and cooling phases take place very slowly, and the total sintering process can require several days or longer. Careful control of temperature and temperature changes over time is important to ensure proper sintering and to prevent adverse effects on the workpiece during the process.

As the material is slowly heated, several changes take place. The first is drying, if the workpiece is not already dried. Then, if the material contains water of crystallization, it is driven off at temperatures of 350 to 600 °C (660 to 1100 °F). Considerable shrinkage may take place during sintering because material moves to fill the open spaces between grains.

Ceramics that do not produce a great deal of glassy phase during firing mature by the process of sintering. This is the densification by solid-state diffusion at elevated temperature driven by reduction of surface area. Several processes or stages can occur, including evaporation-condensation and surface and bulk diffusion. Usually, a small amount of sintering additive (sintering aid) is introduced to accelerate the sintering process by introducing liquid-phase sintering and/or accelerating diffusion rates. Other additives function as grain growth inhibitors, preventing the competitive process of crystal growth (large crystals may deteriorate strength and other properties). The liquid phase usually forms a thin, glassy layer on the grain (crystal) boundaries of the fired ceramic and has a substantial effect on mechanical, electrical, thermal, and magnetic properties. This grain-boundary film has been identified in virtually all manufactured ceramics—even those with trace amounts of additive or impurities. A careful balance must be struck between pore reduction by densification and grain growth. The sintering additives make ceramics prone to exaggerated grain growth, which may deteriorate properties. It is even possible for large agglomeration pores to grow during firing. During the high-temperature sintering operation, the particles fuse together through diffusion processes.

Sintering is conducted either at ambient pressure or with added pressure in the case of hot pressing or hot isostatic pressing. The final microstructure is developed during thermal processing by sintering, vitrification, or reaction bonding. Sintering takes place by volume, surface, or grain-boundary diffusion and is a solid-state process. Although many industrial ceramic products are sintered without external pressure, high-performance ceramics, because they are extremely sensitive to all flaws, even very small voids, are usually sintered with high external pressures to minimize the amount of porosity.

Vitrification involves the presence of a liquid phase during thermal processing. The liquid phase provides faster diffusion paths and holds the piece together by capillary action during processing. This results in an amorphous or glassy phase being present in the final microstructure. The final microstructure is created by vitrification for systems with less than 99% pure oxide, porcelains, and Si_3N_4 with sintering additives.

In some cases, the thermal processing is aided by adding external pressure during sintering. The pressure can be applied uniaxially in hot pressing or isostatically in hot isostatic pressing. Covalent materials, such as silicon carbide and silicon nitride, and composite systems usually undergo hot pressing. Pressure can also be used to suppress the decomposition of materials, such as in the gas-pressure sintering of Si_3N_4.

The grain structure that results from sintering depends strongly on the grain (powder) size and grain size distribution of the starting material as well as minor additive sintering aids and grain growth inhibitors. One approach to achieving fine ultimate grain size and high strength is to use very fine, uniform starting powder grain size. Some success has been found using uniform nanoscale powders, although such powders do complicate microstructural (final property) control during the materials handling and forming stages. Pore structure and grain growth interact and must be controlled; this can be achieved with selected additives and controlled firing time/temperature (Fig. 10.20) to minimize porosity and control grain growth. The existence of a few larger starting particles, coupled with the effects of additives, can cause exaggerated grain growth. Such grains can act as strength-inhibiting flaws, whether on the surface or in the interior of a ceramic. This can also be a method for introducing elongated particles for toughening.

Hot pressing is used to make cutting tool inserts and other ceramic parts of simple shape. Dry, fine ceramic powder or a powder preform is placed in a graphite die supported by ceramic parts. Pressure is applied at the same time that the ceramic workpiece is heated to the sintering temperature. However, the temperature is lower than that required for regular sintering because the combination of pressure and temperature aids in bonding the particles. The diagram in Fig. 10.21 shows a typical tooling arrangement. The shapes of parts that can be processed in this way are limited, but the parts, after pressing, are finished or nearly so. The process is used with some materials that cannot be densified by sintering without pressure. Microcrystalline ceramics with high strength properties can be

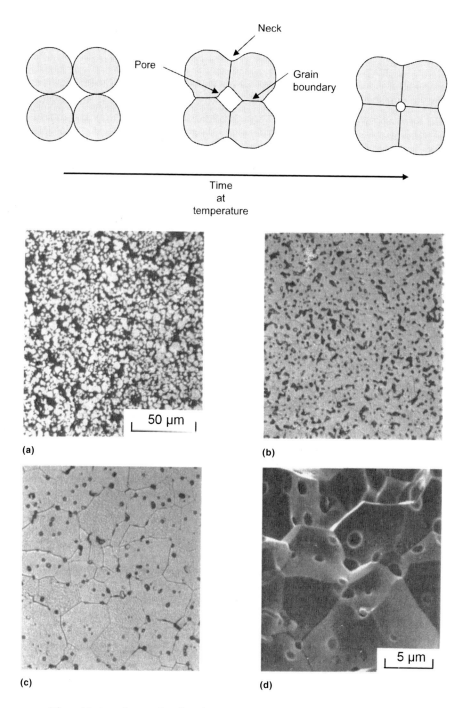

Fig. 10.20 Progressive densification and grain growth at several stages of sintering. (a) Initial stage. (b) Intermediate stage. (c) Final stage. (d) Fracture surface. The fracture surface micrograph shows the desirable placement of spherical pores on grain boundaries in the final stage of sintering. Source: Ref 10.7

produced, and the parts made are used in higher-technology structural applications.

Hot isostatic pressing (HIP) involves the isostatic compaction of powders at an elevated temperature. A water-cooled pressure vessel with an internal high-temperature furnace is used (Fig. 10.22). Pressures reach approximately 310 MPa (45 ksi), and temperatures go up to 2000 °C (3600 °F). Argon, nitrogen, or helium gas is pressurized and acts against the surfaces of the workpiece through a hermetically sealed glass or metal encapsulation. Because of the high temperatures involved, sheet metal, if used for encapsulation, must be refractory. Glass envelopes soften at the

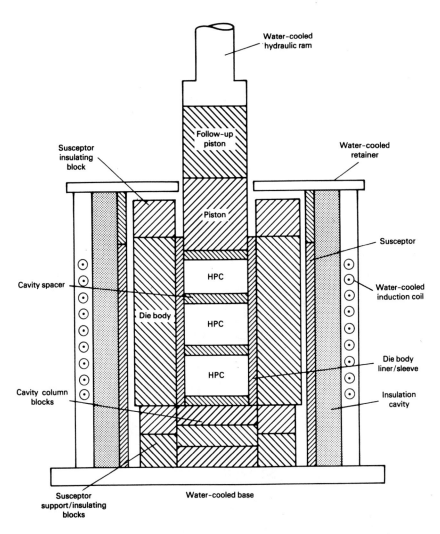

Fig. 10.21 Typical ambient hot press (uniaxial, unidirectional). HPC, hot press cavity. Source: Ref 10.12

temperatures involved but still transmit the pressure to the ceramic material. Pressure and temperature are closely controlled. Electrical resistance heating is usually used. The method is advantageous for producing more complex shapes of parts than by regular hot pressing. Improved, more uniform compaction for critical parts is an important advantage.

Reaction Bonding. Making ceramic parts by reaction-forming mechanisms represents an important alternative to conventional sintering processes. For polycrystalline ceramics made by reaction-forming processes, consolidation between constituent particles occurs by chemical reactions rather than by neck growth between particles. Reaction sintering and forming processes include reaction-bonded silicon nitride and

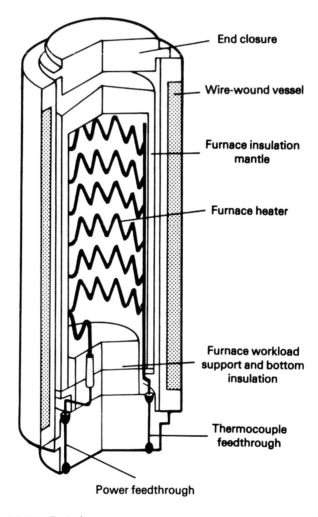

Fig. 10.22 Typical construction of a hot isostatic pressing furnace with a cold pressure vessel wall and internal furnace. Courtesy of Asea Brown Boveri. Source: Ref 10.12

reaction-bonded silicon carbide. In reaction forming, processing temperatures are frequently much lower, and processing times are sometimes much shorter, than in conventional sintering. Microwave heating can reduce processing temperatures even more. One exceptional result is that service temperature limits can substantially exceed processing temperatures. Because many of the reaction-bonded ceramics use gaseous reactants or generate gaseous products, these materials tend to have much higher levels of frequently interconnected, residual porosity. As a result, the properties of reaction-bonded ceramics are generally inferior to sintered, hot-pressed, and HIPed ceramics, as shown in Fig. 10.23.

The reactions used to form ceramics can be difficult to control due to inherent instabilities. Heat management is a significant issue in exothermic reactions, because overheating causes loss of microstructural control and phase chemistry. Achieving single-phase bodies with highly stoichiometric materials is a particularly difficult problem. Even though reactions are normally carried out at relatively low temperatures, the combination of long reaction times and the reactivity of constituents can pose serious detrimental interactions.

For the reaction to proceed to completion, a continuous pore structure must be maintained for diffusion of reactant gases. If the number of pores and pore channels is decreasing, the depth to which the reactions may proceed becomes a limitation. The original density, the particle size of

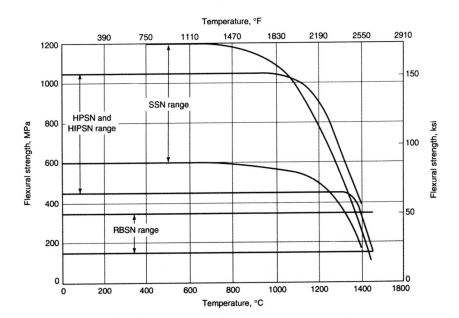

Fig. 10.23 Variation of flexural strength with temperature for various types of silicon nitride ceramics. SSN, sintered silicon nitride; HPSN, hot-pressed silicon nitride; HIPSN, hot isostatically pressed silicon nitride; RBSN, reaction-bonded silicon nitride. Source: Ref 10.7

the initial reactants, and the thickness of the compact become important factors. The greater the density, the less the depth to which the reactions may proceed.

Reaction-Bonded Silicon Nitride. In reaction-bonded silicon nitride, a green body made from silicon powder is nitrided by nitrogen gas. Consolidation in the net-shape fabrication process occurs by filling the void space between silicon particles with the Si_3N_4 reaction product. For Si_3N_4 shapes, several factors are important, that is, particle size and distribution, atmosphere nature, and control, time, and temperature of firing. Complete nitriding is difficult if the initial particle size is >70 μm (200 mesh). Frequently, ≤60 μm (250 mesh) is used for normal ceramic shapes. For high-performance ceramics, however, particle size control is essential, and silicon must be <20 μm, with an average size of 2 μm. Even particles as small as 10 μm can result in inclusions in fired pieces, which can act as flaws and crack initiators.

Frequently, the silicon part is either presintered in argon at approximately 1200 °C (2190 °F) or is partially nitrided to produce a part that can be machined to final dimensions by conventional tooling. After shaping, the part is nitrided to completion, usually in atmospheric-pressure nitrogen gas at temperatures in the 1250 to 1450 °C (2280 to 2640 °F) range. Many reported nitriding schedules employ temperatures in excess of ~1410 °C (2570 °F) (the silicon melting temperature) and times up to 150 h. Residual porosity is typically in the range of 20 vol%, it is usually fully interconnected, and 80% of the pores are smaller than 0.1 μm (4 μin.).

During the reaction of nitrogen with silicon, silicon can melt. This must be avoided if a complete and nondisruptive reaction is to occur. Three basic approaches are used to control the reaction. First, the overall reaction rate is controlled by the flow rate of nitrogen into the furnace. This puts an upper limit on the reaction rate and heat generation within the furnace. Second, the addition of an inert gas such as argon or helium, either in the nitrogen stream or as a cyclic purge, can control the formation of local hot spots within the compact. A third approach is to add a nonreactive material such as prereacted Si_3N_4, SiC, or BN. The heat generation within the body is thereby reduced, and the added material acts as thermal ballast.

Reaction-bonded silicon carbide (RBSC) is the most important example of the processes based on reactions between a porous solid and an infiltrating liquid phase. These processes share the near-net shape and near-net dimension capabilities of gas-phase reaction bonding, as well as the reduced processing temperatures relative to solid-state sintering. Also, they may have significant advantages in the processing rates and material densities that are achievable.

Conventionally, RBSC is made by infiltrating a porous carbon shape with liquid silicon and then allowing the two elements to react to form SiC at temperatures somewhat above the melting point of silicon (~1410

°C, or 2570 °F). This compares very favorably with hot pressing that is carried out at temperatures in excess of 1800 °C (3270 °F) and sintering at absolute temperatures of 2000 °C (3630 °F) or greater. Most RBSC is made with formulations that contain an organic polymer with carbon and α-SiC grain.

Generally, infiltration and reaction processes occur simultaneously and involve kinetics that are extremely complex in detail. The process can be extremely fast, due to good wetting of carbon by liquid silicon and low silicon melt viscosities, and a large exothermic heat of reaction. In fact, processing problems such as thermal stress cracking more often arise from too fast a reaction rate rather than too slow a reaction rate.

Controlling the residual phase content is a particularly important issue for RBSC because SiC exists essentially as a line compound (a compound having a negligible range of stoichiometry). Either excess carbon or silicon will appear in the fired part. Free silicon is particularly troublesome because the molten silicon expands upon freezing, frequently cracking the part. A high free-silicon content can be detrimental to mechanical properties. Failure flaws apparently initiate at and propagate along the weak Si-SiC interface at low temperatures. At elevated temperatures near the melting point of silicon (1200 °C < T < 1410 °C, or 2190 °F < T < 2570 °F, where T is temperature), the silicon phase is ductile and has mixed effects. While the presence of the ductile phase can improve fracture toughness and strength by modest amounts, it can also be the initiation source for cavities that lead to creep rupture at long elapsed times.

An approach for eliminating the residual silicon phase, which substitutes alloyed silicon infiltrants for pure silicon, has recently been investigated. The alloying constituent(s) are selected on the basis of their ability to form refractory silicides that are stable with respect to SiC at high temperatures. Upon reactive formation of SiC, the solutes are rejected into the remaining melt, enriching it until the refractory silicide forms. This approach demonstrates that it is possible to produce materials that are of residual silicon phase and that retain the important processing advantages of liquid-phase reaction bonding.

Finishing

Structural ceramics sometimes require finishing. In general, these operations have one or more of the following purposes: to increase dimensional accuracy, improve surface finish, and make minor changes in part geometry. Additional processing is required where tolerances are tighter than can be achieved by sintering or where a surface must be extremely flat or polished. Diamond grinding is used to provide tight dimensional tolerances. Lapping using abrasive slurries, extremely flat surfaces, and polishing by slurry abrasion will achieve a fine surface finish. Drilling and cutting with diamond-tipped drills and saws is sometimes employed. Ultrasonic machining, laser and electron beam machining, and chemical machining are alternative processes.

10.7 Joining of Ceramics

Mechanical and adhesive joining of ceramics can be useful for low-temperature, low-stress applications. However, to take advantage of the high-temperature capability of ceramics requires some type of diffusion bonding or brazing operation. The highest-temperature-capable joints will be made with straight diffusion bonding. However, because diffusion bonding of ceramics requires high temperatures and pressures, more practical but lower-temperature bonds are often made with metallic interlayers that either do or do not form liquids at the bonding temperature.

Joining Oxide Ceramics

Important joining processes include ceramic-to-ceramic joints and ceramic-to-metallic joints.

Oxide Ceramic-Ceramic Joints. Diffusion bonding is usually performed at temperatures and pressures where at least one component plastically deforms under the applied load so that the surfaces are brought into intimate contact. Although the material that deforms in most cases is either the metal portion of a ceramic-metal joint or an intermediate material placed in the joint, it may also be the ceramic. Coarse-grained oxides are generally difficult to join to themselves and require a material that deforms more readily during bonding, such as a finer-grained material of the same or different composition. The level of dopant also affects bond quality, because higher levels usually result in finer grain size. A combination of cold isostatic pressing and hot isostatic pressing can also be used to join oxides. This requires ceramic green bodies having similar green density and sinterability, in addition to a relatively high green-body strength.

Brazing with filler metals is an attractive process for joining structural ceramics for many applications. Wetting and adherence are the principal requirements for brazing; however, most ceramics are not wetted by conventional brazing filler metals. This problem can be overcome either by coating the ceramic surface with a suitable metal layer prior to brazing (indirect brazing) or through the use of specially formulated filler metals that wet and adhere directly to an untreated ceramic surface (direct brazing).

Indirect brazing involves coating the ceramic in the joint area with a material (usually a metal) that can be wetted by a filler metal that would not wet the untreated surface. Coating techniques include sputtering, vapor plating, and thermal decomposition of a metal-containing compound such as TiH_2. The most widely used brazing technique for joining alumina ceramics is the moly-manganese (Mo-Mn) process, as outlined in Fig. 10.24. The ceramic surface is initially metallized and then brazed to metal using one of various brazing filler metals. The chemical composition and properties of the glassy phase in the moly-manganese layer are very important. Glass or ceramic additions are made to this layer, based on the composition and properties of the ceramic, in order to achieve better compatibility.

During direct brazing, the wetting of an untreated ceramic by a molten metal and the associated adhesion increases with growing affinity of the metal constituents for the elements constituting the solid phase. Thus, chemically oxygen-active metals such as titanium, zirconium, aluminum, silicon, manganese, or lithium, either pure or alloyed with other metals, enhance both wetting of and adherence to oxide ceramics, without the

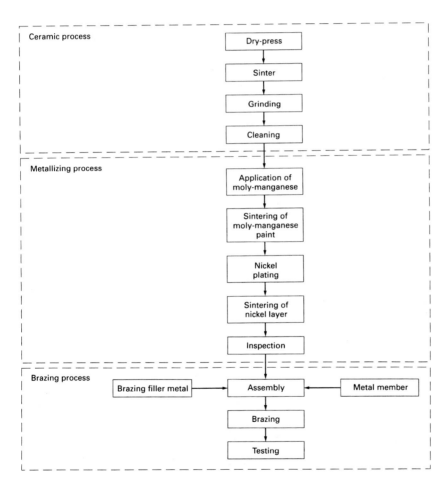

Fig. 10.24 Typical steps in the moly-manganese process. Source: Ref 10.13

need for coating the ceramic surface. Whether the ceramic is an oxide, carbide, or nitride, the active metal reacts with the ceramic surface, forming an interfacial layer that can be wetted by the bulk of the filler metal. Titanium is the most extensively studied and widely used active element addition to filler metals formulated to directly braze high-melting oxide ceramics. The critical interfacial reaction product in the case of oxide ceramics brazed with titanium-containing filler metals is either TiO or Ti_2O_3, with appreciably higher adhesion in systems that result in the formation of TiO.

There are four primary variables in direct brazing of ceramics: active element content of the filler metal, brazing time, brazing temperature, and the surface condition of the ceramic. In general, as the primary variables are increased, the strength of a joint increases to a peak value and then decreases, as shown schematically in Fig. 10.25. This behavior can be explained on the basis of the formation and growth of the reaction products at the interface between the ceramic and filler metal. These products are a necessary part of the wetting and adherence process, but if the reaction products grow excessively, the bond may be weakened. It is assumed that flaws are created by the stresses and strains associated with the volume changes that occur when the reaction products are formed and grow, and that these flaws result in the observed degradation of joint strength.

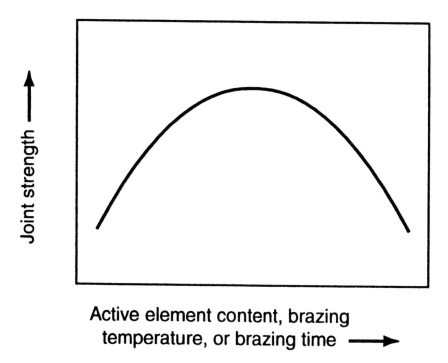

Fig. 10.25 Relationship between the joint strength and the primary brazing variables for structural ceramics joined by active filler metals.
Source: Ref 10.13

Oxide Ceramic-Metal Joints. Both ceramic-ceramic and ceramic-metal joints have been fabricated using diffusion bonding. Most of the work has been on ceramic-metal joints in which the metal portion of the joint is either a bonding material added to the interface between two ceramic components or the second member of a bimaterial joint. For instance, alumina has been joined to a nickel-base superalloy at a pressure of 150 MPa (20 ksi) and a temperature of 1300 °C (2370 °F). Ceramic-ceramic diffusion bonding and a variation on this process, in which ceramic powder compacts are simultaneously sintered and bonded (sinter-bonding process), have also been developed.

Two types of interaction have been observed in diffusion bonding of ceramic-metal joints. In the first type, bonding is driven by a decrease in surface energy as a new interface is formed to replace the two original surfaces. This physical interaction is indicated by an abrupt transition in microstructure between metal and ceramic at the interface. Examples of this type of bond are those between alumina bonded under high vacuum to aluminum, copper, nickel, platinum, and niobium. However, under certain conditions of temperature and excess oxygen activity, reaction products can form at some of these interfaces.

In the second type of solid-state bond between ceramics and metals, chemical reaction further lowers the energy of the system and increases the strength of the bond. Chemical reaction product layers are observed in the joints after bonding. These reaction products generally cause stronger bonds; however, weakening of the joint may occur if the reaction products grow to excessive thicknesses during bonding or in service. Diffusion bonds between oxide materials and metals that result in the formation of interfacial reaction products include alumina with aluminum, aluminum and magnesium with SiO_2 and ZrO_2, nickel with MgO, and nickel and titanium with Al_2O_3.

The temperature required for diffusion bonding usually ranges from 0.5 to 0.8 of the absolute melting point of either the ceramic, the metal being bonded to the ceramic, or an interlayer that is added at the interface to promote bonding. A wide range of applied pressures has also been used, ranging from 20 kPa to 200 MPa (3 psi to 29 ksi) and higher. The surface finish of the mating materials is also a critical variable, because the success of the process depends on the attainment of intimate interfacial contact. If all other variables are equal, rough surfaces require greater pressures to force the surfaces into contact for bonding.

Joining Nonoxide Ceramics

Conventional joining research and development for ceramics has concentrated on oxide ceramics. The use of the molybdenum-manganese metallizing method is well established in industry for joining alumina. However, this method cannot be used for nonoxide ceramics, because it is based on the reaction between glass-forming compounds in the paste and the glass grain-boundary phase of the alumina ceramic. A glass phase suitable for this reaction does not exist in nonoxide ceramics. Thus, a number of alternative joining techniques have been developed for nitride and carbide ceramics.

Nonoxide Ceramic-Ceramic Joints. Joining a silicon-base ceramic to itself can be performed using metallic interlayers. For example, laminated interlayers of nickel and nickel-chromium alloy have been used to join silicon nitride to itself. Laminated interlayers can produce higher bond strengths compared to a single interlayer. The main advantage of using metal interlayers is that it becomes possible to reduce the temperature, pressure, and time required for joining. However, with interlayers, it is generally not possible to have both heat resistance and oxidation resistance in a joint at the same time. Direct diffusion bonding without any metallic interlayer gives joints that are stable above 1000 °C (1830 °F); however, this technique is impractical for many situations.

Hot isostatic pressing without any interlayer under isostatic pressures greater than 100 MPa (15 ksi) can give a nearly ideal joined interface. Complete contact at the interface can be achieved, and the interfacial region may have the same structure as the matrix. When the silicon nitride contains a large amount of sintering additive, the additive may concentrate in the interface region during joining, which can reduce high-temperature strength. Uniaxial hot pressing is another way to achieve solid-state joining without an interlayer. In the hot pressing process, the pressures are limited to much lower values than in hot isostatic pressing, because the uniaxial pressing can easily cause large deformations of components. The high processing costs of solid-state bonding may prevent its wide application, but several cost-saving improvements have been reported. Microwave heating of the interfacial region promotes joining and has been used for silicon nitride. This method does not require conventional furnaces, and joining can be completed in a short time. Joint strengths of up to 500 MPa (70 ksi) have been achieved.

Silicon carbide is a difficult material to join by solid-state methods, even by hot isostatic pressing. A silicon braze or TiC interlayer, plus aluminum foil, has been used to join sintered silicon carbide and reaction-bonded silicon carbide, respectively. These two materials have also been bonded without any interlayer. Joining by reaction sintering of silicon carbide further reduces the joining temperature. In this technique, a powder mixture of silicon carbide and carbon is used as the joining interlayer and is siliconized at 1450 °C (2640 °F). A similar method uses capillary infiltration of molten silicon into tape-cast SiC-C powder precursor interlayers to produce reaction-bonded silicon carbide interlayers. Both methods achieve bond strengths above 300 MPa (40 ksi) that are maintained at temperatures of 1400 °C (2550 °F) and higher.

Nonoxide Ceramic-Metal Joints. The interfacial microstructures of silicon nitride and metals are generally divided into three types, based on their reaction processes. The first type is the interface made by active metal brazing, which is one of the most popular joining processes for ceramic-to-ceramic or ceramic-to-metal systems. The second type is formed by a eutectic melting reaction between silicon-base ceramics and metals. Silicon-base ceramics react with some transition metals to form eutectic liquids. The method uses this eutectic reaction by optimizing the reaction conditions. The third type is formed by complete solid-state reactions under pressure.

Active metal brazing materials, which contain a small weight percent of active elements such as titanium, are available for silicon-base ceramics. Several kinds of active filler metals have been examined. Silver-copper eutectic alloy with a small weight percent of titanium or zirconium, as well as copper-titanium eutectic alloy, and pure aluminum and aluminum-titanium alloy have been examined for use with silicon nitride. Silver-copper-titanium alloy, nickel-titanium alloy, and aluminum have been explored for use with silicon carbide.

Brazing with the Ag-Cu-Ti alloy is a typical process that is accomplished without any metallization. The constituents of the joint are fixed in a furnace, and a slight load is applied. They are then held at the brazing temperatures for Ag-Cu-Ti filler metals, which range from 800 to 950 °C (1470 to 1740 °F). The brazing atmosphere is either a vacuum below approximately 0.0133 Pa (10^{-4} torr) or an inert gas, such as high-purity argon. The maximum interfacial strength obtained using the Ag-Cu-Ti filler metals exceeds 500 MPa (72.5 ksi), as measured by the four-point bend test.

Eutectic reactions can also be used for joining. Most metallic materials react with silicon ceramics to form eutectic liquids between the silicon and the metal. These reactions can be used for joining. The eutectic reaction temperatures, which are primarily determined from the two-component phase diagrams of metal-silicon systems, are approximately 1100 °C (2010 °F) for nickel and its alloys and approximately 1200 °C (2190 °F) for iron and its alloys with silicon nitride. For silicon carbide, the temperature ranges are lowered by 50 to 100 °C (90 to 180 °F) by the presence of carbon, and the reactions are much more extensive than in silicon nitride. The presence of oxygen in the reaction atmosphere may lower the temperature of liquid formation. Reaction layers are formed at the interface in these systems. The reaction products are mainly metal silicides when silicon nitride is joined. High-temperature strengths of silicon nitride joints, one of the critical properties required for high-temperature applications, are shown in Fig. 10.26.

During solid-state bonding, refractory metals such as molybdenum, niobium, tantalum, zirconium, and tungsten react in the temperature range between 1300 and 1500 °C (2370 and 2730 °F) with silicon ceramics by

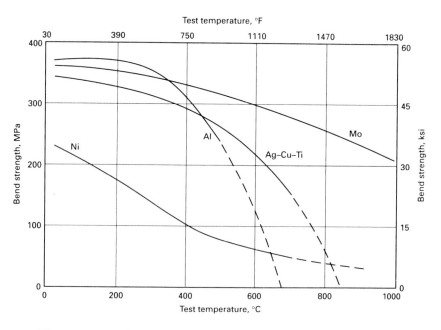

Fig. 10.26 High-temperature strength of silicon nitride joints fabricated using various interlayers. Source: Ref 10.13

solid-state processes to form strong interfaces. The maximum interfacial strength with solid-state bonding is achieved by the optimization of time, temperature, and pressure. It is possible to achieve an interfacial strength above 300 MPa (45 ksi) for silicon nitride/refractory metal systems. In solid-state bonding, the pressure has a great influence on the completeness of the contact at the interface. Isostatically applied high pressure is the most effective means for obtaining full contact.

ACKNOWLEDGMENTS

Portions of the material for this chapter came from "Structural Ceramics" by G.L. DePoorter, T.K. Brog, and M.J. Readey in *Properties and Selection: Nonferrous Alloys and Special-Purpose Materials,* Vol 2, *ASM Handbook,* ASM International, 1990; and "Effects of Composition, Processing, and Structure on Properties of Ceramics and Glasses" and "Design for Ceramic Processing," both by V.A. Greenhut in *Materials Selection and Design,* Vol 20, *ASM Handbook,* ASM International, 1997.

REFERENCES

10.1   P.A. Thrower, Bonding, *Materials in Today's World,* Revised Edition, McGraw-Hill, Inc., 1992, p 25–38

10.2   I. Birkby, Ed., *Ceramic Technology International,* Sterling Publications Ltd., 1992, p 58

10.3   G.L. DePoorter, T.K. Brog, and M.J. Readey, Structural Ceramics, *Properties and Selection: Nonferrous Alloys and Special-Purpose Materials,* Vol 2, *ASM Handbook,* ASM International, 1990

10.4   Structural Applications for Advanced Ceramics, *Engineered Materials Handbook Desk Edition,* ASM International, 1995

10.5   "Ceramic Materials for Light-Weight Ceramic Polymer Armor Systems," CeramTec product literature, CeramTec-ETEC GmbH

10.6   Design Considerations for Advanced Ceramics for Structural Applications, *Engineered Materials Handbook Desk Edition,* ASM International, 1995

10.7   V.A. Greenhut, Effects of Composition, Processing, and Structure on Properties of Ceramics and Glasses, *Materials Selection and Design,* Vol 20, *ASM Handbook,* ASM International, 1997

10.8   G.D. Quinn, Fracture Mechanism Maps for Advanced Structural Ceramics: Part 1, Methodology and Hot-Pressed Silicon Nitride Results, *J. Mater. Sci.,* Vol 25, 1990, p 4361–4376

10.9   V.A. Greenhut, Design for Ceramic Processing, *Materials Selection and Design,* Vol 20, *ASM Handbook,* ASM International, 1997

10.10 R.K. Govila, J.A. Mangels, and J.R. Baer, Fracture of Yttria-Doped, Sintered, Reaction-Bonded Silicon Nitride, *J. Am. Ceram. Soc.,* Vol 68 (No. 7), 1985

10.11 Forming and Predensification of Ceramics, *Engineered Materials Handbook Desk Edition,* ASM International, 1995

10.12 Densification and Sintering of Ceramics, *Engineered Materials Handbook Desk Edition,* ASM International, 1995

10.13 Joining Ceramics, *Engineered Materials Handbook Desk Edition,* ASM International, 1995

SELECTED REFERENCES

- *American Ceramic Society Bulletin,* American Ceramic Society, Westerville, OH
- C.B. Carter and M.G. Norton, *Ceramic Materials: Science and Engineering,* Springer, New York, 2007
- *Ceramic Forum International,* Bauverlag GmbH, Weisbaden, Germany
- *Ceramic Industry,* Business News Publishing Co., Troy, MI
- *Ceramic Science and Engineering Proceedings,* American Ceramic Society, Westerville, OH
- Y.-M. Chiang, D.P. Birnie, and W.D. Kingery, *Physical Ceramics,* John Wiley & Sons, Inc., New York, 1997
- R.A. Flinn and P.K. Trojan, *Engineering Materials and Their Applications,* 5th ed., John Wiley & Sons, Inc., New York, 1995

- J. Hlavac, *The Technology of Glass and Ceramics,* Elsevier Scientific Publishing Company, New York, 1983
- J.T. Jones and M.F. Berard, *Ceramics—Industrial Processing and Testing,* Iowa State University Press, 1972
- W.D. Kingery, H.K. Bowen, and D.R. Uhlmann, *Introduction to Ceramics,* 2nd ed., John Wiley & Sons, 1976
- H.P. Kirchner, *Strengthening of Ceramics,* Marcel Dekker, Inc., New York, 1979
- R.L. Lehman et al., Materials, Chap. 12, *CRC Mechanical Engineering Handbook,* CRC Press, 1997
- M.N. Rahaman, *Ceramic Processing,* CRC Taylor & Francis, Boca Raton, FL, 2007
- J.S. Reed, *Principles of Ceramic Processing,* 2nd ed., John Wiley & Sons, 1995
- D.W. Richerson, *Ceramics—Applications in Manufacturing,* Society of Manufacturing Engineers, Dearborn, MI, 1989
- D.W. Richerson, *Modern Ceramic Engineering,* 2nd ed., Marcel Dekker, 1992
- D.W. Richerson, *Modern Ceramic Engineering: Properties, Processing, and Use in Design,* 3rd ed., CRC Taylor & Francis, Boca Raton, FL, 2006
- S.J. Schneider, Ed., *Ceramics and Glasses,* Vol 4, *Engineered Materials Handbook,* ASM International, 1991
- F. Singer and S.S. Singer, *Industrial Ceramics,* Chemical Publishing Company, New York, 1963
- S. Somiya, Ed., *Advanced Technical Ceramics,* Academic Press, San Diego, CA, 1989
- *Uhlmann's Encyclopedia of Industrial Chemistry,* 5th ed., VCH, 1986

CHAPTER **11**

Ceramic-Matrix Composites

MONOLITHIC CERAMIC MATERIALS contain many desirable properties, such as high moduli, high compression strengths, high-temperature capability, high hardness and wear resistance, low thermal conductivity, and chemical inertness. The high-temperature capability of ceramics makes them very attractive materials for extremely high-temperature environments. However, due to their very low fracture toughness, ceramics are limited in structural applications. While metals plastically deform due to the high mobility of dislocations (i.e., slip), ceramics do not exhibit plastic deformation at room temperature and are prone to catastrophic failure under mechanical or thermal loading. They have a very low tolerance to cracklike defects, which can result either during fabrication or in service. Even a very small crack can quickly grow to critical size, leading to sudden failure.

While reinforcements such as fibers, whiskers, or particles are used to strengthen polymer- and metal-matrix composites, reinforcements in ceramic-matrix composites are used primarily to increase toughness. Some differences in polymer-matrix and ceramic-matrix composites are illustrated in Fig. 11.1. The toughness increases afforded by ceramic-matrix composites are due to energy-dissipating mechanisms, such as fiber-to-matrix debonding, crack deflection, fiber bridging, and fiber pull-out. A notional stress-strain curve for a monolithic ceramic and a ceramic-matrix composite is shown in Fig. 11.2. Because the area under the stress-strain curve is often considered as an indication of toughness, the large increase in toughness for the ceramic-matrix composite is evident. The mechanisms of debonding and fiber pull-out are shown in Fig. 11.3. For these mechanisms to be effective, there must be a relatively weak bond at the fiber-to-matrix interface. If there is a strong bond, the crack will propagate straight through the fibers, resulting in little or no energy absorption. Therefore, proper control of the interface is critical. Coatings

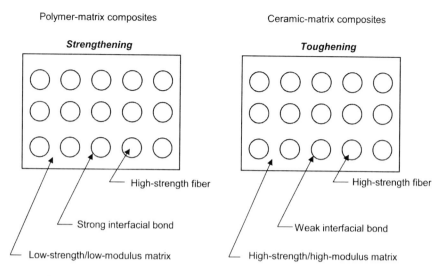

Fig. 11.1 Comparison of polymer-matrix composites with ceramic-matrix composites. Source: Ref 11.1

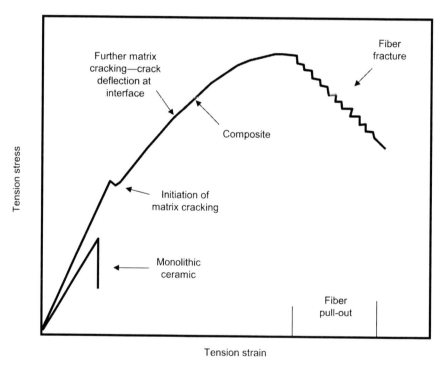

Fig. 11.2 Stress-strain curves for monolithic ceramics and ceramic-matrix composites. Source: Ref 11.1

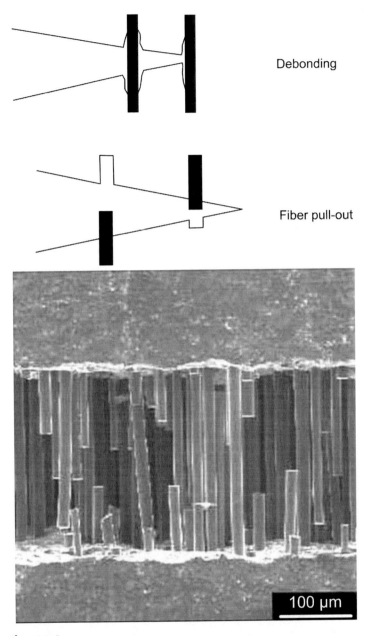

Debonding

Fiber pull-out

Fig. 11.3 Crack dissipation mechanisms. Source: Ref 11.1

are often applied to protect the fibers during processing and to provide a weak fiber-to-matrix bond.

Ceramic-matrix composites (CMCs) can be broadly classified into three types: discontinuously reinforced ceramic-matrix composites, continuous fiber ceramic composites, and carbon-carbon composites. The advantage of ceramic-matrix and carbon-carbon composites is that they can operate reliably at very high temperatures in harsh environments (Fig. 11.4). The need for high-temperature materials spans the range of industries, from aerospace to pulp and paper, and is often driven by the desire to increase energy efficiency and/or reduce emissions. Current metals-based technology can produce materials that are stable to approximately 1000 °C (1830 °F), and superalloys can be used to slightly higher temperatures if thermal barrier coatings and cooling systems are used. However, further technological advances in high-temperature materials will probably not come from the use of metals. Components produced from ceramic-based materials can operate at temperatures that are hundreds of degrees above the melting point of superalloys, and carbon-base materials are stable to temperatures approaching 3000 °C (5430 °F).

Applications for CMCs fall into four major categories:

- Cutting tool inserts
- Wear-resistant parts
- Aerospace and military applications
- Other industrial applications, including engines and energy-related applications

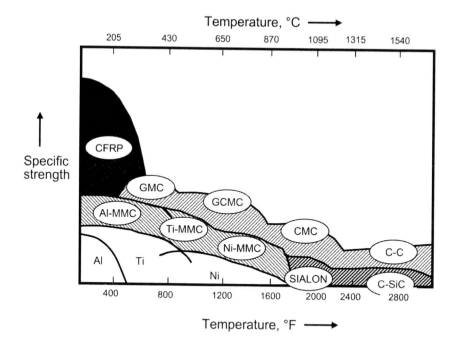

Fig. 11.4 Relative material temperature limits. CFRP, carbon fiber-reinforced plastic; GMC, glass-matrix composite; MMC, metal-matrix composite; GCMC, glass-ceramic-matrix composite; CMC, ceramic-matrix composite; C-C, carbon-carbon; SiAlON, silicon-aluminum-oxynitride. Source: Ref 11.1

11.1 Discontinuously Reinforced Ceramic-Matrix Composites

Discontinuously reinforced CMCs are designed to retain the attractive properties of monolithic ceramics, while enhancing their reliability for structural applications. In discontinuously reinforced CMCs, discrete particles (p), platelets (pl), whiskers (w), or fibers (f) are incorporated into a ceramic or glass matrix to enhance composite performance. Discontinuous reinforced CMCs are characterized by the following:

- Their processing methodology is similar to that of monolithic ceramics, and therefore, any increases in the cost of the composite relative to the monolithic should be small.
- Their mechanical and physical properties are approximately isotropic.
- Their fracture toughness is typically 25 to 100% higher than the corresponding monolithic ceramic.

However, even though the possible increases in fracture toughness are relatively large with the addition of discontinuous reinforcement, discontinuously reinforced CMCs remain similar to monolithic ceramics in that they will fail catastrophically when a crack or flaw grows beyond a critical size. Because strength is typically a secondary concern in these materials, most discontinuously reinforced CMCs are designed for enhanced toughness, with little change in strength relative to the monolithic ceramic. In addition, it is important to recognize that while these various toughening mechanisms are active at low to intermediate temperatures, almost all of them can be predicted to degrade during prolonged exposure to temperatures above ~1100 °C (~2010 °F).

Any changes in the toughness of the composite depend on the specific toughening mechanism(s) that are activated by the addition of the reinforcement, and the magnitude of the change can be difficult to predict. Possible toughening mechanisms for these materials include crack deflection, crack bridging, pull-out, microcracking, crack pinning, and phase transformation. In most instances, the toughening observed in a discontinuously reinforced CMC is the result of the combination of two or more of these toughening mechanisms.

The ceramic matrix in a discontinuously reinforced CMC can either be an oxide or nonoxide material and, depending on the processing techniques employed, may also contain some residual metal. Several important matrix materials are listed in Table 11.1. The more common oxide matrices include polycrystalline alumina (Al_2O_3), mullite ($3Al_2O_3$-$2SiO_2$), and glasses. However, alumina and mullite are by far the most common oxide-matrix materials because of their thermal and chemical stability and their compatibility with most of the common reinforcement materials.

Nonoxide-matrix materials tend to have superior structural properties to their oxide counterparts, and they have better corrosion resistance in some environments. However, they are sensitive to oxidizing environments at elevated temperatures. The more common nonoxide matrices are silicon carbide (SiC) and silicon nitride (Si_3N_4). Of these, SiC is the more widely used, although Si_3N_4 may be selected when higher strength is needed.

The reinforcement phase in a discontinuously reinforced CMC can have a variety of morphologies and is typically composed of one of the following: SiC, Si_3N_4, AlN, TiC, TiB_2, BN, or B_4C. The improvement in fracture properties as a result of discontinuous reinforcement in Al_2O_3 and SiC is illustrated in Table 11.2. Because of its stability in a wide variety of ceramic matrices, and because of its commercial availability in a variety of morphologies, SiC is the most widely used reinforcement phase. However, for applications that require long-term exposure to high temperatures, an oxide reinforcement may be a better choice. Silicon carbide is susceptible to degradation in oxygen-containing environments at temperatures above 1200 °C (2190 °F), and even when used as a reinforcement at these temperatures, it must be protected by some sort of coating system to ensure reliable service.

Table 11.1 Select ceramic-matrix materials

Matrix	Modulus of rupture, ksi	Modulus of elasticity, 10^6 psi	Fracture toughness, ksi · in.$^{1/2}$	Density, g/cm³	Thermal expansion, 20^{-6}/°C	Melting point, °F
Pyrex glass	8	7	0.07	2.23	3.24	2285
LAS glass-ceramic	20	17	2.20	2.61	5.76	...
Al_2O_3	70	50	3.21	3.97	8.64	3720
Mullite	27	21	2.00	3.30	5.76	3360
SiC	56–70	48–67	4.50	3.21	4.32	3600
Si_3Ni_4	72–120	45	5.10	3.19	3.06	3400
Zr_2O_3	36–94	30	2.50–7.70	5.56–5.75	7.92–13.5	5000

Note: Values depend on exact composition and processing. Source: Ref 11.1

Table 11.2 Room-temperature strength and fracture toughness of alumina and silicon carbide discontinuously reinforced ceramic-matrix composites

Composite system	Vol% reinforcement	Fracture strength		Fracture toughness	
		MPa	ksi	MPa · m$^{1/2}$	ksi · in.$^{1/2}$
99.5% Al_2O_3	0	375	54.4	4–5	3.6–4.5
Al_2O_3 + SiC (p)	20	520	75.4	4–5	3.6–4.5
Al_2O_3 + SiC (pl)	30	480	69.6	7.1	6.5
Al_2O_3 + SiC (w)	30	700	101.5	8.7	7.9
Al_2O_3 + SiC (w)	20	620	89.9	7–8	6.4–7.3
Al_2O_3 + TiC (p) + SiC (w)	10 TiC (p)	690	100	9.6	8.7
SiC	0	480	69.6	4–5	3.6–4.5
SiC + TiB_2 (p)	16	480	69.6	8.9	8.1
SiC + TiC (p)	15	680	98.6	5.1	4.6

Reinforcement: p, particulate; pl, platelet; w, whisker. Source: Ref 11.2

In addition to toughness, the incorporation of ceramic whiskers into a ceramic matrix can result in an improvement of other material properties, such as hardness and thermal conductivity. Increases in hardness translate to better wear resistance, although this can be tempered by a localized decrease in the fracture toughness at the microscale. Increases in thermal conductivity, combined with toughening mechanisms that resist the coalescence of thermal shock-induced cracks into critical flaws, result in enhanced thermal shock resistance relative to the monolithic ceramic.

Because of their superior properties and their relative ease of manufacture, whisker-toughened CMCs have had more commercial success to date than any of the other ceramic-based composite systems. Alumina toughened by SiC whiskers (Al_2O_3/SiC_w) is used as cutting tools for nickel-base superalloys (Fig. 11.5). The advantages of these CMCs in this application include increased tool lifetimes, reduced downtime, improved surface finish, lower maintenance costs, and increased production rates. Although the Al_2O_3/SiC_w composites cost more to manufacture than the more traditional carbide tools, their increased cost is more than compensated for by their superior performance.

11.2 Applications for Discontinuously Reinforced Ceramic-Matrix Composites

Cutting Tools. Both particulate-reinforced and whisker-reinforced Al_2O_3 have found use as cutting tool inserts. The incorporation of SiC whiskers (20 to 45 vol%) into an Al_2O_3 matrix with subsequent hot pressing results in a composite with significantly improved toughness. The microstructure of SiC whisker-reinforced Al_2O_3 is shown in Fig. 11.6. The

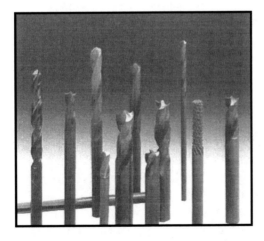

Fig. 11.5 SiC_w/Al_2O_3 composite cutting tools. Courtesy of Greenleaf Corporation

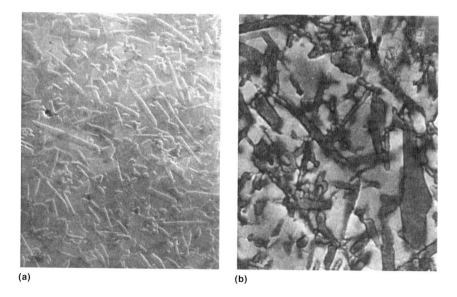

(a)   (b)

Fig. 11.6 Microstructure of SiC whisker-reinforced Al_2O_3 composite. (a) Original magnification: 2100×. (b) Original magnification: 5000×. Source: Ref 11.3

whiskers, which are small fibers of single-crystal SiC approximately 0.5 to 1 μm in diameter and 10 to 125 μm long, have a higher thermal conductivity and a lower coefficient of thermal expansion than Al_2O_3. This improves thermal shock resistance.

The SiC whiskers in the Al_2O_3 matrix also produce a twofold increase in fracture toughness. The fracture toughness is enhanced by the occurrence of whisker pull-out. A close examination of the fracture surface at high magnification will reveal not only a clear indication of the whiskers randomly dispersed throughout the matrix but also the obvious hexagonal voids where whiskers have actually been pulled out during the fracture process (Fig. 11.7). A large amount of energy is required to pull the whiskers out, and this greatly inhibits crack propagation.

The Al_2O_3-SiC_w composite tool grades are used for rough turning of alloy and hardened steels (32 to 65 HRC), ductile irons (150 to 300 HB), and chilled irons (50 to 65 HRC). Their primary application area, however, is for roughing, finishing, and milling of difficult-to-machine age-hardened nickel-base superalloys. As shown in Fig. 11.8, much higher cutting speeds and longer tool life can be achieved when machining these materials with whisker-reinforced Al_2O_3.

Wear-Resistant Materials/Applications. Particulate-toughened ceramics, such as the zirconia-toughened ceramics, are filling applications such as bearings, bushings, precision balls, valve seats, and die inserts, where their friction and wear characteristics have improved both

Fig. 11.7 Fracture surface of a SiC whisker-reinforced Al_2O_3 ceramic. Note hexagonal voids or holes due to whisker pull-out upon fracture. Original magnification: 950×. Source: Ref 11.3

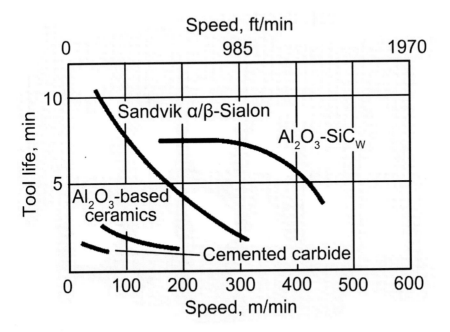

Fig. 11.8 Tool life of ceramic, ceramic-matrix composite, and cemented carbide materials when machining Inconel 718 (feed of 0.2 mm/rev; depth of cut of 2 mm). Source: Ref 11.4

performance and cost. One example is zirconia-toughened alumina, or ZTA (Al_2O_3-ZrO_2), where Al_2O_3 is considered the primary or continuous (70 to 95%) phase. These types of ZTA compositions have been used in transportation equipment where they must withstand corrosion, erosion, abrasion, and thermal shock. Other applications for ZTA include cutting tool inserts and its use as an abrasive grinding medium. Zirconia-toughened alumina has also seen some use in thermal shock applications. Extensive use of monoclinic ZrO_2 can result in a severely microcracked ceramic body. The ZTA microstructure allows thermal stresses to be distributed throughout a network of microcracks, where energy is expended opening and/or extending microcracks, leaving the bulk ceramic body intact.

An Al_2O_3-matrix composite reinforced with coarse SiC particles has been optimized for applications requiring slurry erosion resistance. Wear test data for the Al_2O_3-SiC_p composite and other ceramic and metallic materials are compared in Fig. 11.9. Silicon carbide particle-reinforced Al_2O_3 composites of this type are used in a range of applications, including slurry pump components, hydrocyclone liners, chute liners, and materials handling systems. Several examples of wear components that are available commercially are shown in Fig. 11.10.

Aerospace Applications. While most CMCs under development for aerospace structural applications are continuous fiber CMCs, the most

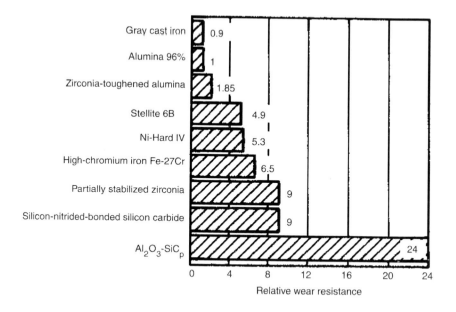

Fig. 11.9 Relative slurry erosion wear performance of metallic, ceramic, and ceramic-matrix composite (Al_2O_3-SiC_p) materials. This evaluation compared measured material losses in a slurry pot test in which 14 mm (0.54 in.) diameter by 60 mm (2.4 in.) long test pins were rotated at 10 m/s (33 ft/s) for 20 h in a slurry of 40 wt% 300 to 600 μm SiO_2 particles in neutral water. Source: Ref 11.5

Fig. 11.10 Examples of wear-resistant Al_2O_3-SiC_p components produced for use in applications requiring resistance to slurry erosion. The pump housing is approximately 640 mm (25 in.) in diameter. Large wear rings are produced up to 1.2 m (48 in.) in diameter. Source: Ref 11.3

notable aerospace application of discontinuously reinforced CMCs is the Space Shuttle thermal protection system (TPS). A key to the success of the Space Shuttle orbiter is the development of a fully reusable TPS capable of being used for up to 100 missions. The key element of the TPS is the thousands of ceramic tiles that protect the shuttle during reentry. The orbiter and the temperatures reached during reentry in a typical trajectory are shown in Fig. 11.11. During reentry of the shuttle into the Earth's atmosphere, its surface reaches 1260 °C (2300 °F) where the ceramic tiles are used. Even hotter regions (up to 1650 °C, or 3000 °F) occur at the nose tip and the wing leading edges, where reinforced carbon-carbon (RCC) composites must be employed. The materials chosen for various areas of the TPS are shown in Fig. 11.12.

The ceramic tiles are made from very high-purity amorphous silica fibers ~1.2 to 4 μm in diameter and 0.32 cm (0.125 in.) long, which are felted from a slurry and pressed and sintered at ~1370 °C (2500 °F) into blocks. Two tile densities are used: 144 kg/m³ (9 lb/ft³) (LI-900) and 352 kg/m³ (22 lb/ft³) (LI-2200). The LI-900 depends on a colloidal silica binder to achieve a fiber-to-fiber bond, whereas LI-2200 depends entirely on fiber-to-fiber sintering.

The silica tiles offer many advantages. The tile is 93% void; thus, it is an excellent insulator, having conductivities (through the thickness) as low as

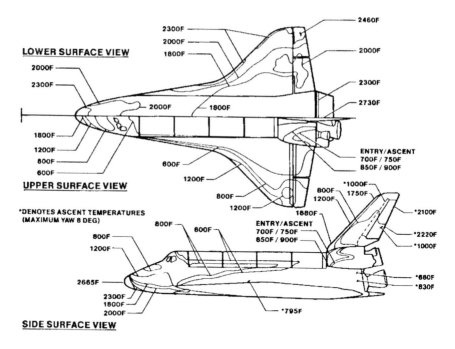

Fig. 11.11 Space Shuttle orbiter isotherms for a typical trajectory. Source: Ref 11.6

0.017 to 0.052 W/m · K (0.01 to 0.03 Btu · ft/h · ft^2 · °F). The low coefficient of expansion of amorphous silica, as well as the low modulus of the tile, eliminates thermal stress and thermal shock problems. Because of the very high purity (99.62%), devitrification is limited, avoiding the high stresses associated with the expansion or contraction of the crystalline phase (cristobalite). Silica has high-temperature resistance; it is capable of exposure above 1480 °C (2700 °F) for a limited time. Because it is an oxide, further protection is unnecessary, whereas RCC and refractory metals must have oxidation-protective coatings.

Heat-Resistant Industrial Applications. A CMC containing SiC particles in an Al$_2$O$_3$ matrix has been developed for high-temperature applications, such as furnace and heat exchanger components. They use very fine SiC particles (<10 μm), and the parent metal and processing conditions are selected to optimize high-temperature stability and mechanical characteristics. Tests have shown that high-temperature strength is unaffected by holding at temperatures at least up to 1500 °C (2730 °F) for 1000 h or more. Prototype heat exchanger tubes and furnace components have been tested successfully, including a rig test of a burner tube involving temperature cycling to 1530 °C (2790 °F). Examples of a heat exchanger and furnace components are shown in Fig. 11.13.

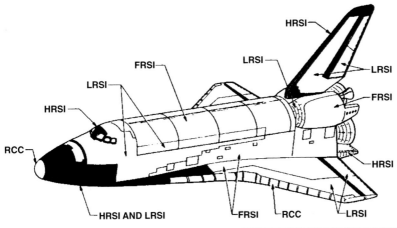

Material generic name	Material temperature capability, °C (°F)(a)	Material composition	Areas of orbiter
Reinforced carbon-carbon (RCC)	to 1650 (3000)	Pyrolized carbon-carbon, coated with SiC	Nose cone, wing leading edges, forward external tank separation panel
High-temperature reusable surface insulation (HRSI)	650–1260 (1200–2300)	SiO$_2$ tiles, borosilicate glass coatong with SiB$_4$ added	Lower surface and sides, tail leading and trailing edges, tiles behind RCC
Low-temperature reusable surface insulation (LRSI)	400–650 (750–1200)	SiO$_2$ tiles, borosilicate glass coatong	Upper wing surface, tail surfaces, upper vehicle sides, OMS(b) pods
Felt reusable surface insulation (FRSI)	to 400 (750)	Nylon felt, silicone rubber coating	Wing upper surface, upper sides, cargo bay doors, sides of OMS(b) pods

(a) 100 missions; higher temperatures are acceptable for a single mission. (b) Orbital maneuvering (OMS) engines

Fig. 11.12 Thermal protection system materials for the U.S. Space Shuttle. More than 30,000 ceramic tiles are included in the system. Other materials making up the system are reinforced carbon-carbon composites (44 panels and the nose cap) and 333 m^2 (3581 ft^2) of felt reusable surface insulation. Source: Ref 11.6, 11.7

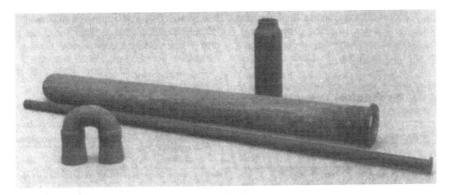

Fig. 11.13 Heat exchanger and furnace components made from an Al$_2$O$_3$-SiC$_p$ composite. Source: Ref 11.5

11.3 Continuous Fiber Ceramic-Matrix Composites

Continuous fiber CMCs can be produced that have the high hardness, high strength, high melting point, chemical inertness, good wear resistance, and low density common to monolithic ceramics. However, unlike monolithic ceramics, continuous fiber CMCs can be designed to be strain tolerant and therefore have a much lower probability of failing catastrophically in a high-stress environment. The development of continuous fiber CMCs for application at high temperatures is not without its challenges, however. Because the fibers and fiber-matrix interfacial regions of the most successful continuous fiber CMCs tend to be either carbides or nitrides, they can exhibit poor stability at elevated temperatures when exposed to oxygen-containing environments. The performance of continuous fiber CMCs depends critically on the integrity of these regions, and thus their degradation can be catastrophic. In addition, because they require specialized materials and processing techniques, continuous fiber CMCs are expensive to produce relative to either monolithic ceramics or discontinuously reinforced fiber CMCs.

Ceramic-matrix materials for continuous fiber CMCs include SiC, Si_3N_4, $MoSi_2$, Al_2O_3, mullite, yttrium-aluminum-garnet, and spinel. However, SiC and Al_2O_3 are the most commonly used matrix materials. Silicon-carbide-matrix continuous fiber CMCs are designed for applications where high thermal conductivity, low thermal expansion, light weight, and good corrosion and wear resistance are required. Alumina-matrix continuous fiber CMCs are better in applications where a high tolerance to salt corrosion, molten glass, or oxidation is required in combination with light weight and high thermal shock resistance. In addition, a variety of glass and glass-ceramic matrices can be used in applications that do not require structural stability above the softening temperature of the glass, for example, ~600 °C (1100 °F) for a borosilicate glass or ~1150 °C (2100 °F) for a high-silica glass. For these applications, the relative ease of processing glass-matrix continuous fiber CMCs can result in a significant cost savings.

The fiber reinforcement is typically either SiC or carbon, although a number of other oxide and nonoxide fibers are currently under development and several are commercially available. The properties of several ceramic fibers can be found in Table 11.3. To act as a reinforcement, the fibers must have the following characteristics: high strength, environmental stability (particularly in oxygen-containing environments), creep resistance, and a coefficient of thermal expansion that is close to that of the matrix. In addition, the fibers must have a small-enough diameter to allow weaving. Currently, it is oxidation and creep resistance that are the limiting factors for most fiber reinforcements. In the absence of oxygen, carbon fiber is the best choice for most applications. However, carbon fibers suffer from oxidation at temperatures as low as 350 °C (660 °F). Silicon carbide fibers have good strength up to 1000 °C (1832 °F) but begin to creep at higher temperatures. In addition, all SiC fibers will oxidize at temperatures greater than 1200 °C (2190 °F) and have a maximum-use temperature of not more than 1400 °C (2550 °F). Current-generation oxide fibers, such as Al_2O_3, are polycrystalline materials with silica at their grain boundaries. As a result, alumina-base fibers begin to creep at temperatures as low as 900 °C (1650 °F) and retain almost no strength at temperatures above 1150 °C (2100 °F). Thus, although alumina and other oxide fibers have considerable potential as continuous fiber CMC reinforcement because of their oxidative stability, their poor high-temperature mechanical properties currently limit their usefulness. The application of single-crystal oxide fibers, currently under development, may be the solution to this problem of high-temperature stability.

The fibers within a continuous fiber CMC laminate can be oriented in a variety of ways, depending on the performance requirements of the specific application. A unidirectional (0°) alignment of fibers results in the highest level of property translation efficiency but at the expense of intralaminar and interlaminar strength. A two-dimensional, planar, interlaced fiber architecture results in better intralaminar properties; however, the interlaminar strength is limited by the properties of the matrix due to the lack of through-the-thickness reinforcement. A three-dimensional, integrated-fiber architecture results in enhanced interlaminar properties by providing reinforcement in the through-the-thickness direction. The type of fiber architecture that is best suited for a particular application will depend on the performance requirements of that application.

The mechanical behavior of a continuous fiber CMC is determined to a large degree by three factors: the strength of the reinforcing fibers, the characteristics of the fiber-matrix interface, and the residual stress present in the composite due to thermal expansion mismatch between fibers and

Table 11.3 Properties of selected continuous ceramic fibers

Fiber	Composition	Tensile strength, ksi	Tensile modulus, 10^6 psi	Density, g/cm³	Diameter, mils	Critical bend radii, mm
SCS-6	SiC on C monofilament	620	62	3.00	5.5	7.0
Nextel 312	$62Al_2O_3$-$14B_2O_3$-$15SiO_2$	250	22	2.7	0.4	0.48
Nextel 440	$70Al_2O_3$-$2B_2O_3$-$28SiO_2$	300	27	3.05	0.4–0.5	...
Nextel 480	$70Al_2O_3$-$2B_2O_3$-$28SiO_2$	330	32	3.05	0.4–0.5	...
Nextel 550	$73Al_2O_3$-$27SiO_2$	290	28	3.03	0.4–0.5	0.48
Nextel 610	99 α-Al_2O_3	425	54	3.88	0.6	...
Nextel 720	$85Al_2O_3$-$15SiO_2$	300	38	3.4	0.4–0.5	...
Almax	99 α-Al_2O_3	260	30	3.60	0.4	...
Altex	$85Al_2O_3$-$15SiO_2$	290	28	3.20	0.6	0.53
Nicalon NL200	57Si-31C-12O	435	32	2.55	0.6	0.36
Hi-Nicalon	62Si-32C-0.5O	400	39	2.74	0.6	...
Hi-Nicalon-S	68.9Si-309.C-0.2O	375	61	3.10	0.5	...
Tyranno LOX M	55.4Si-32.4C-10.2O-2Ti	480	27	2.48	0.4	0.27
Tyranno ZM	55.3Si-33.9C-9.8O-1Zr	480	28	2.48	0.4	...
Sylramic	66.6Si-28.5C-2.3B-2.1Ti-0.8O-0.4N	465	55	3.00	0.4	...
Tonen Si_3N_4	58Si-37N-4O	360	36	2.50	0.4	0.80

Source: Ref 11.1

matrix. High-strength fibers are important because once a matrix crack is initiated and extended, load is transferred from the matrix to the fibers in the wake of the crack. As a result, the ultimate load-bearing capability of the composite is determined by the load-bearing characteristics of the fibers. The level of toughening imparted by the fibers depends on how well the interface between matrix and fibers is engineered. In an ideal composite, the interface will be strong enough to transfer load from the matrix to the fiber yet weak enough to allow debonding of the fiber in the path of a crack.

Interfacial, or interphase, coatings are often required to protect the fibers from degradation during high-temperature processing, aid in slowing oxidation during service, and provide the weak fiber-to-matrix bond required for toughness. The coatings (Fig. 11.14), ranging in thickness from 0.1 to 1.0 μm, are applied directly to the fibers prior to processing, usually by chemical vapor deposition (CVD). The CVD produces coatings of relatively uniform thickness, composition, and structure, even with preforms of complex fiber architecture. Carbon and boron nitride (BN) are typical coatings, used either alone or in combination with each other. Frequently, in addition to the interfacial coatings, an overcoating is also applied, such as a thin layer (~0.5 μm) of SiC that becomes part of the matrix during processing. The SiC overcoating helps to protect the interfacial coating

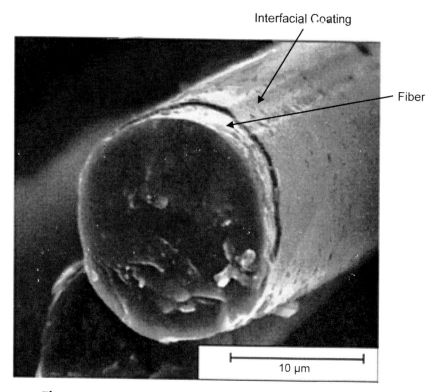

Fig. 11.14 Interfacial coating. Source: Ref 11.1

from reaction with the matrix during processing. Because carbon and BN interfacial coatings will degrade in an ambient environment, the overcoating is usually applied immediately after the interfacial coating. During service, the overcoating also acts to protect the fibers and interfacial coatings from aggressive environments, such as oxygen and water vapor. The interfacial and overcoating are sometimes repeated as multilayer coatings to provide environmental protection layers in the presence of in-service-generated matrix cracks.

In addition to chemical bonding at the interface, residual stresses from thermal expansion mismatch between the matrix and fibers can contribute to the mechanical strength of the fiber-matrix interface and thus influence composite toughness. The room-temperature mechanical properties of several commercially available continuous fiber CMCs are shown in Table 11.4, and the thermal properties of the same composites are given Table 11.5.

Although ceramic materials are known for their high-temperature stability, characteristics of a polycrystalline ceramic microstructure make the current generation of materials susceptible to deformation and creep at temperatures as low as 900 °C (1650 °F). This is true both for the matrices and for the fibers that are used to produce continuous fiber CMCs. This susceptibility results from the presence of grain boundaries and the tendency for relatively low melting phases, which may be introduced either intentionally or unintentionally during processing. When these grain-boundary phases begin to soften, the grains can slide relative to one another, resulting in the initiation of creep cracks and the deformation of

Table 11.4 Room-temperature mechanical properties of several commercially available continuous fiber ceramic-matrix composites

Fiber matrix system	Fiber architecture	Volume fraction fibers, %	Tensile strength x-y		z		Tensile modulus x-y		z		Compressive strength x-y		z		Compressive modulus x-y		z		Shear strength	
			MPa	ksi	MPa	ksi	GPa	10⁶ psi	GPa	10⁶ psi	MPa	ksi	MPa	ksi	GPa	10⁶ psi	GPa	10⁶ psi	MPa	ksi
C/SiC	3D	...	69	10	48	7	83	12	48	6.9	503	72.9	73	10.6	110	16	16	2.3	103	14.9
SiC/Al$_2$O$_3$	2D	37.5	246	35.7	...	...	144	21	...	...	550	79.7	...	...	84	12.2	...	...	63	9.1
SiC/SiC	2D	31.2	215	31.2	...	...	140	20.3	...	...	502	72.8	...	...	145	21	...	...	30	4.3

3D, three-directional; 2D, two-directional. Source: Ref 11.2

Table 11.5 Thermal properties of selected commercially available continuous fiber ceramic-matrix composites

Fiber matrix system	Fiber architecture	Volume fraction fibers, %	Thermal expansion(a), 10⁻⁶/K x-y	z	Thermal diffusivity(b), 10⁻⁶ m²/s x-y	z	Thermal conductivity(b), 10² W/m · K x-y	z	Specific heat(b) J/kg · K
C-SiC	3D	...	4.27	3.10	6.4	5.8	11.6	10.5	1351
SiC-Al$_2$O$_3$	2D	37.5	5.8	...	...	...	5.7	...	...
SiC-SiC	2D	35	5.2	...	1.7	...	...	...	1500

3D, three-directional; 2D, two-directional. (a) From 20–1500 °C (68–2730 °F). (b) Measured at 1000 °C (1830 °F). Source: Ref 11.2

the bulk material at temperatures much lower than the softening point of the matrix. Because of grain-boundary creep, current-generation oxide fibers have a maximum-use temperature of approximately 1000 °C (1830 °F), rendering them unsuitable for many continuous fiber CMC applications. Depending on the amount of oxygen present in the material, SiC fibers can also exhibit significant creep in the same temperature range. As a result, poor fiber stability is considered one of the major barriers to the high-temperature application of continuous fiber CMCs, and research emphasis is now focusing on the potential of single-crystal oxide fibers as a possible reinforcement material.

In addition to high-temperature deformation, environmental susceptibility is a problem for many of the nonoxide-matrix and reinforcement materials used in continuous fiber CMCs. Both SiC and Si_3N_4 form a SiO_2 surface layer upon exposure to an oxidizing environment, which is protective to temperatures approaching 1200 °C (2190 °F). However, decomposition reactions at higher temperatures render the oxide coating nonprotective, and volatilization of the material can be problematic. This high-temperature decomposition occurs even in nonoxide fibers embedded within the composite matrix. In the case of fibers, cracks in the matrix can expose the reinforcement to an oxygen-containing environment. Alternatively, fast diffusion of oxygen along grain boundaries, or along interfaces, can cause internal volatilization of the reinforcement, resulting in cavitation, bloating, and/or blistering. Finally, oxidation of the protective boron nitride or carbon coating typically applied to fibers can result in the formation of a strong interfacial bond that will embrittle the composite.

As a result of these limitations, it appears that oxide-oxide continuous fiber CMCs, which include single-crystal reinforcement, have the highest potential for structural applications at temperatures greater than 1400 °C (2550 °F). However, issues with grain-boundary creep of the matrix and thermal shock resistance have yet to be resolved.

11.4 Applications for Continuous Fiber Ceramic-Matrix Composites

The high-temperature stability, corrosion resistance, and toughness of continuous fiber CMCs make them suitable for a wide range of applications in both the aerospace and industrial sectors. Many of the subsequent applications surveyed were brought about by The Continuous Fiber Ceramic Composite Program supported by the Department of Energy Office of Industrial Technologies. This was a collaborative effort between industry, national laboratories, universities, and the U.S. government. Examples of potential applications for CMCs include:

* *Turbine engine components.* This includes combustor liners, shrouds, seals, vanes, blades, and other parts used in gas turbines, including

utility, industrial, and aeronautical engine applications. Both SiC-SiC and Al_2O_3-SiC composites are under development for such applications. Various processing methods, including melt infiltration, chemical vapor infiltration (CVI), polymer impregnation and pyrolysis, and directed metal oxidation, are used to produce these components.

- *Radiant burner screens.* SiC-SiC reverberatory screens for natural gas burners are being produced by CVI. These burners are used in industrial applications, such as paper or paint drying, chemical processing, metal treating, and glass forming. The single-layer porous mesh screen is mounted directly above the burner surface.

- *Immersion heater tubes.* SiC-SiC composite burner tubes (Fig. 11.15) have been developed for melting aluminum in casting foundries. The part is not wetted or chemically attacked by the molten aluminum. The processing approach involves the application of a water-based ceramic slurry onto a mandrel that is filament wound with SiC fiber. The green preform is nitride bonded (thermally treated in a nitrogen atmosphere), which results in a material composed of SiC fibers in a matrix of SiC and Si_3N_4.

- *Hot gas filters.* Al_2O_3-Al_2O_3 composites are being evaluated for use as hot gas filters in advanced coal-fired energy applications. Removing coal ash particles from the exhaust gas of a coal-fired pressurized fluidized bed combustor protects the downstream equipment, such as

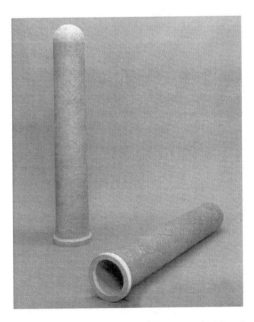

Fig. 11.15 Immersion tubes for molten aluminum holding furnaces made from a filament-wound continuous fiber ceramic-matrix composite. Courtesy of Textron Systems. Source: Ref 11.3

Fig. 11.16 Fastrac rocket engine showing the location of the turbopump that contains the continuous fiber ceramic composite blisk shown in Fig. 11.17. Source: Ref 11.3

gas turbine engines, from damage, enabling the generation of energy with reduced emissions and with higher efficiency. Sol-gel and slurry infiltration methods are used to produce these filters.

- *Turbine disks for rocket engines.* Integrally bladed disks ("blisks") for rocket engine turbopumps have been produced from carbon-fiber-reinforced SiC. Figure 11.16 shows the Fastrac engine developed by the National Aeronautics and Space Administration (NASA) and the locations of the turbopump and blisk. The blisks were produced by CVI of a woven preform, which was subsequently machined and thermally treated to complete densification. The final turbopump assembly is shown in Fig. 11.17.

- *Exhaust nozzle flaps and seals.* Ceramic-matrix composites have successfully entered service as exhaust nozzle flaps and seals in the F414 engine, now used in the Navy F-18 E/F aircraft (Fig. 11.18). The exhaust temperature of the F414 is over 80 °C (145 °F) higher than for the F404 engine used in the previous version of the F-18. As a result, the metal flaps and seals were failing in tens of hours. The CMC parts consist of a Nicalon SiC fiber with an inhibited carbon matrix. A thick SiC overcoat and glaze provide protection from oxidation. There are 12 flaps and 12 seals per engine, and the seals are attached to metal backing plates with metal rivets and a zirconia overcoat. The seals are

Fig. 11.17 Turbopump assembly showing the carbon-fiber-reinforced SiC blisk (right) and a metal inducer and impeller mounted on its shaft. The use of continuous fiber ceramic-matrix composites for this application increases the temperature capability up to 1370 °C (2500 °F). Source: Ref 11.3

Fig. 11.18 Exhaust nozzle of an F414 engine on an F-18 E/F aircraft, showing the twelve sets of ceramic-matrix composite (CMC) flaps and seals. The white areas on the seals are a zirconia overcoat for mechanical fasteners. Over an order of magnitude increase in life has been obtained with the CMC flaps and seals. Source: Ref 11.8

subjected to the highest temperatures, and the flaps must support the largest mechanical loads. Further, the flaps must survive a high thermal gradient, and the CMC is subjected to rubbing with the back face of the seal. The CMC flaps have a useful life that is at least double the design requirement of 500 h.

Other potential applications for continuous fiber CMCs include heat exchanger tubes, armor, pipe hangers for petroleum refining, submersible pump housings for chemical processing, parts for ethylene cracking reformers, parts exposed to methyl chloride and dimethylchlorosilane, heat treating furnace fans, hot particle separators in municipal waste incinerators, and low-heat-rejection diesel engine exhaust valve guides that are self-lubricating.

11.5 Carbon-Carbon Composites

A number of advantages—including stable mechanical properties to temperatures greater than 2000 °C (3630 °F), high strength-to-weight ratio, high thermal conductivity, good thermal shock resistance, good abrasion resistance, and good corrosion resistance—make carbon-carbon (C-C) composites a candidate material for virtually any application requiring structural integrity at very high temperatures. In addition, C-C composites are biocompatible, and this characteristic, combined with the ability to tailor the directional properties of these materials, makes them attractive for use in various load-bearing prosthetic devices. However, C-C composites have at least two drawbacks that prevent their widespread application. The first is that C-C composites are extremely susceptible to degradation by oxidation at temperatures as low as 350 °C (660 °F). As a result, the long-term use of these materials as structural components at high temperature in oxygen-containing environments is impossible without some form of protective coating. The second drawback is that production costs can be quite high, making C-C composites prohibitively expensive for many applications. Because of these drawbacks, use of C-C composites to date has been limited primarily to military and space applications. However, C-C composites have also been successfully utilized as commercial aircraft brakes, where their superior performance outweighs their increased costs.

Carbon-carbon composites are composed of carbon fibers embedded in a carbon matrix. The microstructural characteristics, and therefore the physical properties, of the carbon matrix can vary widely depending on the starting materials used and methods of composite fabrication. For example, the carbon matrix can be approximately isotropic, composed of small, randomly oriented crystallites of poorly graphitic or turbostratic carbon. The matrix can also be strongly anisotropic, consisting of relatively large crystallites of highly graphitized carbon. More often, however, the matrix microstructure is a combination of these two extremes and includes some porosity and microcracking. The level of porosity is determined by the methods of manufacture, whereas the matrix microcracking most often results from the strong anisotropy of carbon in thermal expansion.

The carbon fibers within a C-C composite are most often continuous, with a variety of orientations. Available fiber architectures include unidirectional, bidirectional (two-dimensional) laminates, three-directional (three-dimensional) orthogonal weaves, or multidirectional weaves and braids. The fiber architecture selected depends on the properties desired in the composite. Multidirectional C-C composites containing reinforcing fibers in three or more directions are advantageous because their fiber architecture can be designed to accommodate the specific requirements of the final component. Unidirectionally reinforced C-C composites, on the other hand, tend to have very high on-axis strength and fracture toughness, but at the expense of their off-axis mechanical properties. In addition to fiber architecture, C-C performance also depends on the physical properties of the fibers themselves, as well as on the characteristics of the interface between fibers and matrix. The carbon fibers used in C-C composites range from the isotropic variety, with low elastic moduli and modest strengths, to highly oriented, high-performance fibers, with high elastic moduli and strengths between 3 and 5 GPa (0.4 and 0.7 msi). Most of these fibers are derived from polyacrylonitrile resin, although petroleum pitch or rayon may also be used as fiber precursors.

In C-C composites, the strain to failure of the carbon matrix is typically much lower than that of the reinforcing fibers, and the matrix is frequently microcracked as a result of thermal expansion mismatch stresses created during processing. As a result, it is frequently the characteristics of the interfacial bonding between matrix and fiber that govern the mechanical properties of the C-C composite. When the bonding between fibers and matrix is strong, cracks that form in the matrix are propagated across the fiber-matrix interface, resulting in fiber failure. The result is brittle failure of the composite, with the ultimate strength of the composite controlled by the low strain to failure of the matrix. On the other hand, weak interfaces between fibers and matrix allow matrix cracking to occur without crack propagation through the fibers. Intact fibers bridge the matrix cracks and maintain load-bearing capability until increasing load finally initiates fiber fracture and failure, resulting in a much tougher composite. Although the mechanical performance depends on weak fiber-matrix interfaces, other physical properties are enhanced by strong fiber-matrix interfaces. As a result, fiber treatments and fabrication methods are varied to tailor the interfacial characteristics of the C-C composite to match the requirements of the specific application.

Fiber architecture, fiber type, matrix precursor, and processing methodology all contribute to determine the physical properties of C-C composites. As a result of the large number of variables involved, C-C composites can be produced with a wide range of properties. Nonetheless, all C-C composites have the following basic characteristics in common: low thermal expansion that increases with temperature, good strength that increases with temperature in inert environments, and thermal conductivity that decreases with temperature. A comparison of the room-temperature mechanical properties of a fine-grained monolithic graphite with unidirectional and three-directional C-C composites is shown in Table 11.6. The data clearly indicate the benefits of adding carbon fibers to the carbon matrix. Depending on the specific fiber architecture, all room-temperature mechanical

Table 11.6   Room-temperature mechanical properties of graphite and carbon-carbon (C-C) composites

Material	Elastic modulus		Tensile strength		Compressive strength		Fracture energy, kJ/m²
	GPa	10⁶ psi	MPa	ksi	MPa	ksi	
Graphite	10–15	1.46–2.17	40–60	5.8–8.7	110–200	16–29	0.01
Unidirectional C-C composite	120–150	17.4–21.7	600–700	87–101	500–800	72.5–116	1.4–2.0
Three-directional C-C composite	40–100	5.8–14.5	200–350	29–50.7	150–200	21.7–29	5–10

Source: Ref 11.9

properties are improved. Unidirectional fibers provide dramatic increases in the strength of the material in the principal fiber direction. While the addition of a three-dimensional network of fibers results in more modest increases in room-temperature strength, there is a large increase in fracture toughness.

The mechanical and thermal properties of several C-C composites with different fiber architectures are given in Tables 11.7 and 11.8, respectively. Room-temperature mechanical properties are many times higher when measured in the principal axis directions of both the unidirectional and fabric laminate composites. On the other hand, the woven orthogonal (three-dimensional) composite has more isotropic mechanical properties.

Generally, the elastic modulus that is measured in the principal fiber direction of C-C composites reflects the fiber modulus according to the law of mixtures. Strength utilization of the fibers is typically between 25 and 50%, depending on how the composite is processed. However, as is common with fiber-reinforced composites, the shear properties of the C-C composites tend to be relatively low and therefore will usually have a strong influence on materials selection and design. The same is true for the tensile properties of one- and two-dimensional composites in directions normal to the principal fiber directions. These low properties are the direct result of the weak fiber-matrix interfaces that are designed to prevent composite brittleness and low fracture strengths parallel to the fiber directions.

Table 11.7   Room-temperature mechanical properties of selected carbon-carbon composites

Construction	Fiber volume fraction			Tensile strength				Tensile elastic modulus				Compressive strength				Compressive elastic modulus			
				x		z		x		z		x		z		x		z	
	x	y	z	MPa	ksi	MPa	ksi	GPa	10^6 psi	GPa	10^6 psi	MPa	ksi	MPa	ksi	GPa	10^6 psi	GPa	10^6 psi
Unidirectional (one-dimensional)	0.65	0	0	1000	145	2.0	0.29	260	37.7	3.4	0.5	620	89.9	...	...	250	36.2	...	...
Fabric laminate (two-dimensional)	0.31	0.30	0	350	50.7	5.0	0.72	115	16.7	4.1	1.59	150	21.7	...	...	100	14.5	...	...
Woven orthogonal (three-dimensional)	0.13	0.13	0.21	170	24.6	300	43.5	55	7.97	96	13.9	...	...	140	20.3	...	...	90	13.0

Source: Ref 11.10

Table 11.8 Thermal properties of selected carbon-carbon composites

Construction	Fiber volume fraction			Coefficient of thermal expansion(a), 10^{-6}/K		Thermal conductivity(b), W/m · K	
	x	y	z	x	z	x	z
Unidirectional (one-dimensional)	0.65	0	0	1.1	10.1	125	10
Fabric laminate (two-dimensional)	0.31	0.30	0	1.3	6.1	95	4
Woven orthogonal (three-dimensional)	0.13	0.13	0.21	1.3	1.3	57	80

(a) Thermal expansion measured from room temperature to 1650 °C (3000 °F). (b) Thermal conductivity measured at 800 °C (1470 °F). Source: Ref 11.10

Carbon-carbon composites are unique in that their mechanical properties do not degrade with increasing temperature until 2200 °C (3990 °F). Increases in the tensile strength and decreases in the elastic modulus of the composites (measured in the principal fiber directions) as the temperature increases are typical of carbon fibers. However, it is the properties of the matrix, and of the matrix-fiber interface, that dictate the effect of temperature on the shear, cross-fiber tensile, and compressive strengths of the composites. Generally, these properties also improve with increase in temperature, a phenomenon that is attributed to the closing of matrix microcracks as the temperature is increased. Creep behavior has received only minimal attention to date; however, the steady-state creep rates of most C-C composites are predicted to be at least 4 orders of magnitude lower than that of most technical-grade ceramics.

As with the mechanical properties in C-C composites, the bulk thermal expansion and thermal conductivity can be highly anisotropic in a unidirectional composite, reflecting the large anisotropy in these properties in single-crystal graphite, but approximately isotropic in the three-dimensional composite (Table 11.8). The thermal expansion of C-C composites tends to increase with increasing temperature, although the value remains low relative to other ionic and covalently bonded materials. The thermal conductivity, on the other hand, tends to decrease with increasing temperature. The combination of good strength, relatively high thermal conductivity, and low thermal expansion makes C-C composites very resistant to thermal shock.

The retention of mechanical properties to high temperatures, combined with a resistance to damage by thermal shock, makes C-C composites an ideal candidate for many high-temperature structural applications. However, the lack of reliable oxidation protection for these materials has been, and continues to be, a serious limitation to their widespread application. Depending on its crystal structure, carbon can become susceptible to oxidation at temperatures as low as 350 °C (660 °F), especially when the environment contains atomic oxygen. This susceptibility to oxidation increases with increase in temperature until approximately 800 °C (1470 °F), where the rate of oxidation is limited only by the diffusion rate of oxygen to the carbon surface. Because the oxides of carbon are gaseous, oxidation results in significant material loss and serious degradation in material performance. As a result, some form of oxidation protection is essential to the long-term reliable use of C-C composites at all but the lowest temperatures. External coatings are the most direct, and to date the most effective, method of oxidation protection for these materials. However, the thermal expansion characteristics of C-C composites make establishing and maintaining coherent and adherent coatings extremely difficult. In addition, coating defects are unavoidable with current coating technology, and thus, the addition of some mechanism for coating self-healing and/or internal oxidation protection is an absolute necessity for these materials.

The primary coating candidate for oxidation protection at temperatures below 1200 °C (2190 °F) is SiC, because it forms a surface layer of SiO_2 that is highly oxidation resistant, it is chemically compatible with carbon, and it has a relatively low coefficient of thermal expansion. The temperature limitation for SiC in oxidizing environments results from a reaction between SiC and SiO_2 that renders the SiO_2 nonprotective, and this leads to the rapid erosion of the SiC coating. Coatings of Si_3N_4 can also be used but have essentially the same temperature limitations as SiC. Silicon carbide and Si_3N_4 are usually applied to the C-C composite surface in combination with a boron-containing inner coating to protect the carbon from the inevitable formation of coating cracks. Oxidation of this inner coating through the cracks results in the formation of a sealant glass that flows into and closes the cracks and thus prevents the oxidation of the underlying composite.

Oxidation protection for C-C composites above ~1500 °C (2730 °F) requires the use of either a noble metal or a highly refractory ceramic coating as a primary oxygen barrier. The most attractive noble metal for carbon protection at high temperatures is iridium, which has a melting point of 2440 °C (4425 °F). In addition, iridium has a very low oxygen permeability to 2100 °C (3810 °F), is nonreactive with carbon below 2280 °C (4135 °F), and is an effective carbon diffusion barrier. However, there are also disadvantages to iridium, including a susceptibility to erosion by volatile oxide formation, a lack of adherence to carbon, and a thermal expansion incompatibility with C-C composites. The use of refractory ceramic coatings for high-temperature oxidation protection is limited by the high oxygen permeabilities of the refractory oxides. All of the refractory carbides, nitrides, borides, and silicides oxidize rapidly at temperatures above 1750 °C (3180 °F), and most oxidize at significantly lower temperatures. As a result, refractory ceramic coatings cannot provide long-term oxidation protection to C-C composites without the identification of effective additives that can act as oxygen barriers. Problems such as chemical compatibility with carbon at high temperature and thermal expansion mismatch also need to be solved.

11.6 Applications for Carbon-Carbon Composites

Carbon-carbon composites are used for applications requiring ablation and high-temperature stability (i.e., rocket nozzles and exit cones) or wear resistance at moderate temperatures (i.e., aircraft brakes). Two notable applications of C-C composites are the nose cone, wing leading edges, and forward external tank separation panel on the Space Shuttle (Fig. 11.12) and improved aircraft brakes.

Space Shuttle C-C. The fabrication process for the Space Shuttle C-C components is a multistep process, typical of the infiltration and pyrolysis technology used to produce C-C composites. As shown in Fig. 11.19, initial material lay-up is similar to that used for thermoset composite parts. Plain weave carbon fabric, impregnated with phenolic resin, is layed-up on a fiberglass/epoxy tool. Laminate thickness varies from 19 plies in the external skin and web areas to 38 plies at the attachment locations. The part is vacuum bagged and autoclave cured at 149 °C (300 °F) for 8 h. The cured part is rough trimmed, x-rayed, and ultrasonically inspected. The part is then postcured by placing the part in a graphite restraining fixture, loading it into a furnace, and heating it to 260 °C (500 °F) very slowly to avoid distortion and delamination. The postcure cycle alone can take up to 7 days.

The next step is initial pyrolysis. The part is loaded in a graphite restraining fixture and placed in a steel retort, which is packed with calcined coke. The part is slowly heated to 816 °C (1500 °F) and held for 70 h to facilitate conversion of the phenolic resin to a carbon state. During pyrolysis, the resin forms a network of interconnected porosity that allows volatiles to escape. This stage is extremely critical. Adequate volatile escape paths must be provided, and sufficient times must be employed to allow the volatiles to escape. If the volatiles become entrapped and build up internal pressure, massive delaminations can occur in the relatively weak matrix. After this initial pyrolysis cycle, the carbon is designated reinforced carbon-carbon-0 (RCC-0), a state in which the material is extremely light and porous, with a flexural strength of only 21 to 24 MPa (3.0 to 3.5 ksi).

Densification is accomplished in three infiltration and pyrolysis cycles. In a typical cycle, the part is loaded in a vacuum chamber and impregnated with furfural alcohol. It is then autoclave cured at 149 °C (300 °F) for 2 h and postcured at 204 °C (400 °F) for 32 h. Another pyrolysis cycle is then conducted at 816 °C (1500 °F) for 70 h. After three infiltration/

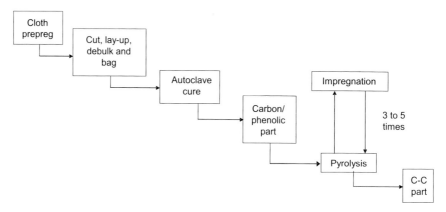

Fig. 11.19 Fabrication sequence for Space Shuttle carbon-carbon (C-C) parts. Source: Ref 11.1

pyrolysis cycles, the material is designated RCC-3, with a flexural strength of approximately 124 MPa (18 ksi).

To allow usage at temperatures above 1982 °C (3600 °F) in an oxidizing atmosphere, it is necessary to apply an oxidation-resistant coating system. The coating system consists of two steps: applying a SiC diffusion coating to the C-C part, and applying a glassy sealer to the SiC diffusion coating. The coating process (Fig. 11.20) starts with blending of the constituent powders, consisting of 60% SiC, 30% Si, and 10% Al_2O_3. In a pack cementation process, the mix is packed around the part in a graphite retort. The retort is loaded into a vacuum furnace, where it undergoes a 16 h cycle that includes drying at 316 °C (600 °F) and then a coating reaction up to 1650 °C (3000 °F) in an argon atmosphere. During processing, the outer layers of the C-C are converted to SiC. The SiC-coated C-C part is removed from the retort, cleaned, and inspected. During cooldown from 1650 °C (3000 °F), the SiC coating contracts slightly more than the C-C substrate, resulting in surface crazing (coating fissures). This crazing, together with the inherent material porosity, provides paths for oxygen to reach the C-C substrate.

To obtain increased life, it is necessary to add a surface sealer. The surface sealing process involves impregnating the part with tetraethylorthosilicate (TEOS). The part is covered with a mesh, placed in a vacuum bag, and the bag is filled with liquid TEOS. A five-cycle TEOS impregnation process is then performed on the bagged part. After the fifth impregnation cycle, the part is removed from the bag and oven cured at 316 °C (600 °F) to liberate hydrocarbons. This process leaves silica (SiO_2) in all of the microcracks and fissures, greatly enhancing the oxidation protection.

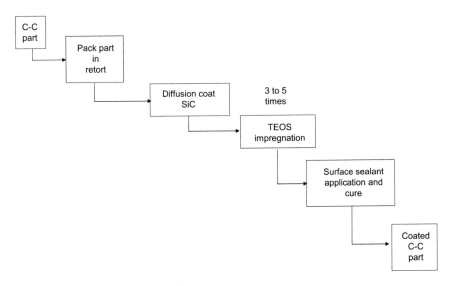

Fig. 11.20 Coating sequence for Space Shuttle carbon-carbon (C-C) parts. TEOS, tetraethylorthosilicate. Source: Ref 11.1

Improved aircraft brakes became necessary as aircraft became heavier and faster. Aborted takeoffs that were terminated at maximum ground speed were uncertain with steel disk brakes. The brakes, if successful in stopping the aircraft, were often destroyed because of warping or melting due to the intense heat generated during the stop. Carbon-carbon composites have a high melting point, are resistant to thermal shock, and have excellent friction and wear characteristics. They can be fabricated in shapes and sizes that are suitable for brake applications. Carbon-carbon brakes, designed to take advantage of these useful properties, can provide superior stopping capability, survive several abortive stops, and last far longer than conventional aircraft brakes. Furthermore, this superior performance is accomplished with a significant weight savings.

Disk brakes for aircraft are composed of a number of disks, half of which are keyed to the nonrotating brake mechanism. The other half rotate with the wheel to which they are keyed. Braking is accomplished by forcing the disks together, at which time friction is converted to heat, which must be dissipated. This requires a material that is resistant to thermal shock, stable to very high temperatures, and has low thermal expansion and good thermal conductivity. In addition, the material should have a friction coefficient of approximately 0.3 to 0.5 for good stopping performance. Carbon-carbon composites have all of these important properties, along with a density of approximately 1.9 g/cm^3 (0.069 $lb/in.^3$), which provides nearly four times the stopping power of copper or steel brakes. Some advanced aircraft require 820 to 1050 kJ/kg (350 to 450 Btu/lb) of the carbon-carbon brake to stop the aircraft. High-performance automobiles require only 300 to 520 kJ/kg (130 to 220 Btu/lb). Carbon-carbon brakes with fibers in the plane of the flat surface have low wear and good frictional characteristics. Thus, carbon-carbon disks are made with layers of carbon fabric or with random fibers arrayed parallel to the braking surface.

Special modifications to the carbon-carbon brake system include the incorporation of refractory borides to avoid temporary reduction in the coefficient of friction after a period of idleness. The friction loss is probably caused by water absorption, and it can be overcome operationally by applying brake pressure several times before takeoff. The borides eliminate the need for this preconditioning. Metal caps on the teeth of the rotating disks are used on some systems to reduce wear on the teeth and to increase thermal conductivity.

11.7 Processing of Ceramic-Matrix Composites

Ceramic-matrix composites can be processed either by conventional powder processing techniques, used to process polycrystalline ceramics, or by newer techniques developed specifically for CMCs.

Cold Pressing and Sintering

Cold pressing of a matrix powder and fiber mixture followed by sintering is a natural extension from conventional processing of ceramics. Shrinkage is a common problem associated with sintering of most ceramics. This problem is exacerbated when a glass or ceramic matrix is combined with a reinforcement material. Thus, after sintering, the matrix generally shrinks considerably, and the resulting composite exhibits a significant amount of cracking. One of the reasons for high shrinkage after sintering is that fibers and whiskers, that is, reinforcements with high aspect ratio (length/diameter), can form a network that may inhibit the sintering process. Depending on the difference in thermal expansion coefficients of the reinforcement and matrix, a hydrostatic tensile stress can develop in the matrix on cooling, which will counter the driving force (surface energy minimization) for sintering. Thus, the densification rate of the matrix will, in general, be retarded in the presence of reinforcement.

Hot Pressing

Hot pressing is frequently used in a combination of steps or in a single step in the consolidation stage of CMCs. Hot pressing is an attractive technique because the simultaneous application of pressure and high temperature can significantly accelerate the rate of densification, resulting in a porosity-free and fine-grained compact. An example of a common hot-pressed composite is SiC whisker-reinforced Al_2O_3, used in cutting tool applications.

A common variant of conventional hot pressing is the slurry infiltration process. It is perhaps the most important technique used to produce continuous fiber-reinforced glass and glass-ceramic composites. The slurry infiltration process involves two main stages: incorporation of the reinforcing phase into a slurry of the unconsolidated matrix, and matrix consolidation by hot pressing.

A schematic of the slurry infiltration process is shown in Fig. 11.21. The first stage involves some degree of fiber alignment, in addition to incorporation of the reinforcing phase in the matrix slurry. The slurry typically consists of the matrix powder, a carrier liquid (water or alcohol), and an organic binder. The organic binder is burned out prior to consolidation. Wetting agents may be added to ease the infiltration of the fiber tow or preform. The fiber tow or fiber preform is impregnated with the matrix slurry by passing it through a slurry tank. The impregnated fiber tow or preform sheets are similar to the prepregs used in fabrication of polymer-matrix composites. The impregnated tow or prepreg is wound on a drum and dried. This is followed by cutting and stacking of the prepregs and consolidation by hot pressing. The process has the advantage that the prepreg layers can be arranged in a variety of stacking sequences, for example, unidirectional, cross-plied (0°/90°/0°/90°, etc.), or angle-plied

$(+\theta/-\theta/+\theta/-\theta$, etc.). An optical micrograph of a transverse section of a uni-directional alumina-fiber/glass-matrix composite is shown in Fig. 11.22(a), while the pressure and temperature schedule used during hot pressing is shown in Fig. 11.22(b).

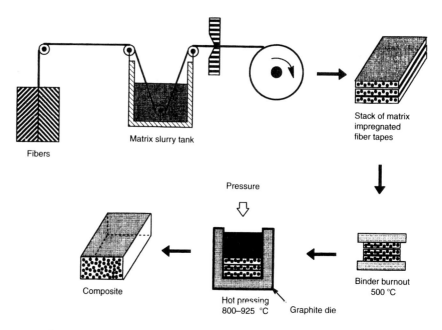

Fig. 11.21   Schematic of the slurry infiltration process followed by hot pressing. Source: Ref 11.11

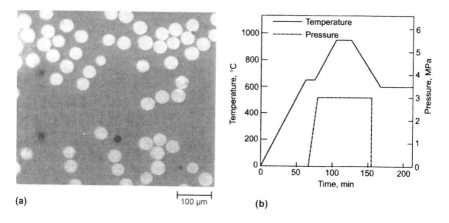

Fig. 11.22   Unidirectional alumina-fiber/glass-matrix composite formed by slurry infiltration followed by hot pressing. (a) Light micrograph of transverse section (some porosity can be seen in this micrograph). (b) Pressure and temperature schedule used during hot pressing of this composite. Source: Ref 11.11

As mentioned previously, the slurry infiltration process is well suited for glass- or glass-ceramic-matrix composites, mainly because the processing temperatures for these materials are lower than those used for crystalline-matrix materials. The hot pressing process does have the limitation of not being able to produce complex shapes. Application of a very high pressure during hot pressing can also easily damage the fibers and decrease the strength of the composite. The fibers may also be damaged by mechanical contact with refractory particles of a crystalline ceramic or from reaction with the matrix at very high processing temperatures. The matrix should have as little porosity as possible in the final product, because porosity in a structural ceramic material is highly undesirable. Thus, it is important to completely remove the fugitive binder and use a matrix powder particle smaller than the fiber diameter. The hot pressing parameters are also important. Precise control within a narrow working temperature range, minimization of the processing time, and using a pressure low enough to avoid fiber damage are important factors in final consolidation. Fiber damage and any fiber-matrix interfacial reaction, along with its detrimental effect on the bond strength, are unavoidable attributes of hot pressing.

In summary, the slurry infiltration process generally results in a composite with fairly uniform fiber distribution, low porosity, and relatively high strength. The main disadvantage of this process is that one is restricted to relatively low-melting- or low-softening-point matrix materials.

Whisker-reinforced CMCs are generally made by mixing the whiskers with a ceramic powder slurry, dried, and hot pressed. Sometimes hot isostatic pressing rather than uniaxial hot pressing is used. Whisker agglomeration in a green body is a major problem; mechanical stirring and adjustment of the pH level of the suspension (matrix powder-whiskers in water) can help minimize this. Addition of whiskers to a slurry can result in very high viscosity. Also, whiskers with large aspect ratios (>50) tend to form bundles and clumps. Obtaining well-separated and deagglomerated whiskers is of great importance for reasonably high-density composites. Use of organic dispersants and techniques such as agitation mixing assisted by an ultrasonic probe and deflocculation by a proper pH control can be useful.

Most whisker-reinforced composites are made at temperatures in the 1500 to 1900 °C (2730 to 3450 °F) range and pressures in the 20 to 40 MPa (3 to 6 ksi) range. A scanning electron micrograph of a hybrid composite, consisting of SiC fibers (Nicalon) and whiskers in a glass-ceramic matrix, is shown in Fig. 11.23.

Reaction Bonding

Reaction bonding processes similar to the ones used for monolithic ceramics can also be used to make CMCs. These have been used mostly with silicon carbide or silicon nitride matrices. The advantages are:

Glass ceramic matrix
with whiskers

Nicalon
fiber

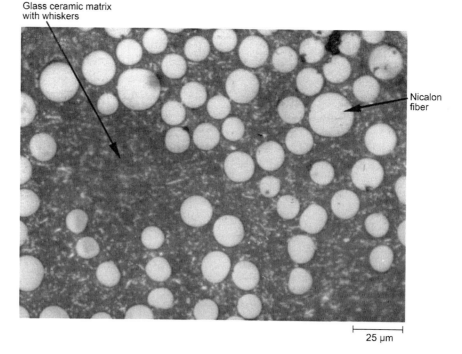

25 μm

Fig. 11.23 Scanning electron micrograph of a hybrid composite consisting of SiC fibers (Nicalon) and whiskers in a glass-ceramic matrix. Source: Ref 11.11

- Little or no matrix shrinkage occurs during densification.
- Large volume fractions of whiskers or fiber can be used.
- Multidirectional, continuous fiber preforms can also be used.
- The reaction bonding temperatures for most systems are generally lower than the sintering temperatures, so that fiber degradation can be minimized.

One great disadvantage of this process is that high porosity is hard to avoid, because the reaction bonding reactions evolve gaseous by-products.

A hybrid process involving a combination of hot pressing with the reaction bonding technique can also be used. The flow diagram for this process is shown in Fig. 11.24, while a micrograph of a composite (SCS-6 fiber/Si_3N_4) made by this process is shown in Fig. 11.25. Silicon cloth is prepared by attrition milling a mixture of silicon powder, a polymer binder, and an organic solvent to obtain a "dough" of proper consistency. This dough is then rolled to make a silicon cloth of desired thickness. Fiber mats are made by filament winding of silicon carbide with a fugitive binder. The fiber mats and silicon cloth are stacked in an alternate sequence, debinderized, and hot pressed in a molybdenum die and in a nitrogen or vacuum environment. The temperature and pressure are adjusted to produce a handleable preform. At this stage, the silicon matrix is converted to silicon nitride by transferring the composite to a nitriding furnace between 1100 and 1400 °C (2010 and 2550 °F). Typically, the silicon nitride matrix has approximately 30% porosity, which is not unexpected in reaction-bonded silicon nitride. Note also the matrix density variations around fibers in Fig. 11.25.

Infiltration

Infiltration of a preform made of a reinforcement can be done with a matrix material in solid (particulate), liquid, or gaseous form. As shown in Fig. 11.26, liquid infiltration is very similar to liquid polymer or liquid metal infiltration. Proper control of the fluidity of liquid matrix is the key to this technique. It yields a high-density matrix with no porosity in

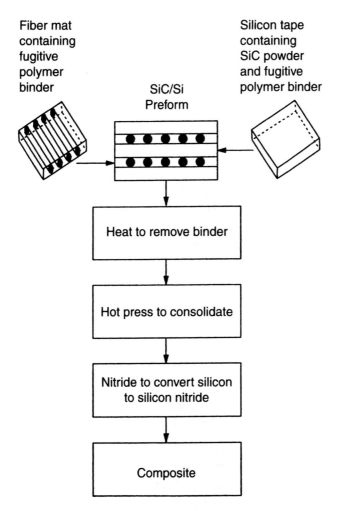

Fig. 11.24 Flow diagram of the reaction bonding process for processing SCS-6 fiber/Si₃N₄ composites. Source: Ref 11.11

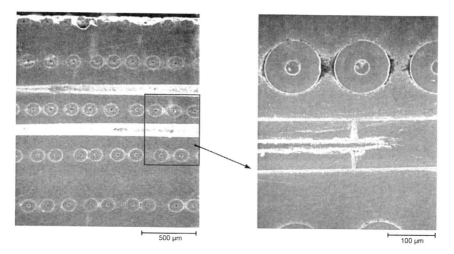

500 µm

100 µm

Fig. 11.25 Microstructure of reaction-bonded SCS-6 fiber/Si_3N_4 composite showing uniform fiber distribution and small amounts of residual porosity around the periphery of the large-diameter fibers. Source: Ref 11.11

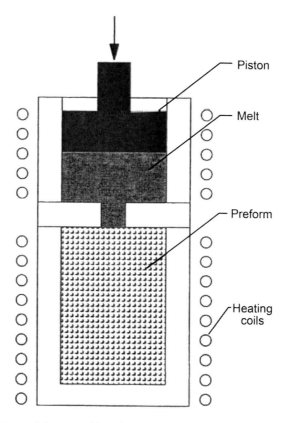

Piston

Melt

Preform

Heating coils

Fig. 11.26 Schematic of liquid infiltration processing. The technique is very similar to liquid polymer or liquid metal infiltration. Source: Ref 11.11

the matrix. Almost any reinforcement geometry can be used to produce a virtually flaw-free composite. However, the temperatures involved are much higher than those used in polymer or metal processing. Processing at such high temperatures can lead to deleterious chemical reactions between the reinforcement and the matrix. Thermal expansion mismatch between the reinforcement and the matrix, the rather large temperature interval between the processing temperature and room temperature, and the low strain to failure of ceramics can add up to a formidable set of problems in producing a crackfree CMC. Viscosities of ceramic melts are generally very high, which makes the infiltration of preforms rather difficult. Wettability of the reinforcement by the molten ceramic is another item to be considered. A preform made of reinforcement in any form (for example, fiber, whisker, or particle) having a network of pores can be infiltrated by a ceramic melt by using capillary pressure. Application of pressure or processing in vacuum can aid in the infiltration process.

Assuming that the preform consists of a bundle of regularly spaced, parallel channels, one can use Poissuelle's equation to obtain the infiltration height, h:

$$h = \sqrt{\frac{\gamma r t \cos\theta}{2\eta}} \qquad \text{(Eq 1)}$$

where r is the radius of the cylindrical channel, t is the time, γ is the surface energy of the infiltrant, θ is the contact angle, and η is the viscosity. Note that the penetration height is proportional to the square root of time and inversely proportional to the viscosity of the melt. Penetration will be easier if the contact angle is low (i.e., better wettability) and the surface energy (γ) and the pore radius (r) are large. However, if the radius of the channel is made too large, the capillarity effect will be lost.

The advantages of different melt infiltration techniques can be summarized as:

- The matrix is formed in a single processing step.
- A homogeneous matrix can be obtained.

The disadvantages can be summarized as:

- High melting points of ceramics mean a greater likelihood of reaction between the melt and the reinforcement.
- Ceramics have higher melt viscosities than metals; therefore, infiltration of preforms is relatively difficult.
- Matrix cracking is likely because of the differential shrinkage between the matrix and the reinforcement on solidification. This can be minimized by choosing components with nearly equal coefficients of thermal expansion.

Directed Metal Oxidation

Directed metal oxidation (DIMOX), or reactive melt infiltration, uses liquid aluminum that is allowed to react with air (oxygen) to form alumina (Al_2O_3) or with nitrogen to form aluminum nitride (AlN). A schematic of the DIMOX process is shown in Fig. 11.27. The first step in this process is to make a preform. In the case of a fibrous composite, filament winding or a fabric lay-up can be used to make a preform. A barrier to stop growth of the matrix material is placed on the preform surfaces. In this method, a molten metal is subjected to directed oxidation; that is, the desired reaction product forms on the surface of the molten metal and grows outward. The metal is supplied continuously at the reaction front by a wicking action through channels in the oxidation product. For example, molten aluminum in air will become oxidized to aluminum oxide. To form aluminum nitride, molten aluminum is reacted with nitrogen. The reactions are:

$$Al + air \rightarrow Al_2O_3$$
$$Al + N_2 \rightarrow AlN$$

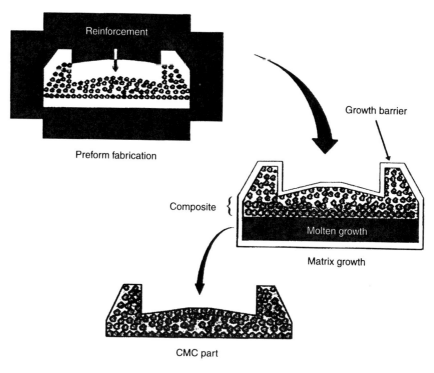

Fig. 11.27 Schematic of the directed metal oxidation process. CMC, ceramic-matrix composite. Courtesy of Lanxide Corporation. Source: Ref 11.11

The end product in this process is a three-dimensional, interconnected network of a ceramic composite plus approximately 5 to 30% of unreacted metal. When filler particles are placed next to the molten metal surface, the ceramic network forms around these particles. A fabric made of a continuous fiber can also be used. The fabric is coated with a proprietary coating to protect the fiber from highly reducing aluminum and to provide a weak interface, which is desirable for enhanced toughness. Some residual aluminum (6 to 7 wt%) remains at the end of the process. This must be removed if the composite is to be used at temperatures above the melting point of aluminum (660 °C, or 1220 °F). On the other hand, the presence of a residual metal can be exploited to provide some fracture toughness.

The process is potentially low cost because near-net shapes are possible. Also, good mechanical properties (strength and toughness) have been reported. The main disadvantages of this process are that it is difficult to control the chemistry and produce an all-ceramic matrix by this method; there is always some residual metal that is not easy to remove completely. It is also difficult to envision the use of such techniques for large, complex parts, such as those required for aerospace applications.

Liquid Silicon Infiltration

Liquid silicon, or one of its lower-melting point alloys, is used to infiltrate a fiberous preform to form a silicon carbide matrix. The fibers must contain an interfacial coating, along with a SiC overcoating, to protect them from the liquid silicon. Before infiltration, a fine grained silicon carbide particulate is slurry cast into the fiber preform. After removal of the slurry carrier liquid, melt infiltration is usually done at 1398 °C (2550 °F) or higher and is usually complete within a few hours. The liquid silicon bonds the silicon carbide particulates together and forms a matrix that is somewhat stronger and denser than that obtained by chemical vapor infiltration. Because the resulting matrix can contain up to 50% unreacted silicon, the long-term use temperature is limited to approximately 1204 °C (2200 °F). The amount of unreacted silicon can be reduced by infiltrating the preform with carbon slurries prior to the silicon infiltration process. However, some unreacted or free silicon will always be present in the matrix.

In another variant of the liquid metal infiltration process, liquid silicon is reacted with unprotected carbon fibers to form a silicon carbide matrix. After the initial step of using prepreg to form a highly porous carbon-matrix part, liquid silicon is infiltrated into the structure, where it reacts with the carbon to form silicon carbide along with unreacted silicon and carbon. Because the carbon fibers are intended to react with the liquid silicon, no coatings are used on the fibers. However, the poor oxidation resistance of the carbon fibers means that the entire part will require an oxidation-resistant coating.

The liquid metal infiltration processes have several advantages: they produce a fairly dense SiC-base matrix with a minimum of porosity, the processing time is shorter than for most CMC fabrication processes, and the dense and closed porosity on the surface can often eliminate the need for a final oxidation-resistant coating. The major disadvantage is the high temperatures (>1400 °C, or 2550 °F) required for liquid silicon infiltration that expose the fibers to possible degradation due to the high temperatures and corrosive nature of liquid silicon. In addition, the temperatures can be even higher due to the possibility of an exothermic reaction between silicon and carbon.

Chemical Vapor Infiltration

When chemical vapor deposition (CVD) is used for infiltration of rather large amounts of matrix material in fibrous preforms, it is called chemical vapor infiltration (CVI). Common ceramic-matrix materials used are SiC, Si_3N_4, and HfC. The preforms can consist of yarns, woven fabrics, or three-dimensional shapes. A filament-wound Nicalon tube and a braided Nextel tube before CVI and after CVI are shown in Fig. 11.28.

Chemical vapor infiltration has been used extensively for processing near-net shape continuous fiber CMCs. Approximately half of the commercially available carbon-carbon composites today are made by CVI. Chemical vapor infiltration can be thought of as a bulk form of CVD, which is widely used in depositing thin coatings. The process involves deposition of the solid matrix over an open-volume, porous, fibrous preform by the reaction and decomposition of gases. An example of a CVI

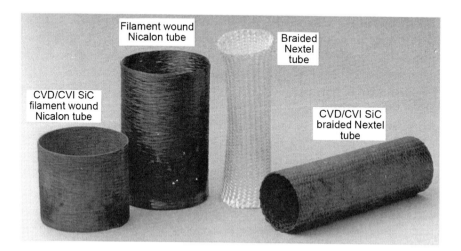

Fig. 11.28 Filament-wound Nicalon tube and a braided Nextel tube before and after being processed by chemical vapor deposition/chemical vapor infiltration (CVD/CVI). Courtesy of Thermo Electron Corporation. Source: Ref 11.11

reaction is the deposition of titanium diboride, which has a melting temperature of 3225 °C (5835 °F) but can be deposited at 900 °C (1650 °F) by CVI:

$$TiCl_4 + 2\ BCl_3 + 5\ H_2 \rightarrow TiB_2 + 10\ HCl$$

The resulting HCl by product is very common in CVI reactions. The solid materials are deposited from gaseous reactants onto a heated substrate. A typical CVD or CVI process would require a reactor with the following parts:

* A vapor feed system
* A CVD reactor in which the substrate is heated and gaseous reactants are fed
* An effluent system where exhaust gases are handled

A simple reactor is shown in Fig. 11.29. One can synthesize a variety of ceramic matrices such as oxides, glasses, ceramics, and intermetallics by CVD. Several examples of ceramic composites fabricated by CVI are shown in Table 11.9. There are two main variations of CVI. Isothermal chemical vapor infiltration (ICVI) relies on diffusion for deposition. The preform is maintained at a uniform temperature while the reactant gases are allowed to flow through the furnace and deposit the solid species. To obtain a uniform matrix around the fibers, deposition is conducted at low pressures and reactant concentrations. However, when the CVI process is carried out isothermally, surface pores tend to close first, restricting the gas flow to the interior of the preform. This phenomenon, sometimes referred to as canning, necessitates multiple cycles of impregnation, surface machining, and reinfiltration to obtain an adequate density. One can

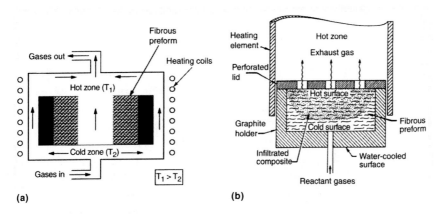

Fig. 11.29 Schematics of chemical vapor infiltration processes. (a) Isothermal chemical vapor infiltration. (b) Forced chemical vapor infiltration. Source: Ref 11.11

Table 11.9   Chemical vapor infiltration reactions for ceramic matrices

Ceramic matrix	Reactant gases	Reactant temperature, °F
C	CH_4-H_2	1800–2200
TiC	TiC_4-CH_4-H_2	1650–1800
SiC	Ch_3SiCl_3-H_2	1800–2550
B_4C	BCl_3-CH_4-H_2	2200–2550
TiN	$TiCl4$-$N2$-H_2	1650–1800
Si_3N_4	$SiCl$-$NH3$-H_2	1800–2550
BN	BCl_3-NH_3-H_2	1800–2550
AlN	$AlCl_3$-NH_3-H_2	1475–2200
Al_2O_3	$AlCl_3$-CO_2-H_2	1650–2010
SiO_2	SiH-CO_2-H_2	400–1110
TiO_2	$TiCl_3$-H_2O	1470–1830
ZrO_2	$ZrCl_4$-CO_2-H_2	1650–2200
TiB_2	$TiCl_4$-BCL_3-H_2	1470–1830
WB	WCl_6-BBr_3-H_2	1550–2900

Source: Ref 11.1

avoid some of these problems by using a forced gas flow and a temperature gradient approach to CVI. Forced chemical vapor infiltration (FCVI) uses a combination of thermal gradients and forced reactant flow to overcome the problems of slow diffusion and permeability obtained in ICVI. This can eliminate, to some extent, the need for multiple cycles. Thus, FCVI processes typically yield much shorter infiltration times while still obtaining uniform densification of the matrix and low residual porosity. As a comparison, a 3 mm (0.12 in.) part infiltrated by ICVI could take several weeks, while the same part infiltrated by FCVI would only take several hours. As is true with all CVI processes, with increasing densification a point of diminishing returns occurs, such that after a certain time the incremental increase in density is not proportional to the time required for deposition.

In FCVI, a graphite holder in contact with a water-cooled metallic gas distributor holds the fibrous preform. The bottom and side surfaces thus stay cool, while the top of the fibrous preform is exposed to the hot zone, creating a steep thermal gradient. The reactant gaseous mixture passes unreacted through the fibrous preform because of the low temperature. When these gases reach the hot zone, they decompose and deposit on and between the fibers to form the matrix. As the matrix material becomes deposited in the hot portion of the preform, the preform density increases, and the hot zone moves progressively from the top of the preform toward the bottom.

When the composite is formed completely at the top and is no longer permeable, the gases flow radially through the preform, exiting from the vented retaining ring. To control deposition, the rate of deposition must be maximized while minimizing density gradients. Deposition reaction rate and mass transport are competing factors, so very rapid deposition results in the exterior of the preform being well infiltrated, while severe density gradients and a large amount of porosity are present within the preform. Very slow deposition rates, on the other hand, require long times and are not economically feasible. A balance between the two factors is required for optimal infiltration.

Consider the process of decomposition of a chemical compound in the vapor form to yield SiC ceramic matrix on and in between the fibers in a preform. For example, methyltrichlorosilane (CH_3SiCl_3), the starting material to obtain SiC, is decomposed between 925 and 1125 °C (1700 and 2060 °F):

$$CH_3Cl_3Si \rightarrow SiC(s) + 3HCl(g)$$

The vapors of SiC deposit as solid phases on and between the fibers in a freestanding preform to form the matrix. The CVI process is very slow because it involves diffusion of the reactant species to the fibrous substrate, followed by outflow of the gaseous reactant products. The CVI process of making a ceramic matrix is a low-stress and low-temperature CVD process and thus avoids some of the problems associated with high-temperature ceramic processing. Using CVI, one can deposit the interfacial coating on the fibers as well as the matrix. For example, for Nicalon/SiC composites with a carbon interface, the carbon layer is deposited first, and then the SiC matrix is infiltrated without changing the preform conditions. The fibrous preforms are stacked layer by layer between perforated plates, through which the gases pass during infiltration. The carbon coating is typically deposited by means of a hydrocarbon gas at approximately 1000 °C (1830 °F) and reduced pressure to protect the fibers.

The graphitic coating on the fibers has a characteristic aligned structure of the basal planes. These basal planes are parallel to the fiber direction but perpendicular to the incoming crack front, so deflection of cracks at the weakly bonded basal planes takes place instead of fracturing the fibers. The softer c-axis of the graphite is also aligned in the perpendicular direction to accommodate the thermal residual stresses that arise from processing. The matrix consists of a nucleation zone in a small region at the coating/matrix interface. After this, long columnar grains perpendicular to the surface of the fiber are seen. The preferred orientation is such that the (111) planes are aligned parallel to the fibers. The grains are composed predominantly of β-SiC with a cubic structure with small disordered regions of α-SiC. For CVI composites reinforced with woven fiber fabrics, the nature of the porosity is trimodal. Macroporosity is found between fiber bundles and between layers of fabric, with pore sizes less than 100 μm. Microporosity occurs between fibers in the fiber bundle, and the pore size is usually on the order of 10 μm. Studies have shown that 70% of the pore volume is in the form of microporosity within the fiber bundle, 25% between the cloth layers, and 5% as holes between layers of the fabric.

This variant of CVI that combines forced gas flow and temperature gradient avoids some of the problems mentioned earlier. Under these modified conditions, 70 to 90% dense SiC and Si_3N_4 matrices can be impregnated in SiC and Si_3N_4 fibrous preforms in less than a day. Under conditions of plain CVI, it would take several weeks to achieve such densities; that is, one can reduce the processing time from several days to less than 24 h. One can also avoid using binders in this process, with their attendant problems of incomplete removal. The use of a graphite holder simplifies the fabrication of the preform, and the application of a moderate pressure to the preform can result in a higher-than-normal fiber volume fraction in the final product. The final obtainable density in a ceramic body is limited by the fact that closed porosity starts at approximately 93 to 94% of theoretical density. It is difficult to impregnate past this point.

Advantages of a CVI technique or any variant thereof include:

- Good mechanical properties at high temperatures
- Large, complex shapes produced in a near-net shape
- Considerable flexibility in the fibers and matrices that can be used (oxide and nonoxide)
- Pressureless process, with relatively low temperatures used compared to the temperatures involved in hot pressing

Among the disadvantages, one should mention that the process is slow and expensive.

Sol-Gel Techniques

Sol-gel techniques, which have been used for making conventional ceramic materials, can also be used to make ceramic-matrix materials in the interstices of a fibrous preform. A solution containing metal compounds—for example, a metal alkoxide, acetate, or halide—is reacted to form a sol. The sol is converted to a gel, which in turn is subjected to controlled heating to produce the desired end product: a glass, a glass-ceramic, or a ceramic. Characteristically, the gel-to-ceramic conversion temperature is much lower than that required in a conventional melting or sintering process. A schematic of a typical sol-gel process for processing CMCs is given in Fig. 11.30. It is easy to see that many of the polymer handling and processing techniques can be used for sol-gel as well. Impregnation of fibrous preforms in vacuum and filament winding are two important techniques. In filament winding, fiber tows or rovings are passed through a tank containing the sol, and the impregnated tow is wound on a mandrel to a desired shape and thickness. The sol is converted to gel, and the structure is removed from the mandrel. A final heat treatment then converts the gel to a ceramic or glass matrix.

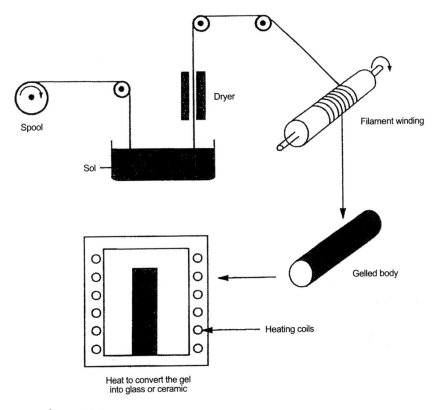

Fig. 11.30 Schematic of sol-gel process. Source: Ref 11.11

Some of the advantages of sol-gel techniques for making composites are the same as those for monolithic ceramics, namely, lower processing temperatures, greater compositional homogeneity in single-phase matrices, and the potential for producing unique multiphase-matrix materials. The sol-gel technique allows processing by low-viscosity liquids, such as the ones derived from alkoxides. Covalent ceramics, for example, can be produced by pyrolysis of polymeric precursors at temperatures as low as 1400 °C (2550 °F) and with yields greater than those in CVD processes. Among the disadvantages of sol-gel are high shrinkage and low yield compared to slurry techniques. The fiber network provides a very high surface area to the matrix gel. Consequently, the shrinkage during the drying step frequently results in a large density of cracks in the matrix. Generally, repeated impregnations are required to produce a substantially dense matrix. The sol-gel technique can also be used to prepare prepregs by slurry infiltration. The sol in the slurry acts as a binder and coats fibers and glass particles. The binder burnout step is thus eliminated because the binder, being of the same composition as the matrix, becomes part of the glass matrix.

Polymer Infiltration and Pyrolysis

Polymeric precursors can also be used to form a CMC. Due to the generally high cost of processing CMCs, polymer infiltration and pyrolysis (PIP) is attractive because of its relatively low cost while maintaining small amounts of residual porosity and minimal degradation of the fibers. Moreover, this approach allows near-net shape molding and fabrication technology that is able to produce nearly fully dense composites. In PIP, the fibers are infiltrated with an organic polymer, which is heated to fairly high temperatures and pyrolyzed to form a ceramic matrix. Due to the relatively low yield of polymer to ceramic, multiple infiltrations are used to densify the composite.

Polymeric precursors for ceramic matrices allow the use of conventional polymer composite fabrication technology. Furthermore, by processing and pyrolyzing at lower temperatures (compared to sintering and hot pressing), fiber degradation and the formation of unwanted reaction products at the fiber-matrix interface can be avoided. Desirable characteristics of a preceramic polymer include:

- High ceramic yield from polymer precursor
- Precursor that yields a ceramic with low free-carbon content (which will oxidize at high temperatures)
- Controllable molecular weight, which allows for solvent solubility and control over viscosity for fabrication purposes
- Low-temperature cross linking of the polymer that allows resin to harden and maintain its dimensions during the pyrolysis process
- Low cost and toxicity

Most preceramic polymer precursors are formed from chloro-organosilicon compounds to form poly(silanes), poly(carbosilanes), poly(silazanes), poly(borosilanes), poly(silsesquioxanes), and poly(carbosiloxanes). The synthesis reaction involves the dechlorination of the chlorinated silane monomers. Because a large quantity of the chlorosilane monomers are formed as by-products in the silicone industry, they are inexpensive and readily available. The monomers can be further controlled by an appropriate amount of branching, which controls important properties such as the viscosity of the precursor as well as the amount of ceramic yield.

All silicon-base polymer precursors lead to an amorphous ceramic matrix, where silicon atoms are tetrahedrally arranged with nonsilicon atoms. This arrangement is similar to that found in amorphous silica. High-temperature treatments typically lead to crystallization and slight densification of the matrix, which results in shrinkage. At high temperatures, the amorphous ceramic begins to form small domains of crystalline phase, which are more thermodynamically stable. Silicon carbide matrices derived from polycarbosilane begin to crystallize at 1100 to 1200 °C (2010 to 2190 °F), while Si-C-O (polysiloxanes) and Si-N-C (polysilazanes) remain amorphous to 1300 to 1400 °C (2370 to 2550 °F).

After preform fabrication, the polymer is then cross linked and finally pyrolyzed in an inert or reactive atmosphere (e.g., NH_3) at temperatures between 1000 to 1400 °C (1830 and 2550 °F). The pyrolysis step can be further subdivided into two steps. In the first step, between 550 and 880 °C (1020 and 1620 °F), an amorphous compound of the type $Si(C_aO_bN_cB_d)$ is formed. The second step involves nucleation of crystalline precipitates such as SiC, Si_3N_4, and SiO_2 at temperatures between 1200 and 1600 °C (2190 and 2910 °F). Grain coarsening may also result from consumption of any residual amorphous phase and reduction of the amount of oxygen due to vaporization of SiO and CO. Porosity is typically of the order of 5 to 20 vol%, with pore sizes of the order of 1 to 50 nm. It should be noted that the average pore size and volume fraction of pores decreases with increasing pyrolysis temperature, because the amount of densification and shrinkage becomes irreversible at temperatures above the maximum pyrolysis temperature.

The main disadvantage of PIP is the low yield that accompanies the polymer-to-ceramic transformation and the resulting shrinkage, which typically causes cracking in the matrix during fabrication. Due to shrinkage and weight loss during pyrolysis, residual porosity after a single impregnation is of the order of 20 to 30%. To reduce the amount of residual porosity, multiple impregnations are needed. Reimpregnation is typically conducted with a very low-viscosity prepolymer, so that the slurry may wet and infiltrate the small micropores that exist in the preform. Usually, reimpregnation is done by immersing the part in the liquid polymer in a vacuum bag, while higher-viscosity polymers require pressure impregnation. Typically, the amount of porosity will reduce from 35% to less than 10% after approximately five impregnations.

Significant gas evolution also occurs during pyrolysis. Thus, it is advisable to allow these volatile gases to slowly diffuse out of the matrix, especially for thicker parts. Typically, pyrolysis cycles ramp to 800 to 1400 °C (1470 to 2550 °F) over periods of 1 to 2 days to avoid delamination.

Two extreme cases of polymer-ceramic conversion can be considered. If the volume is not constrained, then diffusional flow will cause the pores to be filled, but a high amount of shrinkage will take place. If the volume is constrained, then shrinkage does not occur, but a large amount of residual porosity is present. It has been reported that for filler-free pyrolysis of poly(silazane) to form bulk Si_3N_4, either a large amount of porosity (>8%) or a large amount of shrinkage (20%) took place.

A typical sequence of processing steps is:

1. Porous SiC or Si_3N_4 fibrous preform with some binder phase is prepared.

2. Fibrous preform is evacuated in an autoclave.
3. Samples are infiltrated with molten precursors—silazanes or polycarbosilanes—at high temperature (507 °C, or 945 °F), and the argon or nitrogen pressure is slowly increased from 2 to 40 MPa (0.3 to 5.8 ksi). The high temperature results in a transformation of the oligomer silane to polycarbosilane and simultaneous polymerization at high pressures.
4. Infiltrated samples are cooled and treated with solvents.
5. Samples are placed in an autoclave, and the organosilicon polymer matrix is thermally decomposed in an inert atmosphere at a high pressure and at temperature in the 530 to 1025 °C (985 to 1880 °F) range.
6. Steps 2 through 5 are repeated to obtain an adequate density. To produce an optimal matrix crystal structure, the material is annealed in the 1025 to 1525 °C (1880 to 2780 °F) range.

Polymer-derived CMCs, similar to carbon-carbon composites, typically have a cracked matrix from processing as well as a number of small voids or pores. The large amount of shrinkage and cracking in the matrix can be contained, to some extent, by the additions of particulate fillers to the matrix, which, when added to the polymer, reduce shrinkage and stiffen the matrix material in the composite. A schematic of filler-free versus active filler pyrolysis is shown in Fig. 11.31. The microstructure of a PIP ceramic-matrix composite, indicating some residual porosity and a clear

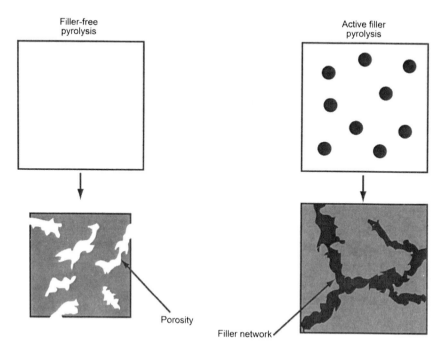

Fig. 11.31 Schematic of filler-free versus active filler pyrolysis. Source: Ref 11.11

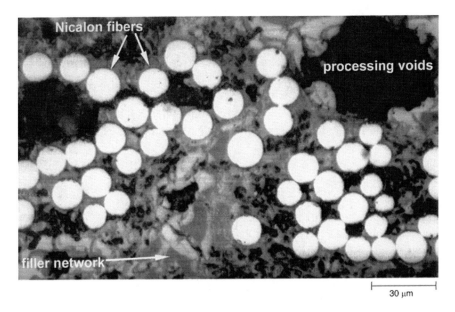

Fig. 11.32 Microstructure of a woven Nicalon/Si-C-O-N-matrix composite with a SiC filler. Small amounts of residual porosity and a clear filler-free network can be seen in the matrix of the composite. Source: Ref 11.11

filler network in the matrix of the composite, is shown in Fig. 11.32. Particulate or whisker ceramics used as fillers in the polymeric matrix can serve a variety of purposes:

- Reduce and disrupt the formation of matrix cracks that form during shrinkage of the polymer
- Enhance ceramic yield by forming reaction products during pyrolysis
- Strengthen and toughen the weak amorphous matrix and increase the interlaminar shear strength of the composite

The filler must be submicrometer in size to penetrate the tow bundle, and the coefficient of thermal expansion of the filler must match that of the polymeric matrix. It should be noted that the filler must not be used in very high fractions, and the slurry should not be forced into the reinforcing fibers because abrasion of the fiber fabric may take place. This is especially true with hard, angular fillers or ceramic whiskers. Typically, the volume fraction of filler is 15 to 25% of the matrix volume fraction. High filler loading may result in an increase in interply spacing and lower volume fraction of fibers. When an "active" filler phase is added to the polymer, it reacts with solid or gaseous decomposition products to form new carbide phases. By controlling the amount of filler, the degree of shrinkage can be controlled.

Fiber architecture can have an impact on part quality. One of the key factors is wetting of the fiber bundles. During pyrolysis, the precursor shrinks around the fibers, so cracks are introduced. For example, two-dimensional woven fabrics seem to have less of a propensity to develop interlaminar cracks than do cross-ply or unidirectional architectures. Satin weaves are preferred over plain weaves because more uniform cracking is achieved, and large cracks between weave crossover points are avoided. Due to the looser nature of the satin weave, better wetting and densification may take place, although the loose nature of the weave also makes it more difficult to handle.

ACKNOWLEDGMENTS

Portions of the text for this chapter came from "Processing of Ceramic-Matrix Composites" by K.K. Chawla and N. Chawla, "Properties and Performance of Ceramic-Matrix and Carbon-Carbon Composites" by C. Powell Doğan, and "Applications of Ceramic-Matrix Composites" by J.R. Davis, all in *Composites,* Vol 21, *ASM Handbook,* ASM International, 2001.

REFERENCES

11.1 F.C. Campbell, *Structural Composite Materials,* ASM International, 2010

11.2 C. Powell Doğan, Properties and Performance of Ceramic-Matrix and Carbon-Carbon Composites, *Composites,* Vol 21, *ASM Handbook,* ASM International, 2001

11.3 J.R. Davis, Applications of Ceramic-Matrix Composites, *Composites,* Vol 21, *ASM Handbook,* ASM International, 2001

11.4 G. Brandt, Ceramic Cutting Tools, *Ceramic Technology International,* I. Birkby, Ed., Sterling Publications Ltd., 1995, p 77–80

11.5 A.W. Urquhart, Directed Metal Oxidation, *Ceramics and Glasses,* Vol 4, *Engineered Materials Handbook,* ASM International, 1991, p 232–235

11.6 L.J. Korb, C.A. Morant, R.M. Calland, and C.S. Thatcher, The Shuttle Orbiter Thermal Protection System, *Ceram. Bull.,* Vol 60 (No. 11), 1981, p 1188–1193

11.7 S.-T. Buljan and V.K. Sarin, Improved Productivity Through Application of Silicon Nitride Cutting Tools, *Carbide Tool J.,* Vol 14 (No. 3), 1982, p 40–46

11.8 Introduction to Composites, *Composites,* Vol 21, *ASM Handbook,* ASM International, 2001

11.9 T. Windhorst and G. Blount, Carbon-Carbon Composites: A Summary of Recent Developments and Applications, *Mater. Des.,* Vol 18, 1997, p 11–15

11.10 J.E. Sheehan, K.W. Buesking, and B.J. Sullivan, Carbon-Carbon Composites, *Ann. Rev. Mater. Sci.,* Vol 24, 1994, p 19–44

11.11 K.K. Chawla and N. Chawla, Processing of Ceramic-Matrix Composites, *Composites,* Vol 21, *ASM Handbook,* ASM International, 2001

SELECTED REFERENCES

- "3M Nextel Ceramic Textiles Technical Notebook," 3M Ceramic Textiles and Composites, 2003
- M.F. Amateau, Ceramic Composites, *Handbook of Composites,* Chapman & Hall, 1998, p 307–332
- P.F. Becher, T.N. Tiegs, and P. Angelini, Whisker Reinforced Ceramic Composites, *Fiber Reinforced Ceramic Composites: Materials, Processing and Technology,* Noyes Publications, 1990, p 311–327
- J.D. Buckley, Carbon-Carbon Composites, *Handbook of Composites,* Chapman & Hall, 1998, p 333–351
- K.K. Chawla, Processing of Ceramic Matrix Composites, *Ceramic Matrix Composites,* Chapman & Hall, 1993, p 126–161
- *Composites,* Ceramics Information Analysis Center, 1995, p 1–31
- G.H. Cullum, Ceramic Matrix Composite Fabrication and Processing: Sol-Gel Infiltration, *Handbook on Continuous Fiber-Reinforced Ceramic Matrix Composites,* Ceramics Information Analysis Center, 1995, p 185–204
- J.A. DiCarlo and N.P. Bansal, "Fabrication Routes for Continuous Fiber-Reinforced Ceramic Composites (CFCC)," NASA/TM-1998-208819, National Aeronautics and Space Administration, 1998
- J.A. DiCarlo and S. Dutta, Continuous Ceramic Fibers for Ceramic Matrix Composites, *Handbook on Continuous Fiber-Reinforced Ceramic Matrix Composites,* Ceramics Information Analysis Center, 1995, p 137–183
- A.S. Fareed, Ceramic Matrix Composite Fabrication and Processing: Directed Metal Oxidation, *Handbook on Continuous Fiber-Reinforced Ceramic Matrix Composites,* Ceramics Information Analysis Center, 1995, p 301–324
- J.E. French, Ceramic Matrix Composite Fabrication and Processing: Polymer Pyrolysis, *Handbook on Continuous Fiber-Reinforced Ceramic Matrix Composites,* Ceramics Information Analysis Center, 1995, p 269–299
- W.J. Lackey and T.L. Starr, Fabrication of Fiber-Reinforced Ceramic Composites by Chemical Vapor Infiltration: Processing, Structure and Properties, *Fiber Reinforced-Ceramic Composites: Materials, Processing and Technology,* Noyes Publications, 1990, p 397–450
- F.F. Lange et al., Powder Processing of Ceramic Matrix Composites, *Mater. Sci. Eng. A,* Vol 144, 1991, p 143–152
- D. Lewis III, Continuous Fiber-Reinforced Ceramic Matrix Composites: A Historical Overview, *Handbook on Continuous Fiber-Reinforced*

Ceramic Matrix Composites, Ceramics Information Analysis Center, 1995

• R.A. Lowden, D.P. Stinton, and T.M. Besmann, Ceramic Matrix Composite Fabrication and Processing: Chemical Vapor Infiltration, *Handbook on Continuous Fiber-Reinforced Ceramic Matrix Composites,* Ceramics Information Analysis Center, 1995, p 205–268

• K.L. Luthra and G.S. Corman, "Melt Infiltrated (MI) SiC/SiC Composites for Gas Turbine Applications," Technical Information Series, GE Research and Development Center, 2001

• T. Mah et al., Ceramic Fiber Reinforced Metal-Organic Precursor Matrix Composites, *Fiber Reinforced Ceramic Composites: Materials, Processing and Technology,* Noyes Publications, 1990, p 278–310

• A. Marzullo, Boron, High Silica, Quartz and Ceramic Fibers, *Handbook of Composites,* Chapman & Hall, 1998, p 156–168

• R.R. Naslain, Ceramic Matrix Composites: Matrices and Processing, *Encyclopedia of Materials: Science and Technology,* Elsevier Science Ltd., 2000

• R. Naslain, Design, Preparation and Properties of Non-Oxide CMCs for Application in Engines and Nuclear Reactors: An Overview, *Compos. Sci. Technol.,* Vol 64, 2004, p 155–170

• State of the Art in Ceramic Fiber Performance, *Ceramic Fibers and Coatings: Advanced Materials for the Twenty-First Century,* National Academy of Sciences, 1998, p 21–36

• R. Zolandz and R.L. Lehmann, Crystalline Matrix Materials for Use in Continuous Filament Fiber Composites, *Handbook on Continuous Fiber-Reinforced Ceramic Matrix Composites,* Ceramics Information Analysis Center, 1995, p 111–136

CHAPTER **12**

Selection Guidelines for Lightweight Materials

THIS FINAL CHAPTER is divided into three parts. The first part discusses the general guidelines for the materials and process selection process. This is only a cursory overview of some of the many methods that have been developed for selecting materials and processes during the various phases of the design process. An excellent overview of these, and other, methodologies that are used is given in Ref 12.1. Reference 12.1 also lists sources where even more detailed information can be found. The second part of the chapter gives some high-level selection guidelines for the lightweight materials covered in this book. Finally, the third part discusses the importance of the automotive sector for the lightweight material industry.

12.1 The Materials Selection Process

Materials and manufacturing processes that convert materials into useful parts are an integral part of engineering design. The enormity of the decision task in materials selection is due to the fact that there are well over 100,000 engineering materials from which to choose. On a more practical level, the typical engineer should have ready access to information on 50 to 80 materials, depending on the range of applications.

The importance of materials selection in design has increased in recent years. The adoption of concurrent engineering or integrated product definition methods has brought materials engineers into the design process at an earlier stage, and the importance given to manufacturing in present-day product design has reinforced the fact that materials and manufacturing are closely linked in determining final properties. Moreover, worldwide competitiveness has increased the general level of automation in manufacturing to the point where materials costs comprise 50% or more of the cost for most products. Finally, materials science has created a variety of new materials and has focused attention on the competition between materials. Thus, the range of materials available to the engineer is much larger than ever before. This presents the opportunity for innovation in design by using these materials in products that provide greater performance at lower cost. To achieve this requires a rational process for materials selection.

With the enormous combination of materials and processes to choose from, the task can be done only by introducing simplification and systemization. The material and process selection becomes more involved as the design progresses. There are three major phases in the design process:

1. *Conceptual design.* At conceptual design, essentially all materials and processes are considered rather broadly. The materials selection methodology and charts developed by Dr. Michael Ashby of Cambridge University, United Kingdom, are appropriate at this stage. Two of these charts—showing strength versus density, and elastic modulus versus strength—are shown in Fig. 12.1 and 12.2, respectively. There are many similar charts for other properties that can be found in Ref 12.2. The decision is to determine whether the part will be made from metal, plastics, ceramic, or composite and to narrow it to a group of materials. The precision of property data needed is rather low. If an innovative choice of material is to be made, it must be done at the conceptual design step because later in the design process too many decisions have been made to allow for a radical change. While the level of fidelity at the conceptual design stage may be low, it is an extremely important phase, as it has been shown that many of the downstream costs are locked in during the early phases of design.

2. *Configuration design.* At the configuration level of design, the emphasis is on determining the shape and approximate size of a part using engineering methods of analysis. Now the designer will have decided on a class of materials and processes, for example, a range of aluminum alloys, wrought and cast. The material properties must be known to a greater level of precision.

3. *Detail design.* At the detail design level, the decision will have narrowed to a single material and only a few manufacturing processes. Here the emphasis will be on deciding on critical tolerances, optimizing for robust design, and selecting the best manufacturing process using quality engineering and cost-modeling methodologies. Depending on the criticality of the part, materials properties may need to be known to a high level of precision (Table 12.1). At the extreme, this requires the development of a detailed database from an extensive materials testing program.

12.2 Performance Characteristics of Materials

The performance or functional characteristics of a material are expressed chiefly by physical, mechanical, thermal, electrical, magnetic, and optical properties. Material properties are the link between the basic structure

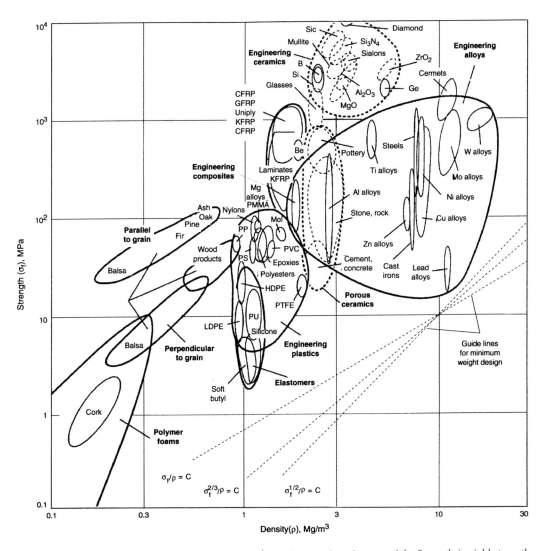

Fig. 12.1 Strength, σ_f, plotted against density, ρ, for various engineering materials. Strength is yield strength for metals and polymers, compressive strength for ceramics, tear strength for elastomers, and tensile strength for composites. Source: Ref 12.2

and composition of the material and the service performance of the part (Fig. 12.3). The goal of materials science is to learn how to control the various levels of structure of a material (electronic structure, defect structure, microstructure, and macrostructure) so as to predict and improve the properties of a material. Not too long ago, metals dominated most of engineering design. Today, the range of materials and properties available to the engineer is much larger and growing rapidly. This requires familiarity with a broader range of materials and properties, but it also introduces new opportunities for innovation in product development. A general comparison of the properties of metals, ceramics, and polymers is provided in

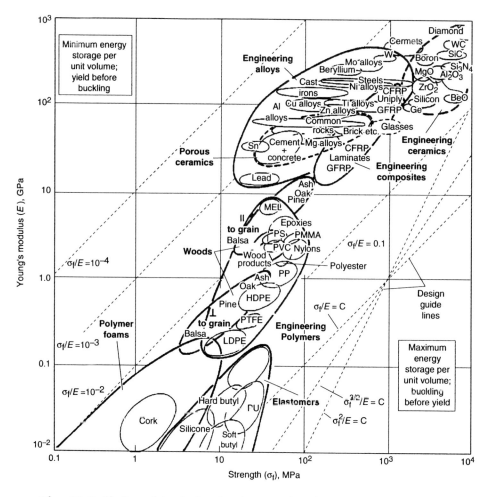

Fig. 12.2 Elastic modulus, E, plotted against strength, σ_f, for various engineering materials. The heavy envelopes enclose data for a given class of material. The diagonal contours show the longitudinal wave velocity. Source: Ref 12.2

Table 12.2, and Table 12.3 gives a listing of the broad spectrum of material properties that may be needed.

An important role of the materials engineer is to assist the designer in making meaningful connections between materials properties and the performance of the part or system being designed. For most mechanical systems, performance is limited, not by a single property but by a combination of them. For example, the materials with the best thermal shock resistance are those with the largest values of $\sigma_f/E\alpha$, where σ_f is the failure stress, E is Young's modulus, and α is the thermal coefficient of expansion. The relationships between standard mechanical properties and the failure modes for materials are shown in Table 12.4. For most modes of failure, two or more material properties act to control the material behavior. Also, it must be kept in mind that the service conditions met by materials

Table 12.1 Examples of materials information required during detail design

Material identification
Material class (metal, plastic, ceramic composite)
Material subclass
Material industry designation
Material product form
Material condition designation (temper, heat treatment, etc.)
Material specification
Material alternative names
Material component designations (composite/ assembly)

Material production history
Manufacturability strengths and limitations
Material composition(s)
Material condition (fabrication)
Material assembly technology
Constitutive equations relating to properties

Material properties and test procedures
Density
Specific heat
Coefficient of thermal expansion
Thermal conductivity
Tensile strength
Yield strength
Elongation
Reduction of area
Moduli of elasticity
Stress-strain curve or equation
Hardness
Fatigue strength (define test methods, load, and environment)

Temperature (cryogenic-elevated)
Tensile strength, yield strength
Creep rates, rupture life at elevated temperatures
Relaxation at elevated temperatures
Toughness

Damage tolerance (if applicable)
Fracture toughness (define test)
Fatigue crack growth rates (define environment and load)
Temperature effects
Environmental stability
Compatibility data
General corrosion resistance
Stress-corrosion cracking resistance

Environmental stability
Compatibility data
General corrosion resistance
Source: Ref 12.3

Environmental stability (continued)
Stress-corrosion cracking resistance
Toxicity (at all stages of production and operation)
Recyclability/disposal

Material design properties
Tension
Compression
Shear
Bearing
Controlled strain fatigue life

Processability information
Finishing characteristics
Weldability/joining technologies
Suitability for forging, extrusion, and rolling
Formability (finished product)
Castability
Repairability
Flammability

Joining technology applicable
Fusion
Adhesive bonding
Fasteners
Welding parameters

Finishing technology applicable
Impregnation
Painting
Stability of color

Application history/experience
Successful uses
Unsuccessful uses
Applications to be avoided
Failure analysis reports
Maximum life service

Availability
Multisource? Vendors?
Sizes
Forms

Cost/cost factors
Raw material
Finished product or require added processing
Special finishing/protection
Special tooling/tooling costs

Quality control/assurance issues
Inspectability
Repair
Repeatability

---

are, in general, more complex than the test conditions used to measure material properties. Usually, simulated service tests must be devised to screen materials for critical complex service conditions. Finally, the chosen material, or a small group of candidate materials, must be evaluated in prototype tests or field tests to determine their performance under actual service conditions.

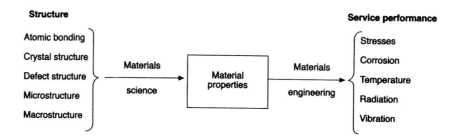

Fig. 12.3 The role played by material properties in the selection of materials.
Source: Ref 12.3

Table 12.2 General comparison of properties of metals, ceramics, and polymers

Property (approximate values)	Metals	Ceramics	Polymers
Density, g/cm³	2 to 22 (average ~8)	2 to 19 (average ~4)	1 to 2
Melting points	Low (Ga = 29.78 °, or 85.6 °F) to high (W = 3410 °C, or 6170 °F)	High (up to 4000 °C, or 7230 °F)	Low
Hardness	Medium	High	Low
Machinability	Good	Poor	Good
Tensile strength, MPa (ksi)	Up to 2500 (360)	Up to 400 (58)	Up to 140 (20)
Compressive strength, MPa (ksi)	Up to 2500 (360)	Up to 5000 (725)	Up to 350 (50)
Young's modulus, GPa (10⁶ psi)	15 to 400 (2 to 58)	150 to 450 (22 to 65)	0.001 to 10 (0.00015 to 1.45)
High-temperature creep resistance	Poor to medium	Excellent	. . .
Thermal expansion	Medium to high	Low to medium	Very high
Thermal conductivity	Medium to high	Medium, but often decreases rapidly with temperature	Very low
Thermal shock resistance	Good	Generally poor	. . .
Electrical characteristics	Conductors	Insulators	Insulators
Chemical resistance	Low to medium	Excellent	Good
Oxidation resistance	Generally poor	Oxides excellent; SiC and Si_3N_4 good	. . .

Source: Ref 12.4

12.3 Relation of Materials Selection to Manufacturing

The selection of a material must be closely coupled with the selection of a manufacturing process. This is not an easy task, because there are many processes that can produce the same part. The goal is to select a material and process that maximizes quality and minimizes the cost of the part. A breakdown of manufacturing processes into nine broad classes is given in Fig. 12.4.

In a very general sense, the selection of the material determines a range of processes that can be used to process parts from the material. The manufacturing methods used most frequently with different metals and plastics are shown in Table 12.5. The material melting point and general level of deformation resistance (hardness) and ductility determine these relationships. The next aspect to consider is the minimum and maximum overall size of the part, often expressed by volume, projected area, or

Table 12.3 Material performance characteristics

Physical properties	Mechanical properties	Thermal properties
Crystal structure	Hardness	Conductivity
Density	Modulus of elasticity	Specific heat
Melting point	Tension	Coefficient of thermal expansion
Vapor pressure	Compression	Emissivity
Viscosity	Poisson's ratio	Absorptivity
Porosity	Stress-strain curve	Ablation rate
Permeability	Yield strength	Fire resistance
Reflectivity	Tension	**Chemical properties**
Transparency	Compression	Position in electromotive series
Optical properties	Shear	Corrosion and degradation
Dimensional stability	Ultimate strength	Atmospheric
Electrical properties	Tension	Salt water
Conductivity	Shear	Acids
Dielectric constant	Bearing	Hot gases
Coercive force	Fatigue properties	Ultraviolet
Hysteresis	Smooth	Oxidation
Nuclear properties	Notched	Thermal stability
Half-life	Corrosion fatigue	Biological stability
Cross section	Rolling contact	Stress corrosion
Stability	Fretting	Hydrogen embrittlement
	Charpy transition temperature	Hydraulic permeability
	Fracture hardness (K_{Ic})	**Fabrication properties**
	High-temperature behavior	Castability
	Creep	Heat treatability
	Stress rupture	Hardenability
	Damping properties	Formability
	Wear properties	Machinability
	Galling	Weldability
	Abrasion	
	Erosion	
	Cavitation	
	Spalling	
	Ballistic impact	

Source: Ref 12.3

weight. Maximum size often is controlled by equipment considerations. Shape is the next factor to consider. The overall guide should be to select a primary process that makes the part as near to final shape as possible (near-net shape forming) without requiring expensive secondary machining or grinding processes. Sometimes the form of the starting material is important. For example, a hollow shaft can best be made by starting with a tube rather than a solid bar. Shape is often characterized by aspect ratio, the surface-to-volume ratio, or the web thickness-to-depth ratio. Closely related to shape is complexity. Complexity is correlated with lack of symmetry. It also can be measured by the information content of the part, that is, the number of independent dimensions that must be specified to describe the shape. Tolerance is the degree of deviation from ideal that is permitted in the dimensions of a part. Closely related to tolerance is surface finish. Surface finish is measured by the root mean square amplitude of the irregularities of the surface. Each manufacturing process has the capability of producing a part with a certain range of tolerance and surface

Table 12.4 Relationships between failure modes and material properties

Failure mode	Ultimate tensile strength	Yield strength	Compressive yield strength	Shear yield strength	Fatigue properties	Ductility	Impact energy	Transition temperature	Modulus of elasticity	Creep rate	K_{Ic}(a)	K_{xxx}(b)	Electrochemical potential	Hardness	Coefficient of expansion
Gross yielding	...	X	...	X	...	...	...	...	...	...	...	...	...	...	...
Buckling	...	...	X	...	...	...	...	...	X	...	...	...	...	...	...
Creep	...	...	...	...	...	...	...	...	...	X	...	...	...	...	...
Brittle fracture	...	...	...	...	...	...	X	X	...	...	X	...	...	...	...
Fatigue, low cycle	X	...	...	...	X	X	...	...	...	...	...	...	...	...	...
Fatigue, high cycle	...	...	...	...	X	...	...	...	...	...	...	...	...	...	...
Contact fatigue	...	...	X	...	...	...	...	...	...	...	...	...	...	...	...
Fretting	...	...	X	...	...	...	...	...	...	...	...	...	X	...	...
Corrosion	X	...	...	...	...	...	...	...	...	...	...	...	X	...	...
Stress-corrosion cracking	X	...	...	...	...	...	...	...	...	...	...	X	X	...	...
Galvanic corrosion	...	...	...	...	...	...	...	...	...	...	...	...	X	...	...
Hydrogen embrittlement	X	...	...	...	...	...	...	...	...	...	...	...	X	...	...
Wear	...	...	...	...	...	...	...	...	...	...	...	...	...	X	...
Thermal fatigue	...	...	...	...	...	...	...	...	...	X	...	...	...	...	X
Corrosion fatigue	...	...	...	...	X	...	...	...	...	...	...	...	X	...	...

Note: An "X" at the intersection of material property and failure mode indicates that a particular material property is influential in controlling a particular failure mode. (a) Plane-strain fracture toughness. (b) Threshold stress intensity to produce stress-corrosion cracking. Source: Ref 12.5

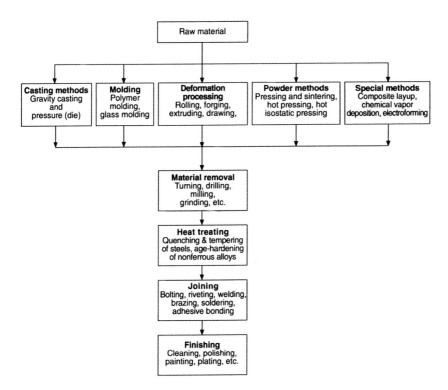

Fig. 12.4 The nine classes of manufacturing processes. The horizontal row contains the primary forming (shaping) processes. The processes in the vertical column are the secondary forming and finishing processes. Source: Ref 12.3

finish (Fig. 12.5). Polymers are different from metals and ceramics in that they can be processed to a very high surface smoothness, but tight tolerances are seldom possible because of internal stresses left by molding and creep at service temperatures. Manufacturing cost increases exponentially with decreasing dimensional tolerance.

Yet another process parameter is surface detail, the smallest radius of curvature at a corner that can be produced. An important practical consideration is the quantity of parts required. For each process, there is a minimum batch size below which it is not economical to go because of the costs of tooling, fixtures, and equipment. Also related to part cost is the production rate or the cycle time, the time required to produce one part. The most commonly used manufacturing processes are evaluated with respect to these characteristics in Table 12.6.

Ashby (Ref 12.2) extended his concepts for materials selection for properties at the conceptual design stage to the development of a set of process selection charts that encompass most of the process attributes described previously. The approach, as with materials selection based on properties, is to search the entire gamut of potential processes to arrive at the small set that should be considered in detail. The approach is illustrated in Fig. 12.6. Figure 12.6(a) shows the relationship between part size and the

Table 12.5 Compatibility between materials and manufacturing processes

Process	Cast iron	Carbon Steel	Alloy steel	Stainless steel	Aluminum and aluminum alloys	Copper and copper alloys	Zinc and zinc alloys	Magnesium and magnesium alloys	Titanium and titanium alloys	Nickel and nickel alloys	Refractory metals	Thermoplastics	Thermoset plastics
Casting/molding													
Sand casting	•	•	•	•	•	•		•		•		X	X
Investment casting		•	•	•	•	•				•		X	X
Die casting	X	X	X	X	•		•	•	X	X	X	X	X
Injection molding	X	X	X	X	X	X	X	X	X	X	X	•	—
Structural foam molding	X	X	X	X	X	X	X	X	X	X	X	•	X
Blow molding (extrusion)	X	X	X	X	X	X	X	X	X	X	X	•	X
Blow molding (injection)	X	X	X	X	X	X	X	X	X	X	X	•	X
Rotational molding	X	X	X	X	X	X	X	X	X	X	X	•	X
Forging/bulk forming													
Impact extrusion	X	•	•		•	•	•		X	X	X	X	X
Cold heading	X	•	•	•	•	•			X		X	X	X
Closed die forging	X	•	•	•	•	•	X	•	•	•		X	X
Pressing and sintering (PM)	X	•	•	•	•	•	X	•			•	X	X
Hot extrusion	X	•			•	•	X	•			X	X	X
Rotary swaging	X	•	•	•	•			•	X	•	•	X	X
Machining													
Machining from stock	•	•	•	•	•	•	•	•			X		
Electrochemical machining	•	•	•	•					•	•		X	X
Electrical discharge machining (EDM)	X	•	•	•	•	•				•		X	X
Wire EDM	X	•	•	•	•	•				•		•	X
Forming													
Sheet metal forming	X	•	•	•	•	•			X		X	X	X
Thermoforming	X	•	X	X	X	X	X	X	X	X	X	•	X
Metal spinning	X	•		•	•	•	•					X	X

Note: •, normal practice; —, less-common practice; X, not applicable. PM, powder metallurgy. Source: Adapted from Ref 12.6

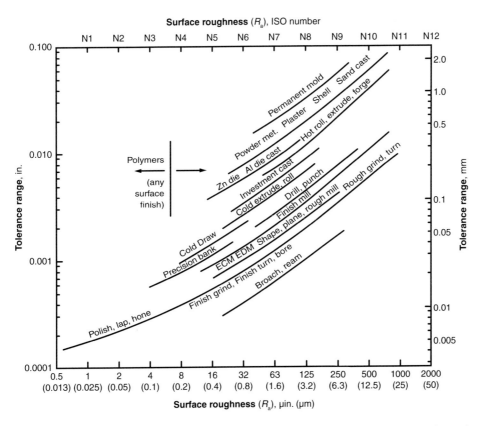

Fig. 12.5 Approximate values of surface roughness and tolerance on dimensions typically obtained with different manufacturing processes. ECM, electrochemical machining; EDM, electrical discharge machining. Source: Ref 12.7

complexity of the shape. Complexity is expressed in bits of information and is given by:

$$C = n \log_2 \left(\frac{\bar{l}}{\overline{\Delta l}} \right)$$
(Eq 12.1)

where C is the part complexity, n is the number of dimensions, $\overline{\Delta l}$ is the geometric mean tolerance, and $\bar{l}$ is the geometric mean dimension.

Simple shapes require only a few bits of information; complex shapes, such as integrated circuits, require very many. The casting for an engine block may have 10^3 bits of information, but after machining the complexity increases both by adding new dimensions (n) and improving the precision (reducing $\overline{\Delta l}$). Casting processes can make parts that vary in size from approximately 1 g to several hundred tons. Machining adds precision by providing parts with a greater range of complexity. Process selection is done by superimposing the envelope of design attributes (Fig. 12.6b). Sometimes the design sets upper and lower limits on the process attribute, region 1, while other times it imposes only upper limits, region 2. Process

Table 12.6 Manufacturing processes and their attributes

Process	Surface roughness	Dimensional accuracy	Complexity	Production rate	Production run	Relative cost	Size (projected area)
Pressure die casting	L	H	H	H/M	H	H	M/L
Centrifugal casting	M	M	M	L	M/L	H/M	H/M/L
Compression molding	L	H	M	H/M	H/M	H/M	H/M/L
Injection molding	L	H	H	H/M	H/M	H/M/L	M/L
Sand casting	H	M	M	L	H/M/L	H/M/L	H/M/L
Shell mold casting	L	H	H	H/M	H/M	H/M	M/L
Investment casting	L	H	H	L	H/M/L	H/M	M/L
Single-point cutting	L	H	M	H/M/L	H/M/L	H/M/L	H/M/L
Milling	L	H	H	M/L	H/M/L	H/M/L	H/M/L
Grinding	L	H	M	L	M/L	H/M	M/L
Electrical discharge machining	L	H	H	L	L	H	M/L
Blow molding	M	M	M	H/M	H/M	H/M/L	M/L
Sheet metal working	L	H	H	H/M	H/M	H/M/L	L
Forging	M	M	M	H/M	H/M	H/M	H/M/L
Rolling	L	M	H	H	H	H/M	H/M
Extrusion	L	H	H	H/M	H/M	H/M	M/L
Powder metallurgy	L	H	H	H/M	H	H/M	L
Key:							
H	>250	<0.005	High	>100	>5000	High	>0.5
M	>63 and <250	>0.005 and <0.05	Medium	>10 and <100	>100 and <5000	Medium	>0.02 and <0.5
L	<63	>0.05	Low	<10	<100	Low	<0.02
Units	μm	in.		Parts/h	Parts		m²

Source: Ref 12.8

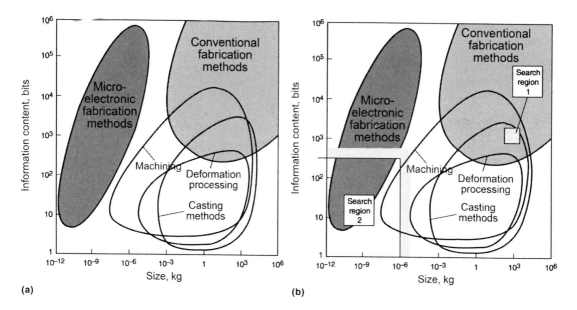

(a)　　　　　　　　　　　　　　(b)

Fig. 12.6 Ashby process selection charts. Process attributes are given on the axes. A given process maps out a characteristic field. (a) Chart for the relationship between part size and complexity of the shape. (b) Same chart with superimposed regions for attributes of a particular design. The design requirements determine the necessary processing attributes, thus establishing a search area. Processes that overlap the search area become candidates for selection. Source: Ref 12.2

selection must be linked with material behavior (Table 12.5) and constraints imposed by material hardness, melting temperature, and flow and fracture properties.

Much attention has been directed in recent years to developing design rules and computer methods for enhancing manufacturability through proper design. The intent is to reduce product cost and increase product quality by integrating the product and process concepts so that the best match is made between product and process requirements so as to ensure ease of manufacture. This approach is called design for manufacturing (DFM). Detailed information about DFM is provided in Ref 12.1, and additional information on DFM can be found in Ref 12.6, 12.9, and 12.10.

12.4 Costs and Related Aspects of Materials Selection

The decision on materials selection ultimately will come down to a trade-off between performance and cost. There is a spectrum of applications varying from those where performance is paramount (defense-related aerospace) to those where cost clearly predominates (household appliances). The total cost of a part includes the cost of the material, the cost of tooling (dies, fixtures), and the processing cost. The unit cost of a part, C, can be expressed by:

$$C = C_m + \frac{C_c}{n} + \frac{C_L}{\dot{n}}$$

(Eq 12.2)

where C_m is the material cost; C_c is the capital cost of plant, machinery, and tooling required to make the part; C_L is the labor cost per unit time; n is the batch size; and $\dot{n}$ is the production rate (parts produced per unit time). This cost equation leads to the familiar plot of a break-even point, a batch size (n) beyond which it is more economical to use a manufacturing process that has higher tooling or capital costs (Fig. 12.7). Below an economical lot size of approximately 60 pieces, the cost of tooling for making investment castings, for example, cannot be offset by savings in material and machining cost.

12.5 Methods of Materials Selection

The act of selecting a material is a form of engineering decision making. It has been shown that in the conceptual stage of design, it is important to consider a wide spectrum of materials. This list is narrowed considerably to enable embodiment or configuration design to be carried out, and it is narrowed still further to a single material for detail design. Clearly, design involves decision making. Many analytical methods have been introduced into engineering design to aid in the evaluation process, which results in a rational decision. These methods are discussed in greater detail in Ref 12.1 and Ref 12.12 to 12.15.

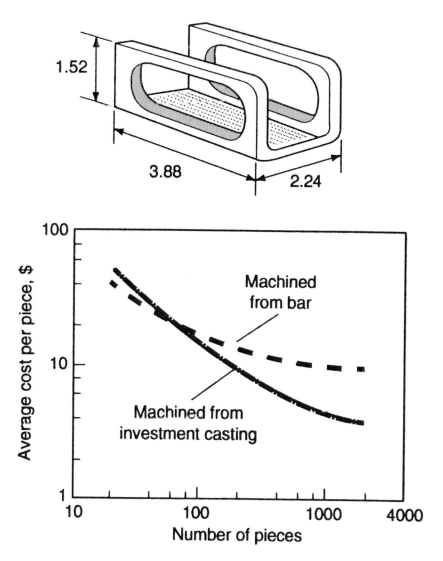

Fig. 12.7 Unit cost (1961 prices) of roller support made by machining from bar stock compared with machining a blank made by investment casting. Material, AISI 8630 steel. Weight of finished part, 0.5 lb. Part dimensions in inches. Source: Ref 12.11

Cost per Unit Property Method

In many materials selection problems, one property stands out as the most dominant service requirement. In this case, a reasonable screening method is to determine how much a material will cost to provide that requirement, that is, the cost of a function.

Following the development due to Ashby (Ref 12.12), it is desired that the material for a solid cylindrical tie-rod of length L to carry a tensile force P with a safety factor S be of a minimum mass. The mass is given by:

$$m = AL\rho \qquad\qquad\qquad \text{(Eq 12.3)}$$

where A is the cross-sectional area, and ρ is the density of the material from which it is made. The design requires that L and P are specified and cannot be changed; that is, they are constraints. However, the radius of the bar is a free variable so long as the bar supports the load P:

$$\frac{P}{A} = \frac{\sigma_f}{S} \qquad\qquad\qquad \text{(Eq 12.4)}$$

where σ_f is the failure strength (yield strength). Eliminating A between these two equations results in:

$$m = (SP)(L)\left(\frac{\rho}{\sigma_f}\right) \qquad\qquad\qquad \text{(Eq 12.5)}$$

Equation 12.5 shows that the performance of the tie-rod is described by three separate factors. The first term in brackets is the functional requirement. The second term is the geometric parameter. The last bracket is the material performance index. For this case of a bar in tension, the performance index M is given by:

$$M = \frac{\sigma_f}{\rho} \qquad\qquad\qquad \text{(Eq 12.6)}$$

The lightest tie-rod that will safely carry the load P without failing is that made from a material with the largest value of the performance index M. A similar calculation for a light, stiff tie-rod leads to the performance index $M = E/\rho$, where E is the Young's modulus. If the design criterion is to minimize cost rather than minimize weight, then the density in Eq 12.5 is replaced by C_ρ, where C is the cost per unit mass. Now M is given by:

$$M = \frac{\sigma_f}{C_\rho} \qquad\qquad\qquad \text{(Eq 12.7)}$$

Other part geometries and types of loading would result in different performance indices. Some of these are given in Table 12.7.

Weighted Property Index Method

Frequently, a design requires that more than one material property or requirement be optimized. The usual approach is to assign a weighting factor to each material requirement or property, depending on its importance.

Table 12.7 Performance indices

Part shape and loading for minimum weight	To maximize strength	To maximize stiffness
Tensile tie-rod: load, stiffness, and length are fixed; section area is variable	σ_f/ρ	E/ρ
Torsion bar: torque, stiffness, and length are fixed; section area is variable	$\sigma_f^{2/3}/\rho$	$G^{1/2}/\rho$
Beam: loaded externally or by self weight in bending; stiffness and length fixed; section area free	$\sigma_f^{2/3}/\rho$	$E^{1/2}/\rho$
Plate: loaded externally or by self weight in bending; stiffness, lenght and width fixed; thickness free	$\sigma_f^{1/2}/\rho$	$E^{1/3}/\rho$
Cylinder with internal pressure: elastic distortion, pressure and radius fixed; wall thickness free	σ_f/ρ	E/ρ

Design goal	Maximize
Thermal insulation: thickness given; minimum heat flux at steady state	$1/\lambda$
Thermal insulation: thickness given; minimum temperature rise after specified time	$C_p\rho/\lambda$
Thermal storage: maximize energy storage for given temperature and time	$(C_p\rho/\lambda)^{1/2}$
Minimum thermal distortion	λ/α
Maximum thermal shock resistance	$\sigma_f/E\alpha$
Pressure vessel: yield before break	K/σ_f

σ_f, failure strength; K, fracture toughness; E, Young's modulus; G, shear modulus; ρ, density; C_p, specific heat capacity; α, thermal expansion coefficient; λ, thermal conductivity. Source: Ref 12.2

Because this is mostly a subjective process, the weighting must be done with care to prevent bias or obtaining the answer you intended. The individual weighted property values are then summed to give a comparative weighted property index.

The importance of a screening property for which go/no-go limits can be set must be kept in mind. Such parameters as weldability and corrosion resistance must be delineated at the outset, and all materials that do not meet the go/no-go criteria are screened out. Other nondiscriminating parameters, such as availability or fabricability, should be used to screen out materials from the weighted property matrix or at least flag them for further deliberation.

The weighted properties index has the drawback of having to combine properties with different units. The best procedure is to normalize these differences by using a scaling factor. Scaling is a simple technique to bring all the different properties within one numerical range. Each property is scaled so that its highest numerical value does not exceed 100.

The scaled value of a property, β, is given by:

$$\beta = \frac{\text{Numerical value of property (100)}}{\text{Largest value in the list}} \qquad \text{(Eq 12.8)}$$

For properties such as cost, corrosion loss, and wear rate, for which a low value is best, the scale factor is formulated as:

$$\beta = \frac{\text{Lowest value of property (100)}}{\text{Numerical value of property}} \qquad \text{(Eq 12.9)}$$

Table 12.8 Using a subjective value to obtain a scaled property

Property	Alternative materials			
	A	B	C	D
Weldability	Excellent	Poor	Good	Fair
Relative rating	5	0	3	1

Source: Ref 12.3

For properties that are not readily expressed in numerical values, for example, weldability, some type of subjective rating is required (Table 12.8). After each material property and requirement is given a scaled value, the weighted property index for the material is simply the sum of the product of the scaled value, β, times the weighting factor, w, for each property:

$$\gamma = \text{weighted property index} = \sum \beta_i w_i \qquad \text{(Eq 12.10)}$$

Cost obviously is an important factor in selecting materials. It can be used as one of the properties, but because it is such an important factor, it may overdominate in the selection of weighting factors. One way to prevent this is to use cost as a modifier to the weighted property index instead of as one of the properties:

$$\gamma' = \frac{\gamma}{C_\rho} \qquad \text{(Eq 12.11)}$$

where C is the material cost per unit mass, and ρ is density.

Limits on Properties Method

In this method of materials selection, the performance requirements are divided into three categories:

- Lower-limit properties
- Upper-limit properties
- Target-value properties

For example, if it is required to have a stiff, light material, then a lower limit is put on Young's modulus and an upper limit is put on density. It has already been shown that this is the approach used when screening a large number of materials with a computer database. After screening, the remaining materials are those whose properties are above the lower limits, below the upper limits, and within the target values of the specified requirements.

To arrive at a merit parameter for each material, the properties are first assigned weighting factors using pairwise comparison. A merit parameter, p, is then calculated for each material using the relationship:

$$p = \left[\sum_{i=1}^{n_i} w_i \frac{Y_i}{X_i} \right]_l + \left[\sum_{j=1}^{n_u} w_j \frac{X_j}{Y_j} \right]_u + \left[\sum_{k=1}^{n_t} w_k \frac{X_k}{Y_k} - 1 \right]_t \qquad \text{(Eq 12.12)}$$

where l, u, and t stand for lower-limit, upper-limit, and target-value properties, respectively. Thus, n_l, n_u, and n_t are the number of lower-limit, upper-limit, and target-value properties; w_i, w_j, and w_k are the weighting factors on lower-limit, upper-limit, and target-value properties; X_i, X_j, and X_k are the candidate material lower-limit, upper-limit, and target-value properties; and Y_i, Y_j, and Y_k are the specified lower limits, upper limits, and target values. Based on Eq 12.12, the lower the value of the merit parameter, p, the more suitable the material. Cost can be treated as an upper-limit property and given the appropriate weight.

12.6 Conclusions

A competent job of materials selection should include consideration and documentation of the following:

- The problem, design, or redesign objective
- The underlying design criteria (primary, secondary, or supporting criteria; manufacturing method considerations; codes and standards)
- Analysis (preliminary concepts, modeling or simulation, optimization and trade-offs, design reliability, economics and cost)
- Alternatives considered, selection criterion, and decision methodology
- Reasons for the selection of the final material and manufacturing process

While some of these methods of materials selection require considerable calculation and may appear tedious at first reading, the growing use of computerized databases can greatly facilitate its use. However, a word of warning is in order. It is very easy for engineers to become quickly enamored with a numerical calculation scheme and lose sight of reality. Care must be taken to properly consider all nonquantifiable factors in the final decision and ensure that the numerical outcome is not unduly influenced by a hasty choice of weighting factors.

Of course, material suppliers want to sell their product and present the most favorable aspects of their material. However, they are amenable to working with their customers to select the right material system for the required application. It is important to seek their help early and often when implementing any new material system. When working with any new material, especially any polymer-based material, it is also important to closely examine the material safety data sheet.

12.7 Lightweight Materials Selection

The lightweight materials covered in this book included metals (aluminum, magnesium, beryllium, titanium, and titanium aluminide), engineering plastics, structural ceramics, and composites (polymer, metal, and ceramic matrix). The following sections give some general guidelines for selecting these materials.

Five lightweight metal alloys were covered in this book: aluminum, magnesium, beryllium, titanium, and titanium aluminide. Of these, two (beryllium and titanium aluminides) are highly specialized metals. Aluminum and magnesium are commodity metals with a wide range of applications. Although titanium is readily available, it is much more costly than both aluminum and magnesium and, as such, has been mainly used in high-value applications, such as aerospace and specialized corrosion-resistant applications.

Aluminum Alloys

For general construction applications, the 5xxx and 6xxx alloys offer the best combination of properties. The 5xxx series of alloys have the highest strengths of the non-heat-treatable alloys. They develop moderate strengths when work hardened, have excellent corrosion resistance even in saltwater, and have very high toughness even at cryogenic temperatures to near absolute zero. They are readily weldable by a variety of techniques, at thicknesses up to 20 cm (8 in.). The 5xxx-series alloys have relatively high ductilities, usually in excess of 25%. The ultimate tensile strength of the 5xxx alloys ranges from a low of 124 MPa (18 ksi) for 5005-O to a high of 310 MPa (45 ksi) for 5456-O. As a result, 5xxx alloys find wide application in building and construction; highway structures, including bridges, storage tanks, and pressure vessels; cryogenic tankage and systems for temperatures as low as –270 °C (–455 °F) or near absolute zero; and marine applications.

The 6xxx alloys are heat treatable to moderately high strength levels, have better corrosion resistance than the 2xxx and 7xxx alloys, are weldable, and offer superior extrudability. With a yield strength comparable to that of mild steel, 6061 is one of the most widely used of all aluminum alloys. Alloy 6063 is widely used for general-purpose structural extrusions because its chemistry allows it to be quenched directly from the extrusion press. Alloy 6061 is used where higher strength is required, and 6071 where the highest strength is required. A unique feature of the 6xxx alloys is their great extrudability, making it possible to produce in single shapes relatively complex architectural forms, as well as to design shapes that put the majority of the metal where it will most efficiently carry the highest tensile and compressive stresses. This feature is a particularly important advantage for architectural and structural members where stiffness-criticality is important. Alloy 6063 is perhaps the most widely used because of its extrudability; it is the first choice for many architectural and structural members.

Where the highest strengths are required, the 2*xxx* and 7*xxx* series of alloys are heat treatable and possess in individual alloys good combinations of high strength (especially at elevated temperatures), toughness, and, in specific cases, weldability. They are not as resistant to atmospheric corrosion as several other series and so usually are painted or Alclad for added protection. In addition, the overaged T7 tempers are available for the 7*xxx* alloys.

The high-strength 2*xxx* and 7*xxx* alloys are competitive on a strength-to-weight ratio with the higher-strength but heavier titanium and steel alloys and thus have traditionally been the dominant structural material in both commercial and military aircraft. However, in some of the newer models of commercial aircraft, carbon-fiber polymer-matrix composites are replacing these alloys. In addition, aluminum alloys are not embrittled at low temperatures and become even stronger as the temperature is decreased without significant ductility losses, making them ideal for cryogenic fuel tanks for rockets and launch vehicles. High-strength 2*xxx* and 7*xxx* alloys can be difficult to fusion weld. The copper content in some of the 2*xxx* and 7*xxx* alloys makes them susceptible to weld cracking during fusion welding operations. The rather newly developed friction stir welding process is being successfully used to weld them.

When high strength is not a consideration but electrical and thermal conductivity are important, the commercially pure 1*xxx* alloys should be used. The primary uses of the 1*xxx* series are applications in which the combination of extremely high corrosion resistance and formability are required, for example, foil and strip for packaging, chemical equipment, tank car or truck bodies, spun hollowware, and elaborate sheet metal work.

Electrical applications are one major use of the 1*xxx* series, primarily 1350, which has relatively tight controls on those impurities that may lower electrical conductivity. As a result, an electrical conductivity of 62% of the International Annealed Copper Standard is guaranteed for this material, which, combined with the natural light weight of aluminum, means a significant weight and therefore cost advantage over copper in electrical applications.

The 3*xxx* series of alloys are often used where higher strength levels than the 1*xxx* alloys are required along with good ductility and excellent corrosion resistance. Additions of magnesium provide improved solid-solution hardening, as in the alloy 3004, which is used for beverage cans, the highest single usage of any aluminum alloys, accounting for approximately ¼ of the total usage of aluminum. The 3*xxx* series of alloys are strain hardenable, have excellent corrosion resistance, and are readily welded, brazed, and soldered. Because of the ease and flexibility of joining, 3003 and other members of the 3xxx series are widely used in sheet and tubular form for heat exchangers in vehicles and power plants.

When an aluminum casting is more appropriate, the default position should be a 3xx.x alloy; that is, there should be a compelling reason for going with one of the other series of casting alloys. The 3xx.x series of castings is one of the most widely used because of the flexibility provided by the high silicon content and its contribution to fluidity, plus their response to heat treatment, which provides a variety of high-strength options. In addition, the 3xx.x series may be cast by a variety of techniques, ranging from relatively simple sand or die casting to very intricate permanent mold, investment castings, and the newer thixocasting and squeeze casting technologies. When the quantities are small, sand casting is probably the most cost-effective. As the quantities become larger, permanent mold and then high-pressure die casting should be considered. Do not assume that die casting can only produce small parts, as shown for the automotive cylinder head and transmission case in Fig. 12.8.

One of the biggest advantages of aluminum alloys is the ease with which they may be fabricated into any form. Often they can compete successfully with cheaper materials having a lower degree of workability. The metal can be cast by any method known. It can be rolled to any desired thickness, down to foil thinner than paper. Aluminum sheet can be stamped, drawn, spun, or roll formed. The metal also may be hammered or forged. Aluminum wire, drawn from rolled rod, may be stranded into cable of any desired size and type. There is almost no limit to the different profiles (shapes) in which the metal can be extruded. An additional advantage of the aluminum alloys is recyclability; they are one of the easiest and mostly widely recycled of the lightweight metals.

Two important enabling fabrication processes for aluminum have been developed in the last ten years. The first, high-speed machining, allows the

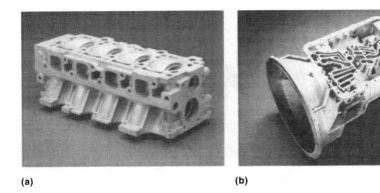

(a)   (b)

Fig. 12.8 Applications for aluminum alloy castings. (a) Alloy 319 automotive cylinder head. (b) Alloy 380 automotive transmission case. Source: Ref 12.16

design of weight-competitive high-speed-machined "assemblies" in which sheet metal parts that were formerly assembled with mechanical fasteners can now be machined from a single or several blocks of aluminum plate. The second process is friction stir welding, which is a solid-state process that operates by generating frictional heat between a rotating tool and the workpiece. The welds are created by the combined action of frictional heating and plastic deformation due to the rotating tool. The advantages of friction stir welding include:

- Ability to weld butt, lap, and T-joints
- Minimal or no joint preparation
- Ability to weld the difficult-to-fusion-weld 2xxx and 7xxx alloys
- Ability to join dissimilar alloys
- Elimination of cracking in the fusion and heat-affected zones
- Lack of weld porosity
- Lack of required filler metals
- No requirement for shielding gases

In general, the mechanical properties are better than for many other welding processes.

Magnesium Alloys

Magnesium alloys are even lighter than aluminum alloys, but they generally have lower strengths and lower moduli. However, when comparing their specific strengths and specific moduli, they become competitive with aluminum alloys. Other potential disadvantages of magnesium alloys include the following:

- Because they have a hexagonal close-packed (hcp) crystalline structure, room-temperature forming of the wrought alloys is rather limited, and most forming operations must be done at elevated temperatures.
- Magnesium alloys are usually more expensive than competing aluminum alloys.
- The corrosion resistance of magnesium alloys has been problematic.

Because the wrought forms are difficult to form, magnesium castings are more widely used. Magnesium castings are used in structural applications because of their low weight and good damping characteristics. Magnesium alloys have a very low viscosity, allowing the metal to flow long distances and fill narrow mold cavities. Their relatively low melting points allow the use of hot chamber die casting. Like aluminum castings, sand casting, permanent mold casting, and die casting are the most widely used processes. Due to their low modulus, magnesium castings have very good damping properties and, as such, are used in high-vibration environments.

The Mg-Ag-rare earth casting alloys with approximately 4 to 5% Y have good elevated-temperature properties. For example, alloy WE43 has a room-temperature tensile strength of 248 MPa (36 ksi) when heat treated to the T6 condition. This alloy maintains a tensile strength of 248 MPa (36 ksi) after long-term aging (5000 h) at 204 °C (400 °F). A relatively new alloy, Elektron 21, offers many of the advantages of WE43; however, the cost is lower and the castability is better. Instead of using yttrium, neodymium and gadolinium are used along with zinc and zirconium. The alloys WE54A and WE43A have high tensile strengths and yield strengths, and they exhibit good properties at temperatures up to 300 and 250 °C (570 and 480 °F), respectively. Alloy WE54A retains its properties at high temperature for up to 1000 h, whereas WE43A retains properties at high temperature in excess of 5000 h. These castings may be a solution when moderately high creep resistance is required.

The impurity elements nickel, iron, and copper must be held to low levels to minimize corrosion. More corrosion-resistant Mg-Al-Zn alloys were developed in the mid-1980s by using higher-purity starting materials and by limiting the amounts of iron (<0.005%), nickel (<0.001%), and copper (<0.015%). The low levels of nickel and copper are controlled by the purity of the starting materials, while the low iron levels are controlled with $MnCl_2$ additions. For example, the high-purity alloy AZ91E, due to its lower iron content, has improved corrosion resistance compared to the earlier alloy AZ91C. Some of the newer casting alloys approach the corrosion resistance of competing aluminum casting alloys.

Even the most corrosion-resistant magnesium alloys should be thoroughly protected in aggressive environments. For severely corrosive environments, for which maximum chemical, salt spray, and humidity resistance are required, specialized paints and coating techniques are used. Good-quality chromate or anodic pretreatments are required. Pretreatments should then be sealed with high-temperature baking organic resins. This is achieved by a process known as surface sealing, in which three coats of thinned resin are applied to a preheated component by spraying or, preferably, by dipping. Epoxy resin systems are preferred.

Titanium Alloys

Titanium is an attractive structural material due to its high strength, low density, and excellent corrosion resistance. However, even though titanium is the fourth most abundant element in the Earth's crust, due to its high melting point and extreme reactivity, the cost of titanium is high. The high cost includes both the mill operations (sponge production, ingot melting, and primary working) as well as many of the secondary operations conducted by the fabricator. The biggest impediment to the wider use of titanium is the high cost of the material itself and the high cost associated with fabrication and assembly. Lower-cost methods of sponge production are actively being pursued. Improved fabrication technologies include newer beta alloys with improved room-temperature formability, superplastic forming, superplastic forming combined with diffusion bonding, and direct metal deposition.

The high strength-to-weight ratio of titanium alloys allows them to replace steel in many applications requiring high strength and fracture toughness. With a density of 4.4 g/cm^3 (0.16 lb/in.^{3}), titanium alloys are only about one-half as heavy as steel and nickel-base superalloys, yielding excellent strength-to-weight ratios. Titanium alloys have much better fatigue strength than aluminum alloys and are frequently used for highly loaded bulkheads and frames in fighter aircraft. When the operating temperature exceeds approximately 130 °C (270 °F), aluminum alloys lose too much strength, and titanium alloys are often required. The corrosion resistance of titanium alloys is superior to both aluminum and steel alloys.

Titanium and its alloys are used primarily in two areas where the unique characteristics of these metals justify their selection: corrosion-resistant service and strength-efficient structures. For these two diverse areas, selection criteria differ markedly. Corrosion applications normally use low-strength commercially pure (CP) titanium mill products fabricated into tanks, heat exchangers, or reactor vessels for chemical processing, desalination, or power generation plants. In contrast, high-performance applications typically use high-strength titanium alloys in a very selective manner depending on factors such as thermal environment, loading parameters, available product forms, fabrication characteristics, and inspection and/or reliability requirements. For example, higher-strength alloys such as Ti-6Al-4V and Ti-3Al-8V-6Cr-4Mo-4Zr are being used for offshore drilling applications and geothermal piping. As a result of their specialized usage, alloys for high-performance applications normally are processed to more stringent and costly requirements than CP titanium for corrosion service.

Historically, titanium alloys have been used instead of iron or nickel alloys in aerospace applications because titanium saves weight in highly loaded components that operate at low to moderately elevated temperatures. Many titanium alloys have been custom designed to have optimal tensile, compressive, and/or creep strength at selected temperatures and, at the same time, have sufficient workability to be fabricated into mill products suitable for specific applications. During the life of the titanium industry, various compositions have had transient usage, but one alloy, Ti-6Al-4V, has been consistently responsible for approximately 45% of industry application. Ti-6Al-4V is unique in that it combines attractive properties with inherent workability, which allows it to be produced in all types of mill products, in both large and small sizes; good shop fabricability, which allows the mill products to be made into complex hardware; and the production experience and commercial availability that lead to reliable and economic usage. Thus, Ti-6Al-4V has become the standard alloy against which other alloys must be compared when selecting a titanium alloy for a specific application. Ti-6Al-4V also is the standard alloy selected for castings that must exhibit superior strength. For elevated-temperature applications, the most commonly used alloy is Ti-6Al-2Sn-4Zr-2Mo + Si. This alloy is primarily used for turbine components and in sheet form for afterburner structures and various high-temperature airframe applications.

Beryllium

Beryllium is a very lightweight metal that has good strength properties and a very high tensile modulus. These properties, combined with its attractive electrical and thermal properties, have led to its use in high-value aerospace electronic and guidance system applications. However, beryllium must be processed using powder metallurgy, which makes it costly. Like magnesium and titanium, its hcp crystalline structure greatly impairs its formability. Finally, beryllium powder and dust are toxic, which further increases its cost through the requirement for a controlled manufacturing environment.

The most noteworthy properties of beryllium are its density, elastic modulus, mass absorption coefficient, and the other physical properties. The specific modulus for beryllium is substantially greater than that of other aerospace structural materials such as steel, aluminum, titanium, or magnesium. Also, beryllium has the highest heat capacity (1820 J/kg · °C, or 0.435 Btu/lb · °F) among metals and a thermal conductivity (210 W/m · K, or 121 Btu/ft · h · °F) comparable to that of aluminum (230 W/m · K, or 135 Btu/ft · h · °F).

Due to its high cost and toxicity problems, beryllium is, and will remain, a niche material for highly specialized applications where no other solutions are available.

Titanium Aluminide Alloys

Titanium aluminide alloys are the least mature of the lightweight metals. Because of their low density and high-temperature strength retention, titanium aluminide alloy development has been driven primarily because of their potential for replacing higher-density nickel-base superalloys in aerospace engine components, primarily in the latter stages of the compressor and turbine sections. Because the gamma (γ) alloys (TiAl) offer the highest potential temperature usage, a significant amount of research and development has been conducted on these alloys.

A significant milestone in the development of these alloys occurred when General Electric certified and implemented TiAl in the new GEnx-1B engine that powers the Boeing 787 Dreamliner that entered service in 2011. The last two stages of the GEnx-1B low-pressure turbine are made from cast gamma-TiAl blades.

Much work remains to be done to further mature these alloys. Improved alloy chemistries, more reliable processing routes, and protective coatings need to be developed.

Engineering Plastics

Unreinforced plastics cannot compete with the lightweight metals on an absolute basis. However, because of their low density, their specific strengths are much more favorable. For example, nylon with a yield strength of only 80 MPa (12 ksi) and a density of 1.08 g/cm^3 (0.04 lb/in.^{3}) has a specific strength equal to a medium-carbon low-alloy steel heat treated to a yield strength of 550 MPa (80 ksi). In addition, engineering plastics offer some unique product benefits. These are usually physical properties, or combinations of physical properties, that allow vastly improved product performance. Most plastics are electrical and thermal insulators. In certain chemical environments, plastics offer outstanding resistance. In addition, certain amorphous plastics are transparent and have a high degree of optical clarity. All plastics are colorable to some extent. Many are capable of a nearly unlimited color spectrum while maintaining high visual quality.

If the application requires maximum efficiency of heat transfer, electrical conductivity, or nonflammable properties, then engineering plastics should not be considered. Applications under constant stress for which a close tolerance must be held must be scrutinized in terms of the effects of creep at the application temperature. The engineer also must be cognizant of the poor ultraviolet (UV) resistance of most, although not all, engineering plastics, and the resulting requirement of a UV absorber.

The first problem an engineer faces when selecting a plastic is trying to figure out what they are getting. For example, when comparing the prices of 6063 aluminum extrusions from different suppliers, the engineer can refer to a handbook to determine the properties; that is, they know what they are getting. This is not true for plastics. There are no industry standards for engineering plastics. Engineering plastics are mainly sold under trade names, and polypropylene from one supplier may be different from polypropylene from another supplier. Plastics are often modified by fillers, plasticizers, flame retardants, stabilizers, and impact modifiers, all of which substantially change mechanical properties. Engineering plastics have a broad range of mechanical properties, and interpreting those listed on a data sheet requires an understanding of the test methods and the method of reporting. Because many of the properties can change dramatically, depending on processing parameters, the engineer must allow for any such changes when selecting materials.

Plastics, unlike most metals, have no true proportional limit, and their stress-strain curves are not linear. Furthermore, the shape of the plastic stress-strain curve is greatly affected by the rate at which stress is applied, and stress-strain behavior is greatly affected by temperature. Because a plastic behaves in a manner that seems to combine elastic solid and highly viscous liquid behaviors, it is said to be viscoelastic. Fatigue is also more complex for plastics. When designing for fatigue resistance, it is necessary to ascertain that the candidate plastic fatigue data were measured using strain rates, temperatures, and deformation modes appropriate for the application.

To select the proper material, the engineer must understand the type of environment in which the plastic part will function, and then determine four critical parameters: operating temperature, stress level, chemical exposure, and adjoining materials. The effects of environmental factors on plastics are unique and sometimes crucial, because plastics are generally more sensitive to environmental factors than other engineering materials. Ultraviolet light causes degradation through a breakdown of molecular weight. To overcome this problem, ultraviolet absorbers are added to plastics. Plastics are also sensitive to chemicals. Alkalis and acids affect them through hydrolysis of polymer chains. The effects of solvents on stress cracking are more difficult to predict, and experimental results should be consulted.

The processing of semicrystalline and amorphous polymers is very different. Semicrystalline polymers must be molten for sufficient flow in processes such as injection molding. In the molten state, the structure is amorphous, and it is when the structure is cooled through the melting point that crystallization takes place. The processing window for amorphous polymers is much wider, because they soften when heated and can attain sufficient fluidity for processing. This softening range is generally greater than the melting range for semicrystalline polymers. This behavior is illustrated in Fig. 12.9, where the tensile modulus of an amorphous and a semicrystalline polymer is plotted against temperature. For amorphous polymers, the modulus is highest at low temperatures, in the glassy region. Here, little intermolecular movement is possible. As the temperature increases, secondary intermolecular forces are overcome, the polymer

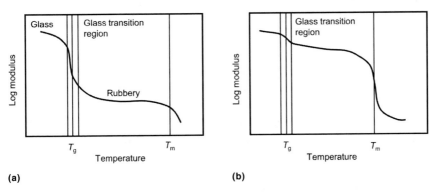

(a) **(b)**

Fig. 12.9 Schematic diagrams showing the effects of temperature on the modulus of (a) amorphous and (b) semicrystalline thermoplastics. T_g, glass transition temperature; T_m, melting temperature

chains have sufficient thermal energy to rotate, and the modulus falls off rapidly through a transition region, characterized by the glass transition temperature, T_g.

The modulus then levels off with a further increase in temperature as the polymer chains remain entangled, although local sections are mobile. In this region, the material can exhibit large elongations under very low load; hence, it is often referred to as the rubbery region. With a further increase in temperature, large-scale molecular movement can take place and the polymer becomes a viscous melt, and the melting temperature, T_m, is often indistinct. Semicrystalline polymers exhibit a similar shape to the curve, although the presence of a T_g depends on the presence of amorphous regions in the microstructure. As the crystalline regions break down, the melting temperature, T_m, is more distinct compared with amorphous polymers. These regions are significant in determining the useful temperature range of thermoplastics. Amorphous plastics, such as acrylonitrile-butadiene-styrene, must be used below the T_g. Semicrystalline polymers are useful above their T_g; for these materials, it is the melting temperature, T_m, that limits their structural stability, although this does depend on the level of crystallinity.

When a thermoplastic is reinforced by fibers, the strength can be greatly increased (Fig. 12.10), although at the expense of some ductility and, sometimes, impact strength. The most commonly used fiber reinforcement is glass, and in articles produced by injection molding, the fibers are of necessity short, little more than a few millimeters in length. This naturally introduces variable anisotropy of properties, and published figures for strength and stiffness frequently fail to take this into account, so that the properties exhibited by real articles may be lower than anticipated. The glass content may vary from 15 to 40% by volume. Fiber reinforcement frequently exacerbates problems of shrinkage and distortion, and where good maintenance of shape and dimensional accuracy is essential, the fibers may be replaced, partially or wholly, with glass beads. The mechanical properties are then, in general, less good, because reinforcement with glass beads is less effective than that produced by fibers.

Some recommended plastics for different applications are shown in Table 12.9. While incomplete, this table should at least provide a beginning. Reinforcements can be used to increase mechanical strength, maximum use temperature, impact resistance, stiffness, mold shrinkage, and dimensional stability.

Generally, resin prices increase with improved mechanical and thermal properties. When there is no clear-cut material of choice, plastics designers generally follow the practice of looking for the lowest-cost material that will meet the product requirements. If there is a reason that polymer is not acceptable, designers start working up the cost ladder until they find one that will fulfill their needs.

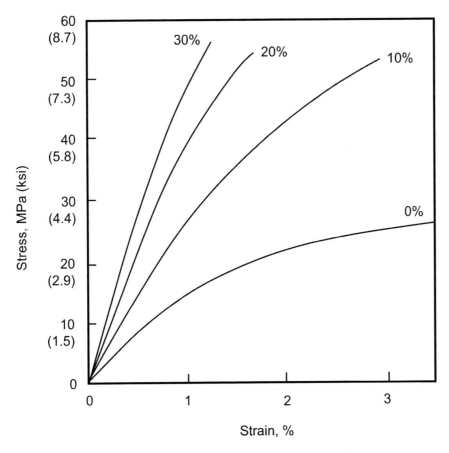

Fig. 12.10 Tensile stress-strain curves for polypropylene with different percentages of glass fibers

In thermoplastics, there are the so-called commodity resins. These are the low-cost resins used in great volume for housewares, packaging, toys, and so on. This group includes polyethylene, polypropylene, polystyrene, and polyvinyl chloride. Reinforcements can improve the properties of these resins at a moderate additional cost. A lower-priced resin with reinforcement will often provide properties comparable to a more expensive resin. It should be remembered that plastic materials are petroleum based, and their costs will fluctuate with the market price for crude oil.

Thermosets usually provide higher mechanical and thermal properties. However, most of the processes used to fabricate thermoset parts are slower and more limited in design freedom than the thermoplastic processes. Furthermore, the opportunity to use 100% of the material that thermoplastics provide is simply not available with thermosets, because the regrind cannot be reused. Recycling possibilities are far more limited for thermosets for the same reason. Nonetheless, glass-fiber-reinforced

Table 12.9 Plastic materials selection guidelines

Property	Recommendation
Abrasion resistance	Nylon
Compressive strength	Polyphthalamide, phenolic (glass), epoxy, melamine, nylon, thermoplastic polyester (glass), polyimide
Dielectric constant (high)	Phenolic, polyvinyl chloride (PVC), fluorocarbon, melamine, alkyd, nylon, polyphthalamide, epoxy
Dielectric strength (high)	PVC, fluorocarbon, polypropylene, thermoplastic polyester, nylon (glass), polyolefin, polyethylene
Dissipation factor (high)	PVC, fluorocarbon, phenolic, thermoplastic polyester, nylon, epoxy, diallyl phthalate, polyurethane
Distortion resistance under load	Thermosetting laminate
Elastic modulus (high)	Melamine, urea, phenolic
Elastic modulus (low)	Polyethylene, polycarbonate, fluorocarbon
Electrical resistivity (high)	Polystyrene, fluorocarbon, polypropylene
Elongation at break (high)	Polyethylene, polypropylene, silicon, ethylene vinyl acetate
Elongation at break (low)	Polyether sulfone, polycarbonate (glass), nylon (glass), polypropylene (glass), thermoplastic polyester, polyetherimide, vinyl ester, polyetheretherketone, epoxy, polyimide
Flexural modulus (stiffness)	Polyphenylene sulfide, epoxy, phenolic (glass), nylon (glass), polyimide, diallyl phthalate, polyphthalamide, thermoplastic polyester
Flexural strength (yield)	Polyurethane (glass), epoxy, nylon (carbon and glass fiber), polyphenylene sulfide, polyphthalamide, polyetherimide, polyetheretherketone, polycarbonate (carbon fiber)
Friction, low coefficient	Fluorocarbon, nylon, acetal
Hardness (high)	Melamine, phenolic, polyimide, epoxy
Impact strength (high)	Acrylic, polycarbonate, acrylonitrile-butadiene-styrene (ABS)
Moisture resistance (high)	Polyethylene, polypropylene, fluorocarbon, polyphenylene sulfide, thermoplastic polyester, polyphenylene ether, polycarbonate
Softness	Polyethylene, silicone, PVC, thermoplastic elastomer, polyurethane, ethylene vinyl acetate
Tensile strength (high)	Glass-filled thermoplastics, glass-filled thermosets, liquid crystalline polymer, polyetheretherketone
Thermal conductivity (low)	Polypropylene, PVC, ABS, polyphenylene oxide, polybutylene, acrylic, polycarbonate, thermoplastic polyester, nylon
Thermal expansion coefficient (low)	Polycarbonate (carbon or glass filler), phenolic (glass), nylon (carbon or glass filler), thermoplastic polyester (glass filler), polyphenylene sulfide (carbon or glass filler), polyetherimide, polyetheretherketone, polyphthalamide, alkyd, melamine
Transparency, permanent (high)	Acrylic, polycarbonate
Weight (low)	Polypropylene, polyethylene, polybutylene, ethylene vinyl acetate, ethylene methyl acrylate
Whiteness retention (high)	Melamine, urea

Source: Adapted from Ref 12.17

thermoset polyester is the material of choice for many severe-environment outdoor applications such as boats and truck housings.

Structural Ceramics

Due to the stability of their strong bonds, structural ceramics normally have high melting temperatures and excellent chemical stability in many hostile environments. Due to the absence of conduction electrons, ceramics are usually good electrical and thermal insulators. However, ceramics are inherently hard and brittle materials that, when loaded in tension, have almost no tolerance for flaws. As a material class, few ceramics have tensile strengths above 172 MPa (25 ksi), while the compressive strengths may be 5 to 10 times higher than the tensile strengths.

Under an applied tensile load at room temperature, ceramics almost always fracture before any plastic deformation can occur. Stress concentrations leading to brittle failure can be minute surface or interior cracks (microcracks), or internal pores, which are virtually impossible to eliminate or control. Fracture toughness values for ceramic materials are much lower than for metals. There is also considerable scatter in the fracture strength for ceramics, which can be explained by the dependence of fracture strength on the probability of the existence of a flaw that is capable of initiating a crack. Therefore, size or volume also influences fracture strength; the larger the size, the greater the probability for a flaw and the lower the fracture strength. The development of toughened ceramics has improved their fracture toughness; however, they must still be considered brittle materials.

The primary properties of interest for structural applications are strength (room and elevated temperature), modulus (elastic, shear, and bulk), fracture toughness, Poisson's ratio, creep and creep-rupture behavior, hardness, tribological properties such as abrasion resistance and friction, chemical resistance, and thermal shock resistance. Optical absorption and index of refraction may also be important where appearance is a consideration, and electrical properties may be important for a high-temperature sensor application. The favorable thermal, chemical, and tribological properties of some of the structural ceramics can also be achieved by the use of ceramic coatings on other materials, such as metals.

The high-temperature properties of advanced ceramics make them attractive for a variety of applications. For instance, the use of ceramic heat exchangers in waste heat recovery has received extensive attention over the last 10 to 15 years. Commercial products now include plate-fin and tubular recuperators. Two similar applications are process heat exchangers and power generation heat exchangers.

Engine components make use of both the temperature resistance and wear resistance of advanced ceramics. Ceramic materials and coatings are currently under development for application in conventional internal combustion engines, adiabatic diesel engines, and advanced gas turbines. The major advantages include reduced inertia, friction, and fuel consumption; lower mass; and improved wear resistance. Sialon, SiC, and silicon nitride are the primary material candidates in these engines.

Ceramics for aerospace applications have traditionally focused on coatings, used for their thermal insulating properties. However, considerable effort has been devoted in recent years to developing ceramic gas turbines for aerospace vehicles. Here the materials issues, especially reliability, are similar to those for the automotive gas turbine.

One of the major advantages of ceramics over other materials is their excellent wear resistance. They have been used in the minerals processing industry for over 30 years. Ceramic metal-cutting tools offer increased productivity, compared to their metallic counterparts, because of higher temperature capability as well as increased wear resistance. Other applications for ceramics that make use of wear resistance include dies for hot extrusion of metal rods and tubes, inserts for table setting, and paper lapping dies, seals, and bearings.

Ceramic materials can increase the performance of bearings and extend the range of their operation to higher temperatures and corrosive environments. Hybrid ceramic bearings (steel rings with ceramic balls or rollers) are now used in machine tool spindles and other high-speed or high-precision equipment. Other applications include turbo-molecular pumps, dental drills, and specialty instruments. Stresses in these bearings are often very high and are localized at or near the surface. The normal contact stress at the surface can be as high as 4 GPa (580 ksi).

Ceramic cutting tools have properties that distinguish them from traditional steel and tungsten carbide cutting materials, including:

- They are more chemically inert.
- They have a higher resistance to abrasive wear.
- They are capable of superior heat dispersal during the chip-forming process.

Collectively, these properties permit the user to increase the rate of metal removal while often obtaining longer tool life. Reduced operating time to produce a finished part, and reduced machine downtime through less frequent insert indexing and replacement, result in an overall improvement in productivity and part cost reduction.

Composites

Composite materials may have discontinuous or continuous reinforcements and matrices of polymers, metals, or ceramics. In general, discontinuous reinforced composites are less expensive than continuously reinforced composites. Polymer-matrix composites are more affordable than metal- or ceramic-matrix composites. Thus, discontinuously reinforced polymer-matrix composites are the mostly widely used.

Polymer-Matrix Composites

If the application is for a low-temperature commercial application, E-glass/polyester should be the default selection. Glass fibers are the most widely used reinforcement due to their good balance of mechanical properties and low cost. E-glass is the most common glass fiber and is used extensively in commercial composite products. E-glass is a low-cost, high-density, low-modulus fiber that has good corrosion resistance and good handling characteristics. Polyesters cure by addition reactions in which unsaturated carbon-carbon double bonds (C=C) are the locations where cross linking occurs. The properties of the resultant polyester are strongly dependent on the cross linking or curing agent used. One of the main advantages of polyesters is that they can be formulated to cure at either room or elevated temperature, allowing great versatility in their processing. Vinyl esters are another choice. Vinyl esters are very similar to polyesters but only have reactive groups at the ends of the molecular chain. This results in lower cross-link densities; therefore, vinyl esters are normally tougher than the more highly cross-linked polyesters. In addition, because the ester group is susceptible to hydrolysis by water and vinyl esters have fewer ester groups than polyesters, they are more resistant to degradation from water and moisture.

If toughness is a major consideration, the aramid fibers are noted for their toughness and damage tolerance. The good energy-absorbing properties of aramid fibers make them useful for ballistics, tires, ropes, cables, asbestos replacement, and protective apparel. Aramid-fiber composites offer good tensile properties at a lower density than glass-fiber composites but at a higher cost than glass-fiber composites. However, their compressive properties are extremely poor, which limits them to tension-dominated designs. In addition, due to the relatively poor fiber-to-matrix bond strength, the matrix-dominated properties, such as in-plane and interlaminar shear and transverse tension, are somewhat lower than that of glass- and carbon-fiber composites. Aramid fibers are normally used with epoxy resin systems. Although lower cost than epoxies, polyesters generally have lower temperature capability, lower mechanical properties, and inferior weathering resistance and exhibit more shrinkage during cure. Epoxies have an excellent combination of strength, adhesion, low shrinkage, and processing versatility. Epoxy resin systems usually cure at 120 or 175 °C (250 or 350 °F), with service temperatures in the range of 80 to 120 °C (180 to 250 °F), respectively.

Carbon fiber offers the best combination of properties but is also more expensive than either glass or aramid. Carbon fiber has a low density, a low coefficient of thermal expansion, and is conductive. It is structurally very efficient and exhibits excellent fatigue resistance. It is also brittle (strain to failure less than 2%) and exhibits low impact resistance. Being conductive, it will cause galvanic corrosion if placed in direct contact with more anodic metals such as aluminum. Carbon and graphite fiber is available in a wide range of strength and stiffness, with strengths ranging from 2 to 7 GPa (300 to 1000 ksi) and moduli ranging from 207 to 1000 GPa (30 to 145×10^6 psi). With this wide range of properties, carbon fiber is frequently classified as high strength, intermediate modulus, or high modulus. The terms *high-strength* and *intermediate-modulus* carbon fibers are a little deceiving. High-strength fibers, with a strength level of approximately 3.45 to 4.14 GPa (500 to 600 ksi) and moduli of approximately 207 to 221 GPa (30 to 32 × 10⁶ psi) were developed before intermediate-modulus carbon fiber. While intermediate carbon fiber does have a higher modulus (276 to 290 GPa, or 40 to 42 × 10⁶ psi), the tensile strength is also higher (4.48 to 5.17 GPa, or 650 to 750 ksi). Standard-modulus polyacrylonitrile (PAN) fibers have good properties with lower cost, while higher-modulus PAN fibers are higher cost because high processing temperatures are required. Tow sizes can range from as small as 1000 fibers/tow up to greater than 200,000 fibers/tow. The cost of carbon fibers is dependent on the manufacturing process, the type of precursor used, the final mechanical properties, and the tow size.

Epoxy-matrix composites are generally limited to service temperatures of approxiamately 121 °C (250 °F). For higher usage temperatures, bismaleimides, cyanate esters, and polyimides are available. Bismaleimide and cyanate ester composites are useful at temperatures up to approximately 177 to 204 °C (350 to 400 °F). For even higher temperatures, up to 260 to 316 °C (500 to 600 °F), polyimides must be used. While cyanate esters and bismaleimides process in a similar manner to epoxies, polyimides are generally condensation-curing systems that require high processing temperatures and pressures. Because the condensation reaction gives off either water or alcohols, polyimides are very prone to developing voids and porosity. In addition, they are brittle systems that are subject to microcracking.

High-modulus pitch-based carbon/epoxy is used in spacecraft applications where extremely lightweight, stiff, and dimensionally stable structures are required. Pitch-based graphite fibers have high values of thermal conductivity, for example, values of 900 to 1000 W/m · K as compared to values of only 10 to 20 W/m · K for PAN-based carbon fibers. These high thermal conductivities are used to remove and dissipate heat in space-based structures. However, all of the properties except for the longitudinal modulus are lower for high-modulus materials compared to the standard high-strength materials. High-modulus materials are also considerably more expensive than the standard-modulus materials. In addition, more expensive cynate ester resins are often specified, because they absorb less moisture and are thus more dimensionally stable and less prone to outgassing in the near-vacuum atmosphere of space. Therefore, high-modulus carbon- and graphite-fiber composites should be restricted in usage to specialized applications.

When examining data sheets for carbon-fiber composites, the unidirectional properties (0° and 90°) are usually the only ones listed. While useful for comparison purposes, most real composite structures have laminate orientations that are quasi-isotropic or near quasi-isotropic. Because the matrix-dependent properties of unidirectional laminates are very low, it is necessary to balance the load-carrying plies fairly evenly in the 0°, ±45°, and 90° directions. In addition, as shown in the Fig. 12.11 envelope, a fair balance of plies in the 0°, ±45°, and 90° directions is required to carry the bearing loads from mechanical fasteners.

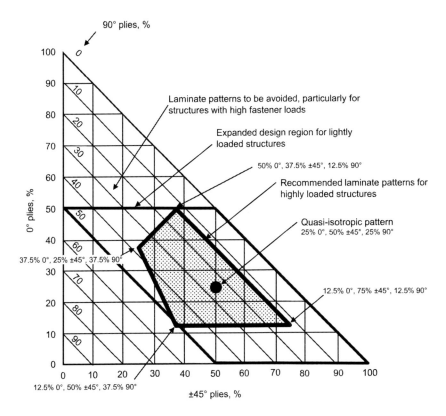

Fig. 12.11 Recommended design envelope for mechanically fastened composites. Note: Lightly loaded minimum-gage structures tend to encompass a greater range of fiber patterns than indicated, because of the unavailability of thinner plies. Source: Adapted from Ref 12.18

A comparison of the unidirectional and quasi-isotropic properties for a high-strength and intermediate-modulus carbon/epoxy composite is given in Table 12.10. In addition to the lower properties that result from cross-plying in the quasi-isotropic laminates as compared to the 0° strengths for the unidirectional laminates, note that while the tension properties of the unidirectional laminates have increased significantly with the development of intermediate-modulus carbon fibers, there has been almost no improvement in the compression strength.

There are a multitude of material product forms used in composite structures. The fibers can be continuous or discontinuous. They can be oriented or disoriented (random). They can be furnished as dry fibers or preimpregnated with resin (prepreg). Not all fiber or matrix combinations are available in a particular material form, because the market drives availability. In general, the more operations required by the supplier, the higher the cost. For example, prepreg cloth is more expensive than dry woven cloth. While complex dry preforms may be expensive, they can translate into lower fabrication costs by reducing or eliminating hand lay-up costs.

Table 12.10 Comparison of unidirectional and quasi-isotropic carbon/epoxy laminates

Property	High-strength carbon/epoxy(a)	Intermediate-modulus carbon/epoxy(a)
0° tensile strength, ksi	290	350
0° tensile modulus, 10^6 psi	19	25
90° tensile strength, ksi	11.5	11.5
90° tensile modulus, 10^6 psi	1.3	1.3
0° compression strength, ksi	190	230
0° compression modulus, 10^6 psi	17	22
Axial tensile strength(b), ksi	85	200
Axial tensile modulus(b), 10^6 psi	8	9
Axial compression strength(b), ksi	85	85
Axial compression modulus(b), 10^6 psi	8	9

(a) 60 fiber vol%. (b) Axial properties are for quasi-isotropic laminates.

If structural efficiency and weight are important design parameters, then continuous reinforced product forms are normally used, because discontinuous fibers yield lower mechanical properties.

Metal-Matrix Composites

Metal-matrix composites (MMCs) offer a number of advantages compared to their base metals, such as higher specific strengths and moduli, higher elevated-temperature resistance, lower coefficients of thermal expansion, and, in some cases, better wear resistance. However, due to their high cost, commercial applications for MMCs are sparse. There are some limited uses for discontinuously reinforced MMCs but almost no current applications for continuously reinforced MMCs.

Most of the commercial work on MMCs has focused on aluminum as the matrix metal. The combination of light weight, environmental resistance, and useful mechanical properties make aluminum attractive. The melting point is high enough to meet many applications yet low enough to allow reasonable processing temperatures. Also, aluminum can accommodate a variety of reinforcing agents. Although much of the early work on aluminum MMCs concentrated on continuous fibers, most of the present work is focused on discontinuously reinforced (particulate) aluminum MMCs because of their greater ease of manufacture, lower production costs, and relatively isotropic properties. The most common reinforcement materials in discontinuously reinforced aluminum (DRA) composites are SiC and Al_2O_3, although silicon nitride (Si_3N_4), TiB_2, graphite, and others have also been used in some specialized applications. For example, graphite/aluminum MMCs have been developed for tribological applications due to their excellent antifriction properties, wear resistance, and antiseizure characteristics. Typical fiber volumes of DRAs are usually limited to 15 to 25%, because higher volumes result in low ductility and fracture toughness. The DRAs exhibit high stiffness, low density, high hardness, adequate toughness at volume percentages less than 25%, and relatively low cost. Discontinuously reinforced aluminum is usually manufactured by melt incorporation during casting or by powder blending and consolidation.

Continuous fiber reinforcements include graphite, silicon carbide (SiC), boron, aluminum oxide (Al_2O_3), and refractory metal wires. Continuous reinforcements usually consist of small multifilaments or much larger monofilaments. Small multifilaments, typically less than 20 μm (0.8 mil) in diameter, usually consist of tows or bundles with thousands of fibers and are derived from textile materials, while monofilaments are much larger, ranging from 100 to 140 μm (4.0 to 5.5 mils) in diameter, and are made one filament at a time by chemical vapor deposition on a small substrate, normally tungsten wire or carbon fiber.

Large, stiff SiC monofilament fibers have been evaluated for high-value aerospace applications. Early work was conducted in the 1970s with boron-reinforced aluminum for aircraft/spacecraft applications. Multifilament fibers, such as carbon and the ceramic fibers Nextel (alumina base) and Nicalon (silicon carbide), have also been used in aluminum and matrices; however, the smaller and more numerous multifilament tows are difficult to impregnate using solid-state processing techniques, such as diffusion bonding, because of their small size and tightness of their tow construction. In addition, carbon fiber readily reacts with aluminum and during processing and can cause these matrices to galvanically corrode during service.

Due to the high cost of continuous fiber aluminum-matrix composites and their limited temperature capabilities, the emphasis has shifted to continuous fiber titanium-matrix composites. Work in the 1990s concentrated on SiC monofilaments in titanium for the National Aerospace Plane. Continuous monofilament titanium-matrix composites offer the potential for strong, stiff, lightweight materials at usage temperatures as high as approximately 815 °C (1500 °F). The principal applications for this class of materials would be for hot structure, such as hypersonic airframe structures, and for replacing superalloys in some portions of jet engines. However, the use of titanium-matrix composites has been restricted by the high cost of the materials, fabrication, and assembly procedures.

The primary MMC fabrication processes are often classified as either liquid-phase or solid-state processes. Liquid-phase processing is generally considerably less expensive than solid-state processing. Characteristics of liquid-phase-processed discontinuous MMCs include low-cost reinforcements, such as silicon carbide particles, low-temperature-melting matrices such as aluminum and magnesium, and near-net-shaped parts. Liquid-phase processing results in intimate interfacial contact and strong reinforcement-to-matrix bonds but can also result in the formation of brittle interfacial layers due to interactions with the high-temperature liquid matrix. Liquid-phase processes include various casting processes, liquid metal infiltration, and spray deposition. However, because continuous aligned fiber reinforcement is normally not used in liquid-state processes, the strengths and stiffness are lower.

Solid-state processes, in which no liquid phase is present, are usually associated with some type of diffusion bonding to produce final consolidation, whether the matrix is in a thin sheet or powder form. Although the processing temperatures are lower for solid-state diffusion bonding, they are often still high enough to cause significant reinforcement degradation. In addition, the pressures are almost always higher for the solid-state processes. The choice of a fabrication process for any MMC is dictated by many factors, the most important being preservation of reinforcement strength, preservation of reinforcement spacing and orientation, promotion of wetting and bonding between the matrix and reinforcement, and minimization of reinforcement damage, primarily due to chemical reactions between the matrix and reinforcement.

Ceramic-Matrix Composites

Carbon-carbon composites are the oldest and most mature of the ceramic-matrix composites. They were developed in the 1950s by the aerospace industry for use as rocket motor casings, heat shields, leading edges, and thermal protection. For high-temperature applications, carbon-carbon composites offer exceptional thermal stability in nonoxidizing atmospheres, along with low densities. Their low thermal expansion and range of thermal conductivities provide high thermal shock resistance. In vacuum and inert gas atmospheres, carbon is an extremely stable material, capable of use to temperatures exceeding 2204 °C (4000 °F). However, in oxidizing atmospheres, it starts oxidizing at temperatures as low as 350 °C (660 °F). Therefore, carbon-carbon composites for elevated-temperature applications must be protected with oxidation-resistant coating systems, such as silicon carbide that is overcoated with glasses. The silicon carbide coating provides the basic protection, while the glass overcoat melts and flows into coating cracks at elevated temperature. In addition, oxidation inhibitors, such as boron, are often added to the matrix to provide additional protection.

Other ceramic-matrix materials include glasses, glass-ceramics, oxides (e.g., alumina, or Al_2O_3), and nonoxides (e.g., silicon carbide, or SiC). The majority of ceramic materials are crystalline with predominantly ionic bonding along with some covalent bonding. These bonds, in particular the strongly directional covalent bond, provide a high resistance to dislocation motion and go a long way in explaining the brittle nature of ceramics. Because ceramics and carbon-carbon composites require extremely high processing temperatures compared to polymer- or even metal-matrix composites, ceramic-matrix composites are difficult and expensive to fabricate.

Reinforcements for ceramic-matrix composites are usually carbon, oxide, or nonoxide ceramic fibers, whiskers, or particulates. Most high-performance oxide and nonoxide continuous fibers are expensive, further leading to the high cost of ceramic-matrix composites. The cost and great difficulty of consistently fabricating high-quality ceramic-matrix composites has greatly limited their applications to date. Reinforcements in the form of particulates and platelets are also used. Silicon carbide, silicon nitride, and alumina are the most commonly used whiskers for ceramic-matrix composites.

The SiC monofilaments are large, so they can tolerate some surface reaction with the matrix without a significant strength loss. However, their large diameter also inhibits their use in complex structures due to their large diameter and high stiffness, which limits their ability to be formed over tight radii. Ceramic textile multifilament fibers, in tow sizes ranging from 500 to 1000 fibers, are available that combine high-temperature properties with small diameters (0.01 to 0.02 mm, or 0.4 to 0.8 mil), allowing them to be used for a wide range of manufacturing options, such as filament winding, weaving, and braiding.

Oxide-based fibers, such as alumina, exhibit good resistance to oxidizing atmospheres, but, due to grain growth, their strength retention and creep resistance at high temperatures is poor. Oxide fibers can have creep rates of up to 2 orders of magnitude greater than nonoxide fibers. Nonoxide fibers, such as SiC, have lower densities and much better high-temperature strength and creep retention than oxide fibers but have oxidation problems at high temperatures. As a class, oxide fibers are poor thermal and electrical conductors, have higher coefficients of thermal expansion (CTE), and are denser than nonoxide fibers. Due to the presence of glass phases between the grain boundaries and as a result of grain growth, oxide fibers rapidly lose strength in the 1205 to 1315 °C (2200 to 2400 °F) range. The oxide-based fibers are typically more strength-limited at high temperatures than the nonoxide fibers; however, oxide fibers have a distinct advantage in having a greater compositional stability in high-temperature oxidizing environments. While fiber creep can be a problem with both oxide and nonoxide fibers, it is generally a bigger problem with the oxide fibers. Fiber grain size is a compromise, with small grains contributing to higher strength, while large grains contribute to better creep resistance.

For temperatures exceeding 980 °C (1800 °F), candidate matrices are carbon, silicon carbide, silicon nitride, and alumina. Although these compositions are possible, the performance requirement that the matrix material have a CTE very close to that of the commercially available carbon-, silicon carbide-, and alumina-base fibers effectively eliminates silicon nitride and alumina as matrix choices for the silicon carbide fibers, and silicon carbide and silicon nitride as matrix choices for the alumina-base fibers. The lack of high thermal conductivity and the availability of oxide-based fibers that are creep resistant for long times above 980 °C (1800 °F) are two factors currently limiting the commercial viability of oxide/oxide ceramic-matrix composites.

Interfacial, or interphase, coatings are often required to protect the fibers from degradation during high-temperature processing, aid in slowing oxidation during service, and provide the weak fiber-to-matrix bond required for toughness. The coatings, ranging in thickness from 0.1 to 1.0 μm, are applied directly to the fibers prior to processing, usually by chemical vapor deposition.

12.8 Lightweight Materials in the Transportation (Automotive) Industry

The automotive industry is arguably the largest and most complex undertaking in industrial history. Its highly evolved production methods satisfy the conflicting demands of price, safety, performance, reliability, emissions, and market appeal. However, due to ever-increasing government regulations on vehicle safety as well as pressure from consumers concerned with fuel economy and environmentalists concerned with emissions, car and truck manufacturers, and their suppliers, must turn to new technologies to help them achieve their goals of making vehicles stronger, lighter, and more efficient.

Worldwide, vehicles consume oil at a rate of nearly 30 million barrels per day. United States vehicle registrations have climbed steadily from 110 million in 1940 to more than 200 million today. Global growth has been much more rapid, expanding from 130 million to more than 450 million during the same time period. *Over the next 50 years, worldwide numbers are likely to increase 3 to 5 times, resulting in 2 to 3.5 billion light vehicle registrations.* While lightweight materials do, and will increasingly, play an important role in all aspects of the transportation industry (rail, ship, and aerospace), it is the ultimate goal of lightweight materials producers throughout the world to increase their penetration into the automotive sector.

Environmental and sustainability consequences of the growing worldwide transportation sector will need to be addressed with new technological solutions. Without these improvements, increased consumption of petroleum will result in increased emissions of greenhouse gases, which contribute to global climate change. Currently, the combustion of fossil fuels in the U.S. transportation sector accounts for one-third of all CO_2 emissions; three-quarters of those emissions are due to road transportation.

Public policy is demanding reductions in fuel consumption and resultant greenhouse gas emissions. European passenger car manufacturers have already taken steps to increase fuel efficiency and thereby reduce CO_2 emissions, averaging 187 g CO_2/km in 1995 to 140 g/km in 2008. More aggressive targets are planned for a further reduction to 120 g/km by 2012. This target emission level would translate to an average fuel economy slightly better than 45 miles per gallon. In the United States, the Energy Independence and Security Act, the so-called Energy Bill, will require the

U.S. auto industry to raise its Corporate Average Fuel Economy standards by 40%, to 35 miles per gallon by 2020.

To achieve these efficiency improvements, automakers are developing a wide array of advanced technologies. These include improvements to engines, drivetrains, transmissions, and aerodynamics and use of hybrid or full electric power systems in conjunction with alternative fuels. However, the most fundamental and effective means of improving efficiency is accomplished by the reduction of vehicle mass. Reducing vehicle size and weight can significantly reduce fuel consumption. Every 10% of weight reduced from the average new car or light truck can cut fuel consumption by 6.9%.

These mass reductions can be accomplished by vehicle design changes, vehicle downsizing, or by lightweight material substitution. Simple vehicle design changes can significantly reduce mass. These changes range from removal of unnecessary systems, such as heated power seating, to more aggressive part integration and optimization techniques. However, consumers may object to the loss of certain "creature comforts" and the phenomenon of "feature creep"—adding more and more functionality year after year, which tends to increase vehicle mass. Reducing vehicle size, or efficiently repackaging to maintain required occupant and cargo volume, is another effective technique to reduce mass. This is also difficult, because market-driven demand for larger vehicles has resulted in increases in the average size of vehicles year after year.

Much promise lies with advanced materials technologies using high-strength, lightweight engineering alloys or composites to reduce vehicle body structure mass that can be amplified throughout the vehicle with lighter drivetrains and subsystems. However, many of these advanced materials have seen very limited use in production due to higher material costs and lack of compatibility with existing manufacturing processes.

A number of different materials are used in manufacturing passenger cars. Some are selected to perform highly specialized functions, while others, such as steel, cast iron, and aluminum, usually perform a more general "structural" function. The estimated material content in a typical American-made passenger car is shown in Table 12.11.

Nearly three quarters of the average vehicle weight is incorporated in its power train, chassis, and body, the bulk of this being ferrous metals. Other structural materials found in an average automobile, and used to a much smaller extent, include aluminum, plastics, or composites. Figure 12.12 shows how the use of aluminum and high-strength steel (HSS) as a percentage of total vehicle mass has been increasing over the past two decades, while the use of iron and mild steel has been declining. Approximately 19% of the mass of the car can be classified as special-function materials that would be very difficult to replace with alternate, lightweight materials. The potential for significant vehicle weight reduction clearly involves replacement of the 68% of the mass constituted by ferrous materials. The

Table 12.11 Material content in typical passenger car

Material	Mass, lb	%
Mild steel	1376	43.7
HSLA steel	259	8.2
Stainless steel	43.5	1.4
Other steels	48	1.5
Total steel	**1726.5**	**54.8**
Cast iron	411.5	13.0
Total ferrous	**2138**	**67.9**
Plastics/composites	245	7.8
Aluminum	177	5.6
Rubber	134.5	4.3
Glass	88.5	2.8
Copper/electrical equivalent	43.5	1.4
Powder metal	26	0.8
Lead	24	0.8
Zinc die castings	16	0.5
Other materials	68.5	2.2
Fluids/lubricants	188.5	6.0
Total	**3149.5**	**100**

Note: HSLA, high-strength, low-alloy. Source: Ref 12.19

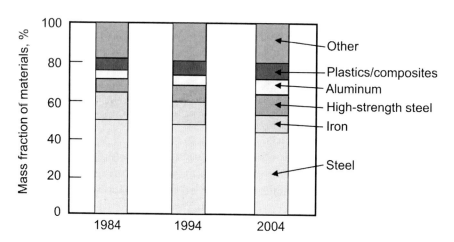

Fig. 12.12 Material composition of average U.S. automobile. Source: Ref 12.20

single largest opportunity for lightweight material substitution lies with the body-in-white, which is made primarily from mild steel. The term *body-in-white* refers to the stage in automobile manufacturing in which the car body sheet metal components have been welded together, but before moving parts (doors, hoods, and deck lids as well as fenders), the motor, chassis subassemblies, or trim (glass, seats, upholstery, electronics, etc.) have been added and before painting.

There are three main structural advanced material group candidates: advanced high-strength ferrous alloys, light metals (aluminum, magnesium, and titanium), and composites, including plastics. The relevant properties of these materials are summarized in Table 12.12 and Fig. 12.13.

Advanced Ferrous Alloys

According to the International Iron and Steel Institute, automotive steels can be classified in several different ways. Common designations include low-strength steels (interstitial-free and mild steels), conventional HSS (carbon-manganese, bake-hardenable, high-strength interstitial-free, and high-strength, low-alloy steels), and newer types of advanced high-strength steel (AHSS), typically classified by their microstructure at room temperature. These AHSS include dual-phase, transformation-induced plasticity, complex-phase, and martensitic steels. Additional AHSS for the automotive market include ferritic-bainitic, twinning-induced plasticity, hot formed, and postformed heat treated steels.

A second classification method important to part designers is strength of the steel, because steel alloys have virtually the same density and elastic modulus throughout the strength ranges. One such system defines HSS as

Table 12.12 Properties and relative costs of lightweight materials for automobiles

Material	Relative density(a)	Yield strength, MPa (ksi)	Tensile strength, MPa (ksi)	Tensile modulus, MPa (ksi)	Relative cost per part(a) (2000)
Mild steel	1.00	200 (29)	300 (44)	200 (29.0)	1.0
High-strength steel (A606)	1.00	345 (50)	483 (70)	205 (29.7)	1.0–1.5
Iron (D4018)	0.90	276 (40)	414 (60)	166 (24.1)	...
Aluminum (AA6111)	0.34	275 (40)	295 (43)	70 (10.1)	1.3–2.0
Magnesium (AM50)	0.23	124 (18)	228 (33)	45 (6.5)	1.5–2.5
Glass-fiber laminate	0.21	...	448 (65)	34 (5.0)	2.0

(a) Relative to mild steel. Source: Ref 12.20

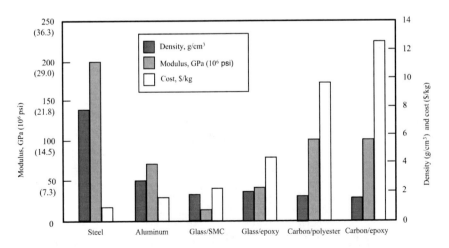

Fig. 12.13 Relative materials properties and costs. SMC, sheet molding compound. Source: Ref 12.21

those having yield strengths from 210 to 550 MPa (30 to 80 ksi) and tensile strengths from 270 to 700 MPa (40 to 100 ksi), while ultra/advanced high-strength steels have yield strengths greater than 550 MPa (80 ksi) and tensile strengths greater than 700 MPa (100 ksi). In addition, many steel types have a wide range of grades covering two or more strength ranges.

A third classification method presents various mechanical properties or forming parameters of different steels, such as total elongation, work-hardening exponent, n, or hole expansion ratio—all measures of formability. As an example, the total elongation for the different metallurgical types of steel is shown in Fig. 12.14.

The principal difference between conventional steels, HSS, and AHSS is their microstructure. Mild steels are low-strength steels with a ferritic microstructure. Drawing-quality and aluminum-killed drawing-quality steels are examples and often serve as a reference base because of their widespread application and production volume. While conventional HSS are single-phase ferritic steels, AHSS are primarily multiphase steels, which contain ferrite, martensite, bainite, and/or retained austenite in quantities sufficient to produce the desired mechanical properties. Some types have a higher strain-hardening capacity, resulting in a strength-ductility balance superior to conventional steels. Other types have ultrahigh yield and tensile strengths and can be bake hardened.

The basic premise in designing with HSS or AHSS is to take advantage of high material strengths relative to conventional steels. The usual result is downgaging a particular part—reducing its wall thickness to save weight while still maintaining functionality and strength. This approach has limitations and does not work as well on parts that are stiffness driven as opposed to strength driven, simply due to the fact that all steels have nearly identical elastic moduli, density, and resulting specific stiffness.

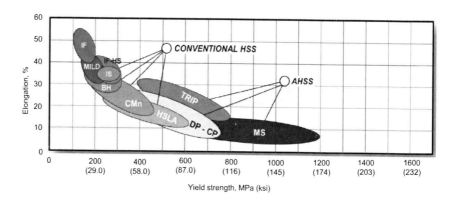

Fig. 12.14 Relationship between yield strength and total elongation for various types of advanced high-strength steels. HSS, high-strength steel; IF, interstitial free; HS, high strength; IS, isotropic; BH, bake hardened; TRIP, transformation-induced plasticity; HSLA, high-strength, low-alloy; DP, dual phase; CP, complex phase; MS, martensitic steel; AHSS, advanced high-strength steel. Courtesy of WorldAutoSteel. Source: Ref 12.22

Very thin gages, if not properly implemented, can introduce new failure modes, such as buckling, and negatively affect noise, vibration, and harshness ride characteristics. Additionally, moving to increasingly stronger materials also limits their formability and hence their range of potential weight-saving applications.

Aluminum Alloys

Aluminum alloys have several properties that are attractive to auto designers. The density of aluminum is one-third that of steel, and certain aluminum alloys have excellent strength, while others exhibit excellent formability. Aluminum is easy to machine, cast, and fabricate and is fairly corrosion resistant. However, aluminum alloys perform poorly in fatigue and are more difficult to weld than conventional steel alloys.

Only 9% of the mass of the average automobile in the United States is aluminum. Most of this is in the form of aluminum castings used mainly in the engine, wheels, transmission, and driveline. Casting is ideal, not only due to the lower melting point of aluminum and subsequent ease of processing but also for the potential of part integration and further mass reduction, if the designer can successfully integrate a number of individual parts into a single casting. Stamped sheet aluminum body panels are not as prevalent; that is, aluminum sheet is more difficult to form than steel and must be handled with care to prevent scratches because it is softer.

Aluminum alloy usage is approximately 145 kg (320 lb) per vehicle. Currently, large-scale use is restricted to the high-value end of the market. Models such as the Audi A8, Jaguar XJ and XK, and Corvette Z06 in addition to niche vehicles such as Aston Martin, Lamborghini, and Ferrari make intensive use of aluminum alloys, mostly for improved power-to-weight ratios on their high-performance vehicles. Body-in-white mass reductions of up to 55% using aluminum may be technically feasible for high-volume production, although the economics of doing so are still uncertain due to the increased material cost.

Magnesium Alloys

Magnesium alloys are the one of the lightest metallic structural materials, 30% less dense than aluminum and 75% lighter than steel. Magnesium is also easier to manufacture and cast, having a lower latent heat (it solidifies faster, and die life is extended) and being easier to machine. However, it has a lower ultimate tensile strength, fatigue strength, modulus, and hardness than aluminum.

The world's automakers already use magnesium alloys for individual components. Die cast magnesium alloys have historically been used where a high level of part integration is feasible, such as for instrument panel beams. Other applications include knee bolsters, seat frames, intake manifolds, and valve covers. Currently, magnesium parts are concentrated in well-protected regions of the car that are not exposed to corrosive environments. Corrosion protection of magnesium becomes a serious issue in exposed applications, for example, body and chassis structures. For such parts, manufacturers can consider proprietary anticorrosion pretreatments available for magnesium alloys, which transform the surface of light alloys into a dense, hard ceramic with outstanding resistance to corrosion and wear.

Newer applications include magnesium/aluminum composite engine blocks. BMW's six-cylinder magnesium/aluminum crankcase is the lightest used in large-scale production, weighing 57% of a comparable gray cast iron block; the weight advantage in comparison to an all-aluminum alloy crankcase is 24%. Novel structural uses of magnesium include the Corvette Z06 magnesium engine cradle, cast as a single-piece component using the newly developed high-temperature magnesium alloy AE44. The 10.85 kg (23.92 lb) engine cradle replaces large, multipiece assemblies and provides a 35% weight savings versus the original aluminum design.

In recent years, the abundance of magnesium in China has had the effect of lowering the raw material cost, at times lower than that of aluminum. However, new processing methods and magnesium alloys are needed before it can become economically and technologically feasible as a major automotive structural material. The low elongation at room temperatures requires that most forming of sheet magnesium alloys be done warm to increase formability. Conversely, this increased formability at high temperatures results in low creep resistance and stress-rupture failure at higher temperatures for typical magnesium alloys.

Per-vehicle magnesium alloy content is expected to grow from 3.5 to 7.3 kg (8 to 16 lb) by 2012. The U.S. Automotive Materials Partnership has forecasted a very ambitious goal of 160 kg (355 lb) per vehicle by 2020. To help achieve this goal, the Department of Energy is sponsoring a joint U.S./Canada/China program to develop a magnesium alloy unibody frontend with the goal of a 50% weight savings compared to conventional steel.

Titanium Alloys

Titanium alloys possess an attractive set of properties, including high specific strength, corrosion resistance, toughness, and specific stiffness; however, their cost limits applications to selected markets. The aircraft industry is currently the single largest market for titanium alloy products due to the exceptional strength-to-weight ratio, elevated-temperature performance, and corrosion resistance. However, this strong dependence of titanium on the aerospace industry has caused titanium production to be very cyclic. The price of titanium alloy sheet fluctuates in step with the aerospace industry cycles, and lead times for product delivery from time of order can be as long as 18 months.

Many automotive systems would benefit from the use of titanium alloy products. Titanium automotive exhaust systems could save as much as 50% of their current mass. Titanium valves and valve springs, connecting rods, suspension springs, wheels, drive shafts, underbody panels, side impact bars, and half-shafts are just some of the automotive applications that could benefit from the use of titanium.

Apart from the cost of manufacturing, technical problems preventing the integration of titanium include wear resistance, a lower modulus than steel, and welding and machining difficulties. Coatings, reinforcements, and compositing can improve wear resistance. The machining difficulties can be reduced by the production of near-net shape parts through powder metallurgy and other methods. The major problem is that titanium costs substantially more than competing materials.

Although the raw cost of titanium ore is significantly more than that of other materials, the difference in the cost of sheet, plate, and most other titanium alloy product forms is largely due to the cost of mill operations or secondary processing. Batch processing methods, such as vacuum arc remelting, are inherently costly but are needed to produce refined titanium alloys. Additionally, thinner plate and sheet gage sections increase the production costs due to the additional batch-type heat treating and rolling steps involved, increasing the amount of production waste and the percentage of total production costs. Processing of titanium from ingot to 2.5 cm (1 in.) plate accounts for 47% of the total cost of the material.

Composites and Plastics

Composites offer high specific material strength and stiffness, along with very high fatigue resistance, allowing for significantly reduced mass while maintaining or even improving component strength and durability and vehicle stiffness. The mechanical properties and degree of isotropy of polymer composites are controllable over a wide range, depending on reinforcing fiber orientation and composition. While 7.5% of total vehicle mass currently is composed of plastic and composite materials, the applications are typically not for the primary vehicle structure. Currently, low-strength and lower-cost composite forms are chosen for their ease of processing and moderate mass reductions.

The most common composites in use today are glass-fiber-reinforced thermoplastic polypropylene, which is applied to rear hatches, roofs, door inner structures, door surrounds, and other trim and interior components. Other types include glass mat thermoplastics, sheet molding compounds (SMCs) made of glass-fiber-reinforced thermoset polyester, and glass-fiber-reinforced thermoset vinyl ester. For instance, SMCs can be stamped into shape with much lower capital costs relative to steel and lower density. Together, these attributes result in overall cost and weight savings.

However, due to the low-performance fibers used, random orientation of the fibers, and low fiber volume fractions, the structural performance and light weight offered by SMC and many of its current rival material systems is only moderate (Fig. 12.15).

Carbon-fiber-reinforced plastic (CFRP) composites are more expensive than their glass-fiber-reinforced cousins, although they offer significant strength- and weight-saving benefits. Advanced CFRP composites are the most logical replacement for steel in vehicle structures where significant weight reduction is desired. Currently, GM's Corvette Z06 features carbon-fiber front fenders resulting in a 6 kg (13 lb) reduction of vehicle mass. Industry analysis shows potential for 60 to 67% body-in-white mass reduction using CFRP. However, the two most widely cited obstacles to the use of carbon composites in automotive structures are the high cost of the raw materials (~$11 to 22/kg versus ~$1.3/kg for steel) and the high labor required to produce advanced composite parts.

The specific crash-energy absorption of CFRP can be 2 to 5 times that of steel. Molding characteristics of composites can provide greater styling flexibility and reduced assembly steps, finish processes, and tooling through parts consolidation and lay-in-the-mold finish coatings, potentially

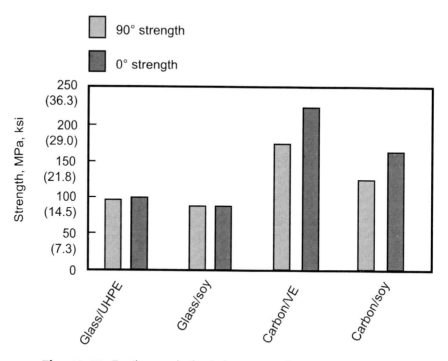

Fig. 12.15 Tensile strength of typical automotive sheet molding compounds with petroleum and soy-based resins. UPHE, ultrahigh-density polyethylene; VE, vinyl ester. Source: Ref 12.23

eliminating CFRP material cost penalties. However, these materials do face manufacturing challenges. Composite parts for structural applications typically suffer from high per unit costs and slow processing cycle times, which confine their use to premium ultralow-volume products. Other technical challenges of CFRP include the lack of infrastructure needed to deliver large quantities of materials and the difficulty in recycling of composites at the vehicle end of life.

Costs

When alternative materials are used to reduce weight, studies have found that on average, weight reduction would cost $2 to $3.50 per kilogram of weight saved. To get a sense of the potential applications of lightweight materials in vehicles and their corresponding manufacturing costs, results from various case studies, most examining a weight reduction of the body-in-white, are presented in Table 12.13.

The point at which mass reduction minimizes vehicle cost and complexity should be determined before the design is finalized with particular structural materials and component choices. Ultimately, these new materials technologies are expected to increase costs, meaning that price increases will be passed on to consumers. However, the impact of such vehicle sales price increases should be somewhat mitigated by the potential savings in fuel costs over the vehicle operating lifespan.

Table 12.13 Incremental manufacturing cost compared to conventional steel alternative

Lightweight vehicle/component	Incremental OEM cost	Weight reduction	U.S. $ per kg reduction	Volume per year
General lightweight vehicle	...	...	2.20–3.70	...
High-strength steel intensive				
Front end	–$13	11 kg	–1.20	...
Sport utility vehicle frame	...	(–23%)	0.68	220,000
Body-in-white	–$32 to 52	52–67 kg	–1.00 to –0.47	225,000
Aluminum intensive				
Vehicle	$661	346 kg	1.91	200,000
Unibody	$537	138 kg	3.88	500,000
Polymer composite intensive				
Body (fiberglass reinforced)	$400	127 kg	3.16	100,000
Body (fiberglass thermoset)	$930	68 kg	13.68	250,000
Body (carbon-fiber reinforced)	...	...	2.20–8.82	...
Body (carbon-fiber reinforced)	$900	196 kg	4.59	100,000
Body (carbon-fiber thermoset)	$728	114 kg	6.39	100,000
Vehicle (carbon fiber)	$2926	444 kg	6.59	250,000
Body (carbon-fiber thermoplastic)	$1140	145 kg	7.86	250,000

Note: OEM, original equipment manufacturer. Source: Ref 12.20

12.9 Summary

Advanced engineering materials can play an important role in improving the efficiency of transportation engines and vehicles. Weight reduction targets are frequently set as a means to increase the fuel economy of vehicles, thereby reducing exhaust emissions. By reducing the mass of the vehicle main body and chassis structure, secondary weight and cost savings can be realized by downsizing subsystems through the mass-decompounding effect. In addition, weight reduction has many significant secondary benefits, improving performance and handling dynamics. Lightweight materials, including AHSS, aluminum, magnesium, and titanium alloys, along with composites, have significant potential to reduce body-in-white mass. With the exception of titanium alloys, Table 12.14 summarizes the highlights and challenges associated with each material.

In the near term, AHSS, aluminum alloys, and possibly magnesium alloys will likely see more use because they are currently the most cost-effective option at large production scales. Titanium alloys and composites, which presently cost significantly more, will likely take a smaller role in weight reduction until material and production costs are decreased in the future.

Targets of 20 of 35% weight reduction commonly referenced in the literature and discussed among policy makers are easily possible by 2035 through the use of these advanced materials, at a cost of $2 to $3.50 per kilogram of weight saved in the average vehicle. Mass reductions of this magnitude could alone result in up to 20% reduction in vehicle fuel

Table 12.14 Summary of alternative lightweight automotive materials

Material	Current use	Merits	Challenges
Aluminum	140 kg/vehicle; 80% are cast parts, e.g., engine block, wheels	Can be recycled. Manufacturers are familiar with metal forming.	High cost of aluminum. Stamped sheet is harder to form than steel. Softer and more vulnerable to scratches. Harder to spot weld; uses more labor-intensive adhesive bonding
High-strength steel	230 kg/vehicle; in structural components, e.g., pillars, rails, rail reinforcements	Makes use of existing vehicle manufacturing infrastructure; OEM support for near-term use	More expensive at higher-volume scale. Lower strength-to-weight ratio compared to other lightweight materials
Magnesium	5 kg/vehicle; mostly thin-walled castings, e.g., instrument panels and cross-car beams, knee boosters, seat frames, intake manifolds, valve covers	Low density, offering good strength-to-weight ratios	Higher cost of magnesium componentsProduction of magnesium in sheet and extruded forms
Glass-fiber composites	Some rear hatches, roof, door inner structures, door surrounds, and brackets for instrument panel	Ability to consolidate parts and functions, so less assembly required. Corrosion resistance. Good damping	Long production cycle time; more expensive at higher-volume scale. Thermosets cannot be recycled.
Carbon-fiber composites	Some drive shafts, bumpers, roof, beams, and internal structures	Highest strength-to-weight ratio, offering significant weight-savings benefit	Same as glass-fiber composites. High cost of fibers ($13–22/kg)

Note: OEM, original equipment manufacturer. Source: Ref 12.20

consumption and associated emissions. Further weight reduction, at a more significant cost, could approach 50% through creative use of lightweight materials, redesign, and downsizing.

ACKNOWLEDGMENTS

Portions of the text for this chapter came from "Overview of the Materials Selection Process" by G.E. Dieter in *Materials Selection and Design,* Vol 20, *ASM Handbook,* ASM International, 1997.

The section on Lightweight Materials in the Transportation (Automotive) Industry came from "Advanced, Lightweight Materials Development and Technology for Increasing Vehicle Efficiency" by D. Codd of KVA Incorporated, 2008.

REFERENCES

12.1 *Materials Selection and Design,* Vol 20, *ASM Handbook,* ASM International, 1997

12.2 M.F. Ashby, *Materials Selection in Mechanical Design,* 3rd ed., Butterworth-Heinemann, 2005

12.3 G.E. Dieter, Overview of the Materials Selection Process, *Materials Selection and Design,* Vol 20, *ASM Handbook,* ASM International, 1997

12.4 V. John, *Introduction to Engineering Materials,* 3rd ed., Industrial Press, 1992

12.5 C.O. Smith and B.E. Boardman, Concepts and Criteria in Materials Engineering, *Properties and Selection: Stainless Steels, Tool Materials, and Special-Purpose Metals,* Vol 3, *Metals Handbook,* 9th ed., American Society for Metals, 1980, p 825–834

12.6 G. Boothroyd, P. Dewhurst, and W. Knight, *Product Design for Manufacture and Assembly,* Marcel Dekker, Inc., 1994

12.7 J.A. Schey, *Introduction to Manufacturing Processes,* McGraw-Hill Book Co., 1987

12.8 E.B. Magrab, *Integrated Product and Process Design and Development,* CRC Press, Inc., 1997

12.9 H.E. Trucks, *Designing for Economical Production,* 2nd ed., Society of Manufacturing Engineers, 1987

12.10 R. Bakerjian, Ed., *Design for Manufacturability,* Vol 6, *Tool and Manufacturing Engineers Handbook,* 4th ed., Society of Manufacturing Engineers, 1992

12.11 *Properties and Selection of Metals,* Vol 1, *Metals Handbook,* 8th ed., American Society for Metals, 1961, p 295

12.12 N. Cross, *Engineering Design Methods,* 2nd ed., John Wiley & Sons, Inc., 1994

12.13 G.E. Dieter, *Engineering Design: A Materials and Processing Approach,* 2nd ed., McGraw-Hill, 1991

12.14 W.E. Souder, *Management Decision Methods for Managers of Engineering and Research,* Van Nostrand Reinhold Co., 1980

12.15 M.M. Farag, *Selection of Materials and Manufacturing Processes for Engineering Design,* Prentice Hall, 1989

12.16 E.L. Rooy, Aluminum and Aluminum Alloys, *Casting,* Vol 15, *Metals Handbook,* 9th ed., ASM International, 1988

12.17 J.I. Rotheiser, Plastics, *Handbook of Materials for Product Design,* 3rd ed., McGraw-Hill, 2001

12.18 L.J. Hart-Smith, Bolted Joint Analysis for Fibrous Composite Structures—Current Empirical Methods and Future Scientific Prospects, *Joining and Repair of Composite Structures,* STP 1455, ASTM International, 2004

12.19 F. Stodolsky, A. Vyas, and R. Cuenca, Lightweight Materials in the Light-Duty Passenger Vehicle Market: Their Market Penetration Potential and Impacts, *Proc. Second World Car Conference,* March 1995 (University of California, Riverside)

12.20 A. Bandivadekar, K. Bodek, L. Cheah, C. Evans, T. Groode, J. Heywood, E. Kasseris, M. Kromer, and M. Weiss, "On the Road in 2035: Reducing Transportation's Petroleum Consumption and GHG Emissions Laboratory for Energy and the Environment," Report LFEE 2008-05 RP, Massachusetts Institute of Technology, July 2008

12.21 A. Lovins and D. Crame, Hypercars, Hydrogen, and the Automotive Transition, *Int. J. Veh. Des.,* Vol 35 (No. 1–2), 2004

12.22 "Advanced High Strength Steel (AHSS) Application Guidelines," Ver. 3, International Iron and Steel Institute, Sept 2006

12.23 "Energy Efficiency and Renewable Energy: Office of Vehicle Technologies, FY2006 Progress Report for Automotive Lightweighting Materials," U.S. Department of Energy, 2007

SELECTED REFERENCES

- J.A. Charles, F.A.A. Crane, and J.A.G. Furness, *Selection and Use of Engineering Materials,* 3rd ed., Butterworth-Heinemann, 1997
- C.A. Harper, *Handbook of Materials for Product Design,* 3rd ed., McGraw-Hill, 2001
- M. Kutz, *Handbook of Materials Selection,* John Wiley & Sons, 2002
- J.K. Wessel, *Handbook of Advanced Materials: Enabling New Designs,* Wiley-Interscience, 2004

Index

A

ABAQUS, 538
abrasive waterjet cutting, 502
acetals, 353–354
acid pickling, 176, 185
acrylic. *See* poly(methyl methacrylate) (PMMA)
acrylonitrile-butadiene-styrene (ABS), 351, 364, 650(T)
acute berylliosis, 221
adhesive bonding
  aircraft applications, 118(F)
  aluminum alloys, 116–119, 118(F)
  applications, 119
  electrical/thermal insulator, 119
  electrochemical corrosion protection, 119
  Kevlar, 117
  magnesium alloys, 170
  Nomex, 117
  polymer-matrix composites, 449–450
  surface preparation, 118
  titanium, 284
  vibration dampers, 119
adhesives
  aluminum alloys, 116–119(F)
  ceramic armor, 518
  epoxies, 365–366, 366
  honeycomb panel construction, 442(F)
  magnesium alloys, 170
  melamine, 366
  node bond adhesive, 366
  plastics, for joining, 380
  thermoplastic joining, 449–450
  types, 380
advanced high-strength steel (AHSS), 663, 664, 664(F), 670
aerospace applications
  carbon fiber, 410, 410(T)
  ceramic-matrix composites, 29–30(F)

discontinuously reinforced CMCs, 578–580(F), 581(F)
  magnesium alloys, 148
  mechanical fastening, 448, 449(F)
  MMCs, 460–461, 463, 657
  polymer-matrix composites, 391
  structural ceramics, 651
  titanium alloys, 644
  titanium casting, 274
age hardening, 1–2, 133
Agena spacecraft, 211
aging
  aluminum alloys, 39, 51, 71–73(F), 107
  artificial, 71, 72
  artificial aging treatments, 72
  magnesium alloys, 157, 165, 168
  natural, 70, 71, 72
  titanium alloys, 271
Airbus, 4, 25, 391, 450
aircraft applications
  adhesive bonding, 117–118, 118(F)
  polymer-matrix composites, 391
  titanium alloys, 225
  titanium machining, 281
  titanium SPF, 269
  titanium SPF/DB, 269
  wrought aluminum alloys, 36(T)
AlBeMet AM 162, 213–214(F,T)
Alclad
  alloy 2024-T3 sheet, 49
  alloy 6013-T6, 51
  aluminum-copper alloys (2*xxx*), 50, 640
  aluminum-zinc alloys (7*xxx*), 640
  applications, 36(T), 37(T), 125
  atmospheric corrosion, 640
  compatibility of metals, 181
  products, 134, 135(T)
  RW, 111
  scalping operation, 57
  soak times, 70

M

Q

X

Y